Günter Tiess

General and International Mineral Policy

Focus: Europe

SpringerWienNewYork

DI Dr. Günter Tiess
University Assistant at the Chair of Mining Engineering
and Mineral Economics, Montanuniversität Leoben, Austria

www.rohstoffpolitik.com

With financial support of
Bundesministerium für Wissenschaft und Forschung in Wien.

This work is subject to copyright.
All rights are reserved, whether the whole or part of the material is concerned, specifically those of translation, reprinting, re-use of illustrations, broadcasting, reproduction by photocopying machines or similar means, and storage in data banks.

© 2011 Springer-Verlag/Wien
Printed in Germany

SpringerWienNewYork is part of
Springer Science+Business Media
springer.at

Product Liability: The publisher can give no guarantee for all the information contained in this book. This does also refer to information about drug dosage and application there of. In every individual case the respective user must check its accuracy by consulting other pharmaceutical literature.

The use of registered names, trademarks, etc. in this publication does not imply, even in the absence of a specific statement, that such names are exempt from the relevant protective laws and regulations and therefore free for general use.

Typesetting: Jung Crossmedia Publishing GmbH, 35633 Lahnau, Germany
Printing: Strauss GmbH, 69509 Mörlenbach, Germany

Printed on acid-free and chlorine-free bleached paper
SPIN: 12445257

With 180 Figures and 87 Tables

Library of Congress Control Number: 2011931324

ISBN 978-3-211-89004-2 SpringerWienNewYork

Preface

The raw materials economy provides an essential contribution to employment and added value and thus is crucial for the European Union industry. Raw materials are the most important link in the value chain of industrial goods' production, which plays a prominent role as a source of prosperity in Europe. A secure supply raw materials for the industry is absolutely necessary. However, a structural change has taken place at the global mineral markets. The old rule of thumb – 20 percent of the world population in Europe, USA and Japan consuming more than 80 percent of the total minerals production – is not valid any more. With the integration of India, the People's Republic of China and other populous emerging countries like Brazil and Russia into the world economy, today more than half of the world's population claims an increasing share in raw materials. Thus the global demand for raw materials stands at the bottom of a new growth curve. It is assumed that by 2030 the worldwide need for raw materials will have doubled.

New estimates by the OECD Development Centre suggest that today's development and emerging countries are likely to account for nearly 60% of world GDP by 2030. Increasing income in these countries means also a growing demand for manufactured industrial goods, which opens new trade options for the export-oriented European industry. Measured by their population, the leading African markets, the so-called "African Lions" Algeria, Botswana, Egypt, Libya, Mauritius, Morocco, South Africa and Tunisia have even overtaken the BRIC states as shown by their economic performance per capita in 2008. China has become the major trade partner of Brazil, India and South Africa; South-South links are of increasing importance as a motor of growth in developing and emerging countries.

Occurrence and consumption of fossil, metallic and industrial minerals are distributed unequally on a worldwide range. Besides that, the access to deposits in Europe has been increasingly limited for various reasons, such as environmental issues or competition for different land uses. Oligopolistic structures in the supplier countries may also lead to market distortions in Europe. Europe is dependent on imports for many raw materials, particularly metals, a fact to which has been paid little attention as yet, but in times of increasing raw materials prices public awareness is growing. Booming prices of fossil fuels and of coke and steel are striking examples of recent years. Recent disturbances on the international raw materials markets were often caused by discrepancies of supply and demand which were also enhanced by rapid volatilities of these market factors. At the same time, developing and emerging countries have increasingly claimed stronger consideration of their interests by political means. Consequences for Europe are uncertainties of supply and problems with availability. Deficiencies of supply in import dependent countries like Europe and Japan have recently led to reductions of production. Volatile prices on the international commodity markets impact directly the costs of downstream production sectors and thus influence economy as a whole. Increasing demand together with lacking awareness of the raw materials matter sometimes lead to paradox situations. Mineral security is not a secondary problem of single industrial sectors. It is, on the contrary, a basic prerequisite for the stability of the complete value added chain, relevant for job preservation and thus economic prosperity. Changing conditions on the world market require a proactive economic policy of the European Union and its member states. Further prosperous development of the European economy is crucially dependent on undisturbed and, applying the scale of international competition, profitable supply with mineral raw and base materials.

In view of the described risks and problems, safeguarding the supply with raw materials on the basis of a minerals policy will be a permanent challenge for the European Union and all European countries. Appropriate measures for securing the supply with raw materials have to be taken. A successful coherent minerals' policy will have to take over essential tasks to solve these problems. In particular, enhancement and extension of existing instruments like recycling, substitution and resource efficiency of raw materials can guarantee domestic security of supply and provide a sustainable strategic European minerals policy.

Methods and structure of the essay

This essay focuses on non-energetic mineral raw materials (i. e. metallic minerals, industrial minerals [e. g. magnesium, talc] and construction minerals), as a thorough complementary discussion of energetic minerals would by far exceed the frame of this book.

The matter of minerals policy is characterised by a wide range of aspects and topics; hence a variety of approaches is possible and necessary. At first, the location of mineral occurrences and their distribution in the earth's crust depends on the geological conditions of their respective origin and thus differs widely for metals, industrial minerals and aggregates. Exploration of deposits, exploitation by different mining methods (e. g. subsurface, underground, marine) are the next steps, including mining techniques, feasibility and economic considerations. Land use and access to land as well as licensing procedures have ecological, economic and legal aspects, as do waste disposal and landscape recovery after mine closing. National and international economics, trade policy and industry are further aspects.

The author has analysed the issues relevant to non-energy minerals policy by comparison of economic statistics for demand and supply, imports and exports, examination of engineering technology related to raw materials efficiency, reuse and recycling methods, by discussion of legal conditions in a variety of countries in and outside Europe, and by consideration of European Union economic and planning policies both on EU and on national levels.

Taking into account the European situation – the economic dependence on mineral imports for the industry, parallel to the increasing self-demand of the mineral-rich countries, the price volatilities and uncertain supply – the hypothesis is proposed that only a coherent common minerals policy at a European Union level can cope with the complex and precarious challenges of the years to come.

Chapter 1 provides a general overview of the topic and addresses basic knowledge, thus providing the fundamentals for the treatise.

Chapter 2 underlines the importance of mineral consumption in Europe and emphasizes the variety of functions of raw materials. It is also meant as a contribution to enhance public awareness of raw materials issues.

Chapter 3 discusses the situation of raw materials demand and supply in Europe. The main data sources utilized include World Mining Data ((Mainly) compiled by Weber and Zsak), the British Geological Survey, the UEPG and the U. S. Geological Survey.

Chapter 4 considers the outline of a minerals policy. The concept of raw materials policy is circumscribed and defined scientifically. The need for a European raw materials policy is pointed out and discussed. The objectives of a raw materials policy are placed in the international context. Approaches to the realization of the objectives are discussed referring to the determination of strategies, instruments and concepts.

Chapter 5 resumes the discussion on the basis of the outline of a minerals policy given in Chapter 4. The raw materials policy and economy of selected European countries is analysed by various criteria, focussing on the national level. European Union member countries and non-member countries are discussed.

Chapter 6 focuses on the European Union, referring to the outline of a minerals policy as shown in Chapter 4. Firstly, critical remarks to the status of minerals policy of the EU are made.

Chapter 7 discusses the options to establish a coherent European minerals policy and covers targets, determination of strategies and designing concepts.

Chapter 8 completes the discussion and lists fundamental remarks and theses for the implementation of a coherent European minerals policy.

The Appendix shows a variety of mineral political approaches and strategies of several non-European countries.

Concluding remark

In November 2008, the Commission of the European Union published the so-called "Communication from the Commission to the European Parliament and the Council: The raw materials initiative – meeting our critical needs for growth and jobs in Europe" (COM (2008)699). In this Communication the political intention to establish an EU raw materials strategy is declared. The Member States are expected to report the implementation of appropriate measures to the Commission by 2010.

Thus the Communication and this book aims at the same objective, the relevance of European security of raw materials supply. Although methods and structure of both show certain similarities, they differ in some points. The Communication of the European Commission makes an analysis of supply and demand of non-energy raw materials and as a policy response, postulates an integrated strategy. This is based on three pillars: access to raw materials on world markets in undistorted conditions; foster a sustainable supply of raw materials from European sources, reduce the EU's consumption of primary raw materials. At last, the way forward is described and a proposal to launch a European Raw Materials Initiative is made. The Commission is to report to the Council in two years on the implementation of the raw materials initiative.

This book is proposed to be a useful complementation to the Communication of the European Commission (COM (2008)699). Its purpose is to make further suggestions and propose options for the establishment of a coherent common European minerals policy, thus making another contribution to the security of raw materials supply in Europe.

As this is a very extensive topic any additional remarks and further information is always welcome to the author.

Acknowledgements

I am grateful to Prof. Zach Agioutantis at the Department of Mineral Resources Engineering, Technical University of Crete for taking the time to review the whole book and for his numerous comments and suggestions. I would like to thank Dr. Bodo Schirmer for his professional contributions. Also, many thanks to Prof. Horst Wagner, Prof. Peter Moser and Dr. Thomas Drnek.

For assistance in completing the book, I also thank Ms. Anna Werner, Ing. Mag. Karim Karman and Ms. Irene Leischner.

Last but not least, also many thanks to my family.

Table of contents

Chapter 1 Introduction 1
  1.1 Raw materials 1
  1.2 Minerals for the economy 2
  1.3 Importance of the extractive industry in the value chain 6
  1.4 Characteristics of the extractive industry 8
    1.4.1 Mining – Initial production 8
    1.4.2 Access to minerals 13
    1.4.3 Competitiveness of the extractive industry 15
  1.5 Mineral markets 17
    1.5.1 Mechanism of pricing 18
    1.5.2 Demand and supply 21
    1.5.3 Aspects of materials criticality 27
  1.6 Sustainable mineral resource management 29
    1.6.1 General facts 29
    1.6.2 Environmental protection 31
    1.6.3 Resource efficiency 34
    1.6.4 Material efficiency 35
    1.6.5 Recycling 37

Chapter 2 Utilisation of non-energy raw materials 39
  2.1 General facts 39
  2.2 Metallic minerals 40
  2.3 Industrial minerals 43
  2.4 Construction minerals 51

Chapter 3 Demand and supply of non-energy raw materials in Europe 55
  3.1 Historical development of minerals consumption 55
  3.2 European Union – minerals consumption 62
  3.3 Development of international metallic minerals markets 82
  3.4 Future demand of mineral resources – scenarios 103
  3.5 Questions on the security of supply in Europe 110
    3.5.1 Change of paradigm 110
    3.5.2 Availability of raw materials 114
    3.5.3 Supply criticalities 118
    3.5.4 Formation of market controlling mining companies 127
    3.5.5 Conflicts und wars 132

Table of contents

Chapter 4 The concept of a minerals policy 137
   4.1 Definition and terms 137
   4.2 Necessity of a minerals policy 141
        4.2.1 Minerals policy versus market economy 142
        4.2.2 Security of minerals supply 143
        4.2.3 Coordination of interactions 147
   4.3 Objectives of a minerals policy 153
        4.3.1 Targets of minerals supply policies 153
        4.3.2 Different aspects 154
   4.4 Approaches towards a minerals policy 158
        4.4.1 Strategy 158
        4.4.2 Instruments 166
        4.4.3 Conceptions 173

Chapter 5 View of the minerals policies in selected states of Europe ... 181
   5.1 General facts 181
   5.2 EU States 189
   5.3 Non-EU States 323
   5.4 Concluding remarks 346

Chapter 6 EU minerals policy status quo – critical reflections 413
   6.1 General 414
   6.2 Public raw material awareness 414
   6.3 Knowledge basis of raw materials 421
   6.4 Access to minerals outside Europe 423
        6.4.1 Trade policy 423
        6.4.2 Development policy 429
   6.5 Access to minerals inside Europe 435
        6.5.1 Decreasing of exploration 435
        6.5.2 Decreasing of the availability of deposits 438
        6.5.3 Complex permitting procedures 441
        6.5.4 Consultation process of the Raw Materials Initiative 2008   455

Chapter 7 Towards a European minerals policy 459
   7.1 General view 459
        7.1.1 Targets 461
        7.1.2 Strategy 464
        7.1.3 Conception 484
   7.2 Present developments at EU level – The Raw Materials Initiative   492

Chapter 8 Conclusions 521

Chapter 9 References 525

Chapter 10 Appendix 1 – International minerals policy approaches ... 561
    10.1 Alaska 561
    10.2 Brazil 563
    10.3 Canada 563
    10.4 Chile 567
    10.5 China 569
    10.6 Guatemala 570
    10.7 India 571
    10.8 Jamaica 575
    10.9 Japan 578
    10.10 Liberia 579
    10.11 Malaysia 581
    10.12 Pakistan 587
    10.13 Russian Federation 592
    10.14 Sierra Leone 596
    10.15 South Africa 602
    10.16 Tanzania 604
    10.17 USA 608

Chapter 11 Appendix 2 613

List of figures and tables

Figure 1: Illustration of the added value chain raw materials – end product (Data by EC, DG Enterprise and Industry, Commission staff working document, 2007) 6

Figure 2: Diamond deposit near Mirny/Yakutia/Russian Federation (Data by http://www.usmra.com) 10

Figure 3: Development stages of the copper mining project (Data by Gocht 1983) 11

Figure 4: Capital-intensity of economic sectors in EU-13 in 1999–2001 showing the relative position of the non-energetic raw materials industry (Data by EC, DG Enterprise and Industry, Commission staff working document, 2007) 12

Figure 5: Comparison of cash operating costs taking account of credits for sales of other metals in copper mines in Europe and other countries (2004) (Data by EC, DG Enterprise and Industry, Commission Staff Working document, 2007) 16

Figure 6: Quarry in Scotland (Data by http://www.alastairmcintosh.com) 17

Figure 7: International mineral market, Munich 2005 18

Figure 8: Production of aluminium and real aluminium price per ton (Data by RWI Essen, 2006, data from: USGS 2005, USGS 2006) 20

Figure 9: Relative price development of LME metals (Data by Ekdahl 2008) 20

Figure 10: Development of steel consumption (Data by Ekdahl 2008) .. 23

Figure 11: Stock of motor vehicles as a useful indicator for raw materials consumption – displayed with a motor 23

Figure 12: Stock of motor vehicles in the US 1900–2005 (Data by Nötstaller and Wagner, 2007) 24

Figure 13: Producing countries of German raw materials imports (Data by BDI, 2007) 26

Figure 14: Computer (Data by Die Presse, 2008) 27

Figure 15: Securing of raw material supply for a growing economy, society and sound environment (Data by Ekdahl, 2008) 30

Figure 16: Focus of the ETP-SMR 31

Figure 17: Raw materials contribute to environmental protection hybrid cars as an example (Data by Handelsblatt, August 2008) 32
Figure 18: Jet engine (Kugler, 2008) 36
Figure 19: Metal recycling – for example motor vehicles 37
Figure 20: Raw materials demand of an EU-citizen (Data by BGR, 2009 [http://www.bgr.bund.de) 40
Figure 21: Computer – For the production of a computer more than 60 different metals are needed 40
Figure 22: Calcium carbonate – a versatile raw material (Data by http://de.wikipedia.org) 44
Figure 23: Significance of potash (http://www.vks-kalisalz.de) 44
Figure 24: Aggregates – Illustration of the added value chain raw material – end product (Data by Department of Mineral Resources and Petroleum Engineering [UEPG] 2010]) 51
Figure 25: Train station Praterstern, Vienna, Austria 52
Figure 26: ICE high speed route in Germany (http://de.wikipedia.org) 53
Figure 27: Global copper production 1830–2000 (Data by Petterson et al., 2005) 55
Figure 28: Industrial production of the G7 states (Data by Petterson et al., 2005) 56
Figure 29: Correlation of construction raw materials consumption in proportion with population. (Data by EC, DG Enterprise and Industry, Commission Staff Working Document, 2007 [Consumption data based on BGS data of production, import and export of construction raw materials in 2003]) 56
Figure 30: World population from year 0 until 2200 (Data by de.wikipedia.org) 57
Figure 31: Global GDP (Data by de.wikipedia.org) 58
Figure 32: World metal consumption and GDP – given period: 1900–2000. Graph shows growth of global GDP and raw materials consumption (Data by Nötstaller/Wagner 2007) 59
Figure 33: Global concentration of countries in extraction of selected metals. (BDI, 2007) 60
Figure 34: Percentage of refined metal in the EU which originated from external scrap in 2003 (Data by European Commission, DG Enterprise and Industry 2007) 65
Figure 35: Change in EU production (%) of selected metallic minerals between 1993 and 2003 (Data by EC, DG Enterprise and Industry 2007 [Data from BGS]) 65

List of figures and tables

Figure 36: EU's main sources of import for metallic raw materials in 2004 by value (Data by EC, DG Enterprise and Industry, Commission Staff Working Document, 2007) 69

Figure 37: Annual raw materials import in the EU, 1999–2004 (weight in thousand tonnes) (Data by EC, DG Enterprise and Industry, Commission Staff Working Document, 2007) 70

Figure 38: Net amount of import of raw materials in the EU from 1999 to 2004. – Data Source: Eurostat (EC, DG Enterprise and Industry, Commission Staff Working Document, 2007) 70

Figure 39: EU's main sources of import for industrial minerals by value (Data by EC, Commission Staff Working Document, 2007 [Eurostat]) ... 73

Figure 40: Change in EU production (%) of selected industrial minerals between 1993 and 2003 and change in EU's percentage share of global mine production (Data by EC, DG Enterprise and Industry, Commission Staff Working Document, 2007 [data source: BGS, Euromines]) ... 74

Figure 41: Production of some industrial minerals – EU/world (Data by EC, DG Enterprise and Industry, Commission staff working document, 2007) 74

Figure 42: Comparisons of various variables indicating the proportions of the EU economy, the construction housing sector and the aggregates sub-sector in EU-15. (Sources: Bleischwitz/Bahn-Walkowiak, 2006) 77

Figure 43: Production of aggregates 2003–2007 of EU-27 (Data by British Geological Survey, 2009) 78

Figure 44: The trade in gravel in the study area in 2000 (Data by http://internationaal.bouwgrondstoffen.info) 81

Figure 45: World mining production based on the level of development of the producing country (Data by Weber and Zsak, 2008) 84

Figure 46: HWWI Index of the world prices for raw materials Hamburg Archive of International Economics (data until end of 2006), from 2007 continued by the Hamburg Institute of International Economics, Source: Hamburgisches Welt-Wirtschafts-Archiv (HWWA) and Hamburgischen WeltWirtschaftsInstitut (HWWI) 85

Figure 47: Price development of copper, tin and nickel in the US 2002–2009 (Data by UNCTAD 2010) 86

Figure 48: GDP and metal consumption in China 1980–2005 (Data by Nötstaller/Wagner 2007) 87

Figure 49: Trend of demand for raw materials in China (Data by Ekdahl, 2008) 88

List of figures and tables

Figure 50: Copper, nickel, zinc and aluminium prices as an example (Data by Sentient Monitor, 2008) 97

Figure 51: Global copper consumption – raw materials markets in Latin America for financial purposes. Per-capita consumption of copper in the EU-27 is three times as big as in China. (Data by Apel, H., KfW IPEX-Bank GmbH, Lateinamerika Verein e.V., Hamburg, 15.04.2008; Data by International Copper Study Group) 98

Figure 52: Important countries for foreign investments in 2007 (Data by Ernst & Young European Investment Monitor Ernst & Young's 2008 European attractiveness survey (http://arisinvest.ro)) 102

Figure 53: Production and export of copper between 1985–2002 (Data by BGS) 104

Figure 54: Iron ore deposit in Serra dos Carajás, Brazil (Data by Trojer, 2007) 105

Figure 55: Scenario: estimated global steel consumption, averaged growth rate 1890–2005 (3,7%) (Data by Ekdahl 2008; Quelle: BGR) 107

Figure 56: Crude steel production in China and India since 1950 (Data by BDI, 2007) 109

Figure 57: Metals supply of the European Union (Data by Ekdahl 2008) 111

Figure 58: World mining production 1984–2006 (Data by Weber and Zsak (2008) 113

Figure 59: Raw materials production in 2007, several metals including raw materials reserves (Data by Mining Environmental Management, 2008, p. 27. Data based on: USGS Mineral Commodity Summaries 2008) 115

Figure 60: Assumed raw materials reserves in the future (Data by Mining Environmental Management, 2008) 115

Figure 61: Development of iron ore price (Data by VRB, 2008) 120

Figure 62: Iron ore of the most important steel producers, 2005 in Mt (Data by BDI 2007) 123

Figure 63: Extraction of metallic raw materials by political stability of the producing countries in 2005 (Data by Bundesanstalt für Geowissenschaften und Rohstoffe (BGR); World Bank: Worldwide Governance Indicators 2006) 124

Figure 64: Concentration of enterprise in the exploitation of selected raw materials. Allotment of the three biggest mining companies in each case as well as their accumulated interest in the world production in 2005 in % (Data by BDI 2007; Raw Materials Supply Group [RMSG]) 128

List of figures and tables

Figure 65: Movement of iron ore trade transported by sea between 1960 and 2006: 121–725 million tonnes (Data by Ekdahl 2008; Thyssen-Krupp Steel, Eurostat) 129

Figure 66: Norilsk Nickel (Data by http://www. welt. de) 130

Figure 67: Where are the main reserves of metallic raw materials located in future? (Data by Ekdahl 2008 [BGR, USGS]) 131

Figure 68: War in Congo (Data by Presse, November 2008) 133

Figure 69: The Arctic and bordering states 135

Figure 70: Close interlocking of minerals policy, mining legislation and mineral economy 141

Figure 71: Stakeholders relevant for minerals policy (Data by Christmann, 2008) 147

Figure 72: Mechanisms of raw materials policy 152

Figure 73: Chinas mineral strategy (Data by Ekdahl 2008) 164

Figure 74: "Macro-control" of China by state interventions and commercial restrictions at the example of the added value chain of the copper production (Data by BDI 2007 [Eurometaux]) 165

Figure 75: Quarry Rotzloch, Alpnachersee/Switzerland (Data by Kündig et al., 1997) 171

Figure 76: Construction raw materials mining in Switzerland (Data by Kündig et al., 1997) 173

Figure 77: Worldwide supply chain of raw materials and metals for the production of aluminium products in the EU – Sources: European Aluminium Association (EAA) and Organization of European Aluminium Refiners and Remelters (OEA) (Data by EC, DG Enterprise and Industry, Commission Staff Working Document, 2007) 183

Figure 78: National minerals policy – schematic diagram 185

Figure 79: National mineral planning policy – schematic diagram 186

Figure 80: Countries of Europe (Data by http://www.world-atlas.us) .. 188

Figure 81: Mining production in Finland between 1998 and 2008 – selected mineral resources (Data by Weber et al. [WMD 1998–2008]) 189

Figure 82: Production, export and import of chromium (Data by BGS) 190

Figure 83: Pyhäsalmi, copper-zinc mine in Central Finland 196

Figure 84: Mining production in Sweden between 1998 and 2008 – selected mineral resources including aluminium production (Data by Weber et al. [WMD 1998–2008]) 197

Figure 85: Production, export and import of iron ore (Data by BGS) ... 198

Figure 86: Aggregates production in Sweden (Data by Swedish Aggregate Producers Association, 2010) 201

Figure 87: Iron Ore mining Kirunavaara in Sweden (Data by Atzenhofer and Pressler 2007) 202

Figure 88: Cross section of Kirunavaara Mine (Data by Atzenhofer and Pressler (2007), Iron ore mining in Kirunavaara) 202

Figure 89: Mining production in the United Kingdom between 1998 and 2008 – selected mineral resources including aluminium production (Data by Weber et al. [WMD 1998–2008]) 204

Figure 90: Anglesey Aluminium Metal Ltd, aluminium Smelter (Data by http://www.angleseyaluminium.co.uk) 207

Figure 91: Aggregates Production – Great Britain (Data by Minerals Products Association, 2009) 209

Figure 92: Mining production in Ireland between 1998 and 2008 – selected mineral resources (Data by Weber et al. [WMD 1998–2008]) 211

Figure 93: Production, export and import of zinc (Data by BGS) 212

Figure 94: Zinc – lead mines in Ireland 214

Figure 95: Salt and aluminium production in the Netherlands between 1998 and 2008 (Data by Weber et al. [WMD 1998–2008]) 215

Figure 96: Mining production in Poland of selected mineral resource including aluminium production (Data by Weber et al. [WMD 1998–2008]) 227

Figure 97: Production, export and import of zinc (Data by BGS) 229

Figure 98: Facility in Rudna, Kombinat Gorniczo Hutniczy Miedzi (KGHM) 232

Figure 99: Mining production in Germany between 1998 and 2008 – selected mineral resources including iron and aluminium production (Data by Weber et al. [WMD 1998–2008]) 234

Figure 100: Production, export and import of salt (Data by BGS) 238

Figure 101: Sand & Gravel production in Germany (Data by German sand & Gravel Association) 239

Figure 102: "Material queue" (Data by http://www.bv-miro.org) 239

Figure 103: Amberger Kaolinwerke, Werk Caminau (Data by http://www.akw-kaolin.com)................................. 242

Figure 104: Mining production in the Czech Republic between 1998 and 2008 – selected mineral resources (Data by Weber et al. [WMD 1998–2008]) 244

Figure 105: Aggregates consumption/GDP in Czech Republic (Data by Sitensky, Czech Geological Survey – Geofond, Czech Republic, 2010) 247

Figure 106: Mining production in Slovakia between 1998 and 2008 – selected mineral resources including iron and aluminium production (Data by Weber et al. [WMD 1998–2008]) 248

Figure 107: Slovalco (Data by http: //www.slovalco.sk) 249

Figure 108: Production, export and import of magnesite (Data by BGS) 251

Figure 109: Aggregates consumption/GDP in Slovakia (Data by Sitensky, 2010) 253

Figure 110: Erzberg, Steiermark (Source: http: //oepg2008.unileoben.ac) 254

Figure 111: Mining production in Austria between 1998 and 2008 – selected mineral resources (Data by Weber et al. [WMD 1998–2008]) 255

Figure 112: Production, export and import of tungsten (Data by BGS) 256

Figure 113: Organigram of the Austrian Mineral Resources Plan (Phase 1) (Data by Weber 2007) 261

Figure 114: Mining production in Hungary between 1998 and 2008 – selected mineral resources including aluminium production (Data by Weber et al. [WMD 1998–2008]) 263

Figure 115: Per capita aggregates production in Hungary between 1998 and 2008 (Data by Kovács, Hungarian Office for Mining and Geology, 2010) 266

Figure 116: Bauxite mining in Hungary (Data by http.//deutsch.mal.hu) 266

Figure 117: Production, export and import of alumina (Data by BGS) .. 267

Figure 118: Aggregates consumption/GDP in Slovenia (Data by Solar, Slovenian Geological survey, 2010) 269

Figure 119: Mining production in Italy between 1998 and 2008 – selected mineral resources including aluminium production (Data by Weber et al. [WMD 1998–2008]) 271

Figure 120: Production, export and import of feldspar (Data by BGS) .. 274

Figure 121: Furtei Mine (Data by http://www.mindat.org) 277

Figure 122: Mining production in France between 1998 and 2008 – selected mineral resources including aluminium production (Data by Weber et al. [WMD 1998–2008]) 278

Figure 123: Vicat Group, The La Courbaisse quarry and facilities in the Alpes Maritimes, France (Data by http://www.vicat.com) 281

Figure 124: Aggregates consumption/GDP in France (Data by Rodriguez Chavez/Schleifer, 2010) 283

Figure 125: Mining production in Spain between 1998 and 2008 – selected mineral resources including aluminium production (Data by Weber et al. [WMD 1998–2008]) 286

Figure 126: CMR, Lomero –Poyatos (Data by http://www.cambmin.co.uk) 287

Figure 127: La Zarza Copper-Gold Project (Data by http://www.ormondemining.com) 288

Figure 128: Aguablanca nickel-copper sulfide deposit (Data by http://www.lund inmining.com) 289

Figure 129: Production, export and import of fluorspar (Data by BGS) 291

Figure 130: Mining production in Portugal between 1998 and 2008 – selected mineral resources (Data by Weber et al. [WMD 1998–2008]) 296

Figure 131: Neves-Corvo Copper/Zinc mine (Data by http://www.nafinance.com) 297

Figure 132: Mining production in Romania between 1998 and 2008 – selected mineral resources (Data by Weber et al. [WMD 1998–2008]) 302

Figure 133: Gold mining in Romania (Rosia Montana) (Data by http://www.gabrielresources.com) 303

Figure 134: Mining production in Bulgaria between 1998 and 2008 – selected mineral resources (Data by Weber et al. [WMD 1998–2008]) 309

Figure 135: Production, export and import of copper (Data by BGS) ... 310

Figure 136: Kaolin and Bentonite industry in Bulgaria (Data by http://www.kaolin.bg) 312

Figure 137: Mining production in Greece between 1998 and 2008 – selected mineral resources including aluminium production (Data by Weber et al. [WMD 1998–2008]) 315

Figure 138: Production, export and import of bauxite (Data by BGS) .. 316

Figure 139: Larco G.M.M.S.A. 317

Figure 140: S&B Industrial Minerals Company, Voudia bay; Milos Island (Data by http://www. s.andb.gr) 321

Figure 141: Production of selected metallic and industrial minerals in Russia (European part) (Data by Weber et al. [WMD 1998–2008]) ... 323

Figure 142: Production, export and import of nickel (Data by BGS) ... 325

Figure 143: Deposit Stoilensky GOK http://www. nlmksteel.com 326

Figure 144: Mining production in Norway between 1998 and 2008 – selected mineral resources including aluminium production (Data by Weber et al. [WMD 1998–2008]) 328

Figure 145: Production, export and import of titanium (Data by BGS) 329

Figure 146: Mining on the coast of Norway Tinfos Titan & Iron KS – Tyssedal (Data by http://www.tinfos.no) 332

Figure 147: Extraction of sand and gravel in Switzerland (Data by Kündig et al, 1997) 336

List of figures and tables

Figure 148: Copper mine near the city of Bor (Veliki Krivelj Mine) (Data by http://www.euromaxresources.com) 338

Figure 149: Mining production in Albania between 1998 and 2008 – selected mineral resources (Data by Weber et al. [WMD 1998–2008]) 339

Figure 150: Production, export and import of titanium (Data by BGS) 340

Figure 151: Steel works in Elbasan/Albanien (Data by http://www.travelbilder.de) 342

Figure 152: Belo Brdo Mining facilities on the surface (Data by www.kosovomining.org) 344

Figure 153: Global contribution of metallic and industrial minerals production in Europe 396

Figure 154: EU critical minerals (EC, 2010) – Comparison in terms of production, import and export related to the observed countries (19 EU-countries including Non-EU countries Norway and Switzerland) 398

Figure 155: Production concentration of the 'critical' raw materials by source country (EC, DG Enterprise, 2010) 406

Figure 156: Summary and outlook – EU's mining, Trends in global metal mining since 1850 (Data by Ekdahl 2008) 415

Figure 157: Effects of globalization 416

Figure 158: European mining industry on a global scale (Data by Ekdahl 2008) 418

Figure 159: Conflicts of interest in an EU raw materials policy 419

Figure 160: China's share in global consumption, 2005 (Data by Ekdahl 2008) 420

Figure 161: Metal scrap 424

Figure 162: WTO structure (WTO) 425

Figure 163: European Commission, Brussels 429

Figure 164: Mining in development countries, example Africa 430

Figure 165: Africa- Caribbean -Pacific (ACP) raw materials exporting countries 1990–1999 in % of total exports (Data by World Bank, Christman, 2008, Presentation, EU-Parlament) 432

Figure 166: Mineral wealth of Africa including trade partners (Data by Die Presse, 2008) 434

Figure 167: Global exploration 1995–2008 (Data by Ekdahl 2008) 435

Figure 168: Global exploration in 2006 (Data by Ekdahl 2008; RMG, MEG) 437

Figure 169: Reduction of gravel reserves due to other land use utilisations (Data by Kündig et. al. 1997) 439

Figure 170: Aggregates situation in Germany (Data by BGR, 2009) 440
Figure 171: Impact of the EU environmental protection law on the raw materials industry 442
Figure 172: Illustration of the proportion of land of the FFH-areas in the EU-27 (http://ec.europa.eu) 443
Figure 173: Important raw material – metal scrap 449
Figure 174: RHI AG – magnesite extraction (Data by Drnek, 2008) 452
Figure 175: Cement plant Retznei (Ehrenhausen), Lafarge Perlmooser Gmbh (Data by http://www.lafarge.at) 453
Figure 176: EU Commission, Brussels (Data by Wikipedia) 456
Figure 177 : Interrelations of EU minerals policy 488
Figure 178: Organigram chart of a coherent EU minerals policy 491
Figure 179: Raw Materials Initiative research strategy – relevant items (Data by http://ec.europa.eu) 501
Figure 180: http: //asiamining.org 582

Table 1: Summary of manufacturing sectors depending on minerals supply (Data by European Commission [EC], DG Enterprise and Industry, Commission staff working document, 2007) 5
Table 2: Example of the economic importance of EU industries consuming non-metallic minerals (Data by EC, DG Enterprise and Industry, Commission staff working document, 2007. Based on "Good Environmental Practice in the European Extractive Industry: A Reference Guide", with figures for value added and employment updated to 2002–2003 by DG Enterprise and Industry using Eurostat data) 7
Table 3: Ranking of investment decision factors at the exploration and mining investment stage (Data by Otto, 1992) 14
Table 4: Operating costs in mining – comparison of energy costs (electricity and fuel) in proportion to the total operating costs. (Data by EC, DG Enterprise and Industry, Commission Staff Working document, 2007) Comparison of the relative costs of energy (electricity and fuel) as a proportion of overall operating costs. Data source: UEPG and IMA with metallic data derived from minecost.com data 35
Table 5: Production of selected metallic raw materials and examples of the raw material application, (Data by EC, DG Enterprise and Industry, Commission Staff Working Document, 2007 [data: British Geological Survey]) 41

List of figures and tables

Table 6: Summary of production of selected industrial minerals, their main downstream markets and factors identified as affecting demand (Data by EC, DG Enterprise and Industry, 2007 [European Minerals Yearbook 1996–1997; EULA; K+S-estimation; Industrial Minerals Association website (http://www.ima-eu.org]) 46

Table 7: Top three producing regions for selected ores and metals (2004) (Data by Weber and Zsak, World Mining Data, 2006) 61

Table 8: Top three producing regions for selected industrial minerals (Data by EC, DG Enterprise and Industry, 2007 [calculations based on BGS data]) 61

Table 9: Location of metal mining in the EU in 2003 (Data by European Commission, 2007 [Data source: BGS (2005) European Mineral Statistics.]) Austria, Finland, Greece, Ireland, Poland, Portugal and Sweden have metal mining industries that contribute more than 1% to global production of one particular metallic mineral 63

Table 10: Metals which are not mined in the EU, indicating the countries with significant known reserves, and summary of their uses (Data by European Commission, DG Enterprise and Industry, Commission Staff Working Document, 2007) 66

Table 11: Industrial minerals extracted in the EU in 2003 and Member States involved (countries with >2% of world production in bold) (Data by EC, 2007 [Data sources: BGS (2005) European Mineral Statistics; EULA, EuroGeoSurveys (for quartz and silica sand)]) 72

Table 12: Exports of industrial minerals outside the EU – Member States (Data by EC, DG Enterprise and Industry, Commission staff working document, 2007) 76

Table 13: The European Aggregates Industry – Annual Statistics 2008, Quantities in million tonnes. (Data by UEPG, 2010) 79

Table 14: Growth rates of metals consumption and shares of world totals (selected countries) (Data by Crowson, 2008) 83

Table 15: China: Exports of selected mineral commodities in 2008 (Data by USGS, 2009) 90

Table 16: China: Imports of selected mineral commodities in 2008 (Data by USGS, 2009) 92

Table 17: Copper (Data by BGR, 2007) 117

Table 18: European Union import dependence – Imports as percentage of domestic consumption plus exports (Data by Crowson, 2008) ... 120

Table 19: Global demand of the emerging technologies analysed for raw materials in 2006 and 2030 related to today's total world production of the specific raw material (Updated by BGR April 2010) (Data by EC, 2010) 126

Table 20: Principles for an effective and workable mineral planning system (Data by Department of Mining and Tunneling, 2004 [provided by UEPG]) 141

Table 21: Japanese import dependence (Crowson, 2008) 162

Table 22: Instruments of a minerals policy according to Siebert (with supplements of Brandstätter and Tiess) (Data by Brandstätter, 1989) 167

Table 23: Conception of a raw materials policy (the state as a raw materials importer) 174

Table 24: Metallic minerals: production/mining – import – export of commodities of Finland (Data by BGS, 2010 [European Mineral Statistics]) 192

Table 25: Industrial minerals: production/mining – import – export of commodities of Finland (Data by BGS, 2010 [European Mineral Statistics]) 193

Table 26: Metallic minerals: production/mining – import – export of commodities of Sweden (Data by BGS, 2010 [European Mineral Statistics]) 199

Table 27: Industrial minerals: production/mining – import – export of commodities of Sweden (Data by BGS, 2010 [European Mineral Statistics]) 200

Table 28 : Metallic minerals: production/mining – import – export of commodities United Kingdom (Data by BGS, 2010 [European Mineral Statistics]) 206

Table 29: Industrial minerals: production/mining – import – export of commodities of United Kingdom (Data by BGS, 2010 [European Mineral Statistics]) 208

Table 30: Metallic minerals: production/mining – import – export of commodities of the Netherlands (Data by BGS, 2010 [European Mineral Statistics]) 217

Table 31: Industrial minerals: production/mining – import – export of commodities of the Netherlands (Data by BGS, 2010 [European Mineral Statistics]) 219

Table 32: Metallic minerals: production/mining – import – export of commodities of Belgium (Data by BGS, 2010 [European Mineral Statistics]) 222

Table 33: Industrial minerals: production/mining – import – export of commodities of Belgium (Data by BGS, 2010 [European Mineral Statistics]) 224

List of figures and tables

Table 34: Metallic minerals: production/mining – import – export of commodities of Poland (Data by BGS, 2010 [European Mineral Statistics]) 230

Table 35: Industrial minerals: production/mining – import – export of commodities of Poland (Data by BGS, 2010 [European Mineral Statistics]) 231

Table 36: Metallic minerals: production/mining – import – export of commodities of Germany (Data by BGS, 2010 [European Mineral Statistics]) 236

Table 37: Industrial minerals: production/mining – import – export of commodities of Germany (Data by BGS, 2010 [European Mineral Statistics]) 237

Table 38: Metallic minerals: production/mining – import – export of commodities of the Czech Republic (Data by BGS, 2010 [European Mineral Statistics]) 245

Table 39: Industrial minerals: production/mining – import – export of commodities of the Czech Republic (Data by BGS, 2010 [European Mineral Statistics]) 246

Table 40: Metallic minerals: production/mining – import – export of commodities of Slovakia (Data by BGS, 2010 [European Mineral Statistics]) 250

Table 41: Industrial minerals: production/mining – import – export of commodities of Slovakia (Data by BGS, 2010 [European Mineral Statistics]) 252

Table 42: Metallic minerals: production/mining – import – export of commodities of Austria (Data by BGS, 2010 [European Mineral Statistics]) 257

Table 43: Industrial minerals: production/mining – import – export of commodities of Austria (Data by BGS, 2010 [European Mineral Statistics]) 258

Table 44: Metallic minerals: production/mining – import – export of commodities of Hungary (Data by BGS, 2010 [European Mineral Statistics]) 264

Table 45: Industrial minerals: production/mining – import – export of commodities of Hungary (Data by BGS, 2010 [European Mineral Statistics]) 265

Table 46: Metallic minerals: production/mining – import – export of commodities of Slovenia(Data by BGS, 2010 [European Mineral Statistics]) 268

Table 47: Industrial minerals: production/mining – import – export of commodities of Slovenia (Data by BGS, 2010 [European Mineral Statistics]) 269

Table 48: Metallic minerals: production/mining – import – export of commodities of Italy (Data by BGS, 2010 [European Mineral Statistics]) 273

Table 49: Industrial minerals: production/mining – import – export of commodities of Italy (Data by BGS, 2010 [European Mineral Statistics]) 275

Table 50: Metallic minerals: production/mining – import – export of commodities of France (Data by BGS, 2010 [European Mineral Statistics]) 280

Table 51: Industrial minerals: production/mining – import – export of commodities of France (Data by BGS, 2010 [European Mineral Statistics]) 282

Table 52: Metallic minerals: production/mining – import – export of commodities of Spain (Data by BGS, 2010 [European Mineral Statistics]) 290

Table 53: Industrial minerals: production/mining – import – export of commodities of Spain (Data by BGS, 2010 [European Mineral Statistics]) 293

Table 54: Metallic minerals: production/mining – import – export of commodities of Portugal (Data by BGS, 2010 [European Mineral Statistics]) 299

Table 55: Industrial minerals: production/mining – import – export of commodities of Portugal (Data by BGS, 2010 [European Mineral Statistics]) 300

Table 56: Metallic minerals: production/mining – import – export of commodities Romania (Data by BGS, 2010 [European Mineral Statistics]) 305

Table 57: Industrial minerals: production/mining – import – export of commodities of Romania (Data by BGS, 2010 [European Mineral Statistics]) 306

Table 58: Metallic minerals: production/mining – import – export of commodities of Bulgaria (Data by BGS, 2010 [European Mineral Statistics]) 311

Table 59: Industrial minerals: production/mining – import – export of commodities of Bulgaria (Data by BGS, 2010 [European Mineral Statistics]) 313

Table 60: Metallic minerals: production/mining – import – export of commodities of Greece(Data by BGS, 2010 [European Mineral Statistics]) 319

Table 61: Industrial minerals: production/mining – import – export of commodities of Greece (Data by BGS, 2010 [European Mineral Statistics]) 321

List of figures and tables

Table 62: Metallic minerals: production/mining – import – export of commodities of Norway (Data by BGS, 2010 [European Mineral Statistics]) 330

Table 63: Industrial minerals: production/mining – import – export of commodities of Norway (Data by BGS, 2010 [European Mineral Statistics]) 331

Table 64: Metallic minerals: production/mining – import – export of commodities of Switzerland (Data by BGS, 2010 [European Mineral Statistics]) 334

Table 65: Industrial minerals: production/mining – import – export of commodities of Switzerland (Data by BGS, 2010 [European Mineral Statistics]) 335

Table 66: Production of metallic raw materials in selected countries of Europe between 1998 and 2008 (Weber et al., 2004, 2009, 2010) 348

Table 67: Production ranking of selected metallic raw materials – comparison between Europe and the world in 2008 (Weber et al., 2010) 350

Table 68: Production of industrial minerals in selected European countries between 1998 and 2008 (Data by Weber et al., 2004, 2009, 2010) 352

Table 69: Production ranking of selected industrial minerals – comparison between Europe and the world in 2008 (Weber et al, 2010) 356

Table 70: Production of aggregates in selected countries of Europe – comparison based on different features (UEPG 2010) 358

Table 71: European minerals economy (including mining and minerals related industries) – overview (Economy Watch, OECD, USGS, WMD and UEPG) 360

Table 72: Production, import and export of metallic minerals related commodities in selected countries of Europe between 1998 and 2008 (BGS [European minerals statistics]) 368

Table 73: Production, import and export of industrial minerals related commodities in selected countries of Europe between 1998 and 2008 (BGS [European minerals statistics]) 380

Table 74: Summary of different issues of the mineral policy framework in selected European countries countries (Department of Mining and Tunnelling 2004, Land Use Consultants 2010, European mining laws, USGS) 386

Table 75: Comparision of the national minerals strategies of Finland, Germany and France 407

Table 76: Comparison – USGS and EuroGeoSurveys (Data by Christmann, 2008) 436

Table 77: Relevant EU-environmental directives concerning the raw material industry (extract) 441

Table 78: Summary of responses from the non-energetic extractive industry and national geological surveys to the request for information on the current impact of Natura 2000 on permits for minerals extraction (Data by EC, DG Enterprise and Industry, Commission Staff Working Document, 2007) 445

Table 79: Concept of a European raw material policy 485

Table 80: Policies relevant to the EU mineral supply (Data by Michaelis, 1977 with additionally remarks from Tiess) 487

Table 81: EU Raw Materials Initiative – 10 actions (Data by European Commission, 2008b) 494

Table 82: Organisations present at the Luleå Conference 500

Table 83: List of critical raw materials at the EU level (EC, DG Enterprise, 2010) 503

Table 84: Minerals policy – Global and EU relevant activities between 1950 and 2011 (some selections) 517

Table 85: Primary self supply in developed copuntries 1978 – 1980 (in %) (Sources: EC raw material balances [various years], Luxembourg) 613

Table 86: Consolidated EC raw materials balances 1980 613

Table 87: Critical non-ferrous metals (Data by Communication of the Commission dated 7. Feb. 1975 about supply of raw materials for the EC, Supplement 1/75 EC-Bulletin) 614

Chapter 1 Introduction

Raw materials mark the beginning of a value-added chain. In times of increasing globalization their availability is a precondition for prosperity, productivity and development of a country's economy. Raw materials have always been needed for many purposes. Global demand had its origin in the Bronze and Iron Ages and increased exponentially during the Industrial Revolution in the 19th Century. Worldwide demand for raw materials rapidly increased in the 20th Century, caused by the explosive growth of the world population and global economic productivity. In the 21st Century, the information and computer age, the demand for raw materials remains on a high level. Modern jet airlines, automobiles, and mobile phones require a comprehensive cocktail of innumerable raw materials.

1.1 Raw materials

This book specifically covers non-energy raw materials. These include metallic minerals, industrial minerals and construction minerals. Energetic raw materials, like crude oil, natural gas, brown coal, or hard coal are not discussed. In economic geology ores are defined as metal-bearing rocks and mineral mixtures from which metals or metallic compounds can be extracted by technical means and with economic profit. This definition corresponds with mining terminology, although occasionally even unconventional raw materials like evaporite liquor (lithium, magnesium), geothermal waters (zinc), metal accumulating plants (nickel and others), or acidic mine waters (copper) can be processed to metals. In mineralogy, however, the term ore usually refers to an ore mineral such as for example chromite or galena (lead sulfide). As stated in the definition, extractive metallurgy from ores should produce economic profit. This necessitates a minimum mineral content, which can vary depending on the availability of best technological as well as general economic conditions.[1]

Industrial minerals are solid raw materials, which are not used for the extraction of metals and usually are monomineralic, such as talc, mica, or mag-

---

[1] Pohl, W.L. (2005): Mineralische und Energie-Rohstoffe. Eine Einführung zur Entstehung und nachhaltigen Nutzung von Lagerstätten [Mineral and energy raw materials. An introduction to the development and sustainable use of deposits, p. 5].

nesite.² These are used because of certain chemical and/or physical properties for special applications.

Construction minerals are raw materials in the construction industry, which occur as rocks in the geological sense and often are mineral mixtures (e. g. sand and gravel, clay, granite).³

Industrial minerals and construction raw materials along with inorganic salts belong to the group of non-metallic raw materials. These can be used in for a variety of purposes, requirements concerning exploration of deposits, processing, refinement and market research being particularly high. In contrast to the evaluation of ore deposits, it is usually necessary to test processing at semi-industrial standard in order to prove that products will meet all customers requirements.⁴

1.2 Minerals for the economy

Resources are the backbone of every economy.

It is fundamentally important for this book to define mineral raw materials from the mineral economic point of view. According to Fettweis,⁵ the following are counted as mineral raw materials in the broader sense (with the exception of water and air):

1. Minerals and mineral mixtures in solid and liquid state as well as gases that occur naturally and in mining waste and have a practical value as mineral resources or because of their mineral resources content.
2. Materials provided by mining that meet the demand which originates from their practical value.
3. Materials extracted from the air that are in demand, and
4. Secondary and waste materials of mineral origin that are in demand.

Accordingly, Fettweis defines valuable mineral materials as those natural mineral substances, i. e. materials of the inanimate nature, which can be used for purposes of production (including building industry) due to their physical

---

2 Pohl (2005), l.c., p. 232. – Also the term 'construction minerals' is common.
3 Pohl (2005), l.c., p. 232.
4 Pohl, (2005), l.c., p. 232.
5 G. B. Fettweis (1990) in: Siegfried von Wahl, Bergwirtschaft, Band I, Die elementaren Produktionsfaktoren des Bergbaubetriebs, Der Produktionsfaktor Lagerstätte [The basic factors of production of mining operations , Deposit as a production factor] p. 2.

or chemical material properties, energy production or direct consumption for the satisfaction of human needs.[6]

Also Drnek states:[7] Mineral raw materials are used for the production of economic goods, in principle serving three purposes: as a source of energy, as material components of goods in material goods production, and, furthermore, as process additives, for instance fertilizers in agriculture. Raw and base materials represent a factor of production[8] needed by many economic sectors and are part of all material goods.[9] Material goods production refers to the development of infrastructure (energy and water supply), to building industry and trade, as well as the industry of a state as a whole.[10] Thus, raw and base materials are crucial for the competitive position of an entire national economy.

In this connection, the degree of industrialization of a state is of fundamental importance. Obviously, the structural changes of a nations's economy are reflected in the development of the intensity of material use.

It is a fact that in national economies running through the process from primary to secondary sector, the consumption of raw materials increases in the same or even a higher degree than the economic performance, industrialization being a material-intensive process. In advanced economies, the increasing service sector (tertiary sector) causes a gradual decoupling of economic growth and consumption of raw materials, which leads to a decrease of material intensiveness.[11] This applies particularly to the EU countries. Nowadays many European Union countries are mature economies, in which about two thirds of

---

6   Fettweis (1990), l.c., p. 5.
7   Drnek, T. (2008b): Mineralrohstoffwirtschaft/spezielle Mineralwirtschaft (minerals economy/specific minerals economy), Leoben, p. 5.
8   On the term "raw materials as a production factor" see also below.
9   Bundesministerium für Handel, Gewerbe und Industrie (1981): Konzept für die Versorgung Österreichs mit mineralischen Roh- und Grundstoffen (Ministry of Commerce, Trade and Industry (1981): Draft for supply of Austria with mineral raw and base materials), p. 11.
10 Cp. Weber, L. (1997): Mineralrohstoffe als Basis für die Wirtschaft, in: Österreichische Akademie der Wissenschaften [Mineral raw materials as the basis for economy, in: Austrian Academy of Sciences]. Wellmer, F. W., Wagner, M. (2006): Metallic raw materials - constituents of our economy, in: Gleich A., Ayres, R. U., Gößling-Reisemann, S.: Sustainable metals management.
11 Nötstaller, R., Wagner, H. (2007): Überlegungen zum Rohstoffbedarf und zur Rohstoffpolitik, BHM – Berg- und Hüttenmännische Monatshefte [Reflections on resource consumption and resource policy, Journal of Mining. Metallurgical, Material, Geotechnical and Planned Engineering], p.384. – The tertiary sector consists of market and non-market services (trade, hotels and restaurants, transport and communication, financial, insurance, public services, etc).

the economic performance is already attributed to the tertiary sector.¹² Due to these developments material intensity has been decreasing since the 70s.

Despite the change of the economic structure of Europe in favour of the tertiary sector, a high demand of raw materials is still to be expected.¹³ Moreover, many branches in service can only develop complementary to material goods production.¹⁴

Industry continues to be very important to the economy of the European Union because of its contribution to employment and added value. It is the key link in the value-added chain for producing material goods. The question of the economic importance of raw and base material supply cannot be answered without considering the means of integration of the economy into the global economic development, for example considering possibilities for export. In this context, the industrial structure of a state is of great importance as well.

Material goods production still is a central core of the European economy.¹⁵ The total value of output (for example) of the EU-25 steel industry in 2005 was € 220 billions, equivalent to roughly 2,5 % of the total value of industrial production, and the added value amounted to € 49 billions. The sector employs 412 000 people, representing 1,25 % of the total employment in EU manufacturing. With a production of around 210 million tons of crude steel in 2007, the EU represents 16 % of world output and is the second largest producer behind China.¹⁶

The mineral economy for several European Union and non-European Union countries is elaborated more in-depth in chapter 5. At this point, the importance of the European building industry, base material chemistry or steel and metal industry shall be emphasized exemplarily. The engineering industry, for instance, ranks among the most important consumers of metals, most notably steel in various processing forms. The electrical industry, likewise, is a significant consumer and processor of non-ferrous metals, particularly lead, copper, tin, and aluminium. The relatively high share of branches of production with high consumption of raw materials in the European Union, such as the metalworking industry (electrical industry, mechanical engineering, automotive

---

12 This fact can be understood regarding the countries discussed in the chapter 5.
13 Nötstaller/Wagner (2007), l.c., p. 384.
14 Europäischer Wirtschafts- und Sozialausschuss [European Economic and Social Committee] (2006), Stellungnahme des Europäischen Wirtschafts- und Sozialausschusses über "Risiken und Probleme der Rohstoffversorgung der europäischen Industrie" Statement of the European Economic and Social Committee on Risks and problems associated with the supply of raw materials to European industry ", (2006/C 309/16), p. 75.
15 European Economic and Social Committee, l.c., p. 75.
16 http://ec.europa.eu/enterprise/steel/index_en.htm

industry) generates considerable export surpluses (e. g. steel), but also causes a **high dependency of the European economy on raw materials**. Evaluating the future demand for raw materials of the electrical industry, it is to be considered that economic dynamics will lie less in the raw materials-intensive sections (e. g. cable, accumulators) than in the industry for electronic devices.

Table 1: Summary of manufacturing sectors depending on minerals supply (Data by European Commission [EC], DG Enterprise and Industry, Commission staff working document, 2007)

Chemicals
Manufacture of chemicals, chemical products and man-made fibres, e. g. chemical products, pesticides, paints, pharmaceuticals, soaps, detergents and explosives.
Non-metallic products
Manufacture of other non-metallic mineral products, e. g. glass and glass products, ceramics, bricks, tiles, cement and plaster.
Base metals
Manufacture of base metals and fabricated metal products, e. g. aluminium, lead, zinc and fabricated metal products (tools, cutlery, etc.).
Machinery
Manufacture of machinery and equipment, e. g. engines and turbines, pumps, agricultural tractors and domestic appliances.
Electrical/optical
Manufacture of electrical and optical equipment, e. g. office machinery and computers, electric motors, lighting, televisions, optical and photographic equipment.
Transport
Manufacture of transport equipment.

A secure and cost-effective supply of raw materials for the industry is therefore a necessary prerequisite:[17] Consequently, **a minerals policy is of fundamental significance**[18].

---

17 The strategic objective of the EU, which was resolved by the Heads of States and Governments in Lisbon (March 2000), is to be regarded in this context. Accordingly, the EU is to develop the most competitive and dynamic knowledge-based economy in the world capable to create permanent economic growth with more and better jobs and greater social coherence. Thus, the EU should become an "area" in the world whose economic structure is characterized by high resource efficiency.
18 Cp. Anonym (2007): Rohstoffpolitik ist Daueraufgabe und Zukunftssicherung /BDI-Kongress, Metall – Internationale Fachzeitschrift für Metallurgie [Mineral policy is a permanent task and guarantee for the future/International Journal of Metallurgy],

1.3 Importance of the extractive industry in the value chain

The economic importance of a sector in a country's economy is usually measured by its contribution to the gross domestic product (GDP) as well as by the job-creating effects of the respective industry.[19] For an appropriate evaluation of the role of the mining industry from macroeconomic view it is important to consider its special position as the first section of a value-added chain. The **multiplication effect in mining** concerning the production processes of downstream goods directly or indirectly dependent on the mining industry is of utmost significance. Disregard for this fact leads to an incomplete and **misleading image** of the real overall economic importance of mining.[20] In this respect, see figure 1 as well as table 2.

Figure 1: Illustration of the added value chain raw materials – end product (Data by EC, DG Enterprise and Industry, Commission staff working document, 2007)

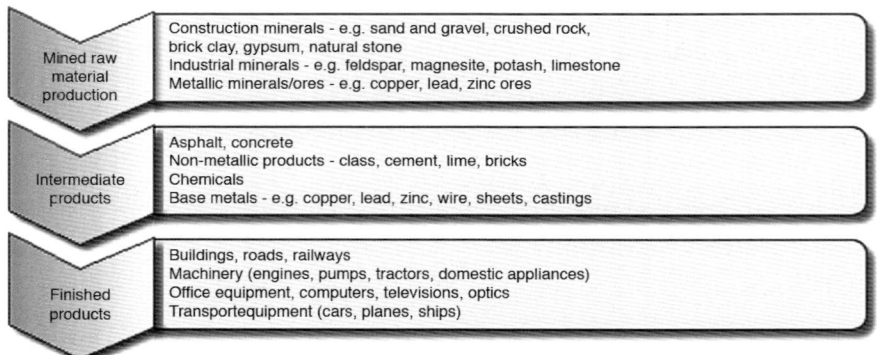

---

pp. 230–231. Reference: Minerals policy contexts are discussed in Chapter 4 (Outline of a raw materials policy).

19 Cp. in this context, the term "structure" according to Molitor (2006): Economic Policy, Munich 2006, sector (p.213): The concept of structure focuses on part units proportional to the complete aggregate or to other part units. An economy is divided into more or less succinctly definable regions, which may vary according to resource inventory, level of development and contribution to the GNP (per capita income). Under functional consideration national economy is divided into sections (industries), each with specific contribution to the overall product (*mining*, energy, transport, agriculture, chemistry, engineering, commerce, etc.).

20 Bundesministerium für Handel, Gewerbe und Industrie ([Austrian] Ministry of Commerce, Trade and Industry) (1981), l.c. p. 11. Wasserbacher, R.. Koller, W., Schneider, H. W., M. Luptáčik, M. (2007): Die volkswirtschaftliche Bedeutung mineralischer Rohstoffe in Österreich, BHM – Berg- und Hüttenmännische Monatshefte (The relevance of raw materials extraction to Austria's economy, Journal of Mining, Metallurgical, Material, Geotechnical and Planned Engineering), pp. 391–396.

Table 2: Example of the economic importance of EU industries consuming non-metallic minerals (Data by EC, DG Enterprise and Industry, Commission staff working document, 2007. Based on "Good Environmental Practice in the European Extractive Industry: A Reference Guide", with figures for value added and employment updated to 2002–2003 by DG Enterprise and Industry using Eurostat data.)

Application	Value added (€ million)	Employment	Mineral content
Construction			
Glass	16 336	375 400	100 %
Ceramic tiles and flags	4 253	94 900	100 %
Bricks, tiles and construction products	3 891	78 300	100 %
Concrete	10 515	256 600	100 %
Cement, lime and plaster	8 717	77 700	100 %
Natural stone production	5 492	189 300	100 %
Non-construction			
Rubber products	17 057	359 400	up to 50 %
Plastic products	55 534	1 310 400	up to 50 %
Paints and varnishes	10 601	179 400	up to 70 %
Paper and paper board	17 429	223 800	up to 30 %
Ceramics used for non-construction uses	6 514	199 100	100 %
Basic chemicals	64 928	584 500	variable
Basic pharmaceuticals	6 812	66 700	variable
Mineral filters			100 %
Sugar			process aid

The non-energetic EU raw materials industry, together with the material goods production and construction, industry contributes considerably to the EU GDP.[21] The following statement by Fettweis (2000) to the significance of

---

21 Anciaux, P. (2005): Die Wettbewerbsfähigkeit des Europäischen Rohstoffsektors, Berg- und Hüttenmännische Monatshefte (The Competitiveness of the European raw material sector, Journal of Mining. Metallurgical, Material, Geotechnical and Planned Engineering), pp 275–278.

mining and mineral raw materials for our society is exactly to the point: "Although global mining contributes on an average only a few percent to the world gross national product, without it the remaining 95 % would not exist."[22]

1.4 Characteristics of the extractive industry

The mining industry shows different characteristics that distinguish it from other industrial sectors and that are important regarding its function of supplying the economy with raw materials. It is substantial for the understanding of the topic to deal with certain terms and contexts.

1.4.1 Mining – Initial production

Mining, like agriculture and forestry, belongs to the initial production.[23] According to Fettweis[24], in order to be able to extract minerals, they must already exist in technically accessible parts of the inanimate nature. To make them available for production, the mining industry has to fulfil three specific sub-tasks:

1. **to locate** deposits of raw materials in the earth's crust,
2. **to mine** these deposits and
3. **to process** the content of the deposits to a mining product suitable for marketing.

The search for minerals can be divided into two stages: prospection (search) and exploration (examination). As "material component" a mineable part of the earth's crust is to be selected (according to certain geological criteria); for the subsequent exploration an already located deposit is available. Mineable or limited mineable deposits are the result of successful exploration.

Extraction is the key process of mining. Hence, a mining company normally is an extraction company. "Material components" of extraction is the deposit field; the product itself is the mined raw material.

---

22 Fettweis, G.B.L. (2000): Über Bergbau und Bergbaukunde im Raum des heutigen Österreich seit 1849, Berg- und Hüttenmännische Monatshefte (About Mining and Mining Engineering in the area of present-day Austria since 1849, Journal of Mining. Metallurgical, Material, Geotechnical and Planned Engineering) 145 (2000), pp. 127–142.
23 Fettweis, Günter B. L. (1994): Zum ökonomischen Prinzip im Bergbau – Besonderheiten und Einordnung in das übergeordnete Rationalprinzip, (On the economic principle in mining – particularities and integration into the superior rational principle) Erzmetall, pp. 23–33.
24 Fettweis (1990), l.c., p. 1.

Since in most cases raw materials are not saleable mining products, they usually have to be processed for enrichment and separation by grain size respectively.

The deposit made available by exploration and its contents are for the mine what materials – be it raw materials, base material or pre-products – are for a processing economy as factors of production.[25]

Location bound deposit

Regarding land use conflicts (see chapter 6.5) it is substantial to address the fact that deposits are location bound.

First the term deposit should be explained. According to Pohl, natural accumulations of usable minerals and rock are called deposits if they can be used for an economic production, considering their size and mineral content.[26] Mineral and rock bodies, which are too small or too poor to be exploited with economic profit, are named occurrences. In common language, occurrences of mineral raw materials are often called deposits, so it is sometimes preferable to speak of usable deposits.[27] Mineral deposits are location bound and exhaustible. The mining site is given by the location of the deposit. The exhaustibility of the deposit means that the mine has to be abandoned upon completion of mining, and exploitation has to be moved to other mining sites.[28] Principally, it is impossible to select an optimal location for mining whilst considering the distance to markets or the availability of infrastructural facilities.[29] Further-

---

25 Fettweis, G. B. L., Brandstätter, W. A., Hruschka, F. (1987): The deposit as a production factor of mining. – Acta Geodaetica, Geophysica et Montanistica Hungarica, Vol. 22, no 3–4, 1987, pp 371–389
26 Pohl (2005), l.c., p. 1.
27 Cp. Murawski, H., Meyer, W., Bonn (2004): Geologisches Wörterbuch. Also: Weber, L. to the term "deposit" in a written statement to the author (2008): Above-average (abnormal) enrichment of one or more elements (element compounds) that can be won with economic benefit.
28 Cp. Wagner, H. (1997): Untersuchung der Versorgung Österreichs mit mineralischen Rohstoffen aus heimischen Vorkommen (Investigation of the supply of mineral raw materials from domestic reserves in Austria) Vol. I–V, Wien, Leoben.
29 To some extent this might establish the construction minerals as an exception: Letouzé distinguishes between absolute and relative location-bound deposits. – Letouzé, G., (1996): Projekt Harmonisierungsmodell, Schritte zu einer bundesweiten Harmonisierung der Materie Mineralrohstoff-Vorsorge, Gutachten zum Fachbereich Rohstoffgeologie (Project harmonization model, steps towards a nation-wide harmonization of mineral raw materials provision, expertise to the Department of Resource Geology), Appendix 2, Wien 1996 In this sense also Solar (2004), in: Department of Mining and Tunnelling, University of Leoben (2004), Minerals Policies and Supply Practices in Europe, Final Report. http://ec.europa.eu/enterprise/policies/raw-materials/files/best-practices/leoben_2004_en.pdf; an extended summary from this report is avail-

more, the assessment of the influence of market prices and production costs is of great importance, as their change might make an inefficient deposit become economically advantageous and vice versa.

Figure 2: Diamond deposit near Mirny/Yakutia/Russian Federation (Data by http://www.usmra.com)

The diamond deposits near Mirny/Yakutia, discovered in 1954, are the second largest in the world (after South Africa). The diameter of the mining site amounts to 1.250 m, with a depth of 525 m. One hundred percent of Russia's diamonds are extracted in Yakutia. Every year 11 000 000 tonnes of kimberlite are produced.

Extractive industry is capital-intensive

Prospection, exploration and exploitation depend on technical circumstances and are capital-intensive (see figure 4). In mining, new capacities are not established very quickly. The building of a new business for the exploitation of mineral raw materials generally takes between five and twelve years, or even up to 20 years for major projects. Realisation is carried out in sections, every phase being based on the results of the preceding one. Accordingly, the range of raw materials on offer is marked by little flexibility.[30]

---

      able under http://ec.europa.eu/enterprise/policies/raw-materials/files/best-practices/leoben_extsum_en.pdf.
30 In particular, conferring to the mining of ores: cp. Gocht (1983), l.c., p. 2. Also: Wagner, M, Huy, D. (2005): Schafft der Strukturwandel in der Nachfrage eine neue Dimension für die Weltrohstoffmärkte, Bundesanstalt für Geowissenschaften und Rohstoffe [Does the structural change in demand accomplish a new dimension to the world raw materials markets? Federal Institute for Geosciences and Raw Materials], Hannover, p. 2: From the discovery of an ore deposit to production, at least 5 years for planning and construction have to be calculated.

1.4 Characteristics of the extractive industry

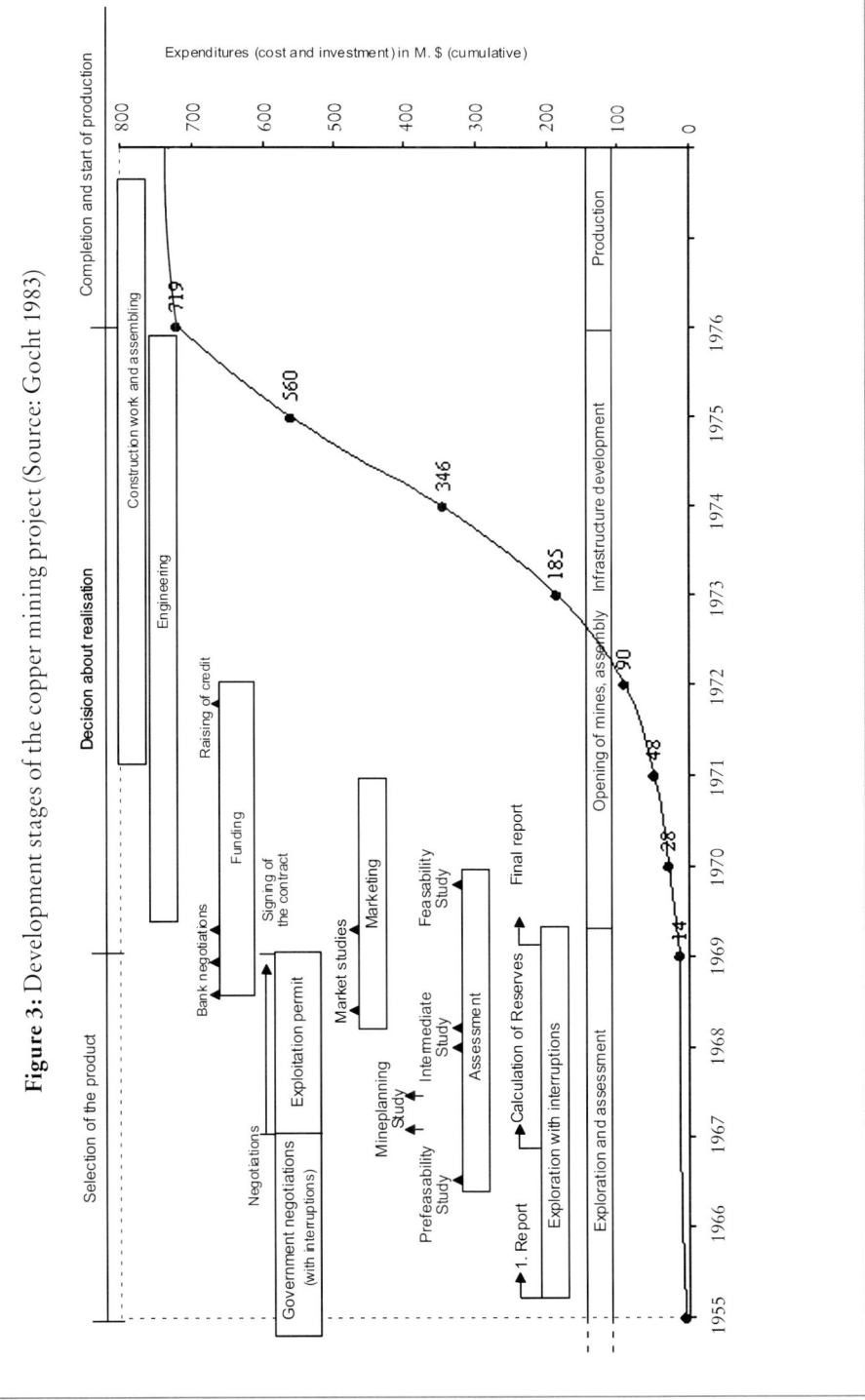

Figure 3: Development stages of the copper mining project (Source: Gocht 1983)

Chapter 1 Introduction

Figure 4: Capital-intensity of economic sectors in EU-13 in 1999–2001 showing the relative position of the non-energetic raw materials industry (Data by EC, DG Enterprise and Industry, Commission staff working document, 2007) Investment in tangibles per person employed – 1999–2001 (1.000 EUR)

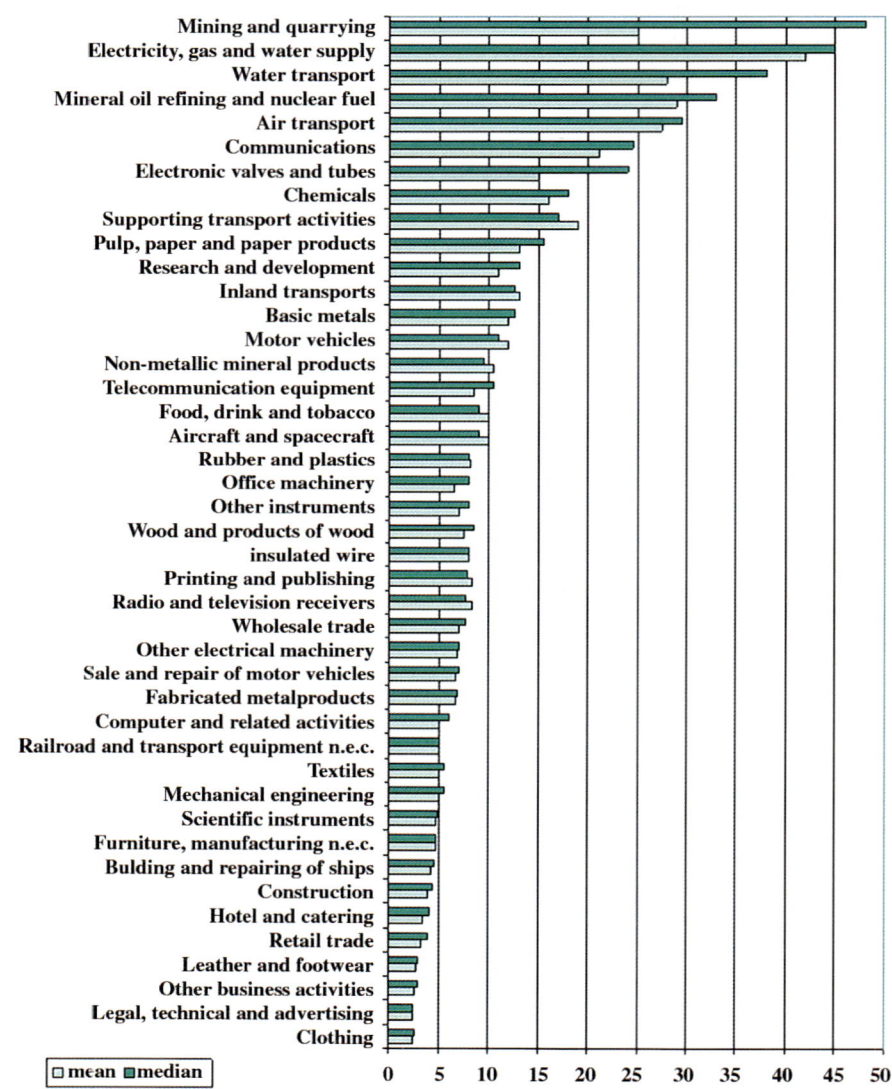

1.4.2 Access to minerals

Unlike other branches of industry, the choice of the business location in mining is determined primarily by geological, market-relevant and legal conditions. The efficiency of exploration and long-term access to raw materials are elementary factors.[31]

A substantial precondition for exploration of raw materials is a basic geological survey and mapping of the state, which usually is a task of the national geological survey.[32]

Prospection and exploration usually are the responsibility of the companies, with increasing participation of international exploration companies. Moreover, in some countries it is common for the state (geological survey) to actively participate in prospection and exploration.

Industrial minerals must possess certain chemical and/or physical properties for special applications. This leads to limitations in the range of products and choice of prospection and exploration areas in this category of raw materials. For businesses, it is more common to carry out exploration near known deposits.[33]

Construction raw materials businesses make use of the geological survey's data. The better the quality of the information available, the more it is applied by companies.[34]

The largest part of the exploration, especially of **metallic raw materials**, is carried out by the so called "junior mining companies" which, upon success, aim to sell the deposits to larger businesses.[35] The administrative effort of such junior mining companies for explorations in the EU is currently considered too much to be competitive.[36]

---

31 EC, DG Enterprise and Industry (2000): "Promoting sustainable development in EU non-energy extractive industry", COM/2000/265 Brussels.
32 In the Annex, there is an overview of the Geological Surveys of more than 20 European countries.
33 EC, DG Enterprise and Industry (2008): Consultation on Raw Material Initiative 2008: The consultation was conducted by DG Enterprise on basis of an on-line questionnaire. This process will be regarded more deeply in chapter 6.1. A part of the results is presented in the appendix.
34 Ibid.
35 Cp. Wellmer, F. W., Dalheimer, M. (1998): Trends in der Rohstoffwirtschaft – Die Rolle der 'Junior'-Firmen und der Berggesetzgebung für die internationale Exploration (Trends in raw materials management – The role of the "Junior" companies and mining legislation for international exploration), Glückauf, pp. 528–534.
36 EC, DG Enterprise and Industry, Consultation on Raw Material Initiative (2008).

There are many factors that have to be taken into account by exploration and mining companies when **deciding** to **invest** in prospection, exploration, and exploitation of mineral raw materials.[37] The Frazer Institute of Annual Survey of Mining Companies 2005/2006 found out that in this connection the **political climate of a region** is becoming more and more important for investment decisions.[38]

Table 3: Ranking of investment decision factors at the exploration and mining investment stage (Data by Otto, 1992)

Exploration stage	Mining-stage	Decision based on:
1	n/a	Geological potential for target mineral
n/a	3	Measure of profitability
2	1	Security of tenure
3	2	Ability to repatriate profits
4	9	Consistency and constancy of mineral policies
5	7	Company has management control
6	11	Mineral ownership
7	6	Realistic foreign-exchange regulations
8	4	Stability of exploration/mining terms
9	5	Ability to predetermine tax liability
10	8	Ability to predetermine environmental obligations
11	10	Stability of fiscal regime
12	12	Ability to raise external financing
13	16	Long-term national stability
14	17	Established mineral titles system
15	n/a	Ability to apply geological assessment techniques
16	13	Method and level of tax levies
17	15	Import/export policies

---

37 Wellmer, F. W., Dalheimer, M., Wagner, M. (2008): Economic evaluations in exploration, Heidelberg: Springer.
38 Problems arise with increasing uncertainties in the interpretation of laws, administrative duplication, and political instability in a region.

Exploration stage	Mining-stage	Decision based on:
18	18	Majority equity ownership held by company
19	21	Right to transfer ownership
20	20	Internal (armed) conflicts
21	14	Permitted external accounts
22	19	Modern mineral legislation

n/a: not applicable.

1.4.3 Competitiveness of the extractive industry

The economy needs safe and cost-efficient raw materials supply and therefore, a competitive extractive industry.

Within the context of this book the competitiveness of the extractive industry can be defined as the ability to ensure medium- and long-term raw materials supply to the economy (ie securing access to raw materials). In other words: The competitiveness of the extractive industry is defined as the ability to make a significant contribution to meeting the demand for minerals required by the European industry and society in terms of quantity, quality and cost of production. The competitiveness is controlled by the quality and quantity of mineral deposits and the political, legal, administrative, social and economic environment in which mineral extraction takes place.

When comparing the raw materials industry to other industrial sectors, a basic difference can be found: raw materials are produced for downstream industries, such as, for example, construction, steel, and chemicals industry; the economic success of the raw materials industry, therefore, is closely connected to the success and the competitiveness of these downstream sectors. Thus, the competitiveness of the raw materials industry has to be seen from a holistic perspective (considering the need for raw materials of these sectors).[39]

---

[39] Cp. The following situation reported by the Presse (Austrian journal), October 2008, p.21: the stock of RHI, a corporation producing refractory products mainly for the steel, cement and glass industries, has depreciated in the wake of the financial crisis of autumn 2008 (after the boom in May 2008 of just under € 34, the RHI share on Oct. 29 was € 12,50), though the company rises extensively (Presse, Oct.29, 2008). The outlook for 2009 was less optimistic: The steel industry reduced its production significantly; further development would depend on employment in the automotive, construction and engineering sector. The growth rates of 2008 will not continue. (Presse, 29.10.2008).

Globalisation is another important factor. Nowadays, mining enterprises increasingly operate in an economy determined by globalisation. This requires an immediate response to permanently changing market conditions and foresight regarding future developments and decisions. First and foremost this is important for the extractive industry, because for the development of new deposits and activation of new facilities investments in the order of several million Euros can be necessary.[40] In addition, it may take years for the original expenditures to amortise and profits to be made.[41]

Figure 5: Comparison of cash operating costs taking account of credits for sales of other metals in copper mines in Europe and other countries (2004) (Data by EC, DG Enterprise and Industry, Commission Staff Working document, 2007)

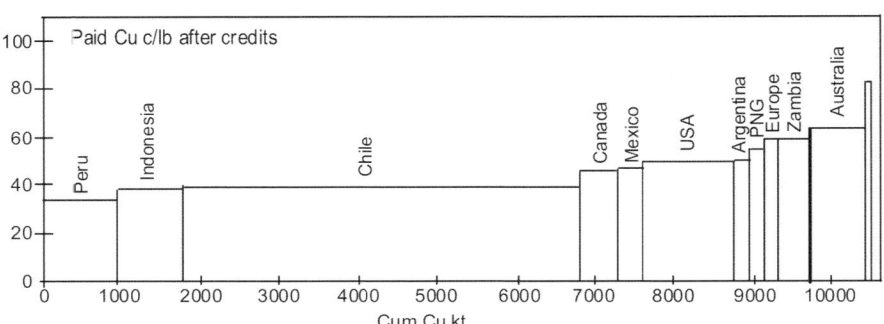

It should be noted that competitive conditions of the industry concerning metallic raw materials and industrial minerals have to be observed on a world scale. For the construction minerals industry usually a local or regional market exists, even though the business of super quarries in coastal regions in north-western Europe (Great Britain and Norway) has established a new dimension.

Metallic raw materials and some industrial minerals are traded on international markets. Raw materials are usually fungible goods (goods which are

---

40 Depending on certain criteria, for example raw materials, mining technology, site conditions.
41 This raises also the issue of investment security. Cp. Tiess, G. (2011): Legal basics of Mineral Policy in Europe, Wien: The validity of the exploration and exploitation permission is essential and standard of a modern mining law.

determined in number and weight and therefore, can be substituted), which means they are negotiable and consequently, can be traded worldwide.[42]

Globally, there are several markets for raw materials of supraregional significance, the three most important ones being Chicago (CME Group), New York (NYMEX), and London (LME). The floated raw materials are divided into energetic raw materials, metals (industrial and precious metals), and agricultural raw materials.

Figure 6: Quarry in Scotland (Data by http://www.alastairmcintosh.com)

1.5 Mineral markets

The encounter of suppliers and consumers for the trading of goods is characteristic for a market. The relation between demand and supply is essential for the formation of prices; the amount of exchanged goods and the extent

---

42 Commodity exchanges and product exchanges can be described as a prototype of the stock market and have their origins in the trade of agricultural products. They originated from market places and trading markets in major terminals and are considerably older than stock exchanges. The London Metal Exchange was already founded in 1877.

of profits are regulated by the market price.[43] Mineral raw materials markets are real assets markets. Strictly speaking, there are separate markets for many raw materials depending on the level of exploitation and quality.

Markets for metallic raw materials are considered world markets because of their good merchantability.

Figure 7: International mineral market, Munich 2005

http://www.mindat.org/munich05/Munich-Show.jpg

The structure of a market is mainly determined by the dispositions of the market participants.

A critical question when analysing markets morphologically is which minimum market share is sufficient to achieve significant market influence (and therefore, price influence) or even market dominance (and therefore, price setting). This share is market-specific and conforms to the competitive situation. Since elasticity of supply concerning price on markets of mineral raw materials is almost always low, a market dominating share is relatively low as well, often between 30 and 40% of the total production.[44]

1.5.1 Mechanism of pricing

In the long-term, raw materials prices are subject to cyclic fluctuations (see fig. 8). Long lasting price depressions result in underdeveloped capacity and a reduction of stocked materials. A rise in prices, however, causes new exploration and prospection in the medium-term which – when the high price period becomes long-term – leads to the development of new mines so that higher pro-

---

43 Klump, R. (2006): Wirtschaftspolitik, Instrumente, Ziele und Institutionen (Economic policy, instruments, objectives and institutions), München, p. 46.
44 Gocht, W. (1983), l.c., p. 83, p. 141.

duction compensates the commodity price.⁴⁵ As already addressed above, the establishment of new production capacity represents a long-term in investment. For the ore mining industry, this requires a minimum time of approximately five years. Consequently, demand for raw materials is usually characterized by marginal flexibility.⁴⁶ From the fundamental principles of the microeconomic theory it is deducted that in the long-term prices of raw materials follow the marginal costs required to make them available, which, in turn, are determined by the costs of extraction and utilization.⁴⁷ Utilisation expenses rise pursuant to the fact that raw materials are no longer available after being exploited and, thus, yield no more profit. The marginal costs of utilization indicate the extent of the present value of future profit the supplier would have to do without if he produced an additional quantity of raw materials today. Utilisation expenses depend very much on whether technological developments, which can affect both the future extraction costs and the demand, or economic developments could increase future demand for such raw materials.⁴⁸

For example, utilisation costs could vanish completely if a complete substitution of the raw material took place overnight. In case of a gradual substitution which could be foreseen today, the present price would be lower than in the case without a possibility for substitution, because this would reduce utilisation costs substantially. A surprising discovery of new, extensive resources leads to a sudden decline of the price of raw materials, thereby reducing utilisation costs. Obviously, there are numerous factors, which can affect the utilization costs. The future development of most of these factors is uncertain. Some factors, like the discovery of new deposits, are inherently unpredictable.⁴⁹

According to the Hotelling rule for exhaustible resources, it is to be expected that the in-situ-price for the mineral – i. e. difference between market

---

45 Wagner, M, Huy, D. (2005), Schafft der Strukturwandel in der Nachfrage eine neue Dimension für die Weltrohstoffmärkte, Bundesanstalt für Wissenschaften und Rohstoffe [Does the Structural change in demand accomplish a new dimension to the world raw materials markets? Federal Institute for Geoscience and Raw Materials, Hannover 2005, p. 2.
46 The income flexibility of raw materials demand describes the change in demand for raw materials in relation to the change in the gross domestic product (GDP) or similar indicators of economic performance. The demand for raw materials is considered to be income flexible if the percentage increase for raw materials per percent increase of GDP equals to 1 or is > 1.
47 Endres, Querner (2000), in: Rheinisch-Westfälisches Institut für Wirtschaftsforschung (Rhine-Westphalian Institute for Economic Research) (RWI Essen), Trends der Angebots- und Nachfragesituation bei mineralischen Rohstoffen (Trends in supply and demand situation for mineral resources), Essen, p. 44.
48 RWI Essen (2006), l.c., p.44.
49 Ibid.

price and exploitation costs, or the profit resulting from the extraction of raw materials – rises gradually.[50]

Figure 8: Production of aluminium and real aluminium price per ton (Data by RWI Essen, 2006, data from: USGS 2005, USGS 2006)

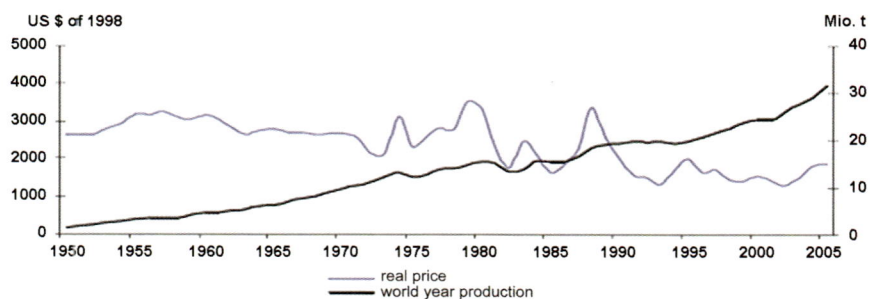

Nevertheless, decreasing price trends for raw materials are to be observed. This is mainly because exploitation costs of these resources could be lowered gradually by improving mining technologies. Additionally, the known mineral reserves are not constant but have risen continuously for many raw materials. That is why the real prices for aluminium, copper or zinc have dropped considerably instead of rising, despite a strong increase of demand. If the future development of exploitation and utilization costs were predictable – which means information about all determinants of future price development was available – an economic model for the exact prediction of mineral prices could be created. Since on one hand the determining factors of the exploitation and utilization costs are extremely complex and on the other hand their future development is uncertain, the future price of such raw materials cannot be predicted precisely.[51]

Figure 9: Relative price development of LME metals (Data by Ekdahl 2008)

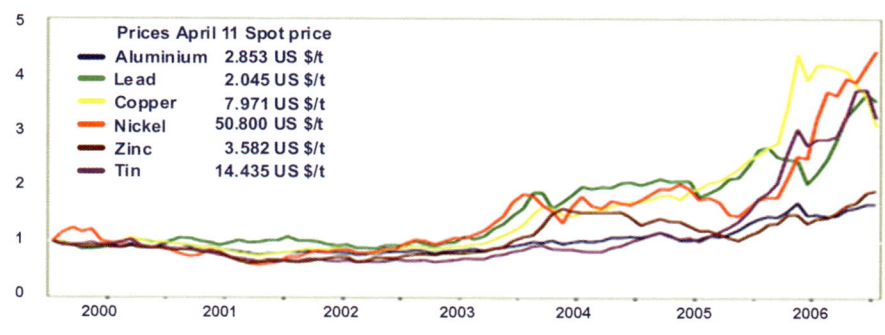

50 Hotelling (1931), in: RWI Essen (2006), pp. 44–45.
51 RWI Essen (2006), l.c., p.45.

Usually predicions about the future price of raw materials are based solely on the variation of cost development in the course of time.[52] This procedure, long established in literature, is based on the assumption that, after an incidental jolt, prices return to their original deterministic trend, which, according to the micro-economic theory, is determined by exploitation and utilization costs. If the presumption of the return trend applies, such time series are called trend stationary.[53]

1.5.2 Demand and supply

On the raw materials market, there are several geological, exploitational and economic factors that influence the **type of products**. They have long-term effects on the development of raw materials production.[54] Apart from that, there are also factors which can trigger short-term effects and often account for market uncertainties and severe price fluctuations.[55]

Determinants for the demand on raw materials markets

The demand for raw materials is basically determined by global economic growth, changes in the industrial structure, price of raw materials, and technological changes.[56] As mentioned earlier, the consumption of raw materials is closely linked to its industrial goods production. This applies to construction, industrial, and metallic raw materials. Since raw materials serve goods production, economic performance forms an essential determining factor of raw materials demand.[57]

Routinely, economic performance is measured by the gross domestic product (GDP) or the gross national product (GNP). In simple terms, the **better**

---

52 Berck, Roberts (1996), Pindyck, Rubinfeld, (1998), Slade (1982), in: RWI Essen (2006), Teil I Part I, Preis-, Angebots- und Nachfragetrends (Price, supply and demand trends).
53 RWI Essen (2006), l.c., p.45.
54 Gocht (1983), l.c., p. 176.
55 Cp. Siebert, Horst [Hrsg.] (1986): Angebotsentwicklung und Preisbildung natürlicher Ressourcen. Schriftenreihe des Energiewirtschaftlichen Instituts (Supply development and pricing of natural resources. Series of the Institute of Energy Economics), 30
56 Cp. Tilton, J.E. Economics of the Mineral Industries (1992), in: H.L. Hartmann (Ed.). SME Mining Engineering Handbook. Society of Mining, Metallurgy, and Exploration.
57 Cp. Nötstaller, R. (2000): Zur Entwicklung der Nachfrage nach mineralischen Rohstoffen – Zusammenhänge und Folgerungen, Berg- und Hüttenmännische Monatshefte (On the evolution of mineral demand - relationships and conclusions, Journal of Mining, Metallurgical, Material, Geotechnical and Planned Engineering), pp. 314–318.

the economic performance of a country, the higher its demand for mineral raw materials. The extent of the GDP depends on the population size and the per-capita income.[58]

In advanced economies, the increasing tertiary sector causes a gradual decoupling of economic growth and consumption of raw materials, which can lead to a decrease of material intensiveness.

At the same time, the per-capita income indicates a country's level of development, as well as the productivity and the level of prosperity of the population. Accordingly, there is a **strong correlation** between the per-capita income and the demand for specific raw materials.[59]

This correlation is predominantly positive, making raw materials demand dependent on the income.[60] It also causes a high specific need for mineral raw materials and energy. This is why **predictions** for future demand are based on several indicators, such as the ratio of energy and metal consumption to the extent of the GDP of a country.[61] Industrial countries with a high GDP tend to consume more than developing countries with a low GDP. A significant dependency can be observed.[62]

---

58 Economic output per capita and countries is published by the World Bank: The World Bank. Size of the Economy, in: World Development Indicators (2001), Washington.
59 Nötstaller, R. (2003), Österreichischer Rohstoffplan, Arbeitskreis 2, Rohstoffwirtschaft und Bergwesen (Austrian Minerals Plan, Working Group 2, Mineral Economy and Mining), Department of Mining and Tunnelling, University of Leoben, Leoben, p. 25.
60 Ibid.
61 Gocht (1983), l.c., p. 183.
62 Cp. u.a. Müller-Ohlsen, L. (1981): Die Weltwirtschaft im industriellen Entwicklungsprozess. Kieler Studien, Tübingen: (World economy in the industrial development process. Kiel Studies, Tübingen): Studies on the relationship between per capita consumption of non-ferrous metals and the degree of industrialization have shown good correlation values. – Cp. also figures presented in Chapter 3.

1.5 Mineral markets

Figure 10: Development of steel consumption (Data by Ekdahl 2008)

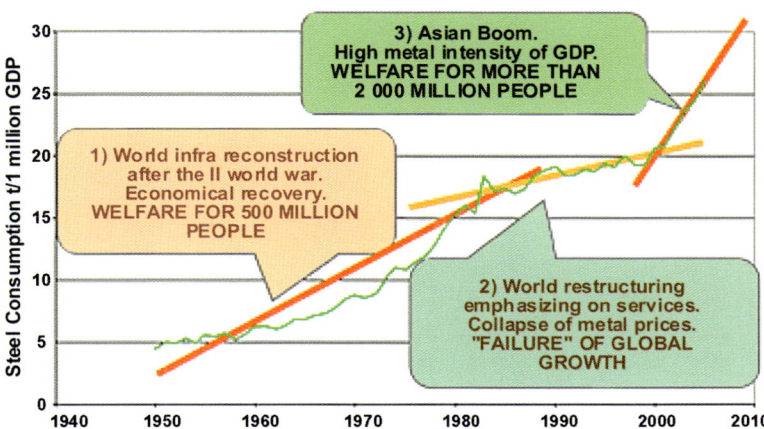

To characterize the raw materials intensity of an economy, the minerals consumption is related to the economic performance. A useful example for this is the stock of motor vehicles of a state, which ranks among the largest consumers of raw materials in every national economy. Thus, this **stock is a useful indicator for the degree of economic development just as for the consumption of raw materials**.

Figure 11: Stock of motor vehicles as a useful indicator for raw material consumption – displayed with a motor

An interesting example is discussed by Nötstaller:[63] Figure 12 shows that the stock of motor vehicles in the US has risen from almost zero at the begin-

---

63 Nötstaller/Wagner (2007), l.c., p. 385f.

ning of the 20th century to more than 800 per 1000 persons at the beginning of the 21st century. In China the stock of motor vehicles at present grows just as rapidly as in the US at the beginning of the past century. The number of motor vehicles rose from 5 per 1000 persons in 1990 to 23,5 in 2005 and therefore, to a level which the US had reached in 1915. In Africa the stock of individual motor vehicles at present amounts to about 3%, in Central and South America to 15%, in Eastern Europe to 30% and in Western Europe to about 70% of the United States value.

The fabrication of a passenger car with an empty weight of 1,5 t demands the equal weight of raw materials, almost all of which are of mineral nature. The stock of motor vehicles in a country increases along with the per-capita income.

Figure 12: Stock of motor vehicles in the US 1900–2005 (Data by Nötstaller and Wagner, 2007)

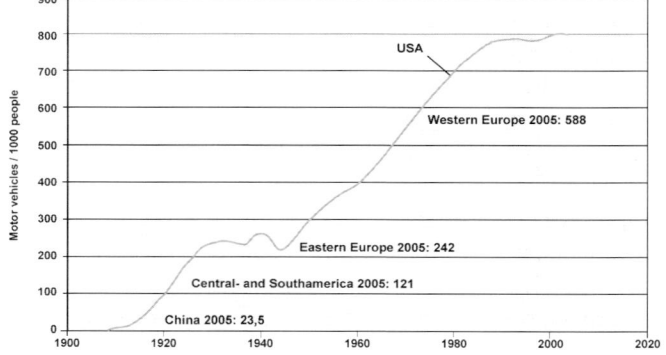

Supply issue

The essential **factors** for long-term development of supply are the **mineral reserves available**, i.e. the amount of reliably estimated profitable resources of the mineral raw materials; the quality of the reserves, the ore grade, the associated minerals; the type of deposits, e.g. all the geometry, depth, and size of the ore body; the paragenetic association of minerals, such as the simultaneous occurrence of copper and cobalt, copper and molybdenum, lead and silver, or zinc and cadmium.[64] Furthermore the following factors are also important for the development of a mineral deposit:

---

64 Gocht (1983), l.c. p. 177.

- The **mining and processing technology available**, including the efficiency of extraction of the valuable minerals, the relative importance of capital and labour, and the operational capability;
- The **infrastructure conditions** in the regions in which the deposits are located, i. e. the accessibility of mines and processing plants;
- The **overall legal and economic conditions** for mining;
- The possibilities for **financing** new mineral exploration and mining projects.

Regional raw materials concentrations and raw materials markets

The geographical spreading of mineral raw material deposits is strongly inhomogeneous. This in particular applies to energy raw materials, metallic raw materials and industrial minerals. In many cases – particularly ores – production is highly concentrated in certain regions. Only three countries produce more than 50 % of the world extraction of some minerals.[65] These concentrations of raw materials do not usually coincide with the regional maximum of demand. The main consumers are mostly not identical with the producing countries. As a result, countries lacking raw materials are highly dependent on states possessing mineral wealth, combined with an **extensive international flow of trade**.[66] Imports of the required raw materials of a state are usually highly diversified to secure delivery. For example, the German industry secures its supply of raw materials by long-term supply contracts with the raw materials producers, by the listed raw materials trade as well as the recycling of waste materials. Germany obtains lead ores from Australia, Ireland, Poland and Sweden; chrome ore is provided by South Africa or Turkey. Argentina, Chile, Indonesia, Papua New Guinea or Peru and Portugal supply Germany with copper ore.[67]

The guaranteed supply of the economy with raw materials on a high level invariably assures that the **condition for the maintenance of the material prosperity is achieved**. In this respect questions about the *raw materials criticality* of a state arise (see section 1.5.3).

---

65 See also Chapter 3.
66 In addition: Bundesministerium für Handel, Gewerbe und Industrie (Austrian Ministry of Commerce, Trade and Industry) (1981), l.c., p.27.
67 Bundesverband der Deutschen Industrie (BDI) (2007), Rohstoffsicherheit – Anforderungen an Industrie und Politik. Ergebnisbericht der of BDI – Präsidialgruppe "Internationale Rohstofffragen" (Minerals security – requirements for industry and policy, Summary report "International raw material issues"), pp. 3–4.

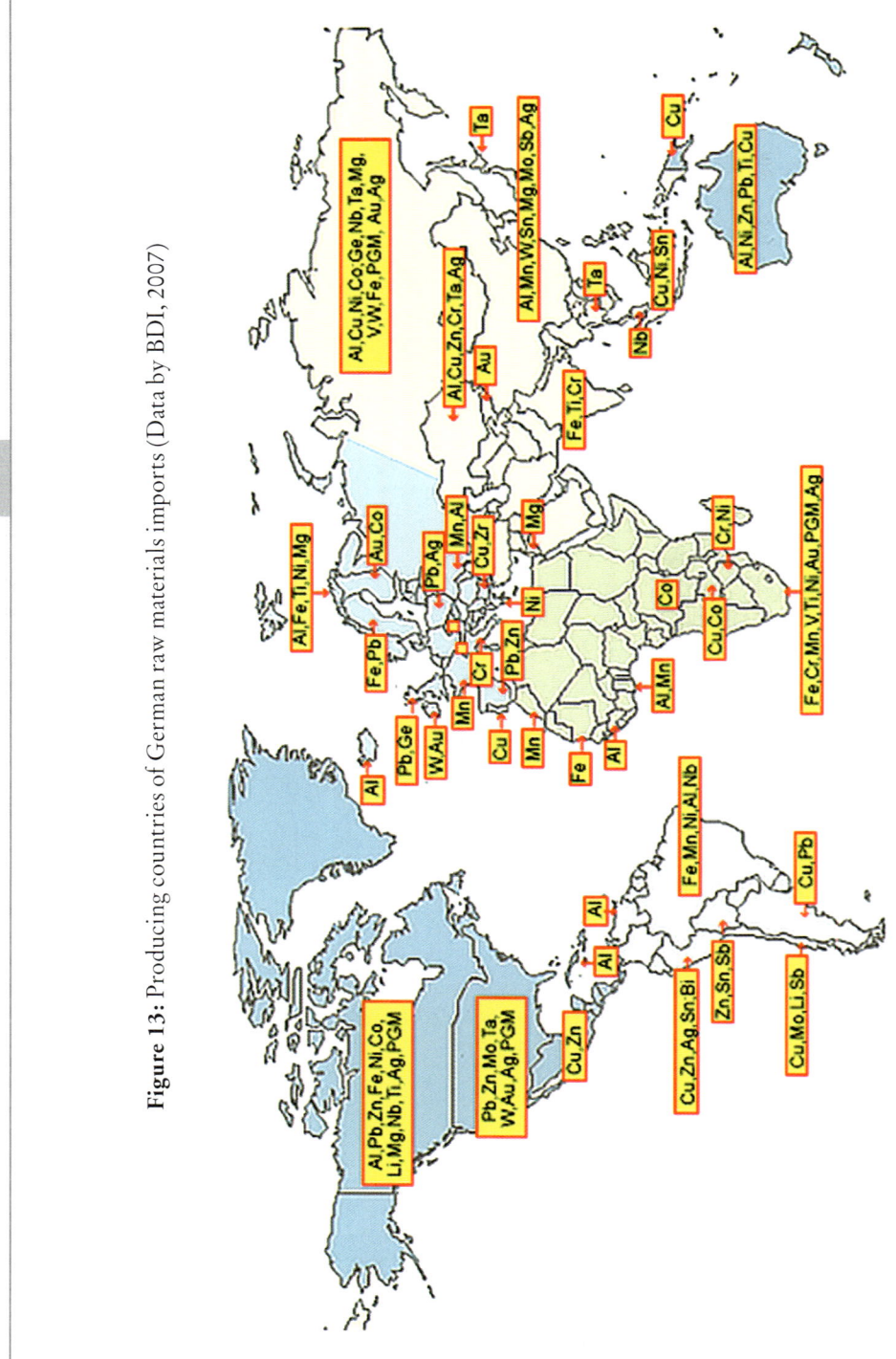

Figure 13: Producing countries of German raw materials imports (Data by BDI, 2007)

1.5.3 Aspects of materials criticality

"**Crucial**" in this sense are not only the areas within which the probability of acute problems of supply is higher than in others, but also those in which the consequences of a supply disruption would be more serious than in others.[68]

The production of technically exacting products increasingly requires critical raw materials, above all high technology metals. If in the eighties of the last century twelve raw materials were needed for the production of computer chips, nowadays their number can be up to 60.

Figure 14: Computer (Data by Die Presse, 2008)

Many important raw materials are produced only in a few countries. China, for example, produces about 95 % of all concentrates of rare earths; these are used for electronic small size units, liquid crystal displays, and high performance magnets. Brazil produces 90 % of all niobium, which is a component of high-strength steel alloys and super alloys for thermally stressed construction units of aero engines. South Africa produces 79 % of all rhodium (for exhaust catalysts of motor vehicles).[69]

Blockages in the supply of raw materials can severely impair the entire downstream industrial production. Raw materials therefore gain an importance which is far beyond their contribution to production or added value respectively.

---

68 Bundesministerium für Handel, Gewerbe und Industrie (Austrian Ministry of Commerce, Trade and Industry) (1981), l.c., p.15.
69 Information from: http://europa.eu/rapid/pressReleasesAction.do?reference=IP/08/1 628&format=HTML&aged=0&language=DE&guiLanguage=en

The foreign trade associated with raw materials and its effects on the balance of payments and current account of a state form an essential criterion for the evaluation of the supply with raw materials. The imports of raw materials of a state often burden the balance of payments considerably.[70] The supply with raw materials is particularly **endangered if**[71]

- international resources or reserves as measured by world consumption are limited;
- consumption in long-term rises faster than the development of new resources;
- an expansion of resources, reserves or production is linked to a high increase in price;
- a substitution in short and particularly in long-term appears to be difficult or impossible;
- reserves and production are in the hand of single or few countries or enterprises,[72] especially if those countries or enterprises could also control substitution materials;
- the political situation in those countries appears unstable or potentially directed against the interests of the consuming country; those countries or enterprises are not dependent on the income from the sale of these products; the routes of transportation appear not safe or efficient.

The effects of a disturbance of the supply vary within the individual raw and base materials, depending on their significance for the processing industry (sensitivity, multiplier effect). To characterize the supply with (and in this sense also the criticality of) raw materials of a state, raw materials statistics and raw materials balances form a substantial condition. As a result, **measures of mineral (planning) policies can be taken for securing minerals supply**. For a statistical display of the supply with raw and base materials of a state not only the quantitative input of domestic and foreign primary or secondary sources, but also a value-based analysis of the particular supply currents is important. A value-based survey of raw and base materials used for consumption does not

---

70   Cp. discussion in Chapter 3 and 5.
71   Bundesministerium für Handel, Gewerbe und Industrie (Austrian Ministry of Commerce, Trade and Industry) (1981), l.c., p. 15. Cp. also Rheinisch-Westfälisches Institut für Wirtschaftsforschung (RWI Essen), Trends der Angebots- und Nachfragesituation bei mineralischen Rohstoffen, Essen, Methode zur Festlegung der kritischen Rohstoffe des Landes Deutschland (Rhine-Westphalian Institute for Economic Research, Trends in supply and demand situation for mineral resources, Essen, p.34f: method of determining the critical resources of Germany.)
72   Exemplarily the raw material Niobium-tantalum ("coltan") can be quoted, shown under point 3.5.5 raw material wars.

only express the economic relevance of a certain raw and base material, but also the adequacy of cost by international comparison.

This is particularly important because at troublefree times and, by international comparison, at least competition-neutral supply with raw materials must be the goal of an effective minerals policy. The quantitative view without appropriate emphasis is insufficient for several reasons. A number of raw and base materials in relatively small quantity are needed urgently for the subsequent treatment or refinement of far larger quantities of other raw and base materials, like alloy metals, stabilizers, catalysts etc. (sensitive raw and base materials). Furthermore, the sufficient and well-priced availability of raw and base materials is of particular importance if they affect any downstream industries. (multiplier effect).[73]

1.6 Sustainable mineral resource management

1.6.1 General facts

A basic principle of use is the **protection of raw materials in the sense of sustainability**.[74] The concept of sustainable development should be applied extensively in the field of mineral economy,[75] considering the characteristics of mineral economy.[76] In the political area as well as in the private sector it is generally agreed that the principle of sustainable development is to be applied in all areas of life and forms the basis of socio-political decision-making process, considering equally the economic, social and ecological aspects.[77] This

---

73 Cp. Bundesministerium für Handel, Gewerbe und Industrie (Austrian Ministry of Commerce, Trade and Industry) (1981), l.c., p. 17.
74 Cp. Hruschka, F. (2008): Aufbau durch Abbau – Gewinnung mineralischer Rohstoffe und Nachhaltige Entwicklung, BHM Berg- und Hüttenmännische Monatshefte (Construction by extraction – Mineral raw materials extraction and sustainable development, Journal of Mining, Metallurgical, Material, Geotechnical and Planned Engineering.)
75 Cp. Wellmer, F. W., Kosinowski, M. (2005): Sicherheit der Rohstoffversorgung unter dem Aspekt der nachhaltigen Entwicklung (Natural resources – security of supply with respect to sustainable development), Berg- und Hüttenmännische Monatshefte (Journal of Mining, Metallurgical, Material, Geotechnical and Planned Engineering), pp. 117–121.
76 See above: Location boundness, unequal distribution and exhaustibility of deposits. Cp. also: Hartung, M. (2007): Rohstoffe und Bergbau - Positionen und Perspektiven, Bergbau: Zeitschrift für Rohstoffgewinnung, Energie, Umwelt (Raw materials and mining - positions and perspectives, Journal of Raw Material Extraction, Energy, Environment), pp. 486–490.
77 Cp. Markandya, A., Mason, P., Friedrich, R., Hacker, M., Gressmann, A., Wagner, H., Nötstaller, R. (2000): SAUNER – Sustainability and the Use of Non-renewable

implies that measures promoting sustainability must not have consequences of competetive distortion. Security of mineral supply should be ensured in such way that economic growth is separated from possible environmental impacts.

Figure 15: Securing of raw materials supply for a growing economy, society and sound environment (Data by Ekdahl, 2008)

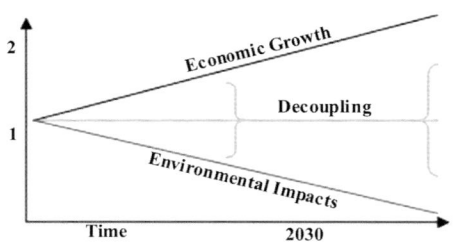

Major challenges for R & D
- ✓ Exploration techniques
- ✓ Zero-impact extraction technologies
- ✓ Mine closure & reclamation
- ✓ Recycling, novel materials
- ✓ Material handling and logistics
- ✓ Life cycle and material flow analysis
- ✓ Environmental accountancy for natural resources

European Technology Platform on Sustainable Mineral Resources (ETP SMR)

In this context, mineral research is of significant importance.[78] In the European Union, the Technology Platform on Sustainable Mineral Resources (ETP SMR) was established in 2005 and officially recognised in 2008. It unites many stakeholders from the mining industry, the research community, regulators, consumers and the civil society around the major technological challenges to the sector, in order to jointly act towards a common vision. It makes great contributions to achieving the goals of the revised Lisbon Strategy and the intentions of the Gothenburg Strategy on Sustainable Development: In its screening of the opportunities and challenges for 27 separate sectors of the EU manufacturing and construction industry, the European Commission identified the European Technology Platform for Sustainable Mineral Resources as one of the sector-specific initiatives for the implementation of the Lisbon agenda.[79]

The European Technology Platform on Sustainable Mineral Resources (ETP-SMR) aims at modernising and reshaping a fundamental pillar of European economy: the European mineral industries. These include oil, gas, coal, metal ores, industrial minerals, ornamental stones, aggregates, smelters as well as technology suppliers and engineering companies. The identified needs for Pan-European collaborative research focuses on a sustainable supply of mineral

---

Resources: Summary final report. Research funded by the European Commission, DGXII Environment and Climate Programme.
78 Further information can be found at www.etpsmr.org
79 www.etpsmr.org

resources to the downstream European industries, also taking into account the decoupling of economic growth from adverse environmental impacts.

Figure 16: Focus of the ETP-SMR

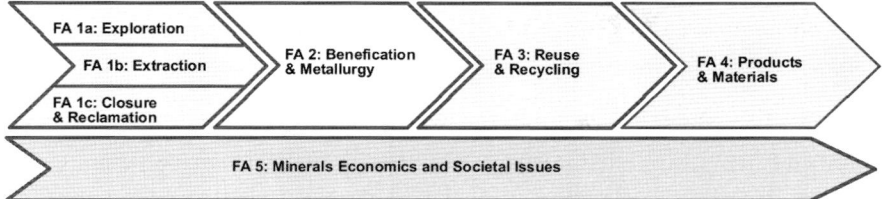

In the EU metal mining companies spent on average 2.3% of their turnover (i. e. € 39 million) on RTD in 2003. However, it should be noted that most RTD in this sector is carried out by independent research Institutes and academia rather than directly through companies. The largest metal mining houses have their own research facilities, but they are located in Australia, South Africa and Canada, and not within Europe. The industrial minerals and construction minerals sectors spent about 1% of their turnover on RTD in 2003.[80]

1.6.2 Environmental protection

Aspects of environmental protection have to be considered comprehensively in the stages of exploration, extraction and processing. The specified emissions have to be observed in each case by technical and organizational means. Furthermore, when planning enterprises, measures for subsequent use of the claimed areas should be stipulated. The objective is the reclamation of mining areas and their integration into the surrounding landscape.

Extraction and processing of raw and base materials produce tailings during extraction, whilst products have to be deposited as waste after a certain time of use. In this connection great importance is assigned to raw materials efficiency and recycling: More efficient exploitation means less mining waste material. Raw material efficiency and recycling are discussed below.

The cycle of extraction of primary raw materials from the earth's crust and the depositing of no longer usable secondary raw materials must be viewed as a closed circle of events in order to guarantee an optimal development of utilization and environment at present and in the future. The most important

---

80 Information from EC, DG Enterprise and Industry, Commission staff working dokument (2007), l.c.

approach to an appropriate formation of this cycle is providing precise criteria for decisions. For this purpose a statistical survey of potential natural landscapes of geogenic and non-geogenic nature and of cycles of already exploited raw materials is required.

Another approach is the promotion of environmental protection based on raw materials. An interesting example for this are rare earth elements. They are needed for catalysts and batteries in hybrid cars, for high performance magnets and fibre optic cables. These metals play a critical role in the development of innovative "environmental technologies" for boosting energy efficiency and reducing greenhouse gas emissions. Hydrogen-fuel based cars require platinum-based catalysts. Electric-hybrid cars need lithium batteries and rhenium super alloys are an indispensible input for modern aircraft production. The demand for hybrid vehicles increases the need of these raw materials. The growth potential for hybrid vehicles is very high because of gasoline prices and the discussion about climate protection.

Figure 17: Raw materials contribute to environmental protection, hybrid cars as an example (Data by Handelsblatt, August 2008)

Rare earth elements are indispensable for the production of hybrid cars. Toyota and Honda intend to considerably increase production. In 2010, Toyota wanted to produce a minimum of one million hybrid cars. World wide production could amount about three million in 2012.[81]

Important progress in environmental protection could be achieved by the development of environmental performance indicators, in order to establish a

---

81 Braune, G. (2008): Begehrte Spezialmetalle – Seltene Erden sind wegen der wachsenden Nachfrage nach Hybrid-Autos gesucht (Special Metals in demand – Rare earths are sought after because of the growing demand for hybrid cars), Handelsblatt (German journal), 12.8.08, No. 155 (www.handelsblatt.com). Further comments: see below.

detailed assessment of the industry's ecological performance, to monitor improvements and to differentiate between various sub-sectors and locations, as influenced by geological conditions and the local ecosystem. Examples of suitable indicators could be resource use, discharges to air and water, and land use. Most importantly, they must satisfy requirements of transparency, relevance, measurability, and analytical soundness.[82]

More comprehensive environmental reporting by the industry could play an important role in the development of indicators. Some companies have already started to develop and use such indicators in their reports. Indicators can only achieve results if they are adapted to common measuring standards, allowing for comparison and evaluation of performance. This approach will provide the degree of objectivity needed to bring about improved dialogue between the stakeholders, a dialogue based on objective analysis which can result in agreed targets for future improvements. Such dialogue can also help to yield greater understanding of the constraints faced by the industry in terms of competitiveness and social development.

Additionally, voluntary initiatives have been taken by industry to foster improvements in environmental performance and to communicate past and on-going achievements. Several companies, especially multinationals in the metallic minerals sector, have adopted environmental and sustainable development policies. Business federations have developed codes of conduct and mission statements for their members, setting out the principles for the sectors' environmental policy. Best practice guides have been developed to illustrate how the industry strives to ensure environmental protection.

The "International Cyanide Management Code" for the manufacture, transport and use of cyanide in the production of gold (ICMC) is a voluntary industrial programme for the gold mining industry to promote responsible management of cyanide used in gold mining, enhance the protection of human health, and reduce the potential for environmental impacts. Companies that become signatories to the Code must have their operations audited by an independent third party to demonstrate their compliance with the Code. In March

---

[82] Meanwhile three international conferences to the topic "Sustainable Development Indicators in the Mineral Industry (SDIMI) were held: SDIMI 2003 in Milos/Greece, SDIMI 2005 in Aachen/Germany and SDIMI 2007 again in Milos. In SDIMI 2003 the so-called Milos Declaration was resolved, quoted in the Proceedings SDIMI in 2007, XV, "A statement of commitment to a sustainable future through the use of scientific, technical, educational and research skills and knowlegde in minerals extraction and utilzation". At the last two conferences the author made important contributions jointly with H.Wagner, S.Šolar, K.Nielsen: SDIMI 2005: Minerals planning policies in Europe; SDIMI 2007: National minerals policy practices: key to minerals supply in Europe.

2007 the Code had 27 signatory companies representing 88 individual facilities and about 36 % of the world's production of gold by cyanidation.[83]

1.6.3 Resource efficiency

The non-renewable nature of most mineral deposits demands that these deposits are extracted as completely as possible and the search for new deposits is an ongoing activity. Improvements in site closures and reclamation techniques made it possible to restore sites for other profitable applications after their production was shut down. Thus, the sustainability and the public image of the industry improved.[84]

Technological development is an ongoing process and requires long term vision and faith in the future of the sector. **Technological changes usually lead to more efficient raw materials usage** (i. e. raw materials input is less) and thus to increased efficiency of resources. This applies in particular to the raw materials group of metals, as well as to the material consumption of goods production.[85] The possibilities for decreasing the demand for construction raw materials are limited because of the need for adequate dimensioning of buildings for static reasons. Further research will need to address the development of

- efficient exploration technologies to find new mineral deposits;
- extraction methods for efficient use of mineral bodies;
- cheaper and more efficient methods of extraction and mineral processing economic extraction of lower quality mineral deposits;
- extraction and mineral processing methods which minimise the environmental impact and damage caused by mining;
- processes which minimise mining waste production;
- technologies which are less mineral resource intensive;
- technological innovations to reduce the specific consumption.[86]

---

83 Total: 88 operating sites, 36% of world gold production.
84 Müller, W.; Werthebach, E.; Schäfer, V. (2000): Rohstoffversorgung und technologischer Fortschritt: Grundelemente einer nachhaltigen Entwicklung in unserer Gesellschaft (Raw materials supply and technological progress: basic elements of sustainable development in our society), Bergbau, pp. 546–548, 550–554.
85 Cp. Antrekowitsch, H.: (2006), Sustainable Technologies in Metal Production and Processing, Berg- und hüttenmännische Monatshefte (Journal of Mining, Metallurgical, Material, Geotechnical and Planned Engineering), pp. 266–269
86 E.g., Introduction of electrolytic tinning for which in contrast to traditional hot dip tinning the amount of required fine tin is strongly reduced.

1.6.4 Material efficiency

Material efficiency means manufacturing products with **less materials** (and consequently with less consumption of raw materials). It can be enhanced by research and development initiatives in each manufactoring sector. In the past decades the raw materials-processing industry succeeded in manufacturing products with less material (and consequently in decreasing consumption of raw materials). Numerous investigations in recent years examine how this continuous process, which already is advanced constantly by the industry considering cost-minimising criteria, can be further encouraged. The expectations of obtaining an increase in efficiency are high.[87]

Research and development as well as innovation are very important for the competitiveness of the European industry.[88] This includes the effort for developing options for improvement of **energy efficiency** in all areas, which also is economically important.[89]

Table 4: Operating costs in mining – comparison of energy costs (electricity and fuel) in proportion to the total operating costs. (Data by EC, DG Enterprise and Industry, Commission Staff Working document, 2007) Comparison of the relative costs of energy (electricity and fuel) as a proportion of overall operating costs. Data source: UEPG and IMA with metallic data derived from minecost.com data.

Sub-sector	Estimated energy costs in the EU as a proportion of overall site operating costs
Construction minerals (aggregates)	3 %
Industrial minerals	11 %–19 %
Metallic minerals (copper and zinc)	15 %–17 %

Example air traffic

Air traffic is still one of the most extensively growing CO_2-emitters, which is why airplanes have to become even more energy efficient and less polluting. Improvements have been mainly through use of new materials. Innovative

---

[87] Cp. BDI (2007), l.c., p. 17.
[88] Cp. Antrekowitsch, H.; Biedermann, H.; Buchmayr, B.; Ebner, F.; Eichlseder, W.; Harmuth, H.; Kepplinger, W.; Kessler, F.; Krieger, W.; Lorber, K.; Ludwig, A. (2006): Universitärer Forschungscluster "Sustainable Technologies in Metal Production and Processing" (STMP), in: Berg- und hüttenmännische Monatshefte (Journal of Mining, Metallurgical, Material, Geotechnical and Planned Engineering), pp. 263–265.
[89] Cp. Ernst, W.G., in: Richards, J.P. (2009): Mining, Society and a Sustainable World, Heidelberg 2009, pp 125–146.

materials result in clearly lighter airplanes, and less weight means less energy input to keep it airborne. This also means less consumption of kerosene and less CO_2-emission. Modern airplanes, like the airbus A380, contain relatively little steel. High-strength materials, aluminium alloys, plastics and carbon fibre reinforced materials have taken over.[90]

New materials also allow for more efficient airplane engines. The engine of the future features about 15 % less consumption of kerosene and less CO_2-emissions. Emissions of nitrogen oxides can decline by 80 % and noise generation can be reduced by 20–25 %.[91] For their construction, materials which clearly withstand higher strain are needed. Hopes are pinned on so-called titanium aluminides. Those are not simply common alloys of metals, but inter-metallic compounds. Certain atoms in a crystal lattice are replaced purposefully and the materials obtain completely new characteristics. The chemical formula for a common inter-metallic material is suggestive of their complexity: Ti-43Al-5 (Nb, Mo)-B. This means that titanium, aluminium, and molybdenum were brought by a definite method into a structure.[92]

Figure 18: Jet engine (Kugler, 2008)

There are many different possible applications of titanium-aluminides. They are used in Formula 1 and aeronautics. In aircraft construction titanium-aluminides

---

90 Kugler, M. (2008): Innovation: Fliegende Materialien, Materialcluster Leoben, Neuartige Werkstoffe wie Titan-Aluminide ermöglichen Dinge, die bisher undenkbar waren (Innovation: Flying materials. The Materialcluster Leoben, Novel materials such as titanium aluminides enable things that were previously unthinkable), Die Presse (Austrian Journal [http://diepresse.com/]), 10.9.2008.
91 Kugler (2008), l.c., quotes Professor Clemens, materials researcher at the University of Leoben, Styria, including Materialcluster).
92 Ibid: Meanwhile in simulations the crystal structures of metals can be calculated. Alloy elements can be inserted into the computer programme and predict the qualities of the material. (Clemens).

will be relevant for application soon: The US General Electrics Group announced the use of titanium-aluminides in jet engines from 2009 onwards.

1.6.5 Recycling

Apart from using primary raw material resources, recycling from secondary resources is gaining more and more importance.[93]

The importance of recycling has been recognised. Through the development of new technologies and processes and logistics systems on the users' side the need for minerals can be reduced substantially.

The **advantage of recycling** is that it contributes to energy efficiency, particularly in the case of metals where production on the basis of secondary raw materials (scrap) is significantly more energy efficient compared to primary raw materials. For example, secondary smelting of aluminium using scrap consumes only 5% of the electricity used compared to primary smelting.[94]

Figure 19: Metal recycling – for example motor vehicles

---

[93] Cp. Prillhofer, R., Prillhofer, B., Antrekowitsch, H. (2008): Verwertung von Reststoffen beim Aluminium-Recycling (utilization of rest materials with the aluminum recycling), Berg- und hüttenmännische Monatshefte (Journal of Mining, Metallurgical, Material, Geotechnical and Planned Engineering), p. 103–108.

[94] http://www.wasteonline.org.uk/resources/InformationSheets/metals.htm. Cp. also Seebacher, H.; Sunk, W.; Antrekowitsch, H. (2006): Recycling von Aluminium in der Automobilindustrie, Fachzeitschrift der Aluminium-Industrie (Journal of the aluminum industry; international journal for industry, research and application, Recycling of aluminum in the automobile industry), p. 24–31.

Recycling must observe certain specified limits because of economic reasons. Recycling is only profitable if the costs of collecting, transporting, preparing and processing are lower than the market price for the concerned raw materials. In some areas of raw materials supply, such as metals, recycling has a significant production share.

The retrieval of metal from high-grade scrap is particularly important as this scrap evolves as by-product of the processing of steel and non-ferrous metals. In this case the collection and transport costs are small, processing is usually not necessary, and the energy needed for remelting is considerably lower than what is needed for the production of ore. Reclaiming (recycling) from scrap causes considerably higher charges. Apart from higher collecting and transport costs, in this case higher recycling costs arise as well.

The rational of metal recycling can be shown using the example of car-recycling: A car consists of numerous components, many of which are reusable. As a rule, 80 % of a car can be reused. Particularly non-ferrous metals can be reclaimed. Cars are shredded in special facilities and the waste is separated in categories: "iron", "non iron" and "plastic waste".

Chapter 2 Utilisation of non-energy raw materials

A brief survey will address the diversity of raw materials use and thus the importance of raw materials in Europe.

2.1 General facts

Raw materials are the first stage of a complex value-added chain. In a time of increasing globalisation they are essential to the performance and to the development and growth potential of a country's economy. This applies to energy sources and to many metallic minerals, industrial minerals and construction raw materials, which are essential primary inputs for the industry.[95]

The **importance of specific raw materials is often only vaguely realised**. This may be due to the fact that raw materials are less important compared to factor allocation as a whole, even though raw materials, unlike other factors of production, can usually not be substituted in the short term.[96]

Raw materials demand of an EU-citizen

The annual demand per capita varies depending on the level of industrialization of a state. Figure 20 describes the average consumption of mineral raw materials of an EU-citizen over the duration of a life of 70 years.

---

[95] Cp. u.a. Hennecke, H. P. (1993): Die Steine- und Erden-Industrie in Europa unter besonderer Berücksichtigung der Kalkindustrie. – Zement, Kalk, Gips, 46. 1993, H.1, S. 1–8 [e.g. The importance of industrial minerals industry for EU economy: Hennecke, HP (1993) Industrial minerals industry in Europe with particular emphasis on lime. – Cement, lime, gypsum, 46 1993, journal no.1, p. 1 to 8]

[96] European Economic and Social Committee (2006), l.c., p.2.

Chapter 2 Utilisation of non-energy raw materials

Figure 20: Raw materials demand of an EU-citizen (Data by BGR, 2009 [http://www.bgr.bund.de]).

Sand & Gravel	307 t		Silica Sand	4.7 t
Brown Coal	158 t		Kaolin	4.0 t
Crushed Rock	130 t		Potash (K_2O)	3.4 t
Petroleum	116 t		Aluminum	1.7 t
Natural Gas (1000 m³)	89,6		Copper	1.1 t
Limestone, Dolomite	72 t		Steel Alloying Metals	0.9 t
Hardcoal	67 t		Sulfur	0.2 t
Crude Steel	39.5 t		Asbestos	0.16 t
Cement	29 t		Phosphate Rock	0.15 t
Rocksalt	12 t		Electricity (MWh)	293.2
Gypsum	8.5 t			

2.2 Metallic minerals

Metals such as iron, aluminium and copper are essential for a broad range of uses, from aerospace technology and electrical engineering to automobile industry.

For the production of a car, more than 40 different raw materials are required, for example aluminium, iron and zinc for the car body, platinum or palladium for the catalyst and copper for the electronics. Rare metals are used in smaller quantities, but they are also indispensable in industrial production because of their application in alloys and super-alloys, which are mostly required in high-tech products.

Figure 21: Computers – For the production of a computer more than 60 different metals are needed

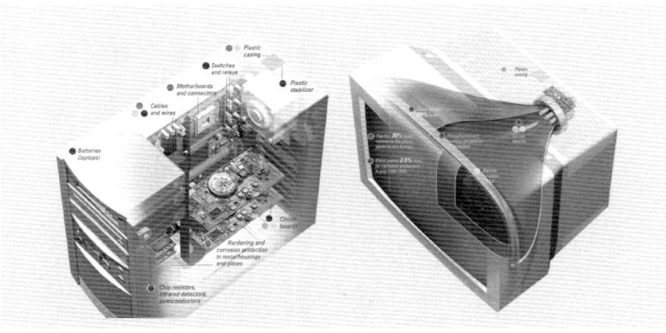

2.2 Metallic minerals

Table 5 provides an overview of important metallic raw materials needed for different raw materials applications in Europe.

Table 5: Production of selected metallic raw materials and examples of the raw materials application, (Data by EC, DG Enterprise and Industry, Commission Staff Working Document, 2007 [data: British Geological Survey])

Metal/Mineral	EU mine production (metal content) in 2003 (tonnes)	EU production as % of global mine production (2003)	Use
Arsenic	2.100	5,7%	Wood preservatives, fertilisers, fireworks, herbicides and insecticides. Alloy in ammunition and solders. Semi-conductors for telecommunication, solar cells and space research
Bauxite	3.251.900	2,1%	Production of aluminium. Important in the automobile industry and also in the building sector, aircraft manufacture, pharmaceutical and hospital equipment, food packaging, high-voltage cables and wires
Cadmium	1.674	9,6%	Batteries, pigments, coating and plating, plastic stabilisers and non-ferrous alloys
Chromium	549.040	3,4%	Chemicals, metals and refractory materials. Used in ferrous and non-ferrous alloys and steel to enhance hardness and resistance to corrosion. Also steel alloys, catalysts, leather processing, pigments and plating of metals
Copper	682.311	5,0%	Production of electrical cables and wires, plumbing, heat-exchangers in fridges, in roofing and building construction, chemicals, pharmaceuticals and electrical machinery, in alloys, alloy castings and electroplated protective coatings.
Gold	15	0,6%	Jewellery, bullion and industrial applications, including high-quality electrical circuits.

Chapter 2 Utilisation of non-energy raw materials

Metal/Mineral	EU mine production (metal content) in 2003 (tonnes)	EU production as % of global mine production (2003)	Use
Iron ore	24.340.028	2,0%	Steel production.
Lead	163.127	5,4%	Lead-acid batteries, sheathing cables and roofing, pigments, glass, ammunition and ceramics.
Lithium	418	3,0%	Ceramics, glass, aluminium production, lubricants and greases, rechargeable batteries and synthetic rubber.
Manganese	48.763	0,2%	Steel production, dry batteries, additives in paints, brick colouring, fertilisers and pet food.
Mercury	770	38,5%	Electrical and electronic uses, production of chlorine and caustic soda and batteries.
Nickel	22.800	1,7%	Production of stainless steel, non-ferrous alloys, steel alloys, foundry products, plating, rechargeable batteries and catalysts.
Selenium	430	25,2%	Glass manufacture, chemicals and pigments, electronics, agriculture and metallurgy.
Silver	1.750	9,3%	Precious metal and photography.
Strontium	152.383	28,3%	Television tubes, magnets and fireworks.
Tin	203	0,1%	Tinplate for food and beverage containers, alloys and solders.
Tungsten	2.096	4,6%	Electrical applications, super alloys, cutting tools for metal working, drilling for oil and gas, mining and construction.
Zinc	816.099	8,5%	Production of galvanising and die-casting alloys. Constituent of brass and bronze. Used to protect steel and as sheets for roofing and rainwater systems. Electric fuel zinc-air battery.

Furthermore, rare earths are increasingly required.[97] Their economic significance surpasses their degree of recognition by far. Rare earths are considered as the precious metals of the 21st century. They primarily find use in communication or environmental technology and, among others, are required for catalytic converters (cerium) and batteries for hybrid-cars, for powerful magnets and fibre optic cables. Erbium for example increases the performance of cables for communication via fibre glass; europium is used as red phosphorus for monitor and television screens and cannot be replaced with any other material.[98]

2.3 Industrial minerals

Industrial minerals cover a wide range of important applications (see table 6). An example is given for calcium carbonate which is a very abundant mineral: Deposits of calcium carbonate rocks, although chemically identical, may come in various forms (chalk, marble, feldspar). Limestone is used in large quantities as mineral fertilizer and as mineral filler in miscellaneous industrial applications (paper, colour, plaster, synthetic material, carpets). Pure calcium carbonate finds environmental application in exhaust gas catalytic converters, as an additive in the glass and the ceramics industry, as well as in cleaning and cosmetics. It can even be found in toothpaste. Calcium carbonate minerals and quicklime are also is also important in cement production (calcium silicate, calcium aluminate).

Furthermore, other raw materials such as silica sand, talc, kaoline and potash should be mentioned. Silica sand is the base material for glass, is vital for the moulds in metal foundry and is used as quartz oscillator in clocks and computers. The applications of talc and kaoline are just as extensive, most notably as fillers and in the cosmetics industry.

Due to changes in nutrition, the importance of potash as a fertilizer has recently increased.[99] While the consumption of meat in Europe remained unchanged over the last three decades, in Asia it grew by a factor of ten. The impact on agriculture is considerable: For the production of one kilogram of beef up to eight kilograms of feed are required. Particularly in Asia, the demand for higher quality of food increases, while a change is also noticeable where a

---

[97] Special metals (e.g. Neodymium); they have already been mentioned in Chapter 1 because of their importance, see also Ebensberger, A., Maxwell, P., Moscoso, C. (2005): The lithium industry: its recent evolution and future prospects. Resource policy, Vol. 30 (2005), 3, pp 218–231.
[98] Braune, G. (2008), l.c.
[99] Vereinigung Rohstoffe und Bergbau (VRB) (2008), l.c., p 17–19.

traditional rather vegetarian nutrition is substituted by more consumption of meat, thus requiring more input of animal feed.

Figure 22: Calcium carbonate – a versatile raw material (Data by http://de.wikipedia.org)

Artistic importance of marble: Large marble deposits in Europe can be found in Austria (Gummern), Norway (Molde), Greece (The parthenon) or in the Italian Carrara, the home of the white Statuario of which Michelangelo made his sculptures. One of the most popular is David by Michelangelo. The Statuario stone is one of approximately 50 different Carrara marbles from the quarries in Carrara in Italy.

Figure 23: Significance of potash (http://www.vks-kalisalz.de)

High-quality mineral fertilizers are created as follows: Raw potash salt from the pit is carried upwards where fine grinding takes place above ground. Depending on the type of raw potash salt, the processing to fertilizers is carried out by a flotation process or by the electrostatic separation method. The method of separation is determined by the structure of the raw salt and by the final product.

All over the world the demand for many agricultural products increases faster than the production. In addition, because of the politically enforced and subsidy supported expansion of biofuel production, capacities for agricultural food production have been reduced in recent years.

As a result, provisioning of agricultural products has fallen, causing a significant rise in prices for food products. Agricultural production would have to increase globally by approximately 2 per cent annually to guarantee the nutrition of the increasing world-population. The amount of arable lands available cannot be expanded significantly. Urbanization and population growth, particularly in Asia, form an obstacle. To improve the yield of agricultural land, sufficient mineral fertilization is necessary.

An overview of the variety of usage of industrial minerals is given below.

Table 6: Summary of production of selected industrial minerals, their main downstream markets and factors identified as affecting demand (Data by EC, DG Enterprise and Industry, 2007 [European Minerals Yearbook 1996–1997; EULA; K+S-estimation; Industrial Minerals Association website (http://www.ima-eu.org)])

Mineral	EU mine production (and as % of global mine production) in 2003	Main uses/downstream industries supplied	Factors affecting demand
Barytes	350.300 tonnes (5,6 %)	Used in oil industry (drilling mud), chemicals, tiles and glass bricks. As a filler in paints, plastics, rubber and inks	There is a growing demand for drilling quality barytes. In Europe the chemical and filler industries account for almost half the barytes consumption. In the USA over 90% of consumption is by the oil industry. Leading world producers: China, India and the USA.
Bentonite and Fuller's earth	3.380.400 tonnes (20%)	Used as a bonding material in preparation of moulding sands for production of iron, steel and non-ferrous castings; as a binding agent in the production of iron-ore pellets; in civil engineering applications as a thixotropic, support and lubricant agent in, for example, diaphragm walls and foundations; as a sealing material in construction and rehabilitation of landfills; in the oils/food markets as a purifier; in agriculture, pharmaceuticals, cosmetics and medical markets; in detergents, paints, dyes and polishes; in cat litter; in paper manufacture and as a catalyst in a range of applications.	A fall in demand for iron-ore pellets in the early 1990's affected demand for bentonite.

2.3 Industrial minerals

Mineral	EU mine production (and as % of global mine production) in 2003	Main uses/downstream industries supplied	Factors affecting demand
Feldspar	5,456,900 tonnes (36,4%)	Ceramics, glass and also paints, plastics and rubber.	The performance of the ceramics and flat-glass industries are closely linked to construction activity and, therefore, general economic development. A very abundant mineral, therefore theoretical reserves are unlimited. Low value and abundance can make transport costs important.
Fluorspar	344,900 tonnes (7,8%)	Production of hydrofluoric acid which is used to produce other chemicals used in production of aluminium, steel pickling, enamelling, glass etching, etc. Glass and ceramics industries (as an opacifier) and metallurgy (fluxing agent for electric steel plants).	Uses have changed rapidly in response to technological developments. Changes in the smelting process for the iron and steel industry and aluminium production were predicted to reduce demand from these sectors. Chinese export quota identified as an issue.
Kaolin	5,080,200 tonnes (22,8%)	Manufacture of paper, ceramics, refractories, rubber, plastics, paint, cement and glass fibres.	Competition from other white minerals, particularly calcium carbonate and talc. Although widely distributed globally, only a small number of deposits are suitable for production of high-quality paper-coating grades (e.g. in the USA, UK, Brazil and Australia

Chapter 2 Utilisation of non-energy raw materials

Mineral	EU mine production (and as % of global mine production) in 2003	Main uses/downstream industries supplied	Factors affecting demand
Industrial limestone	50.000.000 tonnes	Production of lime – used in iron and steel production; cement; flue gas desulphurisation, drinking water treatment, chemicals industry, paper, food and healthcare.	
Magnesite	3.755.600 tonnes (17,5 %)	Manufacture of refractory products (e. g. bricks). Also chemicals, paper and pulp, flue gas treatment and pharmaceuticals. Industries using fused magnesia.	Increasing quality demands of consumers in refractory industries have led to longer kiln life and therefore limited consumption growth. Chinese imports are seen as a problem (anti-dumping duty on imports from China since 1990's and new regulation adopted in May 2006), along with imports from Australia. Extractive industry is in direct competition with identical products derived from chemical processes using brine and seawater as a basic material. If land-based resources were to become depleted, most products could be manufactured from brine and seawater.

2.3 Industrial minerals

Mineral	EU mine production (and as % of global mine production) in 2003	Main uses/downstream industries supplied	Factors affecting demand
Perlite	1.251.100 tonnes (38,5 %)	Formed construction products, plasters, mortars, agribusiness/horticulture, industrial filtration (food, beverages, wine, beer, pharmaceuticals, etc.), cement and thermal insulation products.	Linked mainly to construction activity. Trend towards lighter/insulating products on the construction markets. Although perlite is mainly mined and initially processed in Greece, Hungary and Italy, expansion (secondary processing) it takes place closer to the final markets (all European countries). Logistics play an important role. Europe is a major exporter of perlite to the east coast of the USA, being more competitive than mines located in the west of the USA
Potash	4.690.000 tonnes (16,5 %)	Agriculture (fertiliser production). Also glassware, ceramics, batteries, drilling muds, soaps and detergents, pharmaceuticals and chemicals.	Agricultural land policies (e.g. CAP) affect use of fertilisers. Climate and farming practices also affect use. World markets are predicted to change as lower use of fertilisers in the EU contrasts with increased use in developing countries. Use of potash fertilisers in EU-25 fell from about 7,5 million tonnes in 1979 to approximately 3,5 million tonnes in 2002. Use has increased in developing countries due to the increasing demand for food for a steadily growing population.

Chapter 2 Utilisation of non-energy raw materials

Mineral	EU mine production (and as % of global mine production) in 2003	Main uses/downstream industries supplied	Factors affecting demand
Talc	1,285,400 tonnes (20%)	Paper, plastics and paints. Also agribusiness, pharmaceuticals, ceramics, pesticides and fillers in rubber and asphalt roofing products.	Use of talc in the paper industry is being replaced by precipitated carbonates and kaolin which are cheaper. Production of plastics is using increasing amounts of talc, particularly for automobile construction and household appliances. There is competition from mica and other substitutes. Large reserves within the EU (particularly in France) and globally (China, India, Japan, USA and Brazil). Imports into the EU are mainly from China and Australia.
Salt	46,122,200 tonnes (21%)	Chemical industry (chlor-alkali and other sectors), winter maintenance (de-icing), water treatment (softening and disinfection), food and feed industry and many other uses.	Economic situation of the chemical industry, winter weather, overcapacity within the EU and high imports into the EU. The industry is under pressure from overcapacity, high imports (about 25%), especially from eastern Mediterranean countries but also from South America. High energy costs for production (mining and evaporation) and transportation of salt (about 30 million tonnes) and for the industry's main customer – the chemical industry – with energy-intensive electrolysis.

2.4 Construction minerals

The third raw materials sector of the non-energetic EU raw materials industry is the construction raw materials sector. Construction raw materials (e.g. aggregates) are of manifold use, for example in road, rail and sewer construction as well as in the construction of residential buildings, offices and industrial buildings.

Figure 24: Aggregates – Illustration of the added value chain raw materials – end product (Data by Department of Mineral Resources and Petroleum Engineering [UEPG], 2010)

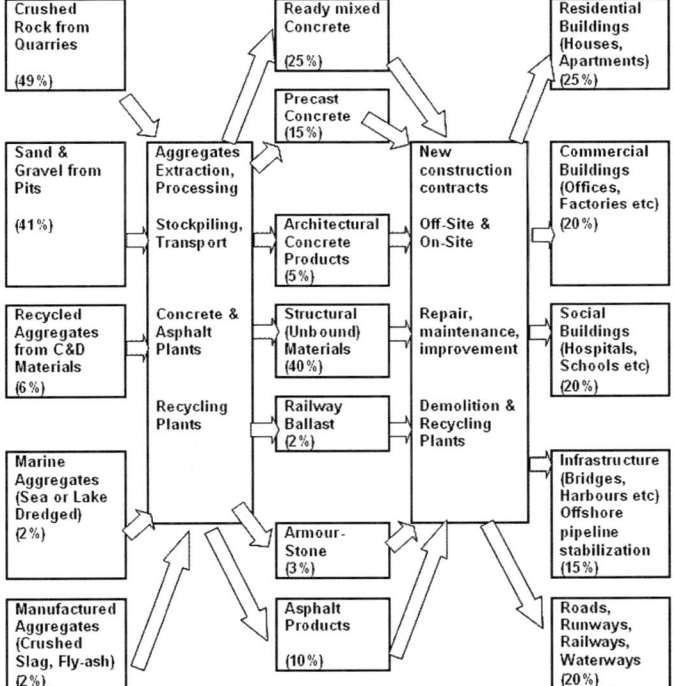

Schematic Diagram for Aggregates, showing sourcing, intermediate products and final usages, with indicative breakdown of percentages by tonnage of aggregates

A single family home with basement contains about 450 tonnes of raw materials. In an 80m² flat approximately 100 tonnes of mineral raw materials have been processed. For the construction of one kilometre of highway about

160.000 tonnes of minerals are needed, for one kilometre of main road about 64.000 tonnes, for one kilometre of country road 32.000 tonnes.[100]

As a conclusion, the construction raw materials industry includes four great divisions:

- Building construction
  - administration buildings, hospitals, schools, swimming pools, prisons, universities, nursing homes, museums, gymnasiums/arenas, football stadiums
- Heavy construction
  - bridges, sewage plants, sewer construction, tunnels, dams, barrages
  - airports
  - roads (including highways, federal and regional routes, community roads, pedestrian areas), railroads
- Commercial construction
  - office buildings, department stores, car parks, television towers, cinemas, halls, workshop buildings
  - stables
- Home construction
  - basements, slabs, walls, roofs
  - stairways
  - garages
  - outside facilities

For example railway

Figure 25: Train station Praterstern, Vienna, Austria

---

100 www.rohstoffforum.at

Figure 26: ICE high speed route in Germany (http://de.wikipedia.org)

An ICE 3 leaving Wahnscheidtunnel.   An ICE 3 between Wiedtal- and Hallerbachtalbrücke, heading for Cologne.

The tracks of the ICE railroad line of German Railways from Cologne to Frankfurt are bedded on a substructure of concrete that consists of 80% aggregate (sand and gravel). However, even the ground under such a "concrete roadway" must be capable of bearing the static and dynamic loads of the trains (including vibrations and oscillations) which travel with more than 300 kilometres per hour, which cause vibrations and oscillations because of their high speed. According to analyses of German Railways, for this particular application an aggregate of sand and gravel lying deep under the tracks is the best solution.[101]

---

101 Die Steine- und Erden-Industrie (2001): Baustoffe 2001, Stein-Verlag Baden-Baden GmbH [Construction Minerals 2001, Stein-Verlag, Baden GmbH].

Chapter 3 Demand and supply of non-energy raw materials in Europe

3.1 Historical development of minerals consumption

The examination (for instance) of a graph of global mine production of copper since 1830, illustrates the remarkable correlation between economic activity and consumption of a basic metallic commodity (figure 27).[102] This graph reflects the electrification of the developed world in the nineteenth century, the two World Wars and the depression periods of the first half of the twentieth century, the post-war reconstruction and the industrialization of Asia, South Africa, and other parts of the world, the impact of rising oil prices in the 1970's and, most recently, wiring the world for the internet. Figure 28 indicates the very good correlation between industrial production and base metal consumption for the world's richest countries.[103]

Figure 27: Global copper production 1830–2000 (Data by Petterson et al., 2005)

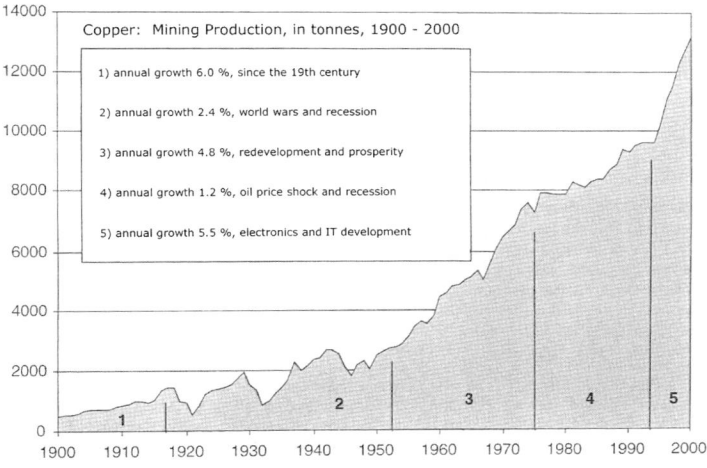

---

102 Petterson, M.G., Marker, B. R., McEvoy, F., Stephenson, M. & Falvey, D. A. (2005): The need and context for sustainable mineral development, in: Sustainable Minerals Operations in the Developing World, Edited by Marker, B. R. et al, p. 5–8.
103 Ibidem. – See also: Svedberg, P., Tilton, J. E. (2006): The real, real price of nonrenewable resources: copper 1870–2000. World development, Vol. 34 (2006), 3, pp. 501–519.

Figure 28: Industrial production of the G7 states (Data by Petterson et al., 2005)

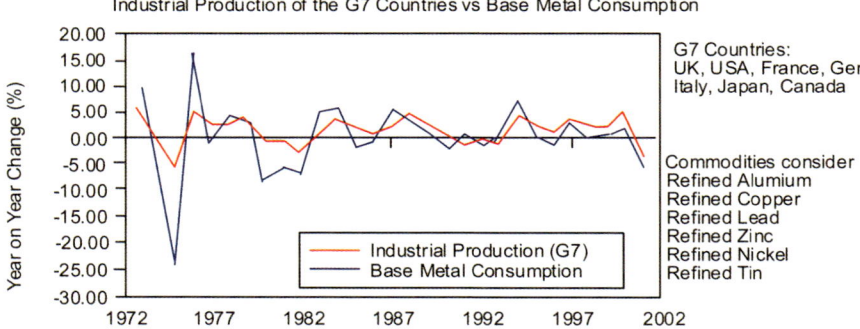

Both graphs demonstrate the importance of the products made from raw materials. The rapid and worldwide increase in demand for raw materials in the 20th century, and partially also in the 21st century, is a phenomenon. The 20th century was in many respects a remarkable episode in human history with special effects on the raw materials industry. This period was marked by an explosive growth of the world population and global economic performance, which lead to a rapid increase in **demand for mineral raw materials of all sorts**.[104]

Figure 29: Correlation of construction raw materials consumption in proportion with population. (Data by EC, DG Enterprise and Industry, Commission Staff Working Document, 2007 [Consumption data based on BGS data of production, import and export of construction raw materials in 2003])

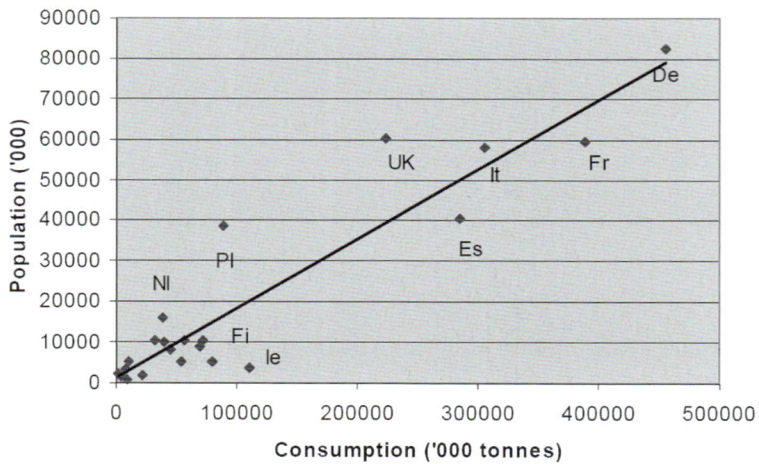

---

104 Nötstaller/Wagner (2007), l.c., p. 383.

3.1 Historical development of minerals consumption

Recent developments as well as relevant projections predict that this trend will continue. Figure 30 shows the historical and the expected growth of the world population. From the graph follows that over many centuries the world population had grown very slowly to one billion at the beginning of the Industrial Revolution, then exploded to more than six billion in the 19th and 20th century. According to pertinent projections, humanity will total about 9,5 billion until the end of this century and stabilize at around 10 billion after 2200. Should this be the case, the world population would have multiplied tenfold in only 300 years.[105]

Figure 30: World population from year 0 until 2200 (Data by de.wikipedia.org)

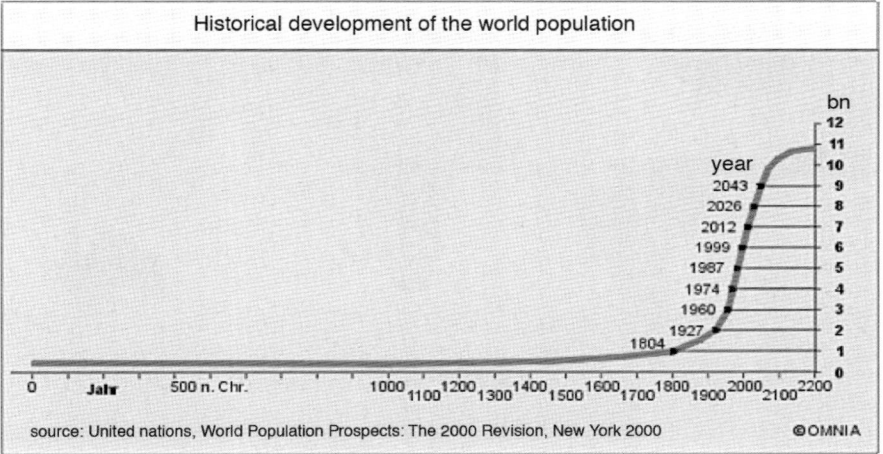

Correlation of economic performance and raw materials consumption

It is essential to discuss the correlation of economic performance and raw materials consumption.[106] The cause of the 20th century's high raw materials consumption is a spectacular growth of the world economic performance coming along with the population development. In the 20th century the global real gross domestic product (GDP) increased by a factor of 18, while the world population quadrupled. The global economic performance grew almost five times as much as the world population; the highest growth was to be observed after World War II.[107] The cause of such an exponential increase was the in-

---

[105] United Nations (2000): World Population Prospects: The 2000 Revision, New York.
[106] The connection between economic achievement and raw materials consumption of a state is already enlarged upon in chapter 1. In the following a reflexion on a global scale takes place.
[107] Nötstaller/Wagner (2007), l.c., p. 383.

dustrialisation of the nowadays highly developed industrial states in Europe, North America and Canada, as well as Japan, Australia, New Zealand and South Korea. The development of infrastructure and goods production coming along with the change from an agrarian to an industrial society led to an enormous rise in demand for mineral raw materials. Thus, **since the end of WW II more raw materials have been consumed than in the whole history of mankind before**.[108]

Figure 31: Global GDP (Data by de.wikipedia.org)

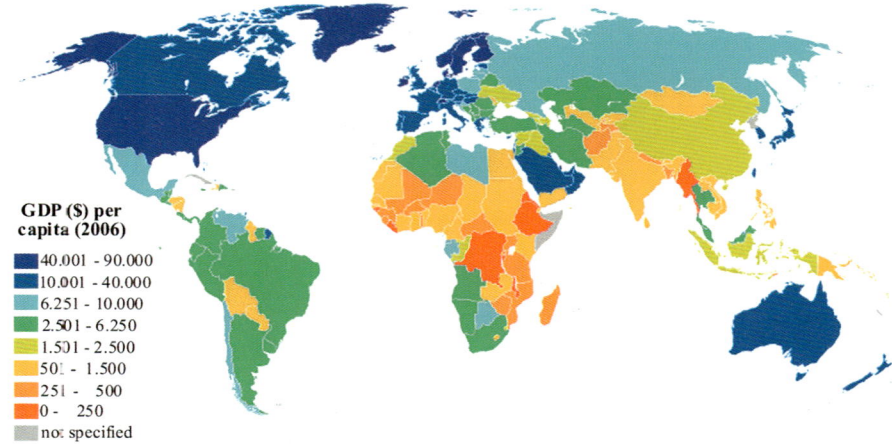

The correlation of the development of the **global** economic performance and raw materials consumption can be demonstrated according to Nötstaller using the example of the industrial metals aluminium, copper, and zinc in figure 32.[109] The examination of the underlying data shows that up until 1960 the demand for metals reacts income elastically, meaning that the rise in raw materials consumption is greater than the rise in economic performance.[110]

While the global GDP rose by a factor of 4,3 from 1900 to 1960, metal consumption of zinc and copper increased by a factor of about 6 and 10 respectively. Metal consumption of aluminium, the industrial production of which started only around 1900, rose by a factor of more than 800. Between 1960 and 2000 the global GDP again increased by a factor of 4,3, whereas metal consumption of zinc and copper only tripled.

---

108 Ibidem. Cp. also Devaney, J., Eden-Green, M., Hargreaves, D. (1994): World index of resources and population, Dartmouth.
109 Nötstaller/Wagner (2007), l.c., p. 384.
110 Ibidem.

Figure 32: World metal consumption and GDP – given period: 1900–2000. Graph shows growth of global GDP and raw materials consumption (Data by Nötstaller/Wagner 2007)

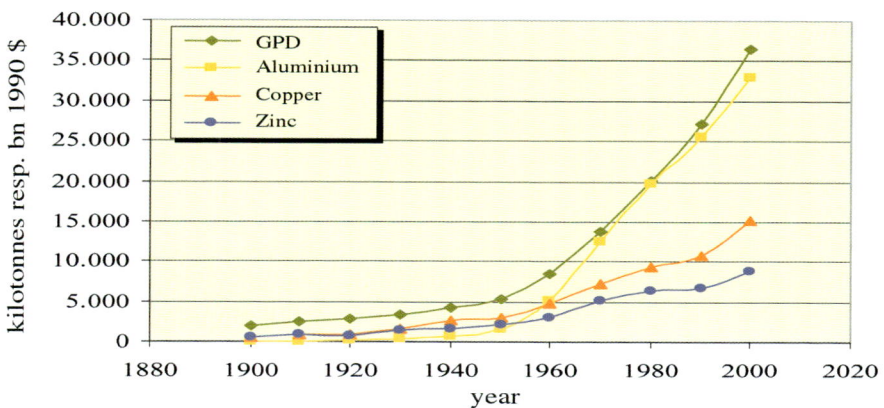

Besides that it is worth to mention that the **worldwide geographical distribution** of certain raw materials deposits (i.e. metallic ores and industrial minerals) is highly **inhomogeneous**. In most cases, the main consuming countries are not the same as the main producing countries. This causes a flow of trade (i.e. development of international raw materials markets). For some raw materials, only three states dominate the world market and possess the majority of known reserves (see figure 33).[111]

Rare earths are a good example.[112] In 2006, global demand for oxides of rare earths was **110.000 tonnes, with about 95% coming from China, amounting to 1,3 billion USD**. Demand is expected to rise to about 185.000 to 195.000 tonnes until 2012.[113] China is going to increase extraction to 125.000 tonnes for its own needs. As China needs an increasing amount of the production to provide its domestic industry, their customers will have to open up new resources on their own. By 2012, the rest of the world will have to extract 60.000 to 80.000 tonnes on their own.[114] Current rare earths supply is coming from just a few sources, the three biggest ones according to TZ Minerals International (TZMI)

---

111 Vgl. Bundesministerium für Handel, Gewerbe und Industrie (1981) (Austrian Ministry of Commerce, Trade and Industry (1981), l.c., p. 27. Also: EC, DG Enterprise and Industry, Comission Staff Working Document SEC (2007) 771, Analysis of the competitiveness of the non-energy extractive industry in EU, p. 39.
112 Concerning relevance of usage of these raw materials: see chapter 1.4 as well as chapter 2.
113 Braune, G. (2008), l.c.
114 Dudley Kingsnorth from the Australian Industrial Minerals Company), in: Braune, G. (2008), l.c.

being in China: the Bayan Obo-mine in Mongolia, the Mianning-Dechang in the Province of Sichuan and scattered deposits in the south of China, particularly near Ganzhou. Only the development of mining companies outside of China can prevent shortages, but possibilities are limited.

Lynas Corp's Mount Weld deposit in Australia is one of the two biggest deposits outside of China. Mount Weld recently started production, which was estimated at 10.500 tonnes for 2009. The second one is Mountain Pass in California belonging to Molycorp, which is part of the Chevron Corporation, from where 20.000 tonnes could soon be produced. The Great Western Minerals Group planned to take up production at Hoidas Lake in Saskatchewan in 2011. Avalon Ventures discovered a significant deposit at Thor Lake in the Northwest Territories.[115]

Figure 33: Global concentration of countries in extraction of selected metals. (BDI, 2007)
Share of the three largest producing countries and their cumulative share of world output in 2005 %

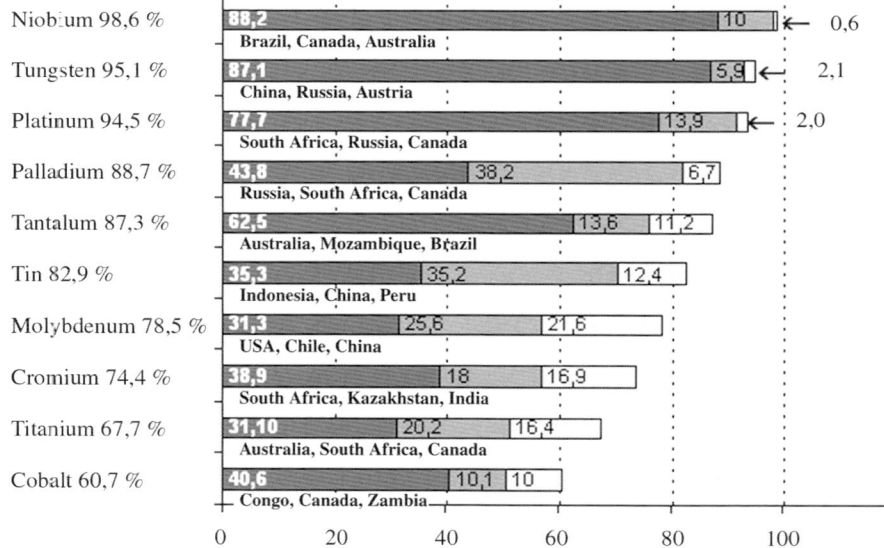

---

115 Braune, G. (2008), l.c.

Table 7: Top three producing regions for selected ores and metals (2004) (Data by Weber/Zsak, World Mining Data, 2006)

	First		Second		Third	
Bauxite	Australia	40%	Guinea	12%	Jamaica	10%
Cadmium	Japan	22%	China	20%	Mexico	12%
Chromium	South Africa	53%	Kazakhstan	18%	India	8%
Copper	Chile	37%	USA	8%	Peru	7%
Iron ore	Brazil	23%	Australia	20%	China	14%
Lead	China	30%	Australia	21%	USA	14%
Manganese	China	24%	Gabon	17%	South Africa	13%
Mercury	EU	43%	Kyrgyzstan	26%	China	23%
Nickel	Russia	24%	Australia	14%	Canada	14%
Silver	Mexico	16%	Peru	15%	Australia	12%
Tungsten	China	87%	Russia	6%	EU	4%
Zinc	China	26%	Peru	14%	Australia	14%

Table 8 is pointing out the top three producing regions for selected **industrial minerals**.

Table 8: Top three producing regions for selected industrial minerals (Data by EC, DG Enterprise and Industry, 2007 [calculations based on BGS data])

	First		Second		Third	
Bentonite	USA	32%	EU	19%	Turkey	7%
Feldspar	EU	36%	China	13%	Turkey	12%
Fluorspar	China	52%	Mexico	17%	EU	8%
Fullers earth	USA	72%	EU	12%	Senegal	4%
Gypsum	EU	24%	USA	16%	Iran	12%
Kaolin	USA	34%	EU	23%	Brazil	19%
Magnesite	China	47%	EU	17%	Turkey	15%
Perlite	EU	39%	China	20%	USA	15%

	First		Second		Third	
Potash	Canada	32%	EU	16%	Russia	16%
Talc	China	46%	EU	20%	USA	13%
Salt	EU	21%	USA	20%	China	16%

3.2 European Union – minerals consumption

The EU is self-sufficient in construction minerals, in particular aggregates, and is a major world producer of gypsum and natural stone. The EU is also the world's largest or second largest producer of certain industrial minerals, such as feldspar, kaolin, salt, though it remains a net importer of most of them. However, the EU is highly dependent on imports of metallic minerals, as its domestic production is limited to about 3% of world production.[116]

Metallic raw materials

The EU (27 members in 2007) is a major user of metallic raw materials; the **global consumption of metallic minerals is about 30%**.[117] The consumption of *certain* metallic minerals accounts for more than 10% of global consumption.[118] For instance, in 1990, the European Union (15 members at the time) shared 22,3% of the world aluminium consumption, in 2007 16,8%. Copper consumption accounted for 28,1% in 1990, 19,9% in 2007. In 1990, the share of zinc consumption was 27,7%; in 2007 it amounted to 20,2%. Finished steel declined from 18,4% in 1990 to 13,4% in 2007.[119]

The main initial markets for metal ores and concentrates in the EU are the **refining and processing sectors which produce semi-finished and finished products for many sectors of the manufacturing industry**.[120] Besides that the EU also has large refining capacities for processing non-ferrous ores and

---

116 Cp. European Commission, DG Enterprise (2008b), l.c., p 3. – Note: *Consumption* of minerals in EU means pruduction of minerals in EU Member States + exports – imports. Many of the needed metallic minerals and also a considerable part of the industrial minerals for the European economy must be *imported* (from different countries all over the world).
117 Ibidem.
118 Chapman, G.R. et. al., British Geological Survey (2003): European Mineral Statistics 1997–2001.
119 Crowson (2008), l.c.: In the same period of time China's share in aluminium consumption rose from 4,5% to 33,3%, copper from 4,7% to 27,1%, zinc from 6,8% to 31,6%, and finished steel from 9,1% to 33,8% of the global consumption.
120 As mentioned in chapter 1, the EU is the second largest producer of iron and steel products, after China.

concentrates and for recycling secondary metals. *Demand* from the manufacturing industry and the construction sector in the EU has been **strong and increasing** in recent decades.[121]

Regarding the **supply offer of metallic minerals**,[122] most of the metallic minerals required by the EU-economy must be imported from countries outside the EU. In recent years, the EU has produced metallic raw materials only to a minor degree, compared to global production. EU mine production of selected metals and the EU share of global mining in 2003 are provided in table 5 (section 2.2). A number of metal ores are extracted within the EU and for some of them, such as copper, lead, silver and zinc; the EU is a relatively important producer. The current distribution of active mines is, however, limited to a relatively small number of Member States (see Table 9).[123]

Table 9: Location of metal mining in the EU in 2003 (Data by European Commission, 2007 [Data source: BGS (2005) European Mineral Statistics.]) Austria, Finland, Greece, Ireland, Poland, Portugal and Sweden have metal mining industries that contribute more than 1% to global production of one particular metallic mineral

Metal ore	Member State
Arsenic	Belgium, France, Germany
Bauxite	France, Greece, Hungary
Cadmium	Belgium, Finland, France, Germany, Italy, Poland, UK
Chromium	Finland
Copper	Cyprus, Finland, Poland, Portugal, Spain, Sweden
Gold	Finland, France, Italy, Poland, Slovakia, Spain, Sweden
Iron	Austria, France, Germany, Slovakia, Spain, Sweden, UK
Lead	Greece, Ireland, Italy, Poland, Spain, Sweden, UK

---

121 EC, DG Enterprise and Industry, Commission staff working document (2007), l.c., p. 47. – See also Chapter 5. Chapter 5 is pointing out the importance of the refining and processing sectors in EU countries and non-EU countries related to metallic minerals. The contribution of these sectors to the added value of a country is essential. Such products are not only produced for the economy of a country but also exported to another EU countries respectively non-EU-countries or even to countries outside Europe.
122 To provide a coherent picture, the following data information mainly is based particularly on the Commission staff working document (2007), published by EC (DG Enterprise).
123 EC, DG Enterprise and Industry, Commission staff working document (2007), l.c., p. 18

Metal ore	Member State
Lithium	Portugal
Manganese	Hungary, Italy
Mercury	Finland, (Spain)
Nickel	Finland, Greece
Selenium	Belgium, Finland, Germany, Poland
Silver	France, Greece, Ireland, Italy, Poland, Spain, Sweden
Strontium	Spain
Tin	Portugal
Tungsten	Austria, Portugal
Zinc	Finland, Greece, Ireland, Poland, Spain, Sweden

Importance of metal recycling

As one of the three sub-sectors (metallic minerals, industrial minerals and construction minerals) **metals provide the highest potential for using recycled materials** (scrap). Many metals, including iron and steel, copper, tin, lead and aluminium, are relatively simple to recycle as they can be melted and recast without losing their important characteristics. Lead scrap (from car batteries), for example, accounts for around 64% of lead consumption in the EU. Recycled aluminium, steel and copper also make significant contributions to total supply within the EU (see figure 34). While recycled metal can make an important contribution to meeting demand, in a growing economy there is a **limit to the extent** to which it can contribute to materials supply. It will be affected by the amount of material originally used (limiting for instance recycling of rare earths) and by its lifetime in use. Metals contained within articles with a short life and high recovery rates will satisfy more of the demand for a particular material than those present in longer lived articles.[124]

Regarding metallic mineral production (of certain metals as listed in table 5 [section 2.2]) in the EU and globally between 1993 and 2003, production of bauxite, cadmium, iron, lead, manganese and mercury all fell. On the contrary, production of chromium, copper, nickel, silver and tungsten increased in 2003 in comparison with 1993 (see figure 35). Despite increases in EU production of some metals, the rate has been lower than the global rate of increase, with the

---

124 EC, DG Enterprise and Industry, 2007, l.c., p. 54.

result that for all but two metals (mercury and tungsten) production in the EU has decreased relatively to global production.[125] Figure 35 points out the change in EU production (%) of (selected) metallic minerals between 1993 and 2003 and the change in the EU's percentage share of global mining production.

Figure 34: Percentage of refined metal in the EU which originated from external scrap in 2003 (Data by European Commission, DG Enterprise and Industry 2007)

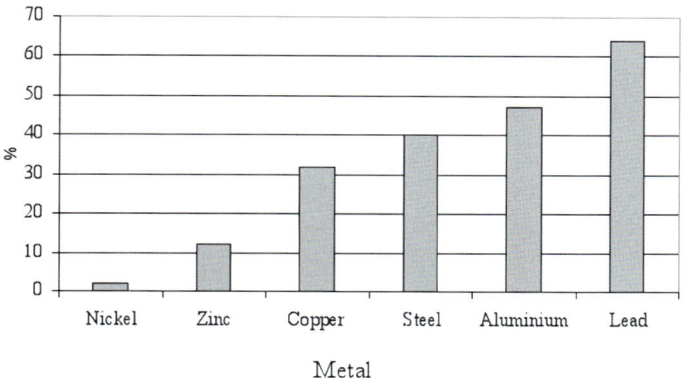

Figure 35: Change in EU production (%) of selected metallic minerals between 1993 and 2003 (Data by EC, DG Enterprise and Industry 2007 [Data from BGS])

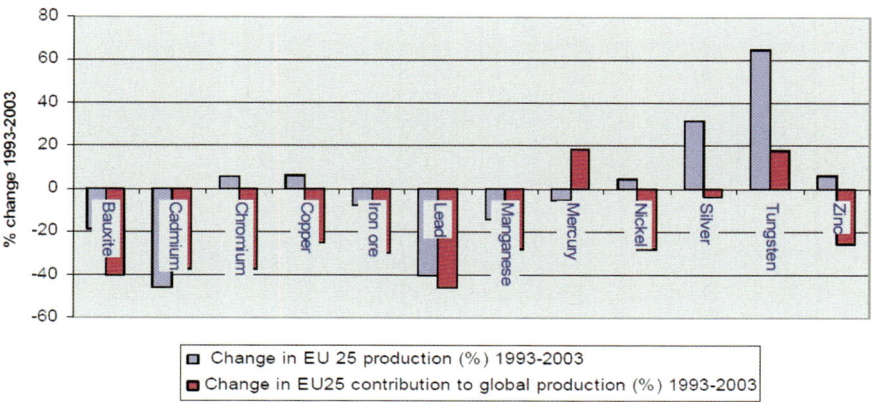

Example: Demand and supply of iron ore/crude steel

Consider the case of iron ore. Since 2003, global iron ore demand has risen at a rate of up to 12% every year, mainly due to the increase in China's consumption, while demand by other countries has increased at a lower rate. Global de-

---

125 EC, DG Enterprise and Industry (2007), l. c., p. 48.

mand is expected to continue growing at an annual rate of about 5% until 2012. In 2006, China increased its domestic iron ore production by 40% and imports by 15%. Due to growing demand for iron ore and continuously rising seafreight prices, yearly iron ore price increases varied from 70% (2005) to 97% (2008).[126] In 1999, the world production of crude steel amounted to 847 million tonnes (Mt), while the European Union (15 members) produced 163 Mt and non-member European countries 17 Mt. China produced 127 Mt, India 26 Mt. By 2007, the total world production had grown to 1.329 Mt whereof the European Union (27 members) produced 198 Mt and non-member European countries 30 Mt, whereas China had increased its production up to 500 Mt and India up to 55 Mt.[127] As mentioned before, EU is the second largest producer of steel products after China; however, as table 5 points out, the EU iron ore production as a percentage of global mine production is only about 2,0% and indicates high import dependency.

Europe's capacity of self-sufficiency through domestic extraction in the case of metallic minerals is very limited. **A number of other metals required by European downstream industries must be imported in the EU** (see also chapter 5).[128] Table 10 illustrates the important uses of many of these metals (particularly for "high-tech" applications) and the global distribution of reserves.[129]

Table 10: Metals which are not mined in the EU, indicating the countries with significant known reserves, and summary of their uses (Data by European Commission, DG Enterprise and Industry, Commission Staff Working Document, 2007)

Mineral	Uses and production	Share of global reserves
Antimony	Alloyed with lead to increase hardness and strength, used in semi-conductors and flame retardants. China accounted for 86% of mine production (2002), although antimony can be produced as a by-product of lead refining. Imported into the EU as ore, metal or oxide.	China (43%), Russia (17%), Bolivia (15%), South Africa (12%), Kyrgyzstan (6%)

---

126 World Steel Association (2008a): International Iron and Steel Institute. Fact sheet Steel and raw materials, http://www.worldsteel.org
127 World Steel Association (2008b): International Iron and Steel Institute. Steel statistics January 2000 (and 2008). Production evoluation of the last months, http://www.worldsteel.org
128 Most of these metallic minerals have not been found in the EU so far in quantities economically viable for extraction.
129 EC, DG Enterprise and Industry (2007), l. c. p 50.

3.2 European Union – minerals consumption

Mineral	Uses and production	Share of global reserves
Beryllium	A lightweight, high-strength metal with high thermal conductivity. Used in electronic components, electrical equipment and aerospace and defence applications. Portugal is thought to possess approximately 0,2% of global reserves.	Brazil (32%), India (15%), China (11%), Russia (11%), Argentina (6%), USA (4%)
Bismuth	Used in pharmaceuticals and as a metal in fusible (low-melting) alloys. Mexico and China accounted for 59% of world production in 2002. Can be produced as a by-product of lead and zinc refining. Bulgaria and Romania both mine bismuth.	China (18%), Australia (16%), Peru (10%), Bolivia (9%), Mexico (9%), USA (8%), Japan (8%)
Boron (Boric oxide)	Glass manufacture (particularly fibreglass) and ceramics. Turkey is the world's largest producer.	Russia (24%), USA (24%), Turkey (18%), China (16%), Kazakhstan (8%), Chile (5%)
Cobalt	Used in steel alloys, super alloys, magnet alloys, batteries and catalysts. Also used in pigments and paint-dryers. Mine production is dominated by just five countries (Zambia, Democratic Republic of Congo, Canada, Russia and Brazil). Belgium and Finland produce significant quantities of cobalt metal from imported ores, while the UK, Finland and France produce significant quantities of cobalt compounds. Often mined as a by-product of other metals (copper, nickel, platinum, silver or zinc). Finland has 0,5% of global reserves.	Congo (44%), Cuba (22%), Australia (15%), Zambia (8%), New Caledonia (5%)
Molybdenum	Used in high-tensile steel to impart hardness, tolerance to high temperatures and resistance to corrosion. It is usually produced as a by-product of copper mining. Production in 2002 was confined to 13 countries, of which the USA, Chile and China accounted for 75% of global production.	USA (49%), Chile (20%), China (9%), Canada (8%), Russia (4%)
Niobium	A soft ductile metal used mainly in special steels and super alloys. Brazil produces approximately 85% of the global total.	Brazil (77%), Russia and other CIS countries (16%)

Mineral	Uses and production	Share of global reserves
Platinum group	Used as catalysts (catalytic converters in cars), in electronics and jewellery. The largest producers in 2002 were South Africa (61%, mainly platinum) and Russia (27%, mainly palladium). Finland is thought to possess approximately 0,1% of global reserves.	South Africa (89%), Russia (9%)
Rare earth elements	A group of 15 metallic elements, of which cerium, lanthanum and neodymium are the most commonly used. Used in automobile catalysts, as metallurgical additives and in glass and ceramics. China produces more than 90% of the global total. Finland and Sweden combined are thought to have relatively small amounts (< 0,01% of global reserves).	China (42%), Russia and the former Soviet Union (18%), USA (17%), Australia (5%)
Rhenium	The main uses are in high-temperature super alloys and petroleum refining.	Chile (52%), USA (15%), Russia (12%), Kazakhstan (8%)
Tantalum	A heavy, very hard, ductile metallic element with a very high melting point (2 996oC) and strong resistance to chemical attack. Used in electronic applications, especially miniature capacitors. Global production dominated by Australia (60%).	Australia (41%), Nigeria (18%), Canada (17%), Congo (11%), Brazil (5%)
Tellurium	Mainly recovered from the anode slimes obtained from the electrolytic refining of copper. Used in iron and steel products, non-ferrous metal alloys, electronics and photoreceptors, catalysts and chemicals, including rubber.	Chile (28%), USA (15%), Zambia (10%), Zaire (9%)
Titanium (ilmenite)	A low-density, strong and corrosion-resistant metal used in the aerospace industry. Most (94%) is used as titanium dioxide as a pigment in paint, plaster, rubber and paper. Finland is thought to possess approximately 0,3% of global reserves.	Australia (25%), South Africa (19%), Norway (12%), Canada (9%), China (9%), Brazil (5%), USA (4%)
(Rutile)		Australia (39%), South Africa (19%), India (15%), Sri Lanka (11%), Sierra Leone (7%), Ukraine (6%)

3.2 European Union – minerals consumption

Mineral	Uses and production	Share of global reserves
Vanadium	A soft ductile metallic element that is highly corrosion-resistant. Mainly used as an additive in steel alloys to which it imparts strength and corrosion resistance. Also used in titanium alloys and as a catalyst. China produces 50% of the global total.	Russia (50%), South Africa (30%), China (20%)

In summary, the EU is highly dependent on imports of metallic minerals, as seen in figures 36, 37 and 38. As an illustration, **168 million tonnes of metallic minerals were imported into the EU in 2004 with a total value of € 10 billion**, compared to the EU's production of some 30 million tonnes; with other words, consumption of 30% of all metallic minerals is contrasted with just 3% of world production.[130] In 2006, the percentage of the EU's self-sufficiency concerning the supply of metallic minerals was rather low: iron 1,99%, chromium 5,36%, tungsten 4,27%, nickel 2,03%, manganese 0,11%; cobalt, molybdenum, niobium-tantalum, vanadium 0% respectively.[131] On the contrary, over 7 million tonnes were *exported* worth € 516 million. In terms of value, Turkey (€ 69 million) and Saudi Arabia (€ 62 million) were the most important destinations. While the level of exports of metallic minerals is considerably lower than imports, there was a 57% increase by weight between 1999 and 2004, with an increase in value of 70%. In terms of weight, exports to Turkey, Saudi Arabia, Egypt, China and Norway were the most important.[132]

Figure 36: EU's main sources of import for metallic raw materials in 2004 by value (Data by EC, DG Enterprise and Industry, Commission Staff Working Document, 2007)

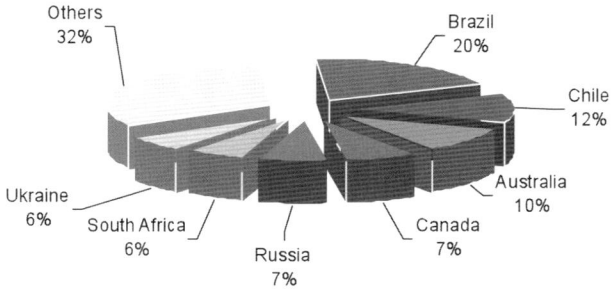

---

130 European Commission, DG Enterprise and Industry, (2008a): Public Consultation on Commission Raw Materials Initiative, Brussels.
131 Weber, L. (2008): Minerals safeguarding activities of the European Union and their impacts on national level, BHM, Journal of Mining, Metallurgical, Material, Geotechnical and Planned Engineering 153 (8), pp. 289–293.
132 EC, DG Enterprise and Industry (2007), l.c., p. 64.

Chapter 3 Demand and supply of non-energy raw materials in Europe

Figure 37: Annual raw materials import in the EU, 1999–2004 (weight in thousand tonnes) (Data by EC, DG Enterprise and Industry, Commission Staff Working Document, 2007)

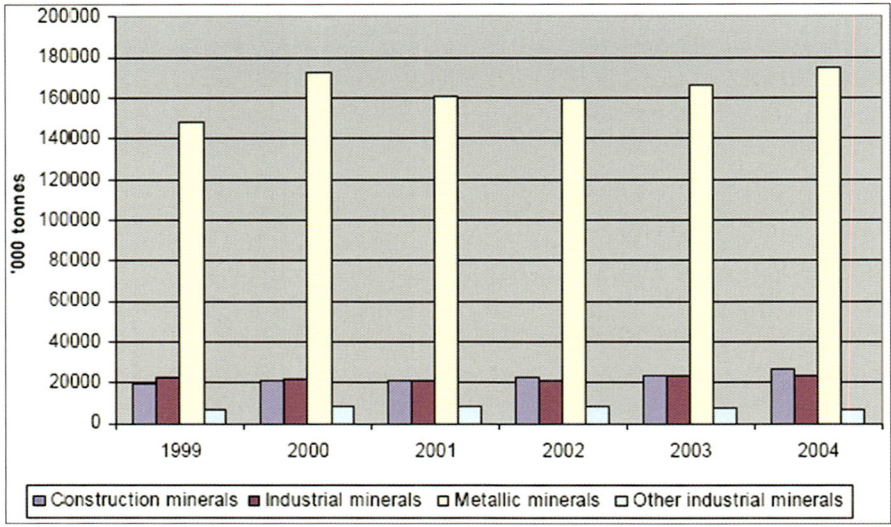

Figure 38: Net amount of import of raw materials in the EU from 1999 to 2004. – Data Source: Eurostat (EC, DG Enterprise and Industry, Commission Staff Working Document, 2007)

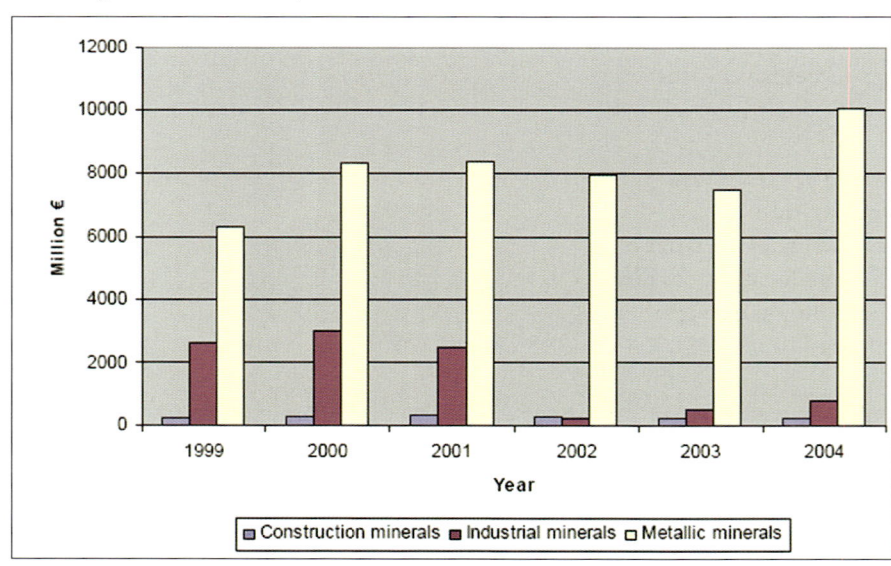

Industrial minerals

The main initial markets for industrial minerals (e.g. magnesite) in the EU are the downstream industries, such as glass, foundry, ceramics, paper, paint and colorants, and plastics industry. As mentioned before, the EU is the world's largest or second largest producer of certain industrial minerals (table 8, section 3.1), though it remains a *net importer of most of them*.[133]

Industrial minerals are extracted in more than 650 mines and quarries and processed in 600 plants in Europe. The sector is present in almost all EU member states, from the north of Scandinavia to the Mediterranean coast. The sector processes an annual volume of some 100 million tonnes, contributing a value of about € 13 billion to the EU's GDP.[134] If downstream industries, such as glass, foundry, ceramics, paper, paint and colorants, and plastics industry are included, these figures are of significantly greater dimension. In 2006, the European Union was the world's largest producer of feldspar (60%), perlite (54%), kaolin (31%), gypsum (23%), salt (22%), and the second largest producer of Fuller's earth, bentonite, talc and potash.[135] The European share of global industrial minerals production is about 20%. Nevertheless, the EU is a net importer of industrial minerals as a whole: In 2004, imports of industrial minerals into the EU amounted to about 18,7 Mt which meant import costs of € 798 million (Figure 37 and 38). On the contrary, 5 million tonnes were exported into countries outside the EU (i.e. Non-EU member states).[136]

Many different types of industrial minerals are extracted within the EU (see table 6 [section 2.3]), however the geographical distribution of suitable resources is very uneven across the EU (see table 11). While the mining industry extracts minerals such as kaolin, industrial limestone and talc in around half of the Member States, others, such as fluorspar, mica, phosphate rock and sulphur, are extracted in just one or two countries. Other industrial minerals, such as iodine, are not mined at all in the EU.[137] Natural variability in the quality and characteristics of a particular mineral found in different parts of the EU and the location of different markets mean that, while a number of Member States might extract the same mineral, they may serve quite different markets. Magnesite extracted in Greece, for example, is predominantly used in agri-

---

133 To provide a coherent picture, the following data information mainly is based particularly on the Commission staff working document (2007), published by EC (GD Enterprise).
134 IMA (Industrial Minerals Association) – Europe, 2009, http://www.ima-eu.org.
135 EC, DG Enterprise and Industry (2008c), Commission Staff Working Document SEC (2008)2741 accompanying COM (2008) 699, Brussels.
136 European Commission, DG Enterprise, 2007.
137 EC, DG Enterprise and Industry, Commission staff working document (2007), l. c., p. 16–17.

culture, whereas the main markets for magnesite produced in other European countries are refractory industries.[138]

Table 11: Industrial minerals extracted in the EU in 2003 and Member States involved (countries with > 2 % of world production in bold) (Data by EC, 2007 [Data sources: BGS (2005) European Mineral Statistics; EULA, EuroGeoSurveys (for quartz and silica sand)])

Industrial mineral	Member State
Barytes	Belgium, France, Germany, Greece, Italy, Poland, Slovakia, Spain, UK
Bentonite and fuller's earth	Cyprus, Czech Republic, Denmark, **Germany**, **Greece**, Hungary, **Italy**, Poland, Slovakia, Spain, UK
Bromine	France, Germany, Italy, Spain, UK
Diatomite	Czech Republic, Denmark, France, Hungary, Italy, Poland, Spain
Feldspar	Czech Republic, Finland, France, Germany, Greece, Italy, Poland, Portugal, Spain, Sweden, UK
Fluorspar	**France**, Germany, Italy, **Spain**, UK
Graphite	Czech Republic, Germany, Sweden
Kaolin	Austria, Belgium, **Czech Republic**, France, **Germany**, Greece, Italy, Poland, Portugal, Slovakia, Spain, **UK**
Industrial limestone/dolomite	Austria, Belgium, Czech Republic, Denmark, Estonia, Finland, France, Germany, Greece, Hungary, Ireland, Italy, Poland, Portugal, Slovakia, Spain, Sweden, UK
Magnesite	**Austria**, **Greece**, Netherlands, Poland, **Slovakia**, **Spain**
Mica	Finland, France, Spain
Perlite	**Greece, Hungary, Italy**, Slovakia
Phosphate rock	Finland
Potash	France, **Germany**, Spain, **UK**
Quartz, silica sand	**Austria, France, Germany**, Greece, **Italy, Poland, Slovakia, Slovenia, Spain, Sweden, UK**
Salt	Austria, Denmark, France, Germany, Greece, Italy, Netherlands, Poland, Portugal, Slovakia, Slovenia, Spain, UK
Sillimanite minerals	France, Spain

---

138 Ibidem. EC, DG Enterprise and Industry, Commission staff working document (2007), l.c., p. 16–17.

3.2 European Union – minerals consumption

Industrial mineral	Member State
Sulphur and pyrites	Finland, Spain
Talc	Austria, Finland, France, Germany, Greece, Italy, Portugal, Slovakia, Spain, Sweden, UK

Figure 39: EU's main sources of import for industrial minerals by value (Data by EC, Commission Staff Working Document, 2007 [Eurostat])

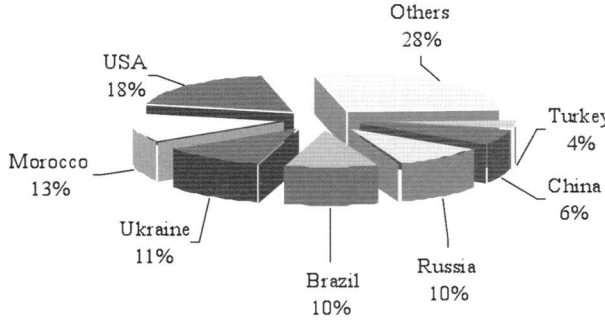

Production trends between 1992 and 2004 for some of the industrial minerals covered in table 6 (section 2.3) can be seen in figure 40. Figure 40 presents recent changes in EU mine production and the contribution that the EU made to global production. The results indicate that EU mine production of bentonite, feldspar, magnesite, perlite, talc and salt was higher in 2003 than in 1993, while production of barytes, fluorspar, Fuller's earth and potash fell. The trend in the EU contribution to world production over the period was upwards for some minerals (bentonite, feldspar, kaolin, perlite and talc) but significantly downwards for others, in particular for barytes, fluorspar, magnesite and potash.[139] The observed trends differ for each mineral type, and in some cases they are the result of a change in demand from downstream sectors. Demand for barytes, for example, is closely linked to the level of oil well drilling activity, while demand for potash, the main source of potassium in fertilisers, is affected by changing agricultural practices in the EU (a general reduction in use) and in developing countries (a general increase). In other cases increased mine production in some developing countries has not only increased the global supply, putting pressure on EU exports, but also led to direct competition within the EU. Chinese magnesite is a good example of this.[140]

---

139 DG Enterprise and Industry, Commission staff working document (2007), l.c., p. 43.
140 Ibid.

Chapter 3 Demand and supply of non-energy raw materials in Europe

Figure 40: Change in EU production (%) of selected industrial minerals between 1993 and 2003 and change in EU's percentage share of global mine production (Data by EC, DG Enterprise and Industry, Commission Staff Working Document, 2007 [data source: BGS, Euromines]) – Figures above zero indicate an increase in EU mine production (light bar) or an increase relative to changes in global production (dark bar) in 2003 compared with 1993. Figures below zero indicate relative decreases.

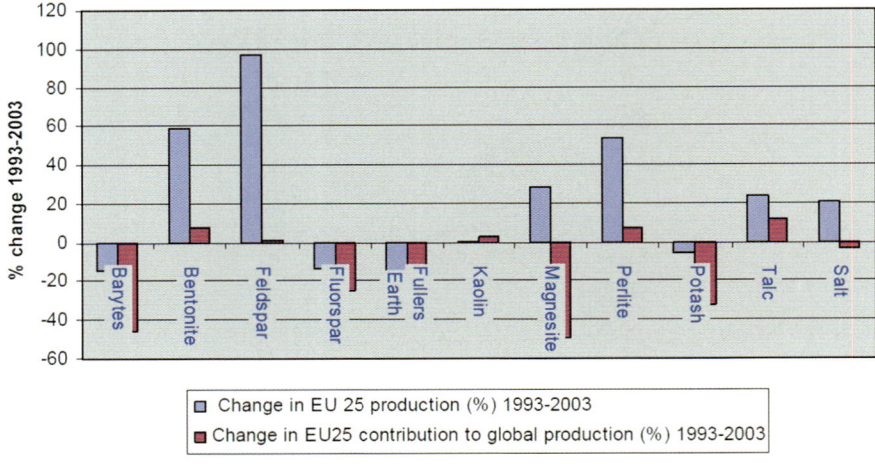

Figure 41: Production of some industrial minerals – EU/world (Data by EC, DG Enterprise and Industry, Commission staff working document, 2007)

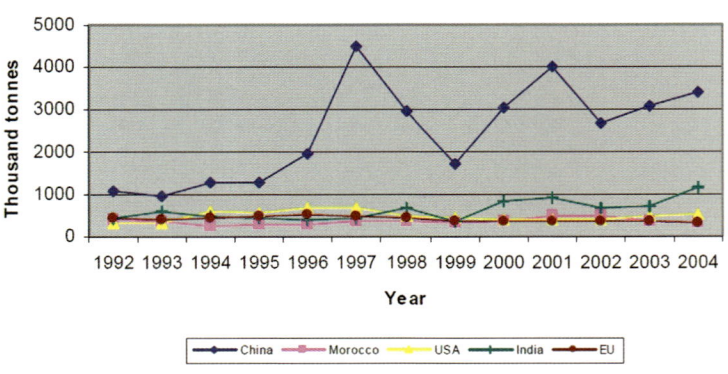

3.2 European Union – minerals consumption

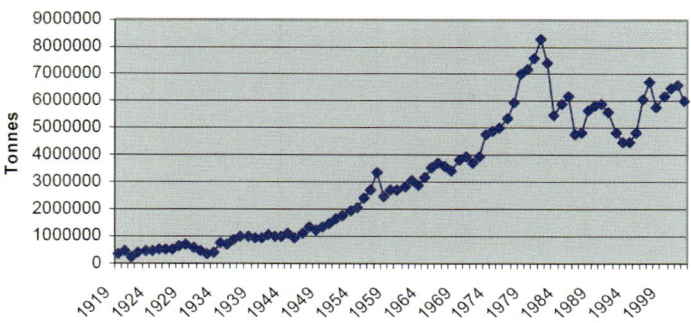

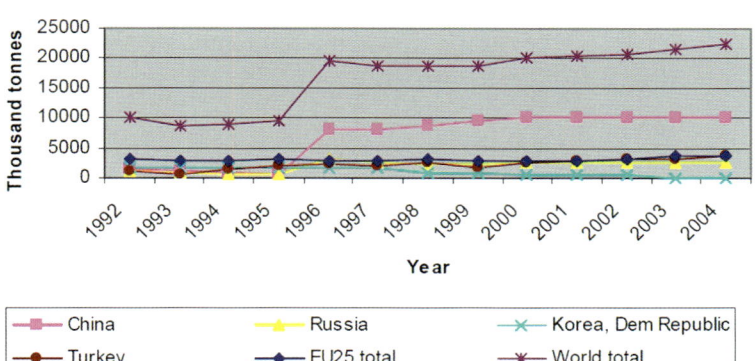

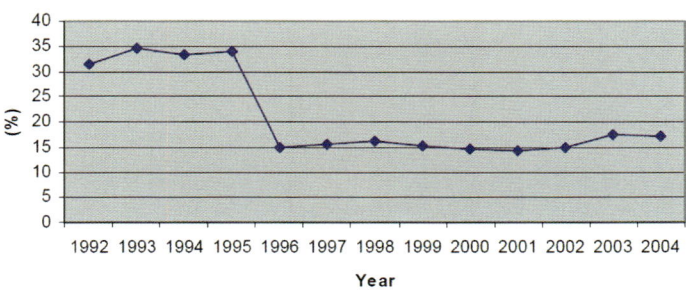

Regarding exports of industrial minerals in countries outside the EU, table 12 indicates that almost 5 million tonnes were exported from the EU in 2004 worth € 509 million. This is an increase of 12% in terms of weight and 33% in terms of value compared with 1999. Almost 20% (by weight) went to Norway, with a value of € 57 million, while Switzerland, Morocco and the USA were other important destinations. There has been a general increase in the level of exports since 1999 with the greatest relative increases in terms of weight being to the USA (281%), China (222%) and Turkey (86%).[141]

Table 12: Exports of industrial minerals outside the EU – Member States (Data by EC, DG Enterprise and Industry, Commission staff working document, 2007)

Country of destination	Weight			Value		
	Thousand tonnes	% of total exports	5 change 1999–2004	Million	% of total exports	5 change 1999–2004
Extra-EU	4 963		12,0	509		32,6
Norway	961	19,4	58,9	57	11,3	25,3
Switzerland	242	4,9	27,0	38	7,5	38,7
Morocco	899	18,1	31,4	34	6,8	52,9
USA	479	9,6	281,1	31	6,2	59,6
Turkey	145	2,9	86,0	30	5,9	63,2
Indonesia	142	2,9	N/A	23	4,5	N/A
China	127	2,6	222,0	23	4,5	291,0
Japan	60	1,2	0,7	18	3,5	27,8
Malaysia	169	3,4	21,1	17	3,4	7,5
Canada	161	3,2	45,0	15	3,0	39,6

Data source: Eurostat

Construction minerals

The construction minerals as the basis of infrastructure development are of particular importance. For construction minerals (in particular aggregates)

---

[141] EC, DG Enterprise and Industry, Commission Staff Working Document (2007), l.c., p. 63.

3.2 European Union – minerals consumption

Europe is self sufficient. More than 3 billion tonnes of sand, gravel and crushed stone are produced annually to meet the demands of the European building industry.[142] Additionally, about 12 Mt of construction minerals (worth € 456 million) were imported into the EU in 2004 (see Figure 37 and 38).[143] A little over 14 Mt of construction minerals were exported from the EU in 2005, worth € 704 million. Almost half (in terms of weight) went to Switzerland, a further 7 % to the USA and 6 % to China. Exports to China increased by almost 400 % (by weight) compared with 1999. However, compared with the total consumption of construction minerals in the EU, the level of exports is very low (0.6 %).[144]

Figure 42: Comparisons of various variables indicating the proportions of the EU economy, the construction housing sector and the aggregates sub-sector in EU-15 (Sources: Bleischwitz/Bahn-Walkowiak, 2006)

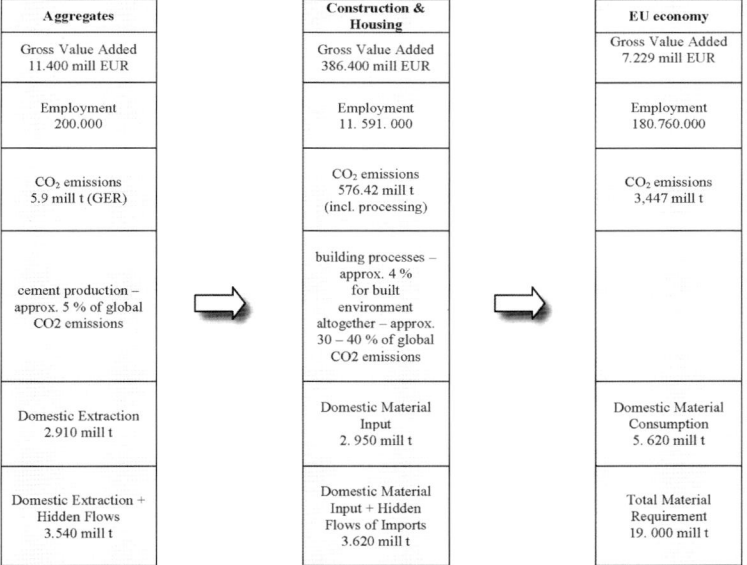

---

142 UEPG, in: Department of Mineral Resources and Petroleum Engineering (2010), l.c.
143 EC, DG Enterprise and Industry, Commission Staff Working Document (2007), l.c., p. 65.
144 EC, DG Enterprise and Industry, Commission Staff Working document (2007), l.c., p. 62 (data: from Eurostat).

Chapter 3 Demand and supply of non-energy raw materials in Europe

The production of construction minerals in Europe appeared widely constant in the years from 1999 to 2004 and ranged between 2.500 to 2.700 million tonnes per year.[145] Countries like UK, Germany, France, Italy and Spain remain in these years in principle in a stable position of being important markets. Demand is slightly increasing at the beginning of 2004 influenced by new EU Member States, particularly Czech Republic, Hungary, Poland, Slovenia and Slovakia. In 2006, production of aggregates came up nearly to 2.900 Mt per year.[146] Figure 43 illustrates the production of aggregates of the EU-27 countries between 2003 and 2007.

Based on extensive analysis, a report (2010) commissioned by UEPG concludes that the demand for aggregates continues to grow with economic development at national and European levels. Empirical evidence shows that advanced economies can demand up to 12 tonnes/capita, though this growth may suffer short-term positive or negative influences from economic boom or recession respectively. Therefore it is reasonable to anticipate that European demand for aggregates will reach 4 billion tonnes in the medium term, driven mainly by economic growth in Central and South-Eastern Europe.[147]

Figure 43: Production of aggregates 2003–2007 of EU-27 (Data by British Geological Survey, 2009)

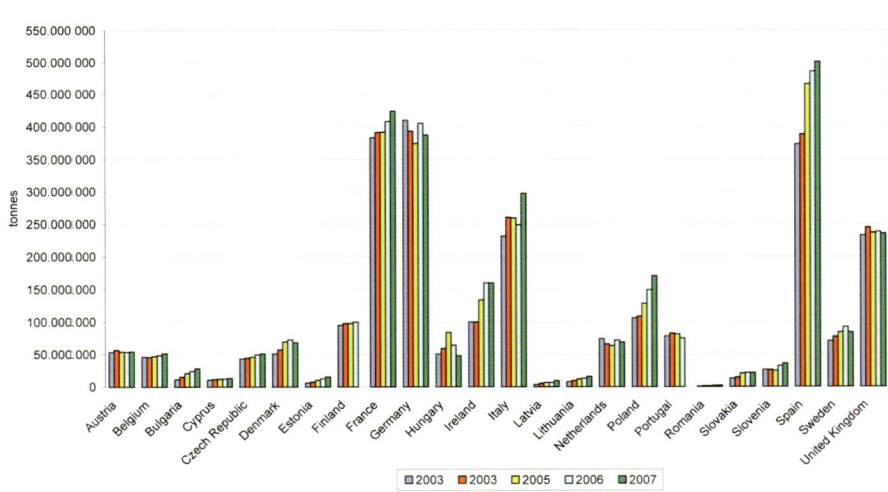

145 Koziol et. Al. (2008), production of aggregates in EU (http://www.min-pan.krakow.pl/Wydawnictwa/GSM2443/koziol-kawalec-kabzinski.pdf.
146 Ibidem.
147 Department of Mineral Resources and Petroleum Engineering (2010): Planning Policies and Permitting Procedures to Ensure the Sustainable Supply of Aggregates in Europe, University of Leoben.

3.2 European Union – minerals consumption

When mounting statistical data of aggregates in Europe, difficulties occur due to overlapping definitions and data (for instance data sources from BGS and UEPG). Present data are fragmented, often inconsistent, partly not covering the whole EU region. Data refer to different definitions and terminology of material groups like construction minerals, building materials or simply minerals, all of them including aggregates to a large extent. Although some studies indicate that the separation of a category like aggregates may be negligible in some cases because the output figures are very close to the total minerals extraction, the problem of inconsistent definitions and data remains at present.[148]

Table 13: The European Aggregates Industry – Annual Statistics 2008, Quantities in million tonnes. (Data by UEPG, 2010)

Country	Total number of producers (companies)	Total number of extraction sites (active quarries and pits)	Sand & gavel (M)	Crushed Rock (Mt)	Marine aggregates (Mt)	Recycled Aggregates (M)	Manufactured aggregates (Mt)	Totals (Mt)
Austria	960	1.290	62	32	0	4	2	100
Belgium	180	253	11	42	4	14	2	72
Bulgaria	200	100	18	22	0	0	0	40
Croatia	367	308	7	22	0	0	0	29
Czech Rep	219	489	27	44	0	4	0	76
Denmark	350	300	43	0	5	0	10	58
Finland	400	2.255	25	60	0	1	0	86
France	1.640	3.050	165	237	7	15	8	432
Germany	2.300	1.510	260	218	0	56	18	552
Greece	300	200	20	20	0	0	0	40
Ireland	150	355	25	25	0	0	0	50
Italy	1.796	2.360	225	135	0	5	3	368
Netherlands	65	225	46	0	54	24	0	124

---

148 Bleischwitz/ Bahn-Walkowiak, 2006. The narrowness of available raw materials data is explicitly mentioned in the Commission Staff Working Document "Analysis of the competitiveness of the non-energy extractive industry" (EC, DG Enterprise and Industry2007). Especially, official statistics for aggregates are rather incomplete. This problem can be traced back, among other causes, to the structure of these sectors: the building raw materials sector, which exhibits a large number of middle, small and very small enterprises in most member states, is particularly concerned. These are not covered by national statistics and thus neither by Eurostat. At the same time, the importance of these sectors is displayed insufficiently. – See also Chapter 6.

Chapter 3 Demand and supply of non-energy raw materials in Europe

Country	Total number of producers (companies)	Total number of extraction sites (active quarries and pits)	Sand & gravel (M)	Crushed Rock (Mt)	Marine aggregates (Mt)	Recycled Aggregates (M)	Manufactured aggregates (Mt)	Totals (Mt)
Norway	690	713	15	52	0	0	0	68
Poland	2.044	1.786	131	49	0	22	1	203
Portugal	350	200	61	15	0	0	17	93
Romania	500	730	18	7	0	1	0	26
Slovakia	170	92	13	21	0	1	0	35
Spain	1.600	2.060	134	244	0	5	1	383
Sweden	985	1.802	19	67	0	7	0	93
Switzerland	350	505	37	5	0	5	0	47
Turkey	770	770	25	290	0	0	0	315
UK	450	781	55	114	12	53	9	243
Total	16.836	22.134	1.441	1.720	81	216	72	3.533

(1) Sand and Gravel: sold production including crushed gravel.
(2) Crushed rock: sold production (excluding crushed gravel).
(3) Recycled Aggregates: materials coming from construction and demolition waste used in aggregates market.
(4) Manufactured aggregates include blast-furnace-slag, electric-arc-furnace-slag, incinerator bottom ash (IBA), pulverised fly ash (PFA) and other industrial and extraction by-products for construction and civil engineering.

In Europe there are countries with a higher potential for sand and gravel extraction, while others possess a higher potential for the production of crushed rocks. Moreover some countries have both mineral categories available. Surely, this can determine the production possibilities and the development of the aggregates market within a country or even between countries. Usually production of construction minerals takes place near metropolitan areas.[149] However, the construction of mega quarries at the coast of Norway and Great Britain denotes a new dimension. This could have particular consequences on parts of Europe (concerning imports and the development of commodity price). Moreover, it is acknowledged that such policies could place additional strain on other countries, and encourage long distance transportation of minerals,

---

149 Association of Swiss hard rock quarries (VHS), which commissioned the study: "Short transport distances limit environmental effects", Sand and Gravel, No. 56 (2002), p. 47. This study has shown that short transport routes are the most effective means of keeping down the cost of transportation and the environmental impact of mineral supply.

3.2 European Union – minerals consumption

something which in general terms appears contrary to the objectives of sustainable development. This development could influence minerals policies and supply practices. It seems that an increased proportion of Europe's (and other countries) demand for aggregates could be met from these sources as other resources diminish and environmental policy is strengthened in the more traditional mining countries. Obviously the environmental impact will be greatest for those areas which have a good access to the sea or are situated along waterways. In all other cases overland transport remains a significant limiting issue.

Figure 44: The trade in gravel in the study area in 2000 (Data by http://internationaal.bouwgrondstoffen.info/ Data in Mt.)

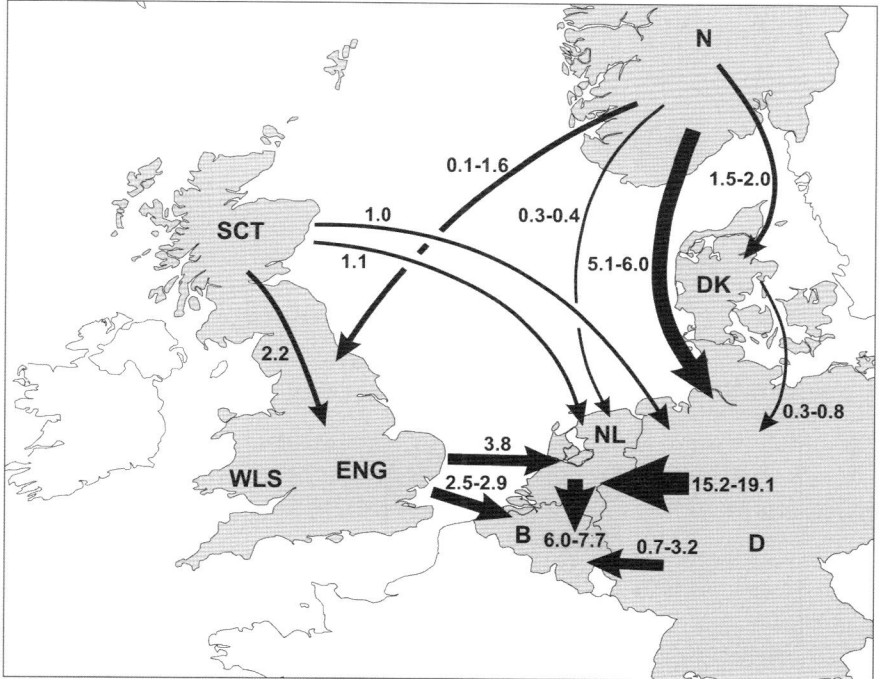

Aggregates and construction industry

Aggregates and construction industry can be considered a moderate growth market in Europe in recent years. Housing and civil engineering are driving forces in the construction industry. Data point out three construction market cycles, the 1st cycle in 1993–1999 and the 2nd cycle in 1999–2006. The European construction market was at a peak in the 2nd cycle in 2006, i. e. 2006 was the culmination of 6 years of development in the construction industry. The highest growth rates were recorded in all countries observed during the period of years 2004–2006. Construction showed the highest growth rates in 2006 and the total European growth reached a peak in the second cycle (3.7%). The next peak of European construction was expected in 2012, with 2007 as the starting year of the third cycle.[150] However, the growth of European aggregates market in 2008–2009 was affected by a financial crisis. Many European banks were involved and it was clear that a long term slump could affect investment levels in Europe in the next years.

The investment outlook for European housing markets and civil engineering construction in 2009 (calculated in 2007 from Euroconstruct) was positive. Regardless of the financial crisis a positive development of housing markets was assumed in a majority of European countries, above all in Spain, Ireland, Finland and Portugal. Also a positive development of civil engineering was assumed: New pressures in various countries stimulated in 2007 a demand for civil engineering works like road construction. Thus, the investment outlook of European countries in the years ahead was optimistic, with growth rates between 3 and 4% annually. The major relative contribution was noticed from the Central and Eastern European countries, with annual growth rates above 10%.[151]

3.3 Development of international metallic minerals markets

The following explanations primarily refer to demand, production and price of **metallic raw materials** as the European Economy is heavily dependent on imports of such raw materials (see section 3.2).

Just as rapidly growing demand, first in a recovering the economically expanding Western Europe, and then in Japan, boosted the rate of growth of demand for metals in the 1960s, so the **Asian countries became the motor of growth from the late 1980s**. The countries of the Pacific Rim led the way, with China and India following in their wake. The impact of differential rate of the economic growth on the consumption of metallic minerals can be seen in table 14, which shows the rate of growth of consumption of metals in the 1990s and 2000s to date, and percentage shares of global usage for selected coun-

---

150 Euroconstruct (2007): construction market, expected growth, http://www.seeurope.net/files2/pdf/rgn1207/6_Expected_Growth.pdf.
151 Ibidem.

tries.[152] Additionally figure 45 provides the world mining production based on the development level of each producing country between 1984 and 2006.

Table 14: Growth rates of metals consumption and shares of world totals (selected countries) (Data by Crowson, 2008)

	Aluminium		Copper		Zinc		Finished Steel	
	1990s	2000s	1990s	2000s	1990s	2000s	1990s	2000s
Growth rates (annual percentage changes)								
US	3,5	−0,7	3,2	−4,7	3,1	−3,4	3,3	−0,5
Japan	−0,6	0,6	−1,3	−0,6	−1,9	1,4	−2,0	1,3
EU (15)	2,1	2,3	2,7	−0,7	2,4	−0,1	2,1	1,6
Asia Pacific (a)	8,5	4,4	9,7	2,2	7,7	1,4	5,4	3,5
Russia	−13,2	6,7	−15,1	21,5	−14,2	6,8	−11,8	8,5
China	13,7	18,4	12,6	13,9	11,7	14,9	8,9	17,6
Other countries	2,3	5,7	1,6	4,6	2,8	3,1	1,3	7,5
World	2,3	5,6	3,0	2,6	2,8	3,5	1,6	6,8
Percentage shares of world consumption								
	1990	2007	1990	2007	1990	2007	1990	2007
US	22,5	15,0	19,9	12,1	15,0	9,3	13,4	9,0
Japan	12,6	5,9	14,6	7,0	12,3	5,2	15,0	6,6
EU (15)	22,3	16,8	28,1	19,9	27,7	20,2	18,4	13,4
Asia Pacific (a)	14,6	6,6	7,1	12,0	7,3	9,1	8,0	9,1
Russia	13,8	2,7	7,8	3,8	9,3	1,8	13,3	3,3
China	4,5	33,3	4,7	27,1	6,8	31,6	9,1	33,8
Other countries	19,7	19,6	21,8	22,0	21,7	22,8	24,5	24,8

---

152 Crowson (2008): Mining Unearthed, London, p. 61.

Figure 45: World mining production based on the level of development of the producing country (Data by Weber and Zsak, 2008)

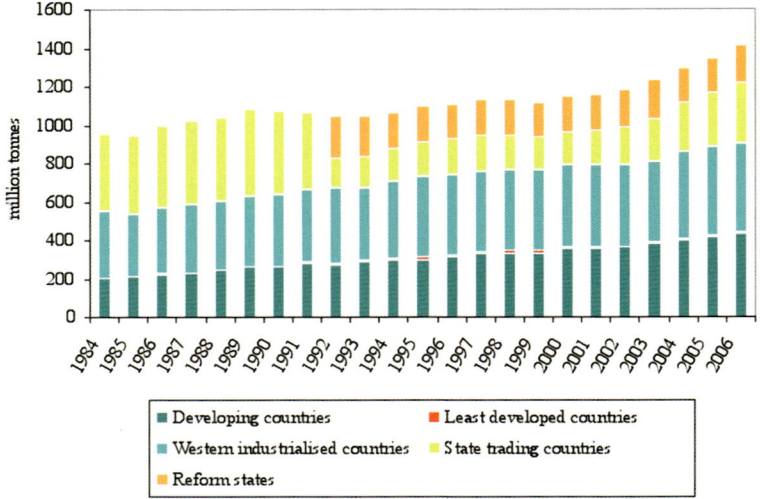

Pricetrend

The commodity price index of the Hamburg Institute of International Economics (Hamburgisches Weltwirtschaftsinstitut [HWWI]) reported that commodity price rose by more than 80% in total between the beginning of 2003 and the end of 2008.[153] In 2006 the overall index was about 20% higher than in the previous year, after it had already risen by more than 28% (2005) and 18% (2004) previously. Not only the price for crude oil, which is in the centre of public interest, but also the metallic raw materials prices rose considerably. From the beginning of 2003 to the end of 2006, the crude oil price increased by 100%, as did prices for iron ore and scrap steel on average (depending on sort and quality); **prices for nonferrous metals even increased by 128%. Prices for some particular metals even rose by 500%.**[154] In 2008 ore prices reached a remarkable level. According to the HWWI commodity price index, raw material prices dropped by 15,8% (due to the financial crisis) compared by the peak in July 2008.

Record prices, net profits and ambitious mine expansion plans in the first half of 2008 for most minerals and metals gave way to price deterioration (fig. 46,

---

153 Linden, E. (2004): Der Bergbau weltweit im konjunkturellen Aufwind – Verkäufermarktbedingungen für Rohstoffe, Bergbau, S. 441–442, 444. [Mining industry worldwide in economic ascent – Sellers' market conditions for raw materials, Mining, pp. 441–442, 444] [Also: BDI (2007), l.c., p. 5].

154 Cp. also Stribrny, B., Vasters, J., Brinkmann, K. (2006): Developments on the international markets of metallic raw materials, World of Metallurgy – Erzmetall, pp. 191–201.

3.3 Development of international metallic minerals markets

47), production cuts and mine closures in the last quarter of 2008 and the first half of 2009. By mid-2009, however, a **slight recovery in hard commodities had begun**, although its sustainability is far from assured. In these uncertain times, consolidation remains an attractive option for many mining companies. Exclusive negotiation at the end of November 2009 between Rio Tinto and Iron Ore Holdings over the latter's Iron Valley deposit in Western Australia, and the recent acquisition of the Australian miner OZ Minerals, the world's second-largest producer of zinc, by CMN, a subsidiary of China Minmetals, can be seen as good illustrations of this merger mania, which is likely to exert an *upward pressure on prices*.[155]

Prices in all major commodity markets, including energy, metals and minerals, agriculture and food, increased sharply in 2007 to reach a peak in 2008, declined strongly from the second half of 2008 and have been on an increasing trend again since the summer of 2009.

Figure 46: HWWI Index of the world prices for raw materials Hamburg Archive of International Economics (data until end of 2006), from 2007 continued by the Hamburg Institute of International Economics, Source: Hamburgisches Welt-Wirtschafts-Archiv (HWWA) and Hamburgischen WeltWirtschaftsInstitut (HWWI)

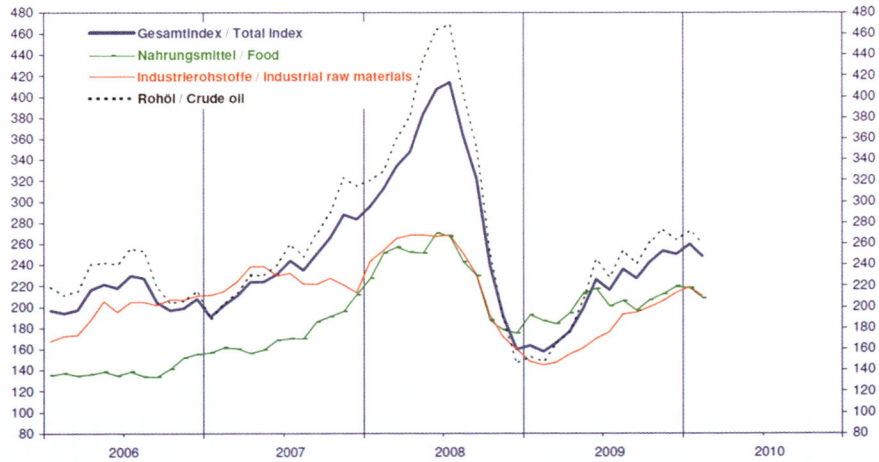

Copyright HWWI = 100 month averages (average for the last month shown is incomplete until the month is over)

---

155 UNCTAD (United Nations Conference on Trade and Development) (2010): Recent developments in key commodity markets trends and challenges (http://www.unctad.org/en/docs/cimem2d7_en.pdf).

At the heart of current developments lies a series of changes in global supply and demand patterns as well as short term shocks in key commodity and raw materials markets. The years 2002 to 2008 were marked by a major surge in demand for raw materials, driven by strong global economic growth, particularly in emerging countries such as China. This increase in demand will be reinforced by the further rapid industrialisation and urbanisation in countries such as China, India and Brazil.

Figure 47: Price development of copper, tin and nickel in the US 2002–2009 (Data by UNCTAD, 2010)

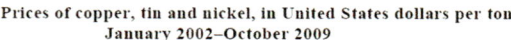
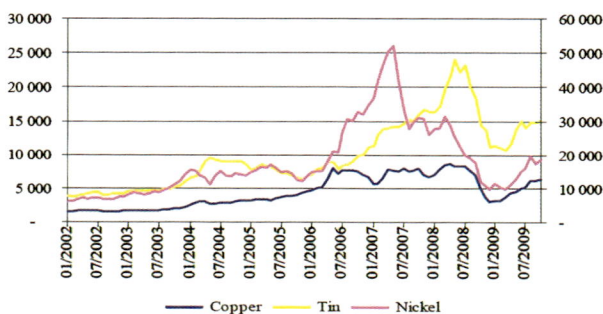

In the following a short description of the development of metallic mineral markets of China, India, South America and Africa is given.

Raw materials markets in China

After a partial deregulation of the market, an economic growth of about 10 % annually was observed from 1980 to 2005.[156] On the international raw materials markets, China is buying metallic raw materials in such quantities that supply shortages, accompanied by an according increase in price, have emerged. China offensively accesses raw materials projects in Africa, Australia, and East

---

156 Cp. Deutsches Institut für Wirtschaftsforschung (1987): China industrialisiert auf breiter inländischer Rohstoffbasis, Wochenbericht/Deutsches Institut für Wirtschaftsforschung, Berlin, S. 289–294 (German Institute for Economic Research (1987): China industrialises on a broad domestic raw material base). – See also: Nötstaller/Wagner (2007), l.c. See also: Ho, P. (2006): Trajectories for greening in China: theory and practice: Development and change, Vol. 37 (2006), 1, pp. 3–28.

Asia.[157] From 1990 to 2004 the Chinese import value of metallic raw materials increased by a factor of 24, whereas its export value of metallic raw materials only increased tenfold.

Whereas for a long time China had been an exporter of raw materials and supplied the world market with cheap raw materials, the country has now become the **largest importer of raw materials in the world**. Over the past decade, China has quadrupled to tenfolded its imports of aluminium, lead, copper and tin. Thus, China has almost doubled its share of world consumption of non-ferrous metals to about 20% during the last six years.[158]

Figure 48: GDP and metal consumption in China 1980–2005 (Data by Nötstaller/Wagner 2007)

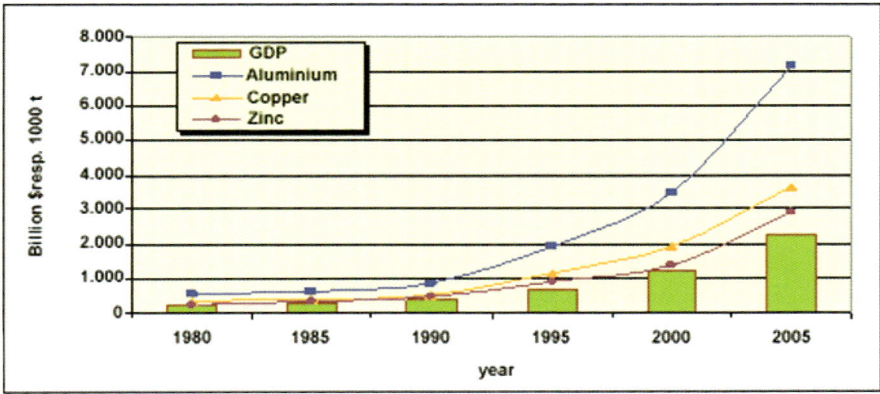

The figure shows the development of the economic performance and the metal consumption in China between 1980 and 2005. While the economic performance increased by a factor of 9 within the observed period of time, metal consumption increased by a factor of almost 10 for copper, 11 for zinc and 13 for aluminium.[159]

As a result of the rapid economic development, China has become the **second largest national economy** and the world's biggest consumer of many raw materials, namely iron ore, copper, zinc, tin, lead, nickel, steel, coal and

---

157 Apel, H., (2008), KfW IPEX-Bank GmbH, Lateinamerika Verein e.V., Hamburg, 15.04.2008.
158 BDI (2007), l.c., p. 7.
159 Similarly, as quick as the metal consumption, the consumption of the building material cement grew by 100 million tonnes in 1980 to 1 billion tonnes in 2005; see Nötstaller/Wagner (2007), l.c., p. 384.

cement. For aluminium, the People's Republic is the largest producer and consumer (BGP, Rohstoffwirtschaftliche Länderstudien, 2008).

Figure 49: Trend of demand for raw materials in China (Data by Ekdahl, 2008)

China is rich in mineral resources and was the world's leading producer of aluminum, antimony, barite, bismuth, coal, fluorspar, gold, graphite, iron and steel, lead, phosphate rock, rare earths, talc, tin, tungsten, and zinc in 2008. It ranked among the top three countries in the world in the production of many other mineral commodities. China was the leading exporter of antimony, barite, coal, fluorspar, graphite, rare earths, and tungsten in the world. The country's demand for chromium, cobalt, copper, iron ore, manganese, nickel, and potash exceeded domestic supply, and **imports were estimated to account for more than 30% of domestic consumption**. Mineral trade increased by 33,5%, which accounted for 25,7% of the country's total trade in 2008.[160]

In 2008, during the largest extent of the global financial crisis, China's economic growth declined to 6,8% in the fourth quarter from 12,6% in the second quarter. Industrial value-added growth also decreased. Much of the slowdown was the result of weak domestic and international demand. In November 2008, the State Council announced a fiscal stimulus package worth about 4 trillion yuan ($ 586 billion) during 2009–10 to promote stable growth in the domestic economy. Also in the last quarter of 2008, the Government announced that 10 major industries in China – automobile manufacturing, electronic information, equipment manufacturing, iron and steel, light industry,

---

160 USGS (U.S. Geological Survey) (Oct. 2009): Minerals Yearbook, China, 2008 (Pui-Kwan Tse).

logistics, nonferrous metals, petrochemical, shipbuilding, and textiles – were to be reformed and upgraded.[161]

In 2008, the Ministry of Land and Resources (MLR) issued the **national plan on mineral resources (2008–15)** which established annual production targets for the most important commodities. The Government projected consumption of 1,5 Gt iron ore; 14 Mt. of aluminum and 7,6 Mt. of copper in 2020. The country was expected to face a shortage of 19 mineral commodities among the 45 major mineral commodities produced. The dependence on imports of copper and potash was expected to be 70%; and iron ore, 40%. During the past several decades of exploration, geologists believed that about one-third of total mineral resources had been discovered in China; most of the undiscovered mineral resources were expected to be located in the western part of the country.[162]

During the past decade, China's rare-earth production accounted for about 90% of the world total. Rare-earth consumption in China had increased steadily. The country consumed about 73.000 t of rare earths in 2007 and 68.000 t in 2008 compared to 19.300 t in 2000. The dominant position China's rare earths in the world became more important because of the wide range of cutting-edge environmental technology, such as wind turbines, low-energy light bulbs, and hybrid cars that depend on rare-earth metals. Owing to an increase in domestic demand, the Government gradually reduced the export quota during the past several years. At the same time, China's companies were looking for rare-earth investments overseas, in Australia.[163]

Production, Export and Import of rare earth and iron are illustrated in 2 Figures in the appendix.

China plays an important part in the automobile market. In 2009, China produced 13,79 million automobile units, of which 8 million units were passenger cars (sedans, sport utility vehicles (SUV), multi-purpose vehicles (MPV) and crossovers), and 3,41 million units were commercial vehicles (buses, trucks, and tractors). In 2009, 7,3 million new cars were sold in China, which is 40% more than in the year before. It is expected that in 2010, about 9 million cars will be sold. As incomes increase the high annual growth rate of private ownership is expected to accelerate. The recent surge in sales has been attributed to the

---

161 Citigroup Global Market Inc., 2008, in: USGS (2009), China.
162 Ministry of Land and Resources, 2008a, b, in: USGS (2009), China.
163 USGS (2009):)Minerals Yearbook, China, 2008.

favorable monetary and fiscal support by the government, which has identified this key sector as one of the ten pillar industries of the Chinese economy.[164]

How much China has to catch up on can be demonstrated through the example of steel consumption: while in Europe the average per-capita consumption of steel is about 450 kg per year, in China it is only 220 kg.[165]

Table 15: China: Exports of selected mineral commodities in 2008 (Data by USGS, 2009)

Commodity	Quantity (metric tons)	Value (thousand $)
Metals		
Aluminum:		
Alumina	44.142	29.558
Metal and alloys:		
Unwrought	841.292	2.136.194
Semimanufactures	1.900.000	6.366.035
Antimony metal, unwrought	9.453	52.680
Barium sulfate	3.840.000	200.991
Copper, metal and alloys:		
Unwrought	102.724	854.036
Semimanufactures	517.522	4.157.241
Iron and steel:		
Pig iron and cast iron	250.000	127.921
Steel:		
Bars and rods	12.620.000	11.540.374
Shapes and sections	3.650.000	3.313.761
Sheets and plates	28.790.000	29.267.799
Tube and pipe	1.340.000	3.004.565
Scrap	204.217	94.796

---

164 http://en.wikipedia.org/wiki/Automobile_industry_in_China (July 19, 2010); Kleine Zeitung, Jan 5, 2010.
165 BDI (2007), l.c., p. 7. Also: Nötstaller/Wagner (2007), l.c., p. 385.

3.3 Development of international metallic minerals markets

Commodity	Quantity (metric tons)	Value (thousand $)
Manganese, unwrought	240.547	852.925
Molybdenum, ores and concentrates	23.626	867.249
Rare earth products	54.963	687.770
Tin, metal and alloys, unwrought	559	11.306
Tungsten, tungstates	5.421	115.602
Zinc:		
Metal and alloys, unwrought	71.320	147.442
Oxide and peroxide	32.779	61.743
Industrial Minerals		
Fluorspar	660.000	190.606
Granite	1.210.000	188.256
Graphite, natural	600.000	163.980
Magnesia, fused	2.270.000	335.326
Talc	690.000	123.469
Mineral Fuels and related materials		
Coal	45.430.000	5.240.265
Coke, semicoke	12.130.000	5.807.369
Petroleum:		
Crude oil	4.160.000	2.979.552
Refinery products	17.030.000	13.665.164

Table 16: China: Imports of selected mineral commodities in 2008 (Data by USGS, 2009)

Commodity	Quantity	Value (thousand $)
(Metric tons unless otherwise specified)		
Metals		
Aluminum:		
Alumina	4.590.000	1.775.694
Metal and alloys, unwrought	260.102	564.138
Semimanufactures	618.620	3.188.836
Scrap	2.150.000	2.540.178
Chromium, chromite	6.480.000	2.714.382
Copper:		
Ore and concentrates	5.190.000	10.440.152
Anode	197.571	1.404.611
Metal and alloys, unwrought	1.702.039	11.660.119
Semimanufactures	934.950	7.567.250
Scrap	5.580.000	5.969.151
Iron and steel:		
Iron ore	443.560.000	60.531.628
Steel:		
Bars and rods	960.000	1.516.407
Scrap	3.590.000	2.467.387
Seamless pipe	1.060.000	4.144.345
Shapes and sections	330.000	370.277
Sheets and plates	12.730.000	16.011.819
Manganese ore	7.570.000	3.469.752
Nickel:		
Ore and concentrates	12.318.022	2.064.439

3.3 Development of international metallic minerals markets

Commodity	(Metric tons unless otherwise specified)	Quantity	Value (thousand $)
Metal		9.600	248.459
Titanium dioxide		250.651	526.263
Industrial Minerals			
Diamond	kilograms	3.388	3.025.786
Fertilizers:			
Compound fertilizers		640.000	402.556
Diammonium phosphate		100.000	126.104
Potassium chloride		5.140.000	2.831.054
Potassium sulfate		110.000	46.309
Urea		67	97
Mineral Fuels and related materials			
Coal		40.400.000	3.509.106
Petroleum:			
Crude oil		178.880.000	129.334.996
Refinery products		38.850.000	30.044.320

Source: General Administration of Customs of the Peoples Republic of China, 2008, China monthly exports and imports, no. 12.

The comparison of import and export data indicates that China has a rising demand of mineral resources and must therefore increase the import.

Raw materials markets in India

India has globally significant mineral resources; its deposits of bauxite and iron ore account for 10%, 4%, and 3% of the world's total resources, respectively. In terms of the relative size of its mineral resources, India's barite resource was the second largest in the world after China; iron ore, the third largest; bauxite (2.3 Gt), the sixth largest. The country's resources of chromium, limestone, and manganese were also among the 10 largest in the world. Of these resources, 1,4 Gt of bauxite are located in the State of Orissa, which also hosts resources of chromium, cobalt, nickel, and titaniferous magnetite. In terms of world production, India was among the leading producers in the world of mica (first); barite, chromium, and talc (second); bauxite and coal (third); iron ore and kyanite (fourth); manganese ore and steel (fifth); zinc (seventh); and aluminum (eighth).[166]

India's mineral industry contributed 1,9 % of the gross domestic product (GDP) and was an important sector of the economy in fiscal year 2008–09. Mineral production in terms of tonnage increased by 2,34 %, and total output in terms of value increased by 7,1 %. The value of mineral exports increased by 17 % and that of mineral imports increased by 15 % compared with those of fiscal year 2007–08. India exported 93 million metric tons (Mt) of iron ore in fiscal year 2006–07, of which 75 % went to China.[167]

In March 2008, the Government approved a new mining policy designed to simplify the country's mining regulations, accelerating the mining lease application process to between 6 and 12 months from more than a year, making mining lease approval automatic for companies that discover the mineral resources, and granting companies the rights to deal with prospecting data.[168]

The mining industry is characterized by a large number of small operational mines. Public-sector companies play a dominant role in mineral production; they control the mining and processing and are the main producers of aluminum, copper, and gold. Small mines are owned mostly by private-sector companies that produce cement and manganese ore.

India is largely self-sufficient in mineral commodities, which are used as primary raw materials in various industries. India exported, in descending order of value, diamond (mostly cut), iron ore, granite, zinc ore and concentrate, chromium, bauxite, and alumina in fiscal year 2007–08. The country imported, in descending order of value diamond (uncut), and other commodities, including copper ores and concentrates, phosphate rock, and sulphur. India exported 3,8 Mt of finished steel products and imported 5,8 Mt.[169]

It is expected that India will need major investment in its mineral industry in the next 5 to 10 years to support and sustain its high rate of economic growth. The value of mineral industry output is expected to reach $ 30 billion and to account for 2,5 % of the GDP within the next 4 years. The country's production of bauxite, alumina, and aluminum is expected to increase owing

---

166 Ministry of Mines, 2009, in: USGS (U.S. Geological Survey) (2010): Minerals Yearbook, India 2008 (Kuo, S.C.), p. 1.
167 Ministry of Mines, 2009, in: USGS (2010), India, p. 1.
168 Industrial Minerals, 2008i, in: USGS (2010), India, p. 1. See also Annex 1.
169 Ministry of Mines, 2009, in: USGS (2010), India, p. 2. See also: Bansal, R. (2010): Iron ore future – the next decade. Journal of mines, metals und fuels, 58 (2010), 3–4, pp. 70–73.

to new mines and plants and expansions of production capacities. The country is likely to achieve a crude steel production capacity of 124 Mt/yr in fiscal year 2011–12 with the start-up of new steel plants and the completion of expansion projects. India has abundant heavy mineral sands. Production of monazite, titanium minerals, and zircon are all expected to increase. The country is expected to become a significant producer of rare earths in the next few years. Owing to mergers and capacity additions, India's output of cement is expected to reach 335 Mt/yr in 5 years.[170]

Production, Export and Import of selected raw materials are illustrated in Appendix 2.

Concerning the development of the economic growth and raw materials demand, India is going to follow suit to China. On the occasion of the 19th World Mining Congress in New Delhi in 2003, the President of India, Abdul Kalam, in his inauguration speech announced the contribution of mining will be significantly raised.[171] The demand for raw materials has risen accordingly. Nonetheless, India is still an exporter of raw materials at the moment. More than half of its iron ore production is exported, because Indian customers alone do not yield sufficient profit.[172] Due to long-term contracts with several states, the country realizes more profit by exporting its iron ore and steel production at global market conditions than from domestic sales.[173] A reversal of the trend might occur in the foreseeable future.[174]

Raw materials markets in Latin America

Latin America (including the Caribbean) consists of 32 independent countries. During the last decades many of them have joined in several different economic associations: Argentina, Brazil, Paraguay and Uruguay are members of the MERCOSUR (Mercado Común del Sur); Bolivia, Colombia, Ecuador, and Peru are members of the Andean Community; El Salvador, Guatemala, Honduras, Nicaragua, Costa Rica and the Dominican Republic ratified the Dominican Republic-Central America Free Trade Agreement (DR-CAFTA)

---

170 U.S. Geological Survey (2010): Minerals Yearbook, India 2008, p. 6.
171 Bundesministerium für Wirtschaft und Arbeit (2008) [Austrian Ministry of Commerce, Trade and Industry (2008)]: Österreichisches Montanhandbuch, 82. Jahrgang, p. 1 [Austrian Mining Handbook, Vol. 82, p.1].
172 Ibid.
173 BDI (2007), l.c., p. 7.
174 Cp. Chapter 3, Section 3.4.

with the USA; Mexico (together with Canada and the USA) is part of the North American Free Trade Agreement (NAFTA).[175]

The Latin American countries make considerable contributions to the production of some mineral commodities. As an example, in 2007 their share of world total production of bauxite was 26%, copper 48%, gold 18%, iron ore 20%, lead 15%, nickel 13%, silver 45%, tin 22%, zinc 22%, hydraulic cement 5%, gypsum 8%, salt 11%.

A number of countries in Latin America are major producers and exporters of metallic and industrial minerals, mineral fuels, and related materials, mostly in crude form. In 2007, record prices for many of the minerals produced in the region encouraged a widespread increase in production. In 2007, Bolivia was one of the top five mine producers of tin in the world. Brazil was the second ranked producer of iron ore in the world and one of the top five mine producers of bauxite and tin. Chile was the leading mine producer of copper and the second ranked producer of refined copper in the world (see figure 53). Peru was one of the top five mine producers of copper, gold, lead, tin and zinc in the world and was among the top five producers of tin metal. Mexico was one of the top five mine producers of lead in the world. The roles of most countries in Latin America and the Caribbean in the global mineral industry were as suppliers of metallic mineral ores and concentrates and of other minerals in crude form, and a few of these countries were significant suppliers of refined mineral products in the world.

Production, Export and Import of selected raw materials are illustrated in Appendix 2.

In 2007, the production and export of mineral commodities, including mineral fuels, accounted for a significant percentage of the gross domestic products (GDPs) and export revenues in many countries in Latin America. During the boom in mineral prices from mid-2003 through the first half of 2008, many countries in the region appeared to enjoy economic benefits, low interest rates and inflation, and expansion of their export volumes. Revenues for already active mineral producers in the region increased exponentially during this timeframe, and many Governments were able to obtain corresponding increases in revenues from the mineral industries in their countries.

The continuing growth in China's (and other emerging economies') demand for minerals affected the GDPs of mineral producing countries in Latin

---

[175] All information regarding raw material markets in Latin America: U.S. Geological Survey (April 2010): Minerals Yearbook, Latin America and Canada, 2007 (By Steven T. Anderson, Omayra Bermúdez-Lugo, Alfredo C. Gurmendi, Alberto A. Perez, Susan Wacaster, Glenn J. Wallace, and David R. Wilbur).

3.3 Development of international metallic minerals markets

America by way of the revenue gained **from exports of minerals and mineral products to China**, which increased its share of exports from many commodity-exporting countries in Latin America from 2000 through 2008. (For example, exports of iron ore accounted for about 30 % of the total value of Brazilian exports to China in 2008.) In 2006, primary commodity (including minerals) exports from Latin America and the Caribbean accounted for about 22 % of all China's (primary) commodity imports.[176]

Mid-year 2008, developments on raw materials markets were mainly positive. Prices for the most important metals were on a new level. Due to the global economic setting and the demand from emerging markets, prices generally remained on a high level. Shortages of particular raw materials are amplified by the direct access (buying into raw materials businesses and projects) of emerging countries.[177] It was also said that the effects of the financial crisis and the fear of recession are completely unclear (which became a fact in autumn 2008).

Figure 50: Copper, nickel, zinc and aluminium prices as an example (Data by Sentient Monitor, 2008)

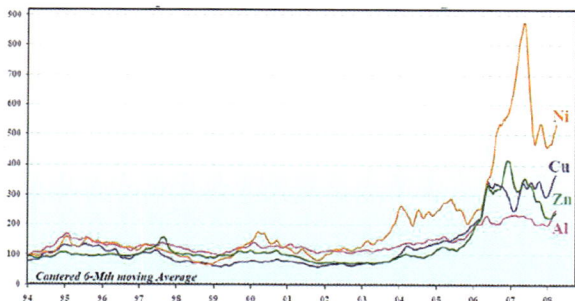

The race for raw materials has begun: Emerging countries still have to catch up on a considerable backlog. Apart from China, other emerging countries are going to follow. The big raw materials demand of emerging markets (compared to industrialized countries) can be shown at the example of copper (see figure 52).[178]

---

176 U.S. Geological Survey (April 2010): Minerals Yearbook, Latin America and Canada, 2007.
177 Apel, H., (2008): Rohstoffmärkte aus Bankensicht, KfW IPEX-Bank GmbH, Lateinamerika Verein e.V., [Raw material markets from banking perspective, KfW IPEX-Bank GmbH, Latin Business Association, incorporated association] Hamburg, 15.04.2008.
178 Apel, (2008), l.c. – See also: Tilton, J.E., Lagos, G. (2007): Assessing the long-run availability of copper. Resource policy, Vol. 32 (2007), 1–2, pp. 19–23 – Gómez, F, Guzmán, J.L., Tilton, J.E. (2007): Copper recycling and scrap availability. Resource policy, Vol. 32 (2007), 4, pp. 183–190 – Guzmán, J.L., Nishiyama, T., Tilton, J.E. (2005): Trends

Chances and risks concerning the investment potential of Latin America's raw materials markets are very diverse. Latin America does have considerable deposits of metallic raw materials (iron ore, copper ore, nickel ore, molybdenum).[179] The risks of investing in Latin America are not low, due to increasing competition of other raw materials producing countries like Russia, South Africa, and in the future also Sub-Saharan Africa (Congo, Angola). Moreover, in Latin American countries economic development and political stability differ widely.[180]

Figure 51: Global copper consumption – raw materials markets in Latin America for financial purposes. Per-capita consumption of copper in the EU-27 is three times as big as in China. (Data by Apel, H., KfW IPEX-Bank GmbH, Lateinamerika Verein e.V., Hamburg, 15.04.2008; Data by International Copper Study Group)

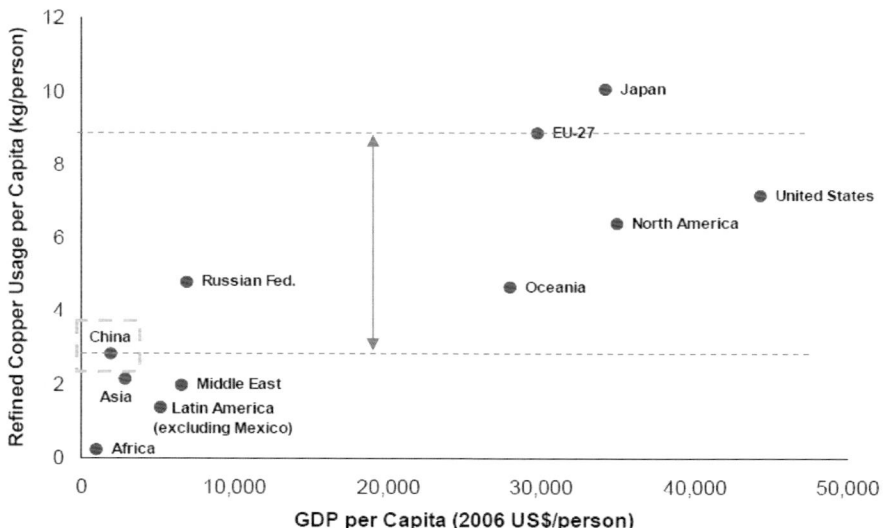

Raw materials markets in Africa

The 56 independent nations and other territories of continental Africa and adjacent islands encompass about 30 million square kilometres, which is more than 20% of the world's total land area, and were home to more than one

---

in the intensity of copper use in Japan since 1960. resource policy, Vol. 30 (2005), 1, pp. 21–27.
179 Ibid.
180 Apel, l.c. Politically stable countries are Chile, Brazil, Mexico, Peru, Colombia. Countries with uncertain political future: Argentina. Countries with protectionist tendencies are Venezuela, Bolivia, Ecuador.

billion people in 2009. For many of these countries, mineral exploration and production constitute significant parts of their economies and remain keys to future economic growth. Africa is richly endowed with mineral reserves and ranks first or second in quantity of world reserves of bauxite, chromite, cobalt, hafnium, industrial diamond, manganese, phosphate rock, platinum-group metals (PGM), soda ash, vermiculite, and zirconium. The mineral industry is an important source of export earnings for many African nations. To promote exports, groups of African countries have formed numerous trade blocs.[181]

Africa accounts for a remarkable share of world total production concerning some mineral commodities, in 2007 above all natural diamond (55%), chromite (50%), cobalt (49%), manganese (33%), gold (20%). Other important commodities are phosphate rock (27%), uranium (17%), bauxite (9%), and copper (5%).[182]

Measured by their population, the leading African markets, the so-called "African Lions" Algeria, Botswana, Egypt, Libya, Mauritius, Morocco, South Africa and Tunisia have even overtaken the BRIC states related to their economic performance per capita in 2008.[183]

The period 2006–07 saw an increase of $ 8 billion in extractive industry investment in Africa's least developed countries, such as Ethiopia and Madagascar, following two successive years of decline. Africa's least developed countries accounted for about 23% of the total foreign direct investment (FDI) inflow to the region, and countries of the **Asia and Pacific region accounted for more than 50% of such investment in Africa**. International corporations from China, India, the Republic of Korea, Malaysia, Singapore, and Taiwan were the top Asian investors to Africa's least developed countries. Interest in African mineral resources also was shown by European and Russian companies.[184]

The debt crisis of the 1980s that affected developing countries around the world lead to new dependence, as financial support and debt-forgiveness became tied to observing Structural Adjustment Programmes (SAPs) in partnership with the World Bank and the IMF. The **South African Mining Charter** of 2002 signalled a change, linking the activities of mining companies directly to broad-based socio-economic development without sacrificing international competitiveness. The Johannesburg Declaration on Sustainable Development,

---

181 USGS (U.S. Geological Survey) (2009), Minerals Yearbook, Africa, 2007 (By Thomas R. Yager, Omayra Bermúdez-Lugo, Philip M. Mobbs, Harold R. Newman, Glenn J. Wallace, and David R. Wilburn)
182 USGS (2009): Minerals Yearbook, Africa, 2007.
183 OECD Development Centre (2010). Perspectives on Global Development 2010..
184 USGS (2009), Minerals Yearbook, Africa, 2007.

ratified at the 2002 Earth Summit, suggested how Africa's mining sector could be used to support development, by recognising that "poverty eradication, changing consumption and production patterns and protecting and managing the natural resource base for economic and social development are overarching objectives of and essential requirements for sustainable development". In the following years, many countries such as the Democratic Republic of Congo, Liberia, Sierra Leone and Zambia began to redraft their mining legislation and re-negotiate contracts.[185]

The African Union (AU) and the **United Nation Economic Commission for Africa (UNECA)** in October 2008 took a major step aimed at ensuring that Africa's mineral resources contribute meaningfully to the continent's development by launching "African Mining Vision 2050". The Vision is based on a knowledge-driven mining sector that catalyses and contributes to the broad-based development of a single African market by 2050 to improve the quality of life of the average African.[186]

Besides the effects of the economic crisis, political risks remain acute, including a sudden change of government, disagreement over the distribution of wealth, and arbitrary transfers of mineral assets to new investors through bilateral government deals linked to infrastructure development, such as certain agreements with China.[187]

While there has been a relative decline during the past decade, there has been a growth in the value created by African mining, and also a shift in terms of the most important metals. For example, gold accounted for more than 50 % of mined metal production in 1984, while it only accounted for some 25 % in 2006. Platinum group metals have become the most important metals, accounting for almost a third of the total.

The struggle for African resources has intensified. There is strong political interest from the European Union, the United States and Japan to secure a stable supply of metals. The free availability of resources can no longer be taken for granted. The Nordic countries of Finland, Sweden and Norway have launched the Mining for Development (M4D) initiative in close co-operation with mainly African countries. This project aims at promoting social and economic development in developing countries all over the world, transferring of Nordic knowledge and skills related to exploration and mining in a wide range

---

[185] USGS (2009), Minerals Yearbook, Africa, 2007.
[186] Pulvermacher, K. (2010): The balance of power. Mining journal special publication – Indaba.
[187] Adams, P. (2010): Africa: a political minefield. Mining journal special publication – Indaba.

3.3 Development of international metallic minerals markets

of areas, from effective regulation of exploration, safe and environmentally-sound operating of mines, developing of new mining technologies to training people in governmental authorities.[188]

Effects on the European industry and political economy

Rising prices on the international raw materials markets in the last years partially led to a massive increase of the production costs in the EU industry.[189] Increased costs for essential input factors have put the profit situation of companies under pressure, especially of businesses for which the acquisition of raw materials is a dominant cost factor. The tightened situation on raw materials markets poses a dynamic challenge in terms of a reduced planning reliability for the companies.[190]

Increasing impact of speculative business investments on the raw materials markets additionally worsens this situation in affected industries.[191] Speculations on commodities exchanges are an important factor of the increase of raw materials prices. The boom of raw materials prices attracted numerous institutional investors and speculators, at least until the taking effect of the world financial crisis in October 2008. Additionally, because of the increased speculation activity volatilities arose. For the prices of many raw materials, fluctuations of 20% and more within a few months were not unheard of. Volatility of price to such an extent is an enormous burden to businesses regarding planning reliability.[192]

---

[188] Ericsson, M. (2010): African countries prepare for the next mining boom. Mining journal special publication – Indaba. – See also: Gylfason, T. (2001): Natural resources, education, and economic development. European economic review, Vol. 45 (2001), 4–6, pp. 847–859.
[189] BDI (2007), l.c., p. 5: These have risen in the period 2002 to 2006 compared to 2001, for example, for steel (oxygen steel) to 175 €/t for zinc to 1118 €/t and copper by as much as € 5230 per tonne.
[190] Hoffmann, Hans-Gerhard (2006): Entwicklungen an den Rohstoffmärkten belasten die Wettbewerbsfähigkeit der deutschen Metallindustrie, [Developments in the raw materials markets strain the competitiveness of the German metal industry] World of Metallurgy – Erzmetall, pp. 216–219.
[191] BDI (2007), l.c., p. 5: The companies operate in a partially highly differentiated market environment. In addition, the market environment in view of globalization is highly competitive. This is compounded by different market structures along the value chain, ranging from oligopoly to a pronounced competition. Therefore, the ability of enterprises to hold the impact of raw materials price fluctuations to a minimum becomes a strategic success or failure factor. The tense situation on the world raw material markets affects not just individual companies or industries, rather it affects the entire industrial value chain from raw materials processing to steel and metal production, up to the plant and machinery and to the automotive, electronics and electrical industry.
[192] Above all in view of the rising costs for the risk hedging, Cp. BDI (2007), l.c., p. 8.

Chapter 3 Demand and supply of non-energy raw materials in Europe

A big part of the raw materials imported in European countries is used for the production of capital goods. The majority of capital goods however are destined to be exported. Export economy traditionally is one of the supporting stands of the economic development of several European states (cp. figure 53).[193]

Even in countries where growing market volume or a shift in production levels along the value added chain led to an obvious turnover gain, cost increases on raw materials markets considerably narrowed this positive effect. For considerable parts of industrial preliminary products renegotiations, even for existing supply contracts, had to be conducted and short-term changes had to be enforced by companies with adequate market power.[194]

Figure 52: Important countries for foreign investments in 2007 (Data by Ernst & Young European Investment Monitor Ernst & Young's 2008 European attractiveness survey (http://arisinvest.ro))

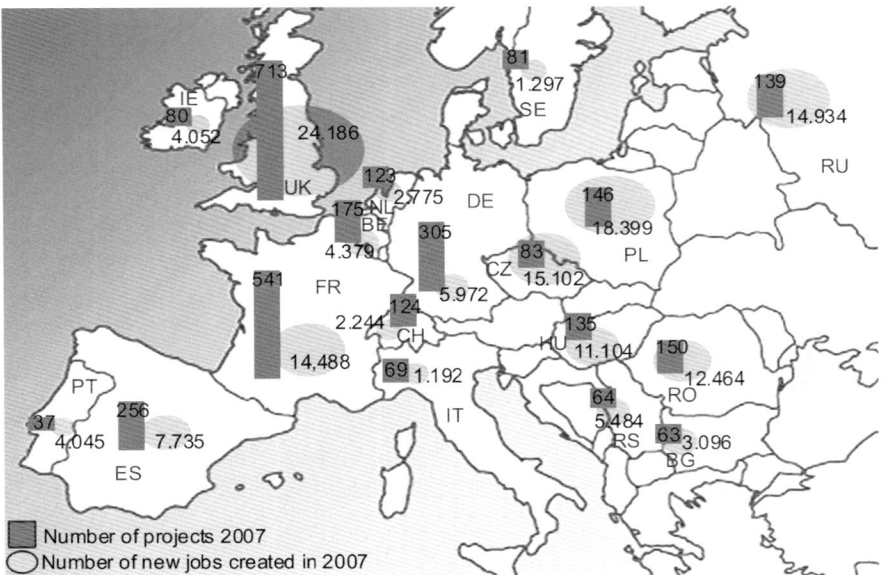

■ Number of projects 2007
○ Number of new jobs created in 2007

*) Job creation figures are based on projects for which the information is available. For more information, please refer to methodology section.

---

193 See also the examples in Chapter 5.
194 Frequently supply contracts at all levels of the industrial added value had to be completed with renegotiation terms or periods of the contract shortened. Cp. BDI (2007), l.c., p. 6.

Furthermore, the topic of metallic secondary raw materials has to be discussed. The degree of recycling adds up to about 40–60% and thus is very relevant for European countries. Compared to primary raw materials, prices have developed similarly in the last years. With *non-iron-metal scrap metal* it has come partly to de facto shortages because of drastic rises in price on account of varied commercial distortion. The worldwide trade in *scrap metal* has expanded considerably over the past few years. Because of the durability of steel products, however, the supply of scrap metal cannot keep pace with demand, which means that the already tight market for scrap metal is set to expand continuously.[195] Scrap metal prices, which tripled between 2002 and 2004, will presumably rise again in the future.[196]

Resume

It is not foreseeable how long the raw materials boom will stay on a high level. Neither do we know if, and to which level, prices will drop intermittently, as is the case in 2008 (financial crisis).[197] However, it is a fact that prices in historical comparison are on a high level. Moreover, the situation will not be the same for all raw materials. Price decreases are predicted to be clearly behind those of previous cycles.[198] The Anglo-Australian mining company Rio Tinto expects the demand for raw materials to double until 2022 at a constantly high price level.[199]

3.4 Future demand of mineral resources – scenarios

In general it can be said that the amount of raw materials available can meet the global demand.[200] The rise in price observed starting in 2000 does not signal an exhaustion of raw materials in medium-term. The fact that the raw materials described, and many other raw materials, are still available in sufficient quantities means that the price rises now being observed do not signal the depletion of resources in the medium term.

However, this does not rule out the possibility of demand and supply shifts, or make price movements random. In the short term, the supply of raw materials is not very flexible owing to the long lead times of capital-intensive

---

195 Because of the close replaceability between secondary and primary raw materials in the field of some metals. Cp. BDI (2007), l.c., p. 5. (Ibidem).
196 European Economic and Social Committee (2006), l.c., p. 75.
197 Cp. the financial crisis, autumn 2008.
198 Wagner/Huy, (2005), l.c., pp. 4–5.
199 http://www.rohstoff-welt.de/news/artikel.php?sid=6910
200 See also section 3.5.2.

exploration projects (see chapter 1). When demand for raw materials is high, it is quite possible that shortages and price rises will occur. The same applies to transport capacity, which also limits the availability of (imported) raw materials. Sufficiency of global reserves and resources may limit the risk of quantitative supply disruptions, but they do not provide a guarantee against marked short-term and medium-term price rises. A complete evaluation of supplier and price risks on international raw materials markets means taking into account political measures, and monopolistic or oligopolistic behaviour of companies with a strong market position.[201]

This is particularly important given that a considerable proportion not just of major energy sources, but also of metal raw materials are concentrated in certain regions of the world and with certain companies, and this concentration has increased significantly since the early 1990s, at least in the case of metals. Thus Chile has almost tripled its share of copper ore production since 1990, and almost 40% of the world's bauxite is produced in Australia.[202]

Figure 53: Production and export of copper between 1985–2002 (Data by BGS)

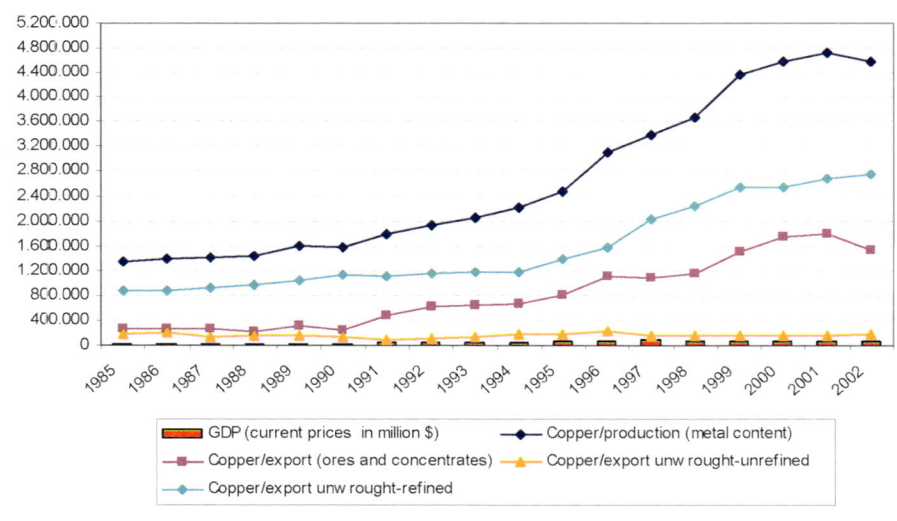

---

[201] European Economic and Social Committee (2006), l.c., p. 75.
[202] European Economic and Social Committee (2006), l.c., p. 75. – See also: Maxwell, P. (2004): Chile's recent copper driven prosperity: does it provide lessons for other mineral rich developing nations? Minerals and energy – raw materials report, Vol. 19 (2004), 1, pp. 16–31.

3.4 Future demand of mineral resources – scenarios

Brazil has also substantially increased its importance as a bauxite supplier, and is now the second-largest bauxite producer, highlighting South America's key role in the production of metal ores. The same goes for iron ores, about 30% of which are produced in Brazil (see figure in appendix).[203] Of the EU Member States, only Sweden is of any importance as an iron-ore producer, but it accounts for only 1,6% of total world production.

Figure 54: Iron ore deposit in Serra dos Carajás, Brazil (Data by Trojer, 2007)

Iron ore deposit in Serra dos Carajás, Amazon region, Brazil, with a volume of about 18 billion tonnes of iron ore. The ore has an iron content of up to 65 percent. The loading device is able to extract the rock directly from the face, so no blasting or drilling is required. A railway line of 900 km length through the jungle was needed to take the ore to the next haven.[204]

In the deep sea there are significant deposits of manganese, copper, nickel and cobalt. There are already legal regulations existing for a future extraction within the Preparatory Commission for the International Seabed Authority and the International Tribunal at the UN. Regarding a shortage of terrestrial deposits here is a significant potential in order to secure a global supply for these metals. However, it should be mentioned that environmentally sound and sustainable and cost-friendly mining methods must be available.

According to an analysis by Skinner (2000), mineral commodity demand could rise in the next 50 years to as much as five times the recent global production.

---

203 Ibid.
204 Trojer M. (2007): Großtagebau, Bachelor Arbeit [Large Opencast Mining, Bachelor Thesis], Leoben.

Skinner makes the point that demand will continue to be driven by a burgeoning population and the **growing aspirations of developing nations**, which are being fuelled by globalisation.[205]

Today, more than half of the world population contributes to the global raw materials demand. In the next 30 years, according to BGR, a **doubling of demand** is to be expected due to the growth of population and economy in today's emerging countries (also see below).[206]

The change from secondary to tertiary sector is accompanied by further consequences: With the declared turning to the production of highly refined (high tech) products of some emerging countries, an increased demand for the raw materials needed for these products comes along.[207]

The huge demand for raw materials is not only caused by China's needs, but it is also the effect of a **growing global economy which reflects itself in the cyclical price movement that is typical for raw materials**. The reasons for economic growth are based usually on globalisation, deregulation, improved communication and improved economic development models.[208]

The present economic revolution in China with an economic growth of 10% annually clearly underlines the affinity between raw materials consumption and economic performance and therefore the relevance of supply security of raw materials supply of a state. The predicted increase of the world population of 50% in this century and the permanently increasing demand for raw materials of a world becoming richer and richer will lead to an **escalating increase of exploration and extraction of raw materials**.[209]

This was confirmed in 2007 already: Turnovers of the 40 largest mining companies worldwide had risen by 32% in 2006, according to PricewaterhouseCoopers. At the same time, running costs, especially for energy, salaries, material, transport, and services, even increased by 38%.[210] A good example for this is the steel industry (see figure 55).[211]

---

205 Skinner, B.J. (2000): Keynote presentation to the 31st International Geological Congress. Rio de Janeiro. Here the BRIC countries are in special focus.
206 BDI (2007), l.c., p. 7; Wagner, M., Huy, D. (2005), l.c., p.1 : Schafft der Strukturwandel in der Nachfrage eine neue Dimension für die Weltrohstoffmärkte [Does structural change in demand create a new dimension in the world commodity markets], Commodity Top News, Fakten – Analysen – Wirtschaftliche Hintergrundinformationen [Facts – Analyses – Economic Background information], 20.09.2005.
207 Cp. Topic "rare earths": about 90% are currently being produced in China.
208 Cp. also 304.
209 Petterson (2005), l.c., p. 5.
210 Rohstoffwelt, 23.06. 2008.
211 Krüger, H. J. (2002): Die Rohstoffversorgung der deutschen Stahlindustrie unter dem besonderen Aspekt der Versorgungssicherheit. – Stahl und Eisen [The raw materials

3.4 Future demand of mineral resources – scenarios

Figure 55: Scenario: estimated global steel consumption, averaged growth rate 1890–2005 (3,7%) (Data by Ekdahl 2008; Quelle: BGR)

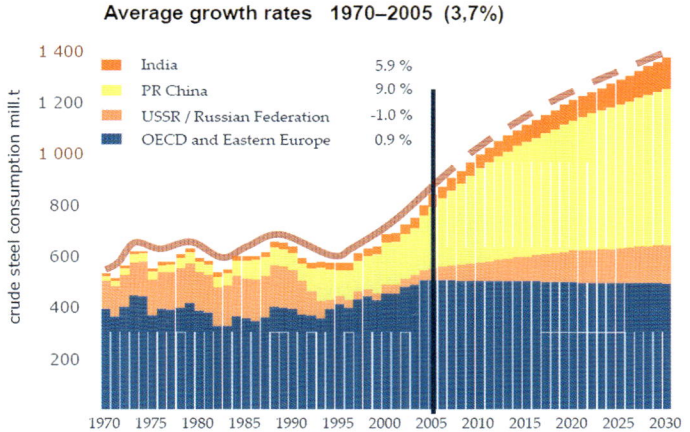

Economic regulations on the development of future global raw materials demand

Following the approach of Nötstaller:[212] Nötstaller starts his analysis with the present structure of the global economy, especially emphasizing differences in income. In 2006, 2,4 billion people (37 % of the world population) in countries with low income have an average per-capita income of 650 US$ per year, 3 billion (47 %) in countries with medium income earn about 3000 US$ annually. One billion people (16 %) in countries with high income earn more than 36.000 US$ per year. The average per-capita income of people in countries with little income is less than 1/50, in countries with medium income less than 1/10 of the comparative value in countries with high income. Correspondingly, there are considerable differences in raw materials consumption: Countries with high income account for half of the world raw materials consumption, the per-capita consumption of mineral raw materials being five times (four for zinc, five for copper, more than six for aluminium, eight for nickel, and five for primary energy) as big as in the rest of the world.

Due to the elasticity of income of raw materials demand, which continues all the way into the peak of the industrialization of political economy,

---

supply of the German steel industry under the particular aspect of supply security. – Steel and Iron], 122. 2002, H. 6, pp. 15 – 20. Gronwald, Leo (2008): Rohstoffversorgung der Stahlindustrie am Beispiel ThyssenKrupp Steel AG [Supply of raw materials for steel industry, at the example of Thyssen Krupp Steel AG], Bergbau (Mining), pp. 318–321.

212 Nötstaller/Wagner (2007), l.c., p. 385–386. In view of the relevance of this topic, Nötstaller's approach is widely accepted and presented.

it is assumed that specific raw materials demand in countries with low and medium income will approach that of countries with high income gradually in the course of further economic development. The speed of this approach, and therefore the growth of demand for raw materials, is determined by the economic growth in the less developed countries. Considerations on the development of raw materials demand can be shown using the example of metallic raw materials based on the predicted world economic performance (reference period: 2005–2030).[213]

For a quantification of the increase of the demand for raw materials coming along with an increased economic performance Nötstaller makes further model assumptions. Simplifying, it is assumed that raw materials demand in OECD member states is saturated already and will not rise any more, and that raw materials demand in non-OECD states reacts continuously income elastic. This proposition seems justified when considering both the development of the demand for raw materials during the industrialization of the OECD member states in the first half of the 20th century and the latest experiences in China. For the forecast period of time, an income elasticity of 1 is assumed, meaning that the demand increases proportionally to the world economic performance.

Finally, it is also assumed that of the total world raw materials consumption in 2005 one half is allotted to the OECD member states and the other to non-OECD states. Based on this set of premises, Nötstaller develops the following scenario: The half of the present raw materials consumption, which falls upon the OECD states, stays unchanged. The other half increases by a factor of 3, equalling the real growth of economic performance in this group of states to 1,5 times the amount of 2005. In 2030 this makes a total of 200% of the starting value, which is a doubling of the raw materials consumption compared to 2005.

A doubling of raw materials demand for the important industrial metals is plausible in view of the relevant predictions for the development of the motor vehicle stock. Accordingly, the global motor vehicle stock will rise from approx. 800 million in 2002 to 2080 million by 2030, whereas in non-OECD

---

[213] Nötstaller/Wagner (2007), l.c., p. 385–386. It concerns the base data set for forecasting of primary energy consumption, which assumes an average annual GDP growth rate of 2,4 % for OECD countries and for the non-OECD countries 5% until 2030. Consequently, the real economic performance of OECD countries will rise from 28 to 50 trillion U.S. $ (reference year 2000) in 2030, while that of non-OECD countries from just under 8 to over 26 trillion U.S. $ and the global total GDP of 36 almost U.S. $ 77 trillion. This means an increase in world economic output by a factor of 2,1 and in the non-OECD countries one of 3,3 during the forecast period. – Cp. Nötstaller/Wagner (2007), l.c., p. 386.

3.4 Future demand of mineral resources – scenarios

countries an increase by 195 million to 1.172 million is expected.[214] The demand for raw materials will rise the strongest in such economies in which the economic performance has the highest growth. According to appropriate model calculations, an above-average increase of the economic achievement till 2030 and afterwards is to be expected particularly in the so-called BRIC states (Brazil, Russia, India and China).[215]

Figure 56: Crude steel production in China and India since 1950 (Data by BDI, 2007)

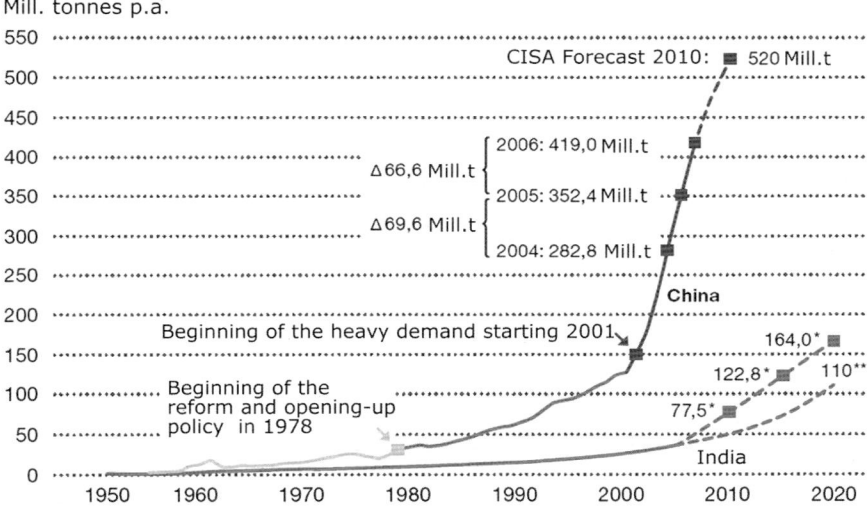

* new projects ** nat. steel policy

The economic performance of the BRIC states, which in 2005 was about 15 % of the GDP in the former G6 states (USA, Japan, Germany, Great Britain, France and Italy), could rise to about 70 % of the comparative value until 2030. Similarly high growth rates of the GDP are also expected for Indonesia, Mexico, and Turkey.[216] Consequently, by the middle of the 21st century the ranking of states according to their economic performance will have changed. China will surpass the USA as the biggest economic power, India will have a GDP in the range of the USA, while the economic performance of the remaining coun-

---

214 Dargay, in: Nötstaller/Wagner (2007), l.c., p. 386.
215 Wilson, D., in: Nötstaller/Wagner (2007), l.c., p. 387.
216 Hawksworth, in: Nötstaller/Wagner (2007), l.c., p. 387.

tries mentioned will reach a level similar to Japan or Germany. In connexion with these changes, regional demand for raw materials will be shifting.[217]

A doubling of the demand for raw materials by 2030 is not to be expected for all groups of raw materials. Especially for many industrial minerals, income elasticity of demand is less than for metals. Neither can we expect an income elastic demand for the quantitatively significant construction raw materials in non-OECD states, as the development of cement consumption in China shows. This means that for construction raw materials a doubling of consumption until 2030 is realistic.[218]

The scenario of Nötstaller is only an option of a possible forecast. In the development of the economies outside the EU and North American energy issues will play a significant role. Fossil fuels and nuclear power will certainly be indispensable in the medium term, but the global environment must increasingly be taken into account. At a possible energy shortage, there will be a drop in consumption of raw materials, particularly in developing countries. Countries in the SADC region of Africa were recently a striking example.

3.5 Questions on the security of supply in Europe

3.5.1 Change of paradigm

A structural change appeared particularly on the raw materials world markets: The rough rule of thumb that 20 % of the population in Europe, USA and Japan consume more than 80 % of the raw materials production is not valid any more nowadays:[219] As above mentioned, with the integration of India, the People's Republic of China and other countries, such as Brazil and Russia (BRIC states) into the world economy, today more than half of the world population participates in raw materials consumption. Worldwide raw materials demand is at the beginning of a new growth curve: Presently, about 30 % of the global raw materials demand and 20 % of the global demand for non-ferrous metals fall upon the People's Republic of China.

---

217 Nötstaller/Wagner (2007), l.c., p. 387.
218 Nötstaller/Wagner (2007), l.c., p. 390.
219 Bundesministerium für Wirtschaft und Arbeit, (2005): Thesen für eine Rohstoffpolitik Berlin. [German Ministry of Economy (2005) Theses for a raw materials policy, Berlin.] Wagner, M., Huy, D. (2005), Schafft der Strukturwandel in der Nachfrage eine neue Dimension für die Weltrohstoffmärkte, [Does structural change in demand create a new dimension to the world commodity markets], Commodity Top News, Fakten – Analysen – Wirtschaftliche Hintergrundinformationen, [Facts – Analyses – Economic Background information] (20.09.2005).

3.5 Questions on the security of supply in Europe

The difficulty of securing the supply of mineral raw materials, as mentioned in Chapter 1, results from the location bound nature of deposits and their exhaustibility on one hand, and the inhomogeneous distribution of deposits and potential areas of use worldwide on the other hand. Europe's problem lies in the regional distribution of raw materials and the discrepancy between the location of reserves and place of consumption.[220] Europe is a region with presently high needs of imports of raw materials (and fossil fuels) and in future, even growing dependence on imports.[221]

Problems concerning access to raw materials from non-EU countries have increased continuously (**external supply issue**, see Chapter 6.1). Additionally, problems concerning the availability of raw materials from domestic deposits have risen considerably in the past years (**internal supply issue**, see also Chapter 6.1).[222] This has to be seen against the background of a growing global economy and an **increased demand for raw materials**, but also a continuous rise in raw materials prices, caused not only by the added demand for raw materials of populous countries like the BRIC states.[223]

Figure 57: Metals supply of the European Union (Data by Ekdahl 2008)

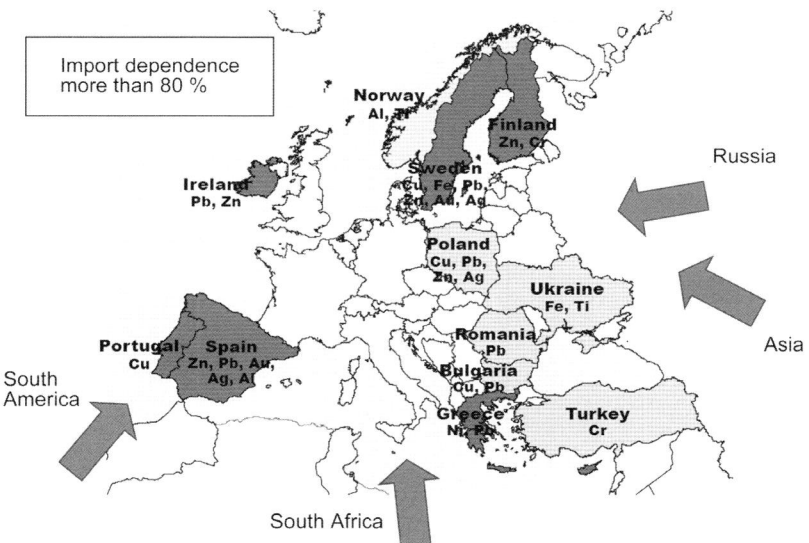

---

220 European Social and Economic Committee (2006), l.c., p. 73.
221 European Social and Economic Committee (2006.), l.c., p. 73.
222 Environmental restrictions, also chapter 6.1.
223 As noted: The main factor for this development is seen in the rapid industrialization of ambitious national economies like Brazil, China and India (where since the 90s a continuous increase of the raw materials needs is to be observed). Since October 2001 the world market prices for many metal raw materials have risen on an average by over 70 %.

An important role is played by raw materials producers who are presently using their entire capacities after having reduced them before, while neglecting new investments, as well as bullish cargo rates due to capacity restraints in transport infrastructure. Since the beginning of the nineties, European enterprises and companies have faced the security of access to external raw materials sources through own interests in foreign mining companies from financially strategic considerations with growing reluctance. With the exception of the worldwide cooperation of European plant manufacturers in the raw materials sector, the trend of withdrawing from mining companies persists in the European mining and raw materials industry.[224]

It is predicted that the surge in growth will continue throughout the Asian region (China, India, Indonesia and others) and significantly determine the situation on the international raw material markets in the long run.[225] Similar tendencies may arise from Russia and Brazil and their connected markets and moreover, the leading African markets, the so-called "African Lions" Algeria, Botswana, Egypt, Libya, Mauritius, Morocco, South Africa and Tunisia which have even overtaken the BRIC states related to their economic performance per capita in 2008.[226] Based on past experience, it is expected that the currently high raw material prices will provoke an increase in exploration and mining activities, whereas also substitution and recycling will be stimulated. This could potentially lead to lower prices again by 2011. However, an uncertain factor in the evolution of prices is the degree of technological advance, which in the past was an important factor in this process.[227]

---

224 BDI, 2007. Cp. the latest development in Germany: Vereinigung Rohstoffe und Bergbau (VRB) (2008): 2008, Positionen und Perspektiven (Positions and outlooks).
225 Wagner/Huy (2005), l.c., p. 5. – See also the above remarks according to Nötstaller (2007), l.c., p. 387. A remark concerning the financial crisis: PRESSE Economist, p. 27: In view of the present world economic crisis since autumn of 2008 there are, according to Roland Berger Strategy Consultants, the following scenarios for progression respectively coping with the crisis: the experts consider possible a "shallow growth depression" which will be overcome by the end of 2009, though their optimism is obviously limited. In this case the national economies of Europe would have to prove very sound, and above all, the BRIC states would have to draw the world economy out of the crisis. In addition, national and international economic pacts as well as the actions of central banks would have to show effect immediately and also restore the confidence of entrepreneurs, investors and consumers. However, a recession that lasts one to two years is assumed most likely. That would mean shrinking the global economy by one to two percent, while the U.S. and Japan would be more affected than Europe. According to Berger, branches which already have structural problems are considered especially crisis-prone. These are the automobile, clothing, chemicals and food industries.].
226 OECD (2010): Perspectives on global development 2010.
227 EC, DG Enterprise and Industry, Consultation Process for Raw Material Initiative (2008).

3.5 Questions on the security of supply in Europe

Generally, raw materials prices, due to the high demand from countries like China and India, are expected to stay on a higher level than before the current boom; the **structural change is expected to manifest sustainably**.[228]

Figure 58: World mining production 1984–2006 (Data by Weber and Zsak, 2008)

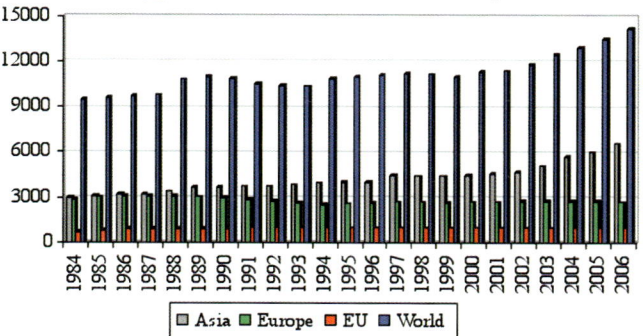

The figure 58 depicts how raw materials production developed differently worldwide. While the change from 1999 to 2000 was 1,3%, from 2003 to 2006 an increase of almost 6% was noted. The biggest increase of production occurred in the Asian region; mining production in Europe declined noticeably due to Eastern Europe's political reorientation. In total, mining production in EU countries is low compared to the world production (BMWA, 2008).

It is hard to assess to what extent the present raw materials boom can lead to a breakout from the existing price structure, even in the current global financial crisis.[229] In the past, warfare and the political and economic change of the world were the factors that coined the raw materials markets. The increased raw materials demand of developing and emerging countries is based on a structural change in the economic-political range similar to certain instabilities in the past. Also it should be mentioned that the (increasing) formation of market dominating mining companies leads to more influencing of prices. A striking

---

228 Wagner, Huy 2005, l.c. – Cp. also Rohstoffwelt, 23.06.2008: The emerging economy of China is the most important factor for the increase in demand for raw materials. The World Bank has raised its forecasts for the growth of the Chinese economy for 2008 to 9,8%. The National Statistics Bureau of China has adjusted its figures for the previous year upwards. Accordingly, China's GDP last year grew by 11,9%, the largest increase since 1994. According to Pricewaterhouse Coopers (PwC), the economy of China will overtake even the U.S. by 2025. India: (BDI, 2007). With the expiry of long-term contracts India may also change very soon from a net raw materials exporter to a net importer. This is also expressed by Ciftci, Ö., Bank ABN Amro in: Presse, 17.06. 2008, Raw materials still have a year-long price boom before them: Upward trend in raw material prices has continued unabated. Raw materials cycles usually last for 15 to 18 years, we are now in year four or five.
229 Cp. also above comments on trend of and demand for raw materials.

example is the production of iron ore: In the future, only few companies might control 70–80 % of the global iron ore extraction and thus could massively influence price development. However, important questions remain:[230] How to improve market information and early warning systems so that they will give adequate signals to market participants and regulators to tame the excessive commodity price booms and busts? Is it possible to forecast more accurately the future demand for various commodity groups, in a way that could form a basis for better-informed investment decisions to achieve more flexible supply responses to changes in demand for commodities?

3.5.2 Availability of raw materials

As mentioned in Chapter 1, raw materials are exhaustible and non-renewable. In recent years, the question of raw materials availability has come up more and more often; it was subject of numerous discussions already in the 1970s and 80s.[231]

Physical compared to technical raw materials availability

Due to geostatistical data and technical considerations in mining there is no indication of imminent physical shortage for the majority of raw materials in the world in the long run. Concerning the future availability of raw materials, the development of the reserve-consumption ratio of different raw materials over time shows that generally no problems in physical availability, no concrete shortages, are to be expected worldwide.[232]

Nevertheless, discussions aiming to tighten the situation are taking place at the moment. Reference is made to the discussion in Mining Environmental Management:[233] production data of the biggest mining producers of certain raw materials in 2007 are contrasted with the total global reserves (figure 59).

---

230 UNCTAD (2010), l.c., p. 7.
231 Cp.. among others Maull, Hanns W. (1984): Western Europe's non-fuel mineral vulnerability: how serious, how vulnerable? Atlantic quarterly, no. 4. 1984, pp 337–358. Wellmer, F. W. (1998): Lebensdauer und Verfügbarkeit energetischer und mineralischer Rchstoffe. Lifetime and availability of energy and mineral resources. Erzmetall, pp. 663–675. Wellmer, F. W. (2003): Die Rohstoffsituation der Welt. The state of natural resources in the world. Erzmetall, pp.705–717.
232 Cp. Rheinisch-Westfälisches Institut für Wirtschaftsforschung (RWI Essen) (2006), [Rhine-Westphalia Institute for Economic Research (RWI Essen) (2006)] l.c., p. 6f.
233 Dixon, K. (2008): Is it possible to predict how long our mineral resources will last and is there anything we can do to slow their inevitable decline?, Mining Environmental Management, July 2008, p. 26.

3.5 Questions on the security of supply in Europe

On this basis, attempts to derive an estimated last for these raw materials are made (figure 60).[234]

Figure 59: Raw materials production in 2007, several metals including raw materials reserves (Data by Mining Environmental Management, 2008, p. 27. Data based on: USGS Mineral Commodity Summaries 2008)

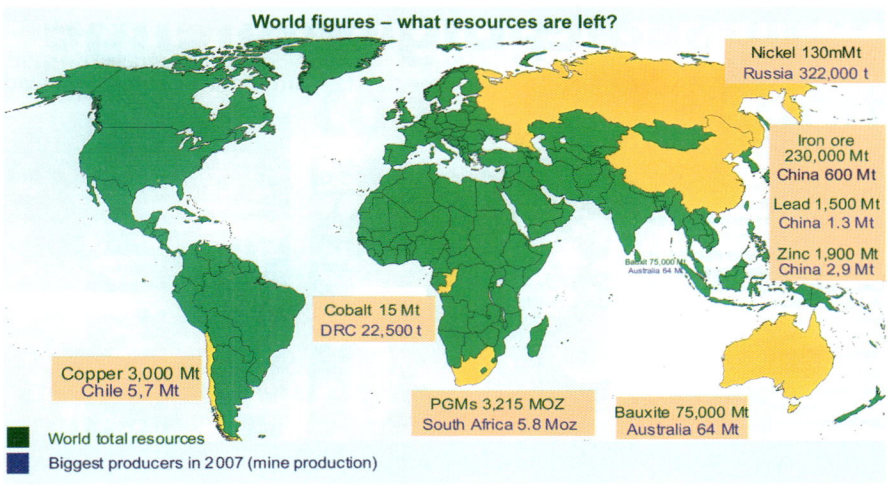

Figure 60: Assumed raw materials reserves in the future (Data by Mining Environmental Management, 2008)

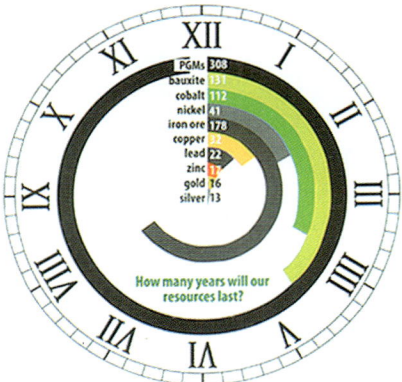

---

234 Ibidem. The reasoning for Figure 60 is as follows: Global reserves divided by world mine production (2007), assuming constant production at 2007 levels and current global reserves. Reserves: the part of the resource that can be economically extracted or produced. (Source referred to: USGS Mineral Commodity Summary 2008).

However, when estimating future raw materials reserves, caution is advisable. According to the US Geological Survey, in 2005 profitably mineable iron ore reserves stood at about 80 billion tonnes of iron equivalent, which is more than 100 times greater than current demand. If deposits that are not currently profitably mineable are included, the total volume of reserves increases to around 180 billion tonnes of iron. Obstructions of the world economical growth by physical raw materials shortage are not to be expected in the foreseeable future; global raw materials reserves are considered sufficient to meet demand despite an increasing market.[235]

Despite these large reserves, it is assumed that iron ore prices will, in future, continue to be high. One reason for that is undoubtedly the dominant market position of three large firms (CVRD, BHP and Rio Tinto) which, together, have a good 40% market share of global iron ore production. Bottlenecks are also expected in maritime transport, leading to increased transport costs and thus to higher ore prices for the European steel industry.[236]

Moreover, the issue of technical raw materials availability should be addressed. In EU third countries, due to the high risk of essential investments, it is not sure whether in future reserves (and access to them) will be available in time, considering political and ecological conditions.[237] On the supply side, technical availability is an important aspect, which is determined significantly by environmental restrictions and conflicts of use, but also by existing mining and investment capacities Environmental restrictions and conflicts of use are dealt with in chapter 6.1. The issue of mining and investment capacities shall be mentioned briefly:

Capacities of mining companies in Europe reached a peak in 1981 – at the end of the last big raw materials price boom. At that time huge capacities were built with the support of governmental programmes. However, despite a continually increasing demand, new investments declined. In recent years, few new investments were made and a rise in output was achieved by expanding existing capacities. The international mining companies' investment behaviour was coined by investing only in the lower third of the cost curve, putting high-cost producers out of business. In the last decades the focus shifted to rich ores and major deposits, which again led to regional accumulations and entailed the accumulation of companies.[238]

---

235 European Social and Economic Committee (2006) l.c., p 74f.
236 European Social and Economic Committee (2006), l.c., p. 74f.
237 Cp. Gocht – already in 1983, l.c., p. 202.
238 Cp. Bundesministerium für Wirtschaft und Arbeit (2005): Thesen zur Rohstoffpolitik, Berlin, [German Ministry of Commerce, Trade and Industry: Theses for a raw materials policy, Berlin] p.3. Also: BDI (2007), l.c., p. 7.

3.5 Questions on the security of supply in Europe

Development on international raw materials markets depicts how the drawbacks of an import dependency can affect the security of raw materials supply and raw materials prices. Due to changes in general framework, a trend reversal might occur. The significance of domestic ore production may increase in EU countries. Reference is made at this point to chapter 5 and the latest development in Ireland, Finland and Sweden; ore exploration is also present in different German federal states. German and foreign companies aim at non-ferrous metals, steel refiners, special and precious metals and rare earths elements. The raw materials processing industry is interested in the raw materials that can be extracted in Germany again.[239]

Table 17: Copper (Data by BGR, 2007)

Usage	Electrical industry, building industry, machine construction, coinage
Supply (2005)	
Worldwide mining production	15,1 Mt (content)
Worldwide definite + probable stock	470 Mt
Regional concentration of production (*)	Chile (35,3%), USA (7,6%), Indonesia (7,1%), Peru (6,7%), Australia (6,2%) Top 5: 62,9%, Top 10: 83,0%
corporate concentration (*)	Codelco (Chile, 12,5%), BHP Billiston (Australien, 8,6%), Phelps Dodge (USA, 6,8%), Grupo Mexico (Mexiko, 5,8%), Rio Tinto (Großbritannien, 5,4%) Top 5: 39,1%, Top 10: 58,1%
Demand (2005)	
Worldwide consumption	16,6 Mio. t
Consumption EU	3,8 Mio. t
Consumption Germany	1,1 Mio. t
Net import	Ore + concentrate 1,1 M t; metal 612.000 t, secondary material 40.100 t, half-finished products – 632.900 t
Important German producer/processor	North German Refinery, KM Europe Metal, numerous small and medium sized processor

---

[239] VRB (2008), l.c., p. 61.

Chapter 3 Demand and supply of non-energy raw materials in Europe

Price development copper, grade A, min. 99% London Metal Exchange (LME), cash, in LME warehouse, prices in USD/t monthly average	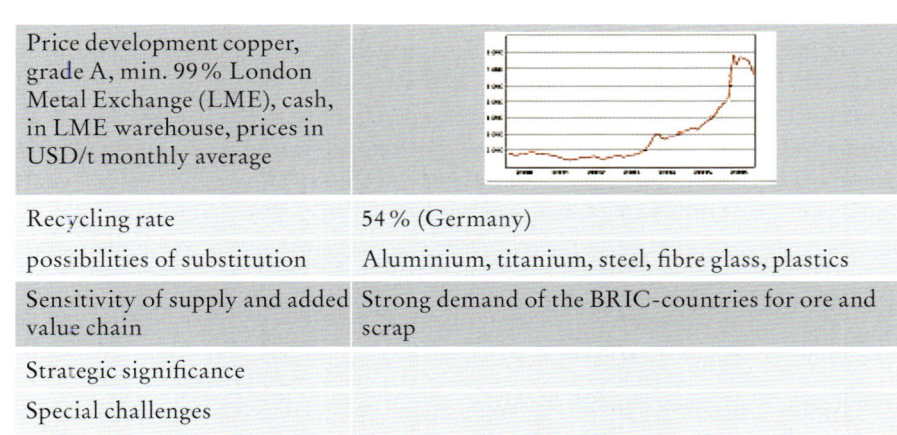
Recycling rate	54% (Germany)
possibilities of substitution	Aluminium, titanium, steel, fibre glass, plastics
Sensitivity of supply and added value chain	Strong demand of the BRIC-countries for ore and scrap
Strategic significance	
Special challenges	

(*) = share of world production in 2005

3.5.3 Supply criticalities

As already mentioned, the potential of technical raw materials availability, along with the criticality of supply, is a primary question.

Historical consideration

In the course of its development, the EU has become more and more dependent on raw materials imports. In 1975 it was already emphasized by the Commission that, especially for metallic raw materials, the degree of dependence of the European Union on imports from third countries ranged between 70 and 100%. These percentages have obviously been increasing alongside the development of industrial activity in Europe.[240] Furthermore it is stated that for a number of raw materials, the Union depends on a very small number of supplying countries due to the geographic distribution of known occurrences. From this two problems arise: in some cases these are countries which might provoke political problems; in other cases there is a danger that the Union, in view of the dominating position of the producing country or countries on the European or global market, must bear hardly acceptable conditions of supply concerning amount and price.[241]

---

240 Commission of the European Communities (1975): The raw materials supply of the Community, Bulletin of the European Communities, Supplement 1/75, p. 5f. Cp. also Vajna, Thomas (1974): Dependence on imports and raw materials policy, Köln : Deutscher Instituts-Verlag, 1974 (Beiträge/Institut der Deutschen Wirtschaft ; 12).
241 Commission of the European Communities (1975)], l.c., p. 5f.

3.5 Questions on the security of supply in Europe

The 1980s and 90s are basically marked by an untroubled raw materials supply situation. The need for raw materials can be met easily and without taking risks on international markets.[242] In the long term this leads to corporate decisions of declining shares in raw materials extraction projects, mainly because raw materials projects, due relatively long payback periods, offer financial earning only after a longer period of time.[243]

Since the late 1990s, a continuous growth of the world economy, considerably affected by the development of emerging markets, has been observed.[244] This leads to remarkable turbulences (price developments) on the international raw materials markets, with notable effects on the EU industry.

In 2004 (as mentioned before), the EU imported more than 160 million tonnes of metallic minerals with a total value of € 10,5 billion, compared to a domestic production of only 30 Mt. The Import dependency rate for many of these minerals ranges from 74% for copper ore, 80% for zinc ore and bauxite, 86% for nickel to 100% for materials as cobalt, platinum, titanium and vanadium.[245] In 2008 it is stated that the EU's options concerning access to strategic raw materials are very limited and connected to a strong increase of raw materials prices.[246]

---

[242] Cp. Bachmann, Hans [et al.] (1980): Rohstoffpolitik der achtziger Jahre zwischen Strategie und Alibi [Raw material policy of the eighties between strategy and alibi], Aussenwirtschaft, Zürich, pp. 287–405.

[243] Bundesministerium für Wirtschaft und Arbeit (2005), Thesen zur Rohstoffpolitik, Berlin, [German Ministry of Commerce, Trade and Industry (2005), Theses for a raw materials policy, Berlin] 9. Februar 2005, p. 2f.

[244] Cp. also Wellmer, F. W.; Dalheimer, M. (1998): Trends in raw materials management – the international supply of raw materials. Glückauf, pp. 319–324. Wellmer, F. W., Wagner, M. (2000): Mineral trends at the beginning of the third millennium, Erzmetall, pp. 569–582. – See also: Eggert, R. G. (2010): Critical minerals and emerging technologies. Issues in science and technology, Vol. 26 (2010), 4, p. 14, ISSN 07485492.

[245] BDI (2007), l.c.

[246] EC, DG Enterprise (2008): Consultation process on raw material initiative 2008: This was carried out by the Directorate General Enterprise, based on an online questionnaire. This process is enlarged on in Chapter 6.1.5. Some of its results are presented in the appendix.

Figure 61: Development of iron ore price (Data by VRB, 2008)

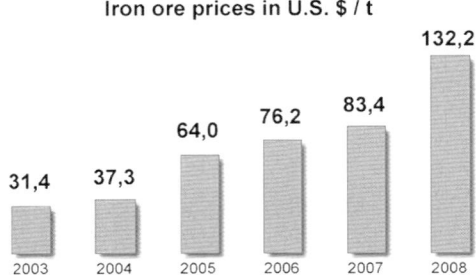

For 2008, iron ore price was assessed at 132,2 US$/t; in 2003 it was 31,4 US$/t. The European steel industry requires about 200 Mt of iron and 85 Mt of scrap iron, whereas EU production is around 24 Mt (about 2 % of the global production). China's share in world iron ore trade rose from 13 % in 1997 to 41 % in 2006; within the past five years it tripled.[247]

Moreover, the EU environment standards are getting more and more restrictive and the availability of deposits decreases. The topic of environment can be displayed using the example of climate policy and emissions trading, which presently implies major problems mainly for the magnesite, cement and steel industry (see chapter 6.1). The issue of deposits availability is particularly relevant for minerals raw materials supply. In summary, fundamental problems concerning the security of supply arise.

Table 18 indicates the (increasing) import dependence of selected important raw materials.

Table 18: European Union import dependence – Imports as percentage of domestic consumption plus exports (Data by Crowson, 2008)

	1979/80	1988	1998	2006
Aluminium	45	48	61	67
Arsenic	na	23	71	21
Asbestos	84	100	46	100
Barytes	18	36	53	54
Bismuth	na	89	100	98
Cadmium (refined)	32	32	25	58

---

247 EC, DG Enterprise (2008): Consultation process on raw material initiative 2008.

3.5 Questions on the security of supply in Europe

	1979/80	1988	1998	2006
Chromium	97	93	47	88
Copper	80	62	69	63
Fluorspar	29	27	52	64
Germanium (refined)	2	9	na	na
Graphite	na	na	71	87
Indium	na	na	31	na
Industrial diamonds	na	na	27	83
Iron Ore	87	90	82	85
Kaolin	na	na	32	40
Lead	44	37	30	26
Lithium	100	100	96	93
Magnesium metal	Over 61	82	89	100
Mercury	91	na	na	92
Nickel	87	85	92	93
Potash	19	26	19	43
Selenium	100	na	70	> 70
Silicon	44	38	68	77
Silver	58	38	na	c. 70
Soda ash	na	na	15	8
Sulphur	31	21	7	6
Talc	na	na	11	19
Tellurium	na	na	63	> 65
Tin	93	77	100	100
Tungsten	78	75	84	> 85
Uranium	70	80	96	98
Zinc	57	71	47	55

Short-term supply interferences

Regarding strategic raw materials, Europe's downstream industries are highly vulnerable by interferences in their supply chains. Temporary shortages in raw materials supply and the accompanying **sudden price increases** can result from unexpected increases in demand, insufficient investment, decisions of oligopolistic groups of enterprises or producing countries and from speculative transactions. These have serious consequences due to the effects on downstream branches of industry (multiplier effect) especially.

Public economic consequences of supply disturbances are the more fatal, the more far-reaching the processing depth and the longer the value added chain. From this follows that the macroeconomic importance of a raw material or a group of raw materials is larger, the more stages of processing it takes to produce the final product.[248] A remarkable example concerning scarcity of raw materials and the rise in price is the Eastern European employer Henkel. Phosphate is needed for the production of cleaning agents; in 2008 a problematic raw materials shortage and a rise in price of 15 % occurred. The reason is striking: Because of the increased meat consumption in emerging countries, more feeding stuff – which uses phosphate as a fertilizer – is needed.[249]

Higher costs on the raw materials side force employers to look for methods of compensation on more competitive markets. Thereby employment in these companies is under pressure to adapt, especially when increasing costs cannot be passed on as price increases. Should the permanently tense situation even result in **parts of the value added chain breaking away**, the impact on employment is obvious.

The dependence on raw materials has serious consequences on the European economy; the impacts of a shortage in raw materials on the industry and production site are enormous. In 2005 costs for raw materials import into Germany amounted to € 77 billion, which was about 12,3 % of total imports. Metallic raw materials, with a value of € 19,3 billion, had a share of about 25 %.[250]

---

248 This fact has already been mentioned in Chapter 1.
249 Die Presse, 19. Mai 2008 (Austrian Journal): Also Thumser, G., Head Eastern Europe, of Henkel, on expensive raw materials. Also cement is used as a raw material for tile glue, raw material access is difficult. Cement has become an extremely scarce product, in Russia alone, 30 million tonnes a year are missing. Regarding the product of soda: Soda is an essential component for the manufacture of glass. With the building boom in 2008, a growing demand for glass facades can be observed, this implies a reduced access for the enterprise. – See also: Keyzer, M. (2010): Towards a closed phosphorus cycle. Economist, Vol. 158 (2010), 4, pp. 411–425.
250 BDI (2007), l.c., p. 6: Calculations show that even taking into account the adaptive responses, the price increases in mineral resources in the period from early 2002 until

EU-external supply risks

The high import dependence of the EU concerning the supply with mineral raw and base materials in several fields can be reduced through specific measures. **Self supply cannot be achieved because adequate/enough deposits do not exist** (in economic scale), apart from possibilities for improvement of extraction from secondary raw materials sources (recycling). The European economy will continue to depend on imports to provide a secure supply with raw and base materials. One example is the EU steel industry (see figure 62).

Figure 62: Iron ore of the most important steel producers, 2005 in Mt (Data by BDI 2007)

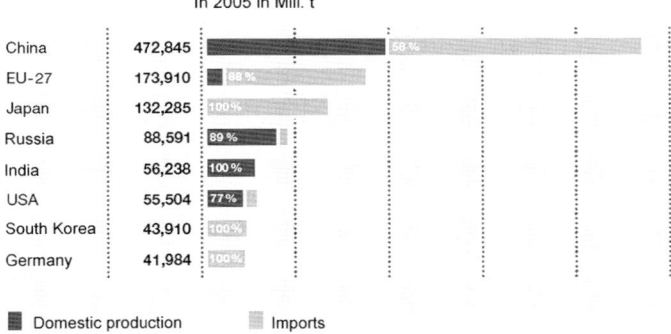

Essential EU-external supply risks are:

- Further scarcity of offer on international markets, naturally or **artificially** caused.[251]

- In the future, exploration and exploitation of deposits worldwide might stay behind the requirements, especially due to **false estimation** of developing countries.[252]

- Tendencies of big EU companies to reduce relevant activities in insecure regions in the world, be it for **political** or for economic reasons can be recognized.

---

late 2006 have caused additional costs of almost € 89 billion for the German industry] See also Wellmer, F.W. (2001): Raw materials and energy – impact of globalization on security of supply in Germany, Mining (Bergbau), pp. 315–321.

251 In this respect, the example of magnesite/export duty for China is to be addressed (information from Dr. Drnek: RHI).

252 Cp. the situation with the ACP countries (mentioned below).

- **Prices** on international raw materials markets could continue to rise and fluctuate strongly. The present financial crisis (October 2008) is an adequate example.
- **Discriminating** treatment against European importers, concerning price in particular. The (increased) formation of market dominating mining companies should be mentioned; for further elaboration see below.

Not only a quantitatively insufficient, but also a not competitively neutral supply of the European Union with raw and base material can have serious consequences also in normal times.[253]

Figure 63: Extraction of metallic raw materials by political stability of the producing countries in 2005 (Data by Bundesanstalt für Geowissenschaften und Rohstoffe (BGR); World Bank: Worldwide Governance Indicators 2006)

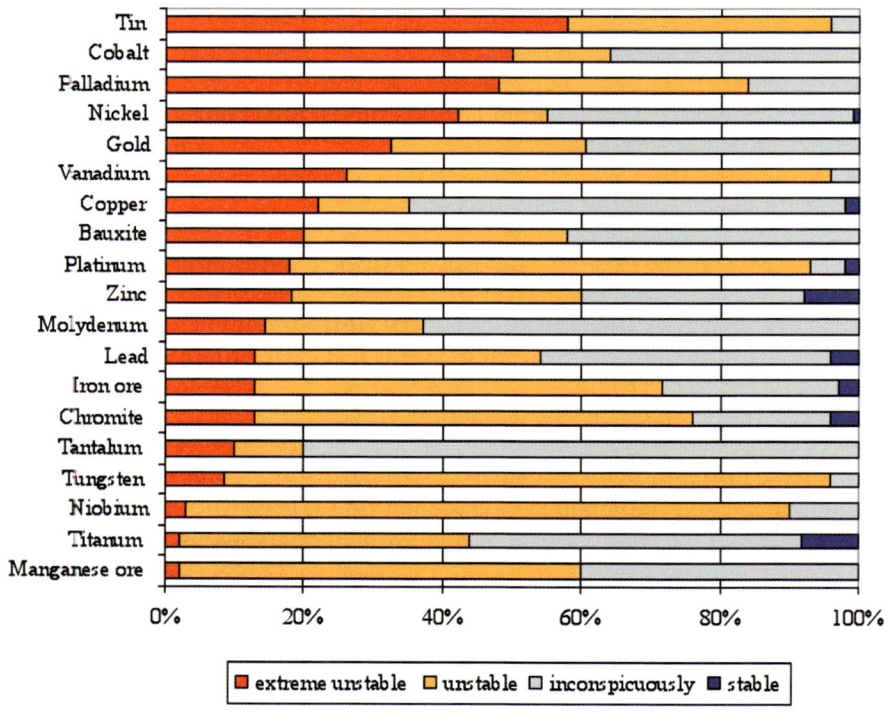

---

253 Cp. also Maull, Hanns W. (1988): Versorgungsrisiken bei "strategischen" Rohstoffen [Supply risks in "strategic" raw materials], Glückauf <Essen>, pp. 572–577. – See also: Hoffmann, H.-G. (2006): Entwicklungen an den Rohstoffmärkten belasten die Wettbewerbsfähigkeit der deutschen Metallindustrie [Development on the commodity markets weigh on the competitivness of the German metal industry] – World of metallurgy – Erzmetall, Vol. 59 (2006), 4, pp. 216–219.

Disturbances of supply can furthermore arise from interruption of international trade routes due to **political and military crises** in the world. Looking at the estimates of the World Bank about the political stability of the raw materials producing countries, one finds that more than half of the world mining industry production originates from **politically unstable** or extremely unstable countries. Regarding metallic raw materials, this situation is even more precarious.[254] **More than 60% of metallic raw materials come from unstable or extremely unstable countries** (World Bank).[255]

EU-internal supply risks

EU-internal extraction of raw and base materials is connected to a set of risks like

- the exhaustion of domestic deposits and the **timely exploration** of new economic deposits staying behind. Here the question arises whether reversal of this trend in Europe might occur.

- **no availability of deposits**. Insufficient and late analysis and assessment of deposits especially in promising regions of Europe and the protection of these deposits. – An indication of poor raw materials planning policy.

- **little elasticity of supply** in raw materials extraction, which is an imminent criterion.[256]

- **insufficient research** for and development of new processes for the search, opening, extraction and processing of mineral raw materials, and processing in the primary and secondary field (innovation). Here the foundation of the European Technology Platform on Sustainable Mineral Resources in 2005 is worth mentioning (see below).

- **undercapitalization** of business in the raw and base materials sector: a disadvantageous financial structure blocks the companies' activities, especially when high-risk or high-cost work has to be performed.

One of the most powerful forces influencing the economic importance of raw materials needed by the European economic in the **future is technological**

---

[254] Bundesministerium für Wirtschaft und Arbeit (BMWA): Österreichisches Montanhandbuch, 2008, 82. Jahrgang (Austrian Mining Handbook, 2008, 82. Issue), p.3: It is reverted to the assessment of the political stability of the WORLD BANK (Kaufmann, D. et al., 2007). This examines in great detail the likelihood of nationalisation or the threat of terrorism in the various producing countries. If the production quantities of the various supplying countries are correlated, it shows clearly that more than half of the world mineral production comes from politically unstable countries.

[255] BDI (2007), l.c., p. 21.

[256] Cp. Chapter 1.

change. In many cases, their rapid diffusion can drastically increase the demand for certain raw materials. Based on a study commissioned by the German Federal Ministry of Economics and Technology, the demand from driving emerging technologies is expected to evolve sometimes very rapidly by 2030.[257]

Table 19: Global demand of the emerging technologies analysed for raw materials in 2006 and 2030 related to today's total world production of the specific raw material (Updated by BGR April 2010) (Data by EC, 2010)

Raw materials	Production 2006 (t)	Demand from emerging technologies 2006 (t)	Demand from emerging technologies 2030 (t)	Indicator[1] 2006	Indicator[1] 2030
Gallium	152	28	603	0,18	3,97
Indium	581	234	1.911	0,40	3,29
Germanium	100	28	220	0,28	2,20
Neodymium (rare earth)	16.800	4.000	27.900	0,23	1,66
Platinum (PGM)	255	very small	345	0	1,35
Tantalum	1.384	551	1.410	0,40	1,02
Silver	19.051	5.342	15.823	0,28	0,83
Cobalt	62.270	12.820	26.860	0,21	0,43
Palladium (PGM)	267	23	77	0,09	0,29
Titanium	7.211.000[2]	15.397	58.148	0,08	0,29
Copper	15.093.000	1.410.000	3.696.070	0,09	0,24

[1] The indicator measures the share of the demand resulting from driving emerging technologies in total today's demand of each raw material in 2006 and 2030
[2] Ore concentrate

---

257 EC, DG Enterprise (2010): Report of the Ad-hoc Working Group on defining critical raw materials, p. 7.

3.5 Questions on the security of supply in Europe

How to define critical raw materials?

An Ad-hoc working group (EC and Member States) has been researching this topic and published in June 2010 a report.[258] With regards to geological availability, the group observes that, as geological scarcity is not considered an issue for determining criticality of raw materials within the considered time horizon of the study, ten years, global reserve figures are not reliable indicators of long term availability. Of greater relevance are **changes in the geopolitical-economic framework** that impact on the supply and demand of raw materials. These changes relate to the growing demand for raw materials, which in turn is driven by the growth of developing economies and new emerging technologies. Moreover, many emerging economies are pursuing industrial development strategies by means of trade, taxation and investment instruments aimed at **reserving their resource base for their exclusive use**. This trend has become apparent through an increasing number of government measures such as export taxes, quotas, subsidies etc. In some cases, the situation is further compounded by a high level of concentration of the production in a few countries.

This report analyses a selection of 41 minerals and metals and puts forward a relative concept of criticality. This means that raw material is labelled *"critical"* when the risks of **supply shortage and their impacts on the economy are higher** compared with most of the other raw materials. Two types of risks are considered: a) the "**supply risk**" taking into account the *political-economic stability* of the producing countries, the level of concentration of production, the potential for substitution and the recycling rate; and b) the "**environmental country risk**" assessing the risks that measures might be taken by countries with weak environmental performance in order to protect the environment and, in doing so, endanger the supply of raw materials to the EU.[259]

3.5.4 Formation of market controlling mining companies

One essential requirement for sufficient supply with raw and base materials is **keeping up functioning raw materials markets**, which can be affected by protectionist measures of single states as well as by the occurrence of accumulations on the offer side, such as monopolies, oligopolies and so on.[260]

---

258 EC, DG Enterprise (2010): Report of the Ad-hoc Working Group on defining critical raw materials.
259 EC, DG Enterprise (2010): Report of the Ad-hoc Working Group on defining critical raw materials.
260 Cp. Vondran, R. (2001): Die Konzentrationswelle erfasst die Rohstoffmärkte. – Stahl und Eisen [The concentration wave reaches the raw material markets. – Steel and Iron], pp. 111–116.

Chapter 3 Demand and supply of non-energy raw materials in Europe

Associations of employers endanger the security of supply with raw and base materials as soon as **anticompetitive market structures arise** and start to settle due to the association. The formation of suppliers' monopolies or oligopolies increases: International developments led to the destabilization of the balance between supply and demand, which until recently determined global raw and base materials trade.

Additionally, the cost pressure of prices going down and the high demand for capital in mining in the 1990s pushed this process of accumulation. As stated in chapter 1, deposits are location bound and distributed regionally.[261]

Mining companies worldwide limited their extraction activity to the lower third of the cost curve. As a result, new and expansion investments of international mining companies were increasingly made in major deposits and deposits rich in ore. The fact that mining production in Europe has clearly declined in the last twenty years tightens the situation even more. In the **last fifteen years many, mainly European, employers have withdrawn from ore mining**, because involvement in this sector no longer seemed profitable.[262]

For a multitude of metallic raw materials, several individual extraction companies hold significant shares in the market. Takeovers or the formation of business groups can lead to a **strong increase in business accumulation**.

Figure 64: Concentration of enterprise in the exploitation of selected raw materials. Allotment of the three biggest mining companies in each case as well as their accumulated interest in the world production in 2005 in % (Data by BDI 2007; Raw Materials Supply Group [RMSG])

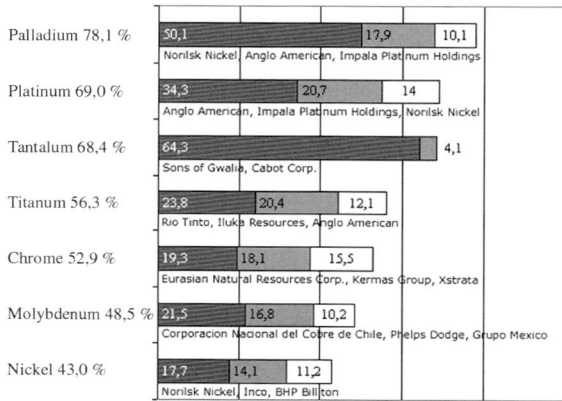

---

261 Cp. Bassani, A. (1993): Steps to a Market Economy. – Paris. – OECD Observer, no 180, February – March 1993. pp. 15–18.
262 BDI (2007), l.c., p. 7. Maybe it could lead again to a slight turnaround, as, for example, developments in Germany show. – Cp. VRB (2008), l.c.

3.5 Questions on the security of supply in Europe

This means that more and more consumers face less and less suppliers from fewer and fewer countries. The remaining suppliers are gaining **market power, which they use for raising prices.** High regional business accumulation is economically and politically dangerous, also because more than half of the metallic raw materials produced worldwide come from politically instable regions; but the risks go beyond the question of political and economic stability of the extracting countries. Apart from the producer's ability to deliver, which, provided a minimum of law and order, should not be obstructed, the willingness to deliver is important as well.[263] The raw materials sector as well as several other sectors in materials industry has suffered from a considerable decrease of suppliers due to the formation of business groups.

A particularly high concentration can be observed on the **iron ore market** as mentioned before. This has led to an increase in iron ore price of 100% during the year 2008, especially for Asian customers.

Originally a further shifting was being considered, particularly as the Australian antitrust authorities have agreed with the acquisition of Rio Tinto by BHP Billiton, and this without any conditions.[264] BHP's takeover bid was 101 billion USD. European and Asian steel producers see the decision of the Australian antitrust authority very critically, though the decisions of the European, Canadian and South African antitrust authorities are still pending.

Figure 65: Movement of iron ore trade transported by sea between 1960 and 2006: 121–725 million tonnes (Data by Ekdahl 2008; ThyssenKrupp Steel, Eurostat)

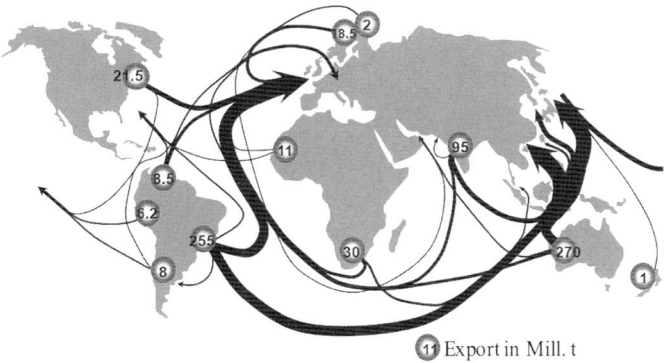

Up to now, three groups have been controlling 71% of iron ore trade by sea route: BHP Billiton, CVRD and Rio Tinto

---

263 BDI (2007), l.c., p.10.
264 R. Hahn (EMFIS), 1.10.2008: Australian antitrust authority approves the takeover of Rio Tinto by BHP Billiton, http://www.rohstoff-welt.de/news/artikel.php?sid=9484. The stocks of Rio Tinto rose after the announcement by 17%.

Dubai World announced the formation of a new branch of business as a response to the enormous growth possibilities in the sector of natural resources worldwide.[265] Dubai Natural Resources World wants to further the growth of its shares along the exploitation chain of natural resources including mining as well as oil, gas and alternative energies. Driving force is the scarcity of raw materials worldwide together with the race for the reserves in developing countries. Furthermore, the permanently increasing raw materials consumption leads to a situation which can no longer assure sustainability.

Parallel to market-induced processes of accumulation some countries pursue a **strategy of developing national conglomerates in the raw materials and plants sector through publicly enforced fusions.** In 2006, the Russian aluminium companies Rusal and Sual merged with Glencore to form the biggest aluminium concern worldwide. The Russian competition authorities approved of this fusion on the grounds that it would strengthen the Russian position as an equal participant in the international economic integration and would increase the nation's influence on the international markets.[266] This process is going on: in 2008, the idea of a merger of the aluminium concern Rusan with Norilsk Nickel came up.[267]

Figure 66: Norilsk Nickel (Data by http://www.welt.de)

Rusal intends to create a new giant in the branch from the fusion of two businesses. It is intended to become global market leader in mining and metals

---

265 Dubai World founds Dubai Natural Resources, World – 06.10.2008, BUSINESS WIRE: The new enterprise aims at long-term investment and sustainable development. Dubai, United Arab Emirates.
266 BDI (2007), l.c., p. 12.
267 28. April 2008, Oberösterreichische Nachrichten, Russian oligarchs battle for Norilsk, "Oleg Deripaska and Viktor Vekselberg want to merge their aluminium company Rusal with Norilsk Nickel to form a true metal giant. Two other Russian oligarchs, however, want to put a spoke in their wheel".

3.5 Questions on the security of supply in Europe

industry within the next ten years for all metals that are dealt on the London Stock Exchange.[268] Moreover, another Russian metal giant might emerge: Metalloinvest and Interros agreed on uniting their metal assets in the future. The aluminium monopolist Rusal has also been invited to cooperate.[269] Should a unification of these three companies actually take place, a metal giant with around 150 billion USD market capitalization, which catches up with the global market leaders BHP Billiton and Rio Tinto, might emerge.[270]

Figure 67: Where are the main reserves of metallic raw materials located in future? (Data by Ekdahl 2008 [BGR, USGS])

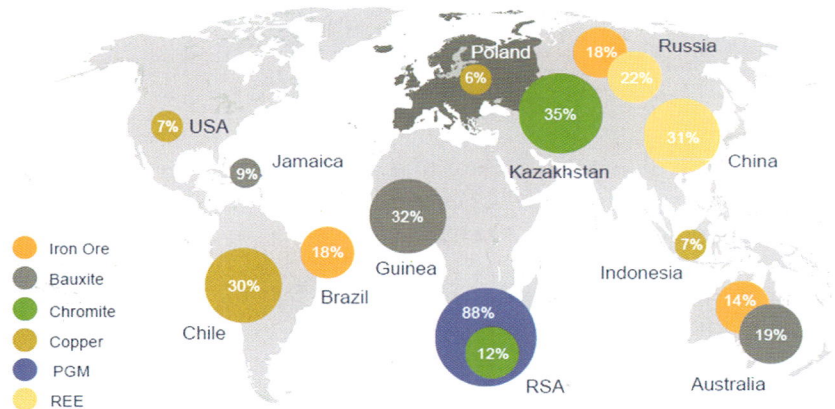

---

[268] 28. April 2008, Wirtschaftsblatt, Rusal will Norilsk bis 2009 voll übernehmen: "Der Aluminiumproduzent Rusal will den Metallproduzenten Norilsk Nickel innerhalb eines Jahres voll übernehmen." [Rusal wants to take over Norilsk completely until 2009: "The aluminium producer Rusal wants to fully take over the metal producer Norilsk Nickel within a year."].

[269] 31. Mai 2008, Die Presse: Ein russischer Metall-Gigant entsteht: "Der erste Schritt zur Bildung eines russischen Metallurgie-Giganten, der Anspruch auf die globale Branchenführerschaft erhebt, ist getan." [A Russian metal giant emerges: "The first step towards the formation of a Russian metallurgical giant which claims global branch leadership is done".] – See also: Schulze, G. (2010): Russlands Metallurgie-Sektor kommt glimpflich durch die Krise. Sinkende Schuldenlast und große Zukunftspläne. Germany Trade & Invest, Berlin, Bonn, Datenbank Länder und Märkte, 18.01.2010. https://www.gtai.de/DE/Navigation/Datenbank-Recherche/Laender-und-Maerkte/(accessed 12.03.2011).

[270] Ibid. There is, however, no indication for rush, because for now Usmanows Metalloinvest will complete its going public. For the merger of Metalloinvest and Potanin's Interros group, a period of five to seven years is estimated.

The trends of development mentioned above as well as the great and increasing negative effects of imports of raw and base materials on the balance of payments suggest making resolute efforts in the EU's supply policy. To a**dapt the usual balance between demand and supply and thus secure future availability**, it takes continuous **innovation and exploration effort as well as the (EU-) political support of global raw materials production**. This includes exploration by Geological Surveys and, for defined projects, the industries of the consuming countries.

Likewise all measures which serve the guarantee of a free development of supply and demand on the raw material markets should be supported by the EU within the scope of a reinforced cooperation with international organisations with lasting effect. The global dimension of this problem was discussed in 2007 by the European Committee.[271]

3.5.5 Conflicts und wars

This issue shall be enlarged upon at last. The subject of conflicts and wars plays a well-established role in history, particularly concerning energetic and metallic raw materials.[272] Both groups of raw materials show a great regional potential of accumulation. Important mineral deposits are to be found in politically problematic regions. For some metallic raw materials almost all of the global production comes from politically unstable countries. In these countries the

---

271 At the same time the importance of securing raw materials for the EU raw materials industry will be highlighted. See below (6.5.4).
272 Cp. among others Zentrales Komitee des Kommunistischen Bundes Westdeutschland (1976): Rohstoffpolitik und Kriegsvorbereitung. [Raw materials policy and preparation for war] – Mannheim : Kühl, 1976. (Kommunismus und Klassenkampf [Communism and class struggle]/Arbeitshefte ; 2). Dams, Theodor; Grohs, Gerhard [Hrsg.] (1977): Kontroversen in der internationalen Rohstoffpolitik: ein Beitrag zur Rohstoffpolitik der Bundesrepublik Deutschland nach UNCTAD IV [Controversies in the international raw materials policy: a contribution to the raw materials policy of the Federal Republic of Germany according to UNCTAD IV]. – München : Kaiser ; Mainz : Matthias-Grünewald-Verlag, 1977 (Reihe Entwicklung und Frieden [Series development and peace] : Materialien ; 7). Mayer, Peter (2006): Macht, Gerechtigkeit und internationale Kooperation : eine regimeanalytische Untersuchung zur internationalen Rohstoffpolitik. [Power, justice and international cooperation: a regime-analytical study on international raw material policy.] – 1. Aufl. Baden-Baden : Nomos, 2006. (Weltpolitik im 21. Jahrhundert ; Bd 13). Roithner, Thomas (2008): Von kalten Energiestrategien zu heißen Rohstoffkriegen? : Schachspiel der Weltmächte zwischen Präventivkrieg und zukunftsfähiger Rohstoffpolitik im Zeitalter des globalen Treibhauses [From cold energy strategies to hot wars for raw materials? : Game of chess of world powers between preventive war and sustainable raw materials policy in the era of a global greenhouse] /Österreichisches Studienzentrum für Frieden und Konfliktlösung [Austrian Study Center for Peace and Conflict Solution] (Hg.). Projektleitung (project management): Thomas Roithner. – Wien (u.a. among others): Lit, 2008.

3.5 Questions on the security of supply in Europe

threat of (civil) war and terror as well as the included risk of the government being overthrown violently are ubiquitous. Furthermore, a nationalization of resources is possible. The Central African region, which is rich in important ore deposits but also suffered from violent wars and civil wars repeatedly, as well as the Central Asian region of the CIS successor states point up these problems.[273]

Above all, Congo can be mentioned as an example:[274] 130 years ago, when the Belgians appropriated the land, the exploitation of the Congolese resources began. The demand for gold, diamonds, copper and Niobium-tantalum ("coltan") has taken many people's lives. In the late 60s mines were nationalized and the mining company Gecamine was founded. Cheap mining rights for foreign concerns secured the dictatorship's acceptance by the West. In 1997 the Democratic Republic of the Congo was founded. Most of the former contracts are still valid, such as those concerning the rich Copper belt (see figure), where, according to the NGO "Southern African Resource Watch" (SARW), one third of the global cobalt reserves and ten percent of all copper occurrences are located.

Figure 68: War in Congo (Data by Presse, November 2008)

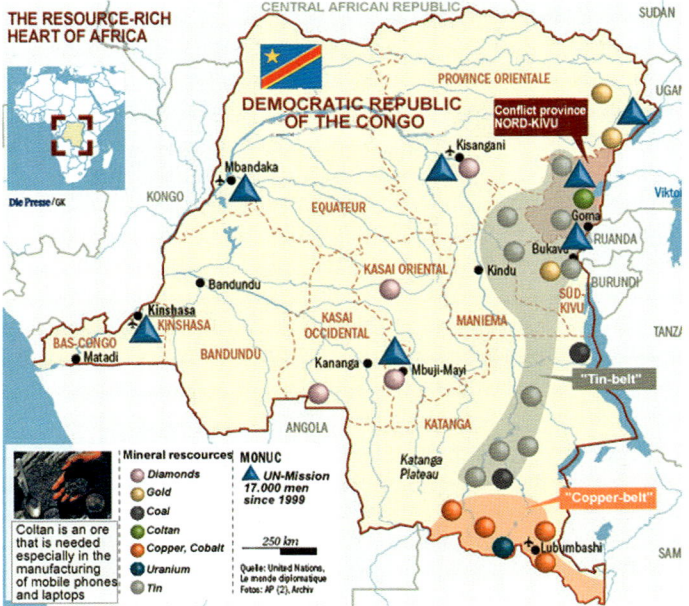

---

273 Aksjuk, Leonid N. (1984): politika razvivajuščichsja stran Afriki [Rohstoffpolitik in den entwickelten Ländern Afrikas]. – Moskva : Izdat. Nauka, Glavnaja Red. Vostočnoj Literatury, 1984.
274 Schwarz, C., Helmar, D. (2008), Geschichte einer Ausbeutung (History of an exploitation), Presse, November 2008.

The raw materials are brought out of the country avoiding the government: The only asphalt road in the region leading to South Africa, is lined by trucks guarded by militia with machine guns. Presently, a revision of all contracts is intended; about 61 agreements will be reviewed by a Commission. Stricter conditions imposed on new contracts displease China, which has acquired mining rights to an enormous extent. According to SARW, Beijing invested billions of dollars in development projects in 2007/08. In return they received mining rights for copper and cobalt valued at 14 billion dollars.

Also in the current conflict in the Congolese province North Kivu a special raw material plays an important role: the crude ore Niobium-tantalum, which is needed for the production of mobile phones and notebooks.[275] Niobium-tantalum (and other raw materials) is a fundamental reason for the power struggle in East Congo. Coltan may again play a role on the global market:[276] Niobium-tantalum extraction in East Congo was identified as conflict-promoting; since then, coltan was no longer sought after on the global market. In December 2008 the biggest tantalum mine in the world, in Wodgina, Australia, was closed. Wodgina so far provided 30 % of the global production. The expected shortage might revive the interest in the Congo. The mining company Talison explains the shutdown by the increasing cost pressure on the part of the electronics industry. This will again favour the tantalum coming from Central Africa, especially from the Democratic Republic of the Congo.[277]

Niobium-tantalum mining there is mainly concentrated in the Lake Kivu region in the Democratic Republic of the Congo. 80 % of the global occurrences are located in North Kivu. Diggers extract concentrate for further smelting from the soil by wet screening and gravity separation. Peak values of these concentrates are between 40 and 45 % tantalite (tantalum oxide Ta_2O_5). One prospector can earn up to 2000 dollars a month; the average annual income in the Congo is 80 dollars. Niobium-tantalum from this region is often mingled with radioactive elements. International companies are involved in the conflicts between the government and the insurgents, who take advantage of the chaos. In exchange for little bribe money, from which they buy weapons, they provide unobstructed export of coltan. Niobium-tantalum is secretly

---

275 The name "Coltan" is derived from the mineral group columbite-tantalite.
276 DOMINIC JOHNSON, http://www.taz.de/1/politik/afrika/artikel/1/coltan-kehrt-auf-weltmarkt-zurueck
277 Tantalum from Central Africa is available at relatively low prices because it is often exploited illegally or without any significant health, safety, environmental and working regulations. It is not possible to compete that. Without Talison the majority of tantalum will come from irregular and unreliable suppliers, much of it from the Congo (so the mining company Talison).

smuggled to Uganda and Ruanda primarily. 130 years after the Belgians appropriated the land conflicts over raw materials have not lessened.[278]

Special developments occurred in **Antarctica** and recently also in the **Arctic**.[279] A study of the United States Geological Survey (USGS) in 2000 showed that about 25 % of the undeveloped fossil fuels worldwide and further strategic raw materials might be found in the Arctic Circle. If these results are correct, the Artic would be as rich in mineral resources as Saudi Arabia or even more.[280] This aspect has gained importance because the climate change makes these deposits accessible: in the last 30 years about 1 million square kilometers of pack ice have melted, thereby making accessible a number of petroleum, natural gas and other minerals (diamonds, gold, silver, lead, copper, zinc) deposits

Figure 69: The Arctic and bordering states

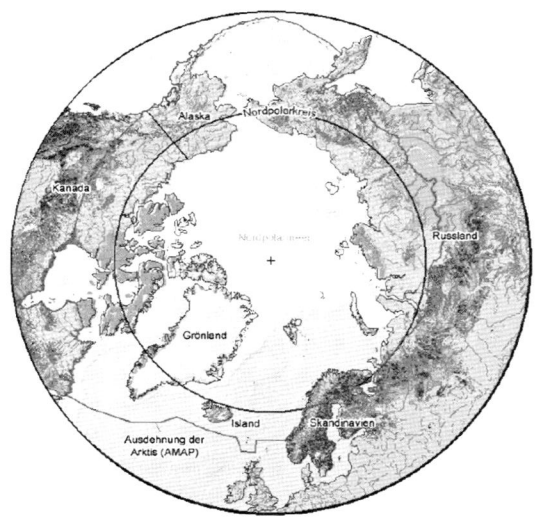

---

278 Schwarz/Helmar (2008), l.c.
279 De Wit, Maarten J. (1985): Minerals and mining in Antarctica. Science and technology, economics and politics. – Oxford: Clarendon Press, 1985 (Oxford science publications). Bermejo García, Romualdo (1990): L'Antarctique et ses resources minérales: le nouveau cadre juridique. – Paris : PUF, 1990. – (Publications de l'Institut Universitaire de Hautes Études Internationales de Genève.). Vakalopoulou, A. (1992): Antarctica: a scientific challenge for Europe. Report drawn up on behalf of the Commission of the European Communities (DG XI) on the scientific, political and legal status of the Antarctic continent. 6 Sept. 1990 / EC, Commission. – Brussels : EC, 1992 (Publication UE/CE) (Publication internationale).
280 Cp. Braune, Gerd: Eiskalter Wettlauf. Bodenschätze in der Arktis [Ice-cold race. Mineral resources in the Arctic]. In: Handelsblatt.com, 01.10.2007, URL: http://www.handelsblatt.com/News/Technologie/Energie-Umwelt/_pv/_p/303081/_t/ft/_b/1326538/default.aspx/eiskalter-wettlauf.html (Zugriff: 17.03.2008).

Another important aspect for the increasing importance of the Arctic is the fact that the Northwest Passage will soon be free of ice during most of the year. This means a shortening of the route of transport between Europe and Japan by 7000 km and by 8000 km between the East Coast of the US and China.[281] Thus, especially the US American and Chinese markets will be provided with raw materials and avoid shortages in strategically important raw materials from politically unstable regions. The bordering states of the Arctic (Russia, USA, Canada, Denmark and Norway, or the European Union) try turn climate change to their advantage in power-politics. The run on territorial claims on the North Pole has already begun, as can be seen from the offensive Arctic policy of the states mentioned.[282]

In **conclusion** it should be stated that the high metal prices in the last ten years not only increased the interest of investors in the raw materials sector, but also that of violent political groups. This became obvious when in 2008 assumedly communist insurrectionists attacked a copper mine worth three billion dollars on the Philippine island Mindanao.[283] The deposit Tampakan, which is the largest mainly undeveloped copper occurrence in Southeast Asia, was attacked twice already in 2008. Although the assaults, which aimed at calling international attention to the actors, did not take any lives, they did cause considerable material damage. Due to the precarious safety situation, the operators Xstrata and Indophil Resources had to go out of business temporarily. The attacks have a negative effect on the economy and provoke fear in the investors. The Philippine military positioned 500 additional soldiers in the region to restore safety. This assault made clear in a spectacular way that raw materials have gained even political significance.

---

281 Kopp, Dominique: Kalter Krieg unter dem Packeis, übers. von Barbara Kleiner [Cold War under the pack ice, translated by Barbara Kleiner]. In: LeMonde diplomatique, 14.09.2007, URL: http://www.monde-diplomatique.de/pm/2007/9/14.mondeText.artikel, a0039.idx,11 (accessed:06.01.2008). – See also: Wegge, Njord (2011): The political order in the Arctic : power structures, regimes and influence. Polar record, Vol. 47 (2011), pp. 165–176. (Published online 11 Jun 2010, DOI:10.1017/S0032247410000331) – Wilson, E. (2007): Arctic unity, Arctic difference: mapping the reach of northern discourses. Polar record, Vol. 43 (2007), 2, pp. 25–133 – Jin, D., Seo, H., Choi, S. (2010): Arctic governance and international organization: a focus on the Arctic council [In Korean]. Ocean and polar research, Vol. 32 (2010), 1, pp. 85–95. ISSN 1598-141x – Udd, J. (2006): Management in action – the frozen north: Arctic mining in Canada. Mining magazine, Vol. 194 (2006), 2, pp. 26–27 – Safonov, Yu. G. (2010): Mineral potential of Russian Arctic : state and efficient development. Russian geology and geophysics, Vol. 51 (2010), 1, pp. 112–120.
282 Cp. Trojer, M. (2007), Über die Wahrscheinlichkeit eines Kalten Kriegs um Gebietesansprüche am Nordpol [About the probability of a cold war for area claims on the North Pole], Raumordnung, Seminararbeit (term paper), Leoben.
283 Pöltner, Wirtschaftsblatt, Juli 2008: Südostasiens größte Kupferlagerstätte ist zum Ziel terroristischer Aktivitäten geworden. [Southeast Asia's largest copper deposit has become the target of terrorist activities.]

Chapter 4 The concept of a minerals policy

The importance of mineral economy as substantial element of the added value of a state has been shown in the first chapter. The **demand and supply situation** of non-energy raw materials in Europe has been discussed in Chapter 3. While demand of mineral resources generally is on a high level, the supply offer of non-energy raw materials in EU Member States is increasingly affected by different issues (external supply risks [supply from politically unstable countries] and internal supply risks [limited access to domestic mineral resources], see Chapter 3). Thus, the importance of minerals policy for **securing the supply** of the **EU economy** with minerals is apparent. In Chapter 4 this topic shall be elaborated in the following order: terms of classification, targets, strategies, instruments and concept.

4.1 Definition and terms

Fact – Minerals policy as part of the economic policy

Minerals policy is part of the economic policy,[284] which again is assigned to the political economy in the scientific sense.[285] In other words: Economic policy is the part of state politics which deals with the shaping of national economy.[286] It seems appropriate to refer to any *state activity* aiming directly at *influenc-*

---

[284] Thesaurus Sozialwissenschaften, http://www.sowiport.de. Siebert, H. (1983): Ökonomische Theorie natürlicher Resourcen, Tübingen [Economic theory of natural resources, Tübingen.]Johansen, Harley E., Matthews, Olen Paul, Rudzitis, Gundars [eds] (1987): Mineral resource development: geopolitics, economics and policy, London. Acciarito, G. (1985): La politica mineraria nel contesto delle relazioni economiche. L'Industria mineraria, p 1–28. Brandstätter, W. (1989): Der Einfluß von Steuern auf die Planung von Bergbaubetrieben, Diss., Leoben [The influence of taxes on the planning of mining operations, Diss, Leoben], p. 22f. Aron, J. (1992): Economic policy in a mineral-dependent economy: the case of Zambia. – University of Oxford, 1992. Drnek, T. (2008b): Spezielle Mineralwirtschaft, Rohstoffpolitik (Allgemeine Ziele), Vorlesungsunterlagen [Special mineral economy, raw materials policy (General objectives), lecture notes], p.1.

[285] Regarding *political economy*: see relevant technical literature. – See also: Gordon, R.L., Tilton, J.E. (2008): mineral economics: overview of a discipline. Resources policy, 33 (2010), 1, pp. 4–11.

[286] Tuchfeldt, in: Gabler Verlag (1984), Gablers Wirtschaftslexikon, 10. Auflage Edition, Wiesbaden.

ing extent, composition or distribution of the national product as economic policy.²⁸⁷ Generally speaking, *economic policy* is a policy including all measures, with which the state *intervenes* regulating and arranging the economy. Economic policy specifies the rules, within those the (to a large extent) privately organized economy can act. This leads to the following **general definition** of minerals policy: A minerals policy can be defined as the *entirety of actions of a State for influencing supply of and demand for mineral resources on its territory and beyond that*.

That implies the **conceptual definition** of (mineral) raw materials policy with reference to "**Minerals policy in Europe**". As the analyses in Chapter 3 (*and* also Chapter 5) shows that demand of minerals for the European economy is at a high level, however the security of minerals supply is affected by external (European high dependency on metallic mineral imports) and internal supply risks, so primarily a **minerals supply policy or minerals security policy** is generated: The *national political interest* in raw materials results from their *position in the economic circuit* and their particularities of production.²⁸⁸ Securing an optimal supply with public (as well as private) goods (raw materials as "limited goods") and (connected with) increasing material prosperity over time are main targets of every realistic economic policy.²⁸⁹ This lead to the following definition:

> A minerals policy is to be defined as a policy to **secure** the (sustainable) **supply** of the economy with (non-energetic) mineral resources.
>
> Regarding the *national and European level*, the following can be stated: A **national minerals policy** can be defined as a policy to **secure** the (sustainable) **supply** of the economy with (non-energetic) mineral resources by the *entirety of actions of a State for influencing the supply of mineral resources on its territory and beyond that*.

---

287 Molitor (2006), l.c., p. 10,11. Economic supporters are all state or state-authorized institutions which responsibly take economic policy decisions. – To obtain a socio economic optimum, five main objectives can be defined: steady growth, increasing per capita real income, full employment, price stability, balance of payments and equitable distribution of income. Cp. also Brandstätter (1989)], l.c., pp. 24, 25.
288 Cp. Siebert, Horst (1981): Strategische Ansatzpunkte der Rohstoffpolitik der Industrienationen nach der Theorie des intertemporalen Resourcenangebots, Institut für Volkswirtschaftslehre und Statistik, Mannheim. (Beiträge zur angewandten Wirtschaftsforschung) [Strategic points of the minerals policy of industrialized nations, on the theory of intertemporal resource supply, Department of Economics and Statistics, Mannheim. (Contributions to applied economic research)].
289 Klump (2006): Wirtschaftspolitik, Instrumente, Ziele und Institutionen, München [Economic policy, instruments, objectives and institutions, Munich].

4.1 Definition and terms

Note: Minerals policies which aim at promoting mineral exports are also an important, encouraging target of a national (and European) mineral policy. However, the supply issue is of particular interest in this book.

A **European minerals policy** can be defined as a policy to **secure the (sustainable) supply** of the EU economy with (non-energetic) mineral resources by *the entirety of actions of the European Union for influencing the supply of mineral resources on its territory and beyond that*.

An effective minerals policy must be embedded in a general economic policy including monetary policies (fees, taxes, revenues, royalties, rents, etc.), so that individual aspects can be evaluated in a larger context.[290] Furthermore, minerals policy is an integral part of a state and has to serve socio-economic objectives, increase the present population's standard of living and public welfare for future generations.[291] A minerals policy, on the basis of appropriate general conditions, needs to ensure that the domestic minerals economy (supply with minerals from domestic resources, internal supply) and the external minerals trade (external supply) contribute to the **gross domestic product of the state** (or a confederation of states) at **optimal cost**.

A policy for securing the supply with minerals should imply establishing a **framework** including **objectives, strategies and action plans**.

The subject of raw materials is rather **complex** and features many links to other branches of politics. The concepts of minerals policy and other state policies involved need to be **coherent** for the reason alone that the former is **part of these policies and must use their instruments**. Description of some component policies relevant for minerals policy:

- Foreign policy: setting objectives of trade and development policy through diplomatic dialogue with Non-EU-States
- Trade policy: securing access to raw materials from Non-EU-States, for example through multilateral contracts
- Development policy: building capacities (cooperation of geological surveys) in Non-EU-States to support political stabilization and the access to mineral deposits/raw materials

---

290 Enzer, H. (1981): Decision making from the administrative viewpoint, Materials and Society, Bureau of Mines, Washington, DC.
291 Cp. Zeller, Jürg R. (1981): Nationale Rohstoffpolitik. – Bern (u.a.) [National raw material policy. – Bern (among others)]: Haupt, 1981 (Beiträge zur Wirtschaftspolitik ; Bd 35) [(Contributions to economic policy, vol 35)]. – See also: Bradley Jr., R. L. (2007): Resourceship: an Austrian theory of minerals resources. Review of Austrian economics, Vol. 20 (2007), 1, pp. 63–90.

- Minerals planning policy: encouraging an active raw materials planning policy, for instance exploration and protection of deposits in context of the land use planning
- Research and technological policy: for instance increasing efficiency of minerals, products and raw materials

From the above follows that a minerals policy is a **cross-section matter**. Establishing a **coherent** raw materials policy requires comprehensive and effective coordination and cooperation between the separate policies. This can be realized by coordination committees on national (European) level. A **solitary view** of partial policies, however, is contradictory and will **not result** in a cost-effective contribution to the gross domestic product of a state (passive raw materials politics).

An **active** minerals **policy** implies encouraging an **raw materials planning policy**, for instance exploration and **protection of deposits** in context of the land use planning. Clear regulations have to be established for mineral authorities which award **mineral rights** to entrepreneurs in order to **secure orderly access** to raw materials.[292] An **active** minerals **policy** also involves creating **stable** mineral rights and a **favourable fiscal policy** (for the entrepreneur). Thus, both access to raw materials and **investment protection** for foreign and domestic investors is ensured.

In summary, through its minerals policy the state contributes to the creation of an **appropriate framework for the** protection and exploitation of its domestic mineral resources (*internal aspect*) and the access to mineral resources needed from countries outside the state (*external aspect*) – in the sense of a **cost-effective input to the gross domestic product**.

---

292 Following context has to be stated: Securing of mineral supply can be defined as the sum of all measures which lead to making deposits availabe for long-term economic exploitation and recycling and to ward off claims by third parties against these objectives. Securing mineral takes place here on the one hand at the public level as a planning tool (in the context of mineral planning policy), on the other hand on the private level in the form of acquisition of property or mining rights. Likewise: mineral supply is carried out by the extractive industries of a country (or the European Community). This can be done from domestic or foreign deposits respectively (or, however, also on the part of the industry of a state by direct import of the raw materials from non-member states).

Figure 70: Close interlocking of minerals policy, mining legislation and mineral economy

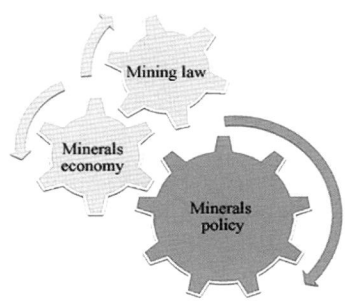

4.2 Necessity of a minerals policy

Minerals policy takes place on several administrative levels: the European Union level, the national level, the regional and the local level. Necessarily, a coherent minerals policy should be based on the EU level, governing the (EU-) external and internal aspects of supply. Furthermore, regarding a **collective** European raw materials policy, a framework of guidelines for establishing national raw materials policies should be created. Such a "framework" implies that the Member States are responsible for the development of their individual raw materials policies themselves.

Table 20: Principles for an effective and workable minerals planning system (Data by Department of Mining and Tunneling, 2004 [provided by UEPG])

	European	National	Regional/Local
Planning	Provide a level playing field based on secure access to mineral resources	Consider minerals as a key resource. Incorporate minerals in land-use planning	Identify and protect reserves of mineral resources
Decision	Include proportionality	Consider public interest. Have a long-term vision. Clarify which authority is in charge	Autonomy from local political pressures. Indicate time length to obtain a permit, or extension of existing permits
Implementation	Monitor best practices across Europe	Promote flexibility by considering local conditions and the specific nature of each project	Give certainty to operators

	European	National	Regional/Local
Evaluation	Assess the results of the transposed directives impacting the minerals industry	Assess reserves of authorised available resources	Number of permit applications Number of refusals

Within a minerals policy the following points should be made clear:

- the **particular role of the state and the private sector**
- the **main focus** of the raw materials policy
- **institutions** significant for the raw materials policy ("partial policies")
- interrelation of different policies (**interactions**)

4.2.1 Minerals policy versus market economy

Even though providing the economy with mineral raw and base materials is primarily a task of the businesses[293], there are essential reasons for the state to control it. Production and consumption of raw and base materials yields relatively serious external consequences.[294] This is because, as noticed above, raw materials are a production factor processed by many economic sectors, which influences all real assets. In other words: the **real value of raw materials in the value-added process of a national economy is crucial** (see chapter 1).

Individual enterprises can externalize economic costs of a supply interruption (by termination of employment). They are not willing to finance preventive measures at the corresponding volume of economic costs for micro economic reasons.

Due to the significance of raw materials in the value-added process of a national economy it is necessary for the state to establish basic conditions for the realization of measures which are of public economic interest.

---

[293] Cp. Linden, E. (1997): Marktrelevante Überlegungen zur Rohstoffversorgung und zu Beteiligungen im internationalen Bergbau, Erzmetall [Market-relevant considerations for raw materials supply and participation in international mining, Erzmetall], pp. 761–768. – See also: Daul, Johannes (2008): Auswirkungen der aktuellen Klimapolitik auf die österreichische Mineralrohstoffindustrie [Impact of current EU climate change policy on the Austrian mineral industry]. BHM – Berg- und Hüttenmännische Monatshefte, Vol. 153 (2008), 8, pp. 296–301.

[294] External effects are costs and benefits which go beyond the account of expenditures and income statement; e.g., interruption of the added value process. – See also: Shields, D. J., Solar, S. V. (2005): Sustainable development and minerals: measuring mining's contribution to society. Geological Society Special publication, 250 (2005), pp. 195–212.

The exploitation of domestic and foreign mineral resources requires long-term planning. Because of the high research and capital expenditure involved and uncertain chances of success the readiness to assume risk is rather marginal at some businesses.

The realization of certain actions that are also of public interest cannot be the responsibility of individual enterprises alone, which is why they have to be supported by public measures.

Due to the relatively high costs of these investigations and the, in the beginning often hardly assessable, chances of success they are on average afflicted with greater risks than investments in other economic sectors.

Therefore such investigations, if they are of public interest, should be supported by the state in order to give an **incentive** for the businesses to also realize more venturous projects. The measures of a minerals policy should be set, above all, where the probability and extent of the risks would without public commitment result in unfavourable effects for the economy not only in quantitative, but also in price aspect.

4.2.2 Security of minerals supply

This "classic initial position" can roughly be characterized as interplay between three supply conceptions:

1) a mainly **protectionist minerals policy** in line with the national interests; 2) a **policy geared to a liberal world trade** circumscribed essentially by the GATT order; 3) a policy influenced by companies[295] concerned with raw materials, international mining companies in particular (industrial policy).[296] This means that the policy of a state can be opposed by the policy of an individual enterprise. The three fundamental objectives of an individual enterprise are (1) maximization of profit, (2) securing economic survival and (3) expansion.[297]

---

295 Cp. Wellmer, F. W., Hennig, W. (2003): Aspects for formulating mineral resources management policies. (Gesichtspunkte für die Formulierung von Unternehmenspolitik im Rohstoffsektor). Erzmetall, p. 3–10.
296 Cp. Boettcher, R. (2003): Thesen zur nachhaltigen Rohstoffsicherung. – [Theses for the sustainable raw materials supply], Bergbau, 54. 2003, H. 5, p. 199–201. – See also: Schächter, Norbert, Johannes, Dieter, Einenkel, Betty (2006): Bergbau und nachhaltige Entwicklung – Ressourcen für heute und morgen [Mining and sustainable development – resources for today and tomorrow]. Bergbau, Vol. 57 (2006), 11, pp. 503–508.
297 Cp. Brandstätter (1988), l.c., p. 24, 25.

Chapter 4 The concept of a minerals policy

The policy of supply with raw materials is manifested in a comprehensive system of rules and complex interrelations (interactions) between the political instances, which are able to determine the situation and the behaviour of contractors and consumers. The entirety of these regulations and measures can be structured according to: [298]

- decision-makers, governments, international organizations, as well as enterprises;
- the tools for realization of these measures, namely as a branch of politics (commercial policy) and definite instruments (custom duties);
- the products concerned and/or aimed at.[299]

Furthermore, it has to be distinguished between **internal** (raw materials planning policy of the state, competition policy, research and environmental policy) and **external** (development, foreign and commercial policy) **economic issues** and in this connection also different options of invention (*influence on supply and demand*).[300]

A realistic minerals policy must take the basic geological and economical considerations into account. The quantity of mineral resources currently accessible in the earth's crust is limited, minerals are non-renewable natural resources, and the regional distribution of the known mineral deposits is uneven. A number of minerals occur as associated products, which means that supply is highly inelastic.[301]

Irrespective of the socio-economic system adopted, almost all planners agree **that industrialization and industrial production** are a source and means of economic welfare, so the process of industrialization still **dominates** globally all concepts of **economic policy**.

The way to maintain the standard of living in industrialised nations and to enhance socio-economic development in Third World countries is to increase the production of industrial goods.[302]

---

298 Michaelis, H.(1976): Europäische Rohstoffpolitik, Bergbau-Rohstoffe-Energie, Schriften über wirtschaftliche und organisatorische Probleme der Gewinnung und Verwertung mineralischer Rohstoffe, Essen [European mineral policies, Mining – Minerals – Energy, scripts on economic and organisational problems of the extraction and exploitation of mineral resources, Essen], p. 22.
299 Here: based on metallic minerals, industrial minerals, construction minerals.
300 See also below: List of instruments to achieve the goals.
301 Cp. above remarks.
302 Gocht (1983), l.c., p. 83, 199–202.

4.2 Necessity of a minerals policy

The global financial crisis in 2008 should also be regarded in this context. Since world-wide recession with substantial effects on the economy is spreading, many states with political-economic concepts tried to steer against this trend and at the same time promote the growth of industrial goods production. So, for example, the Swedish government tried to set the Swedish economic situation in motion, spending 785 million Euros until 2011. This comes up to more than three per cent of the gross national product of Sweden After a growth of 0,8 per cent in 2008, a zero growth was estimated for 2009. Apart from investments in road and railway construction, the building sector, badly hit by the financial crisis, is to profit from the economic stimulus package above all.[303]

France also wanted to take action against the financial crisis through an economic stimulus plan (26 billion Euros), which was meant to create 110.000 new jobs From this plan primarily the automotive industry is to profit, for instance by granting bonuses for environment-friendly cars.

The state wanted to invest effectively into the infrastructure, so the construction of the TGV connection from Lyon to Turin or the Seine Canal in northern France are planned. Also, public enterprises such as mail, railway or energy companies should invest in innovation, developing environmental technologies. The construction industry was supported by the state through the construction of 70 000 residential buildings. On local level the distribution of building licenses should take place unbureaucratically.[304]

The German economy was expected to decrease by about 0,8 % in 2009. In order to compensate the stricken economy of Germany, the state intended to make 23 billion Euros available by 2012. Thus, one million jobs are to be secured and investments for 50 billion Euros may be made possible. For new vehicles bought by mid-2009, the motor vehicle tax was waived. Budget resources for traffic investments and thermal restoration will be increased. For the middle income groups a 15-billion-euro credit program will be started.[305]

In terms of minerals policy, this means that the industrialization process will make the world demand for mineral raw materials increase further. Some

---

[303] Gamillscheg, H. (2008): Finanzkrise – Konjunkturbelebung Schwedens, in: Presse, Economist, 6. 12. 2008, p. 24. [Financial crisis – Sweden's economic recovery, in: Presse, Economist 6. 12. 2008, p. 24].
[304] Ag./r.b. (2008), Finanzkrise – Konjunkturprogramm Frankreich, in: Presse, Economist, 6. 12. 2008, p.27. [Financial crisis – Economic Program of France, in: Presse, Economist, 6. 12. 2008, p.27.]
[305] Ag./b. I. (2008)., Finanzkrise – Konjunkturprogramm Deutschland, in: Presse, Economist, 6. 12. 2008, p.27. [Financial Crisis – Economic Programme of Germany, in: Presse, The Economist]

minerals will become relatively scarce, leading to the exploration and development of new deposits.[306] At the same time an increasing gap will develop between the minerals importing and the minerals exporting countries. Consequently, this means: the exploitation and utilisation of minerals increasingly endangers the environment (particulary if the state of the art level is ignored),[307] mineral costs tend to rise due to more remote and lower grade deposits being opened up, costly measures to protect the environment and increasing taxation imposed by mineral exporting countries. These developments have on one hand led to an **expansion in world trade but on the other caused problems** for the world markets for numerous minerals, in particular:[308]

- Strong, usually short-term **price fluctuations**
- Substantial **fluctuations in the earnings** of producing countries,
- **Small transparency** of mineral markets due to the high level of concentration on the supply side
- **Limitation of access to some markets** due to customs barriers and established trade structures
- **Uncertainty of medium and long-term supply** for the consumer due to a decline in exploration activities in the last years, production problems in exporting countries and export quotas imposed by international cartels.
- Problems concerning the **availability of deposits**

These facts besides others explain the need for a (international) raw materials policy (corrective justice of distribution).

---

306 Described in chapter 3, particularly on the BRIC countries and the leading African markets (so called "African Lions").
307 Cp. Tienhaara, Kyla (2006): Mineral investment and the regulation of the environment in developing countries. - International environmental agreements, Vol. 6,4.2006, pp. 371–394. Bothe, Michael (2005): Environment, development, resources. – Recueil des cours/Académie de Droit International de La Haye, Vol. 318. 2005(2007), pp 333–516. Tiess, G. (2007): Environmental related Aspects of the Non-Ferrous Mining Industry in Europe, BHM Berg- und Hüttenmännische Monatshefte [Monatshefte (Journal of Mining. Metallurgical, Material, Geotechnical and Planned Engineering)], H. 10 p. 309–316.
308 Already emphasized by Gocht (1983), l.c., p.201f. These problems can be described as basic problems of the raw materials policy. They are generally valid. Solving these problems requires a comprehensive interdisciplinary approach (see Section 4.2.3).

4.2.3 Coordination of interactions

Raw materials are a **complex issue**. The exploration, production, processing and transport to the final consumer and furthermore the market-relevant operations, are subject to complicated developments. Also, the numerous interferences of this topic with the interests of nature and environmental protection as well as the space-use conflicts are to be regarded.

Figure 71: Stakeholders relevant for minerals policy (Data by Christmann, 2008)

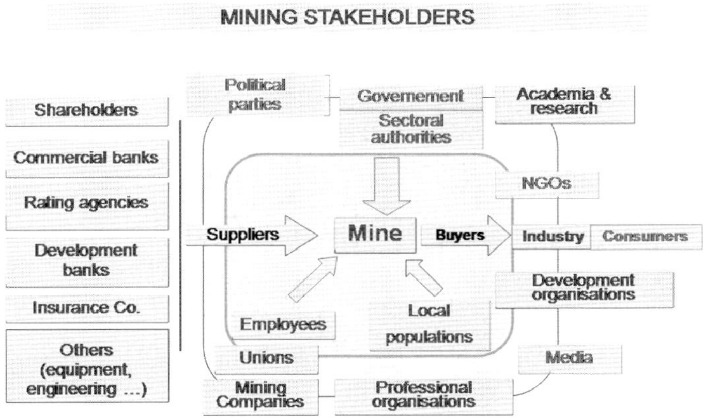

Securing a sustainable supply with raw materials for the national and European economy requires an **interdisciplinary and/or interdepartmental approach**. Such an approach alone can take into account the **complexity of the raw materials subject**. Thereby the following questions arise:

a) What functions (exploration and exploitation of deposits) are to be assigned to which officials (Geological Survey, Ministry of Economics)?

b) Which interactions take place between the officials?

c) How can the interactions between the officials result in sensible synergy effects?

d) How can a comprehensive framework of minerals policy be established for the control and coordination of these interactions?

For this purpose, a list of the **officials and the functions** assigned to them should be made, showing the resulting **interactions**/mechanisms of minerals policy.[309]

---

309 Figure 72 (mechanisms of raw material policy) demonstrates possible interactions.

Demand issue

- **Demand of raw materials:** The need for mineral raw materials results from the public consumption (economy) of a state/confederation of states. An indicator is the degree of industrialization and structuring of a state.

- **Predictions for a state's demand for raw materials:** From the data collected and the developments in recent years as well as international trends and developments on the market predictions can be made.

- **Minerals balance and market analyses:** determining domestic demand for raw materials, raw materials production and raw materials imports against the background of the global market to show up reserves balances and options for development of raw materials supply should be made. The statistical collection of data is carried out by the state (controlled by law) or by external institutions. Several organizations world-wide conduct international raw materials analyses. The need of minerals is determined by the economy due to its different facets.

- Based on a **raw materials balancing critical raw materials** may be identified using the following criteria:
    - import difficulties (high price, not available in sufficient amount)
    - extraction of raw materials limited to few countries which additionally exhibit an insecure or unstable political situation
    - strategic raw materials important for domestic (and export) industry and available only in small amounts world-wide

Supply issue

- **Raw materials imports:** those raw materials that cannot be produced sufficiently to meet domestic demand of the state's economy have to be imported. On the contrary, semi-finished and/or finished products of the manufacturing industry including (imported) raw materials can be exported. Raw materials imports ought to be covered in statistics.

- **Raw materials industry of a state:** is responsible for the sufficiency of production and/or import for supplying the industry. The raw materials produced in a state may also be **exported**. Customs are collected equal to imports. The extraction of domestic deposits is carried out by one or several enterprises. Domestic raw materials may not always find use on the domestic market and thus may also be exported. The raw materials producers usually are the counterpart to the ministry or department responsible for raw materials (the export of commodities is generally not controlled by the state) and form a syndicate to collectively cooperate with governmental authorities and conduct an efficient and cost-effective as well as environmentally sustainable raw materials production.

- The produced or imported raw materials can be sold to the domestic **processing industry** (steel and cement industry). The processing industry, much like the raw materials industry, is an **essential part of the national value-added chain and considerably contributes to the GDP of a state**.[310] Securing the supply with raw materials for the processing industry is elementary. The industry has several possibilities:
  - Securing the supply with raw materials from domestic deposits
  - Securing the supply with raw materials from foreign deposits. Investments in certain regions (exploration, development of foreign deposits) can decrease the dependency from raw materials. The more diversified the raw materials sources, the more secure the supply.
  - A company (or on higher level even a state (like Japan) or a federation of states (USA)) can purchase strategically important raw materials at a favourable price and store them. This storage may be used as a compensation for extra-spendings in times of insecure raw materials prices. On state level storing raw materials can mean a general backup. This task may be under the control of the ministry of economics, the ministry of internal affairs or the ministry of national defense. Short-term trouble with supply lasting up to several months can only be bridged by adequate storage. Establishing and maintaining large storage capacities can serve two purposes: securing supply by means of creating reserves on one hand and stabilizing prices through market interventions for the dilution of major fluctuations in supply and demand.

- **International organizations:** Several international organizations (WTO and World Bank) and forums (World Mining Ministries Forum) deal with the topic of minerals supply policy, which is getting more and more important. The role of these organizations is the construction of international cooperation and networking structures, and in this sense also the consultation of businesses and nations concerning the development of a sustainable and efficient dimensioning of raw materials imports. Further areas of responsibility include the accomplishment or support of relevant projects and studies as well as the supervising of international trade relations.

- **Foreign policy:** Apart from a foreign security policy and cultural policy the ministry of foreign affairs (unless a separate ministry of trade is responsible) plays an important role in foreign trade policy. Its task is the

---

310 Cp. the definition of a minerals policy mentioned under 4.1.

establishment of bi- or multilateral foreign trade relations, a common **trade policy** and the promotion of exports. The foreign ministry furthermore is important for the development policy. It also has to be considered that the supply with raw materials implies external activities (import of strategic raw materials), which are carried out late or are badly coordinated without a foresighted raw materials policy. The biggest part of the global raw materials production comes from politically unstable countries, which is reason enough to study the problems of raw materials supply within a national or European foreign and safety policy. Foreign policy can make use of various instruments to broach the issue of raw materials and steer against a distortion of trade and competition: bilateral discussion of the foreign minister and state secretaries, bilateral and regional summit talks, bilateral mixed commissions and partnership and cooperation agreements.

- One particularity of trade applies to raw materials that are dealt on **commodities exchanges**, London Metal Exchange (LME). The price is determined by the purchasers according to supply and demand. Enormous rises in price in recent years lead to repeated speculation which influenced price in an unprecedented way and duration. Raw materials in- and exports of such mineral raw materials are linked directly to commodities exchange. Many strategic raw materials are not dealt, but sold by extraction companies in fixed-term contracts. On the iron market the three biggest iron producers Vale, Rio Tinto and BHP Billiton determine price and flow of material.

- **Development policy:** Since development policy contributes to economic and political development of developing countries rich in raw materials the stability of these countries might increase. This can lead to the extraction of raw materials which then are available to the global market and whose revenues at the same time support the development of the producing country. Apart from this comprehensive task in development policy it should further contribute to the development of partnering countries and Europe's raw materials security.

- Checking and controlling mineral rights is an essential task of the **ministry responsible for raw materials**; especially awarding and securing mineral rights for entrepreneurs (investment protection, access to raw materials). The relevant legal framework specifies granting procedures. Additionally, several aspects of environmental laws are relevant for such businesses. Raw materials producers are obliged to acquire a mining license, go through different (licensing) procedures and provide the authorities with various evidence and verification, which refer to the exploration of deposits, determination of mineability, the exploitation itself, ceasing of operation and future use of the mine.

- **Fiscal policy:** Taxes and duties from raw materials producers to the state. This may refer to mineral rights, production etc. and is particularly relevant because an attractive fiscal policy crucially influences a business's investments.
- **Geological surveys** play a special role within raw materials planning policy. They issue raw materials maps, information and in some countries they are directly involved in prospection and exploration. On the other hand, nowadays it is common to import exploration companies from abroad to explore domestic deposits, which may be developed and exploited by domestic as well as international companies.
- **Land use planning policy** also plays a special role. Deposits are location-bound and exhaustible, which is particularly important when determining priorities in land use planning if several uses of a certain open space are up for discussion. Thus, spatial planning should not only guarantee a well-planned long-term reservation of mining areas, but also a comprehensive balancing of mining and other kinds of use. Relevant information about the deposit needs to be integrated into spatial structuring.
- **Environmental policy:** The extraction and processing of raw materials in general is very energy-intensive and polluting and thus is subject to energy and environmental law. For the exploration and extraction of raw materials the ministry of the environment is essential. To minimize the effects of mining on the environment an in-depth verification of mining projects by the ministry is needed. Although environmental restrictions might result in numerous problems for the mining industry (inefficient processes, mining prohibitions), environmental law is fundamental for sustainable development. The ministry of the environment is an important controlling body.
- **Research and innovation policy:** It plays an important role for raw materials security, especially for the reduction of raw materials imports. Raw materials imports are a burden in many ways, thinking of (rising) prices and distortion of competition from third states or from political instability. The aim is to increase domestic production by following means:
    - Increasing raw materials efficiency: through efficient research in the field of geophysical, exploration and mining (deeper deposits, maritime mining, etc.) as well as processing can be enhanced.
    - Enhancement of energy and materials efficiency
    - Research in the field of raw materials substitution
    - Research in the field of raw materials recycling

Chapter 4 The concept of a minerals policy

Research can be done by universities and academies as well as by private or public institutes and specialized companies. The ministry in charge could, for example, be the ministry of science. Adequate basic conditions for the training of specialists are of great significance.

Figure 72: Mechanisms of raw materials policy

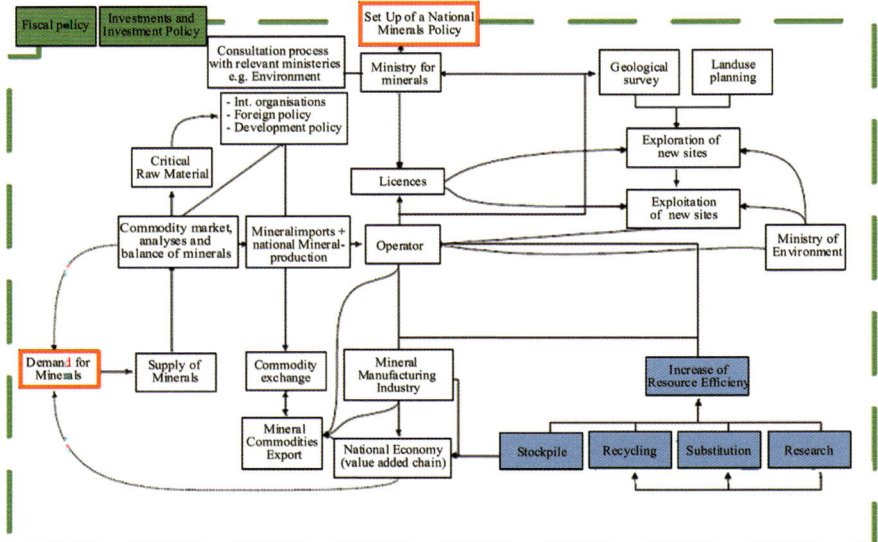

(Highly simplified illustration – Presently intense research with the intention to discuss the connections between these mechanisms and deriving consequences is being done by the Chair of Mining Engineering.)

Above explanations express that **different institutions** can fulfill tasks relevant for minerals policy and therefore have a share in securing raw materials supply. **Many more officials (resp. branches of politics) than assumed so far are of importance**. All of them have their own area of *influence*, especially highlighting interactions between the officials. Coordinating these **interactions** sensibly takes a precise strategy and coherent conception (see below).

In summary the need for an active minerals policy is evident. In the following, reference will be made to raw-materials-political interactions. They will be displayed based on a **general model**, which is a simplified illustration of the course ("mechanisms", "functions", "interactions") of the "issue raw materials policy". On one hand this is meant to display the manifold interactions between the officials, on the other hand it creates a basis for further analysis/research of the topic. Furthermore it should be mentioned that the functions and officials mentioned within this model qualify as "general variable". Thus, the model can be applied to various case studies (national raw materials policies).[311]

4.3 Objectives of a minerals policy

The primary responsibility of minerals policy should guarantee an **optimal**, demand-oriented, cost-effective, timely and environment-friendly, **supply of the economy** with raw materials as well as a fair intertemporal spreading of raw materials reserves and in this sense a cost-optimal contribution to the GDP of a state.[312] A policy of supply with raw materials has to pursue **several** competing, partly complementary and intersecting targets, and to decide priorities and emphases, whereby mid- and long-term supply perspectives are more important than short-term ones.

4.3.1 Targets of minerals supply policies

Primarily relevant is a rather high **security of supply** to prevent endangering the economic growth of a state. The objective of *security of supply* refers to the demand for a timely, demand-oriented and qualitatively suitable allocation of mineral raw materials needed by the economy. This means that shortages in supply and long-term supply disruptions are to be minimized as far as possible.

A **low-cost supply** is relevant for increasing competitiveness. The objective of *economic effectiveness* serves the cost-effect supply with raw materials including the creation and maintenance of fair competitive conditions for domestic producers. For this adequate rules and regulations are required.

Great importance is to be laid upon **rational** (i. e. resource efficient) **use and consumption of raw** materials and in this sense protection of environment as far as possible.

The objective of *environmental compatibility* is in line with the objective of sustainable development.

Another significant topic is transparency of **foreign policy**. It should be conform to the regulations and targets of GATT and contribute to the im-

---
311 See also chapter 7.1.3: this Model is applied there for the presentation of the functionaries at EU level.
312 Drnek (2008b), l.c., p. 1.

provement of trade relations with non-member countries of the European Union, also considering the needs of developing countries.

4.3.2 Different Aspects

Objectives of national mineral policies vary widely reflecting the unique circumstances of each nation. Countries with few mineral resources but requiring substantial mineral inputs – such as Japan, Korea, Taiwan, particularly also the EU countries – will obviously emphasize **different objectives** than a mineral rich non-industrialized nation – such as Papua New Guinea, New Caledonia, or the Democratic Republic of the Congo. Likewise, large mineral producers with substantial internal demand, such as Brazil, China, India and the United States, have their own policy approaches.[313]

Under consideration of the deposit potential and the raw materials strategy of the state it has to be differentiated between raw materials exporters and importers on macroeconomic level.[314]

Objectives of mineral exporting countries

The general economic policy and hence minerals policy objectives in developing countries are primarily geared to economic growth.[315] The export of minerals is viewed as the basis for financing development, foreign exchange earnings help stabilize the national budget and contributing to investment programs.[316] In the following two examples are given: Malaysia and Tanzania.

---

313 Otto, J.M. (1999): Mining, Environment and Development, United Nations Conference on Trade and Development, USA, p. 9.
314 Raw materials exporters are Australia and Canada; increasingly emerging countries such as Indonesia, Jamaica, but also developing countries like Tanzania and Uganda. Raw material importers are, e.g., the EU, Japan, but also China.
315 Cp. Daniel, P. (1990): Economic policy in mineral exporting countries: what have we learned? – University of Sussex, Institute of Development Studies.
316 Other objectives are [Gocht (1983), l.c., p.203]: Diversification of mining production to reduce dependence on individual commodity markets (target: securing revenue). Securing of the domestic supply of mineral raw materials to avoid expenditures of foreign currency for imports (target: autarky). Use of the raw materials sector for the development of rural areas by improving infrastructure and creating new jobs outside of urban areas (target: area development) Minimisation of environmental impacts in the production and processing of mineral resources (target: conservation of ecology). Protection of deposits from improper exploitation and premature abandonment (target: conservation of resources). Obtaining sovereign power of control over the natural resources (target: sovereignty) by state control of production and marketing (target: control), change of ownership structures (target: participation), or promotion of domestic cooperatives.

4.3 Objectives of a minerals policy

The National Minerals Policy of **Malaysia** provides the foundation for the development of an effective, efficient and competitive regulatory environment for the mineral sector.[317] Whilst the trust of the policy is to expand and diversify the mineral sector through optimum exploration, exploitation and utilisation of Malaysia's resources, maximum use of research and development (R&D) and modern technology, emphasis is also given to environmental protection and sustainable development, as well as the management of social impact. The salient features of the policy include security of tenure, favourable fiscal regimes, high priority land use for mining, uniform and efficient institutional framework and transparent guidelines and regulations. The regulatory objectives are

1. To contribute to national and State development by promoting diversification and expansion of the mineral industry.
2. To provide an attractive, efficient, and stable mineral sector regulatory framework
3. To encourage exploration and a beneficial expansion of the mineral industry
4. To provide a stable and conductive fiscal system
5. To accord the mineral industry a high land-use priority in areas open for exploration
6. To enhance the development of domestic expertise for mineral resource development through research, education and training activities
7. To provide environmental protection and management of social impact
8. To provide timely and accurate regulatory, scientific and technical information required by the industry, Federal Government and State Governments, including periodic review and publication of the national policy on the mineral sector

The mineral sector policy of **Tanzania** is designed to address the following national challenges:[318] To raise significantly the contribution of the mineral sector in the national economy and *increase the Gross Domestic Product* (GDP); to increase the country's foreign exchange earnings; to increase government revenues; to create gainful and secure employment in the mineral sector and

---
317 http://www.jmg.gov.my/en/information/legislation-and-policy/doc_details/9-national-mineral-policy-2.html. Further information can be found in the Annex.
318 Ministry of Energy and Minerals (1997): The Mineral Policy of Tanzania (http://www.tanzania.go.tz/pdf/themineralpolicyofTanzania.pdf). Further information is provided in the Annex.

provide alternative source of income particularly for the rural population; and to ensure environmental protection and management. In view of these challenges, the Government's policy of Tanzania for the mineral sector development will aim to attract and enable the private sector to take the lead in exploration, mining development, mineral beneficiation and marketing. The role of the public sector will be to stimulate and guide private mining investment by administering, regulating, and promoting the growth of the sector. Accordingly, the *policy objectives* of the Government for the mineral sector are:

- To stimulate exploration and mining development;
- To regularize and improve artisanal mining;
- To ensure that mining wealth supports sustainable economic and social development;
- To minimize or eliminate the adverse social and environmental impacts of mining development;
- To promote and facilitate mineral and mineral-based products marketing arrangements;
- To alleviate poverty especially for artisanal and small-scale miners.

Objectives of mineral importing countries

A principal goal of the importing industrialized countries is to ensure **guaranteed supplies** of minerals, whereas that of the importing developing countries, to obtain low prices.[319]

- Foreign trade aspect: Guaranteeing the supply of minerals through international mining investments, participation in mining projects abroad, concluding long-term supply contracts or cooperation agreements and diversifying supply sources. Aiming for **cost-effective supply** to maintain the industry's competitiveness.
- Domestic economic aspect: Independence from raw materials suppliers. Maintaining and improving domestic raw materials economy as far as economically justified. Optimizing intensity and kind of exploration and extraction of domestic raw materials as well as the coordination of domestic producers and consumers. Promoting domestic mineral production by supporting mineral exploration programs, subsidizing domestic mining companies and stimulating R&D in the field of substitution possibilities.

---

319 Here also the presentation by Gocht (1983), l.c., p. 204f is applied.

- Developing new technologies to enhance the rational and efficient utilization of raw materials. Enhancing the recycling. Minimizing the environmental impact of the extraction and processing of mineral raw materials.

- Promoting international cooperation between industrial countries and raw materials producing countries for the development of new deposits, the transfer of modern but adapted extraction technologies.

Between raw material exporters and importers an exchanging of ideas takes place. This carries some potential for conflict.[320] The raw material dependent industrialised countries and emerging economies are striving for a long-term safe, adequate and preferably affordable supply. The raw material-exporting countries, however, are primarily concerned with stabilization and continued increases of sales revenues from the export of raw materials and the provision of investment capital for mining projects. The resource-poor developing countries have to rely on international assistance, because they would lack, otherwise, the basis for industrialisation.[321]

The conflict potential can be met by comprehensive planning in advance. The search for **prize mechanisms capable of compromise** is in this respect an essential challenge of an international raw material policy. This has to be seen against the background of an inequitable consumption. The rich world uses far more minerals per head of population than the poor world. For example, in a 77-year lifespan, the average North American will consume around 600 tonnes of primary aggregate and 550 tonnes of fuel, while the average 77-year-old Ethiopian, Bangladeshi or Nepali will have consumed less than 5 tonnes of these commodities.[322] In spite of the unequally distributed demand for minerals, many of the richer countries, particularly the more densely populated areas, are drafting increasingly prescriptive and constraining planning policies and regulations.[323] Environmental designation of areas often excludes mining options. In other words: there is a remarkable demand for raw materials in

---

320 Cp. Bolz, Reinhardt [Hrsg.] (1975): Kooperation oder Konfrontation?: Materialien zur Rohstoffpolitik. – Bonn: Progress Dritte Welt, 1975. [Cooperation or confrontation?: Materials for the Raw material policy. – Bonn: Progress Dritte Welt, 1975] Streit, Manfred E. (1975): Einige alte Überlegungen zu neuerlichen Schwierigkeiten in der internationalen Rohstoffpolitik. – Mannheim: Institut für Volkswirtschaftslehre und Statistik der Universität Mannheim, 1975. [Some early considerations about recent difficulties in the international raw material policy. – Mannheim: Institute of National Economics and Statistics of the University of Mannheim, in 1975.] Donges, J. B. (1976): Kritik der Pläne für eine neue internationale Rohstoffpolitik, Inst. für Weltwirtschaft an der Universität Kiel, 1976. [Criticism of the plans for a new international raw material policy, Inst. for World Economy at the University of Kiel, in 1976]
321 Gocht (1983), l.c., p. 205.
322 Ibid.
323 C.f. EU situation, shown below.

countries, while at the same time, mineral production is shifted to third countries and thus raw material imports increase. This is obviously an inconsistency in raw material distribution policy.

An important objective of an international minerals policy must be to **avoid conflict between producing and consuming countries** over distribution of mineral commodities by setting up a distribution system acceptable to all parties. The latter includes stable production and supply, meeting demand requirements, competitive prices and trade.

4.4 Approaches towards a minerals policy

Security of minerals supply means minimizing economic costs of raw materials imports as well as maximizing profit from domestic raw materials extraction and processing. These **two objectives need to be balanced** in order to avoid shortages of supply and support competitiveness of domestic mining. Adequate strategies, measures and a coherent conception are required.

4.4.1 Strategy

A strategy is a planned long-term intention for an advantageous situation or goal. It aims at the appropriate use of certain means, generally referring to some subordinate objective. Strategy is the "greater plan" or a "basic pattern of actions".

Once the objectives[324] are set and the starting point is clear, the basic conditions for programming measures which promise to achieve the objectives are given.[325] This is the central step in the economic political decision-making process. Basis of the pertinent analysis are the stringent procedural **relations between the means and the objective**, as derived from the general economic theory according to the principle of cause and effect. The highest level of success in achieving the objectives has to be determined by comparison of possible alternatives. Often the optimal measure is a "programme" which uses various means chronologically graduated (**economic political strategy**).[326]

---

324 Here: Securing raw material supply of a state/confederation.
325 Molitor (2006), l.c., p.37.
326 Molitor (2006), l.c., p.37: Whether it be a programme or a single means: the measure with the highest degree of objective rationality is still subject to the proviso that it is compatible with this basic decision for a particular type of economic system, which in our case is with the functional requirements of a market economy regulation.).

4.4 Approaches towards a minerals policy

The general economic political and therefore raw materials political objectives of most **raw materials exporters** are mainly growth-oriented. The strategies for achieving the objectives aim at activities for exploring new deposits through the promotion of prospection and exploration activities.[327] This affects development policy, Geological Surveys and international mining companies. Furthermore, the establishing of an attractive investment/fiscal policy (mainly for foreign investors) should be mentioned.

Strategies regarding raw materials importers

Security of supply for raw materials importers can be realized through: **Domestic economic strategies**, i. e. *raw materials planning* strategies and strategies concerning the *acquisition and usage of raw materials,* through diversification of sources, optimal use of domestic resources, development of cost-saving processing technology and the promotion of recycling and substitution. **Foreign economic strategies,** i. e. strategies for the legal and political coverage of the acquisition of raw materials for the

- maintenance of a permissive **international raw materials trade** and money transfer,
- attainment of **cooperation structures** between industrial countries poor in raw materials and raw materials exporting developing countries, for setting up international cooperation structures between Geological Surveys and research institutions.[328]

Raw materials strategies worth mentioning are pursued by the USA, Russia, Japan and China.[329]

---

327 Gocht (2006), l.c, p. 83, 202–203: Export of mining and mineral oil products is the basis of development finance, the foreign exchange earnings contributing to the stabilisation of the state budget and the implementation of investment programmes. This fundamental significance is attached to mineral exports by the mining developing countries and the smaller group of resource-rich industrialised countries (Canada, Australia, South Africa).
328 Cp. Gocht (2006), l.c., p. 83, 228 f.
329 Michaelis (1976), l.c., p. 39f. OECD (1994): Mining and Non-Ferrous Metals Policies of OECD Countries, Paris (Japan: p. 133ff; USA: p. 219ff). Further relevant mineral strategies are for example: Department of Minerals and Energy, Republic of South Africa (1998): A minerals and mining policy for South Africa. October 1998, Pretoria. http://www.dme.gov.za/minerals/min_whitepaper.stm; China's policy on mineral resources (2003). Chinese Government's Official Web Portal gov.cn, Official publications 2003; India/Ministry of Mines): National mineral policy, 1993, for non-fuel and non-atomic minerals. (Last updated 29/11/2004). http://mines.nic.in/nmp.html; Tanzania National Website [Mineral policy of Tanzania] http://www.tanzania.go.tz/mining.html; Jamaica/Ministry of Agriculture and Lands: The National minerals policy: ensuring a sustainable minerals industry. 2nd draft (for discussion purposes), August 2006. www.

The **United States** have pursued an exemplary strategy of securing their supply with raw and base materials since the 70s. Traditionally, this policy aims at

- creating or maintaining a capacity of producing raw materials and base materials which is sufficient to withstand a political or economic pressure from outside, this policy also includes a limited protection of their own resources; the so-called national defense stock pile, a system of strategic mineral reserves

- comprehensive politically-based efforts to obtain the best possible access to supply sources in other countries, particularly by promotion of direct investments of American companies (on the part of the State) for prospecting, exploration, and exploitation of foreign mineral deposits, which includes an expensive geographical diversification of supply sources.[330]

---

moa.gov.jm/land/minpolicy/national_min_policy_draft_2_sep06.pdf; Government of Canada: The minerals and metals policy of the Government of Canada. Executive summary http://www.nrcan.gc.ca/mms/policy/mmp-e.pdf; Government of Malaysia: National mineral policy. Mineral Sector mission statement (1994) http://www.jmg.gov.my/files/NATIONAL_MINERAL_POLICY.pdf; Republic of Uganda/Ministry of Energy and Mineral Development: The mineral policy of Uganda. September 2000. Executive summary. www.energyandminerals.go.ug/minpol00.pdf; Deutsche Bundesregierung (2007): Elemente einer Rohstoffstrategie der Bundesregierung, Berlin. German Federal Government: Elements of a mineral strategy of the Federal Government.

(In addition, raw material strategies of further countries are also to be mentioned: Department of Mineral and Energy, Republic of South Africa (1998): A minerals and mining policy for South Africa. October 1998, Pretoria http://www.dme.gov.za/minerals/min_whitepaper.stm; China's policy on mineral resources (2003). Chinese Government's Official Web Portal gov.cn, Official publications 2003; India/Ministry of Mines): National mineral policy, 1993, for non-fuel and non-atomic minerals. (Last updated 29/11/2004). http://mines.nic.in/nmp.html; Tanzania National Website [Mineral policy of Tanzania] http://www.tanzania.go.tz/mining.html; Jamaica/Ministry of Agriculture and Lands: The National minerals policy: ensuring a sustainable minerals industry. 2nd draft (for discussion purposes), August 2006. www.moa.gov.jm/land/minpolicy/national_min_policy_draft_2_sep06.pdf; Government of Canada: The minerals and metals policy of the Government of Canada. Executive summary http://www.nrcan.gc.ca/mms/policy/mmp-e.pdf; Government of Malaysia: National mineral policy. Mineral Sector mission statement (1994) http://www.jmg.gov.my/files/NATIONAL_MINERAL_POLICY.pdf; Republic of Uganda/Ministry of Energy and Mineral Development: The minerals policy of Uganda. September 2000. Executive summary. www.energyandminerals.go.ug/minpol00.pdf; German Federal Government (2007): Elements of a commodity strategy of the Federal Government to Berlin.)

330 C.f. also the service of the United States Geological Survey, in particular the worldwide country reports (as they are applied in chapter 5). – Cf: Long, Keith R., Van Gosen, Bradley S., Foley, Nora K., Cordier, Daniel (2010): The principal rare earth elements deposits of the United States – a summary of domestic deposits and a global perspective. U.S. Department of the Interior, U. S. Geological Survey, Scientific Investigations Report 2010–5220 – [Anonymous] (2011): Rare earths are not so rare ; USGS determines

4.4 Approaches towards a minerals policy

The overall strategy to reduce US reliance on uncertain sources of supply of strategic materials is based on a combination of three approaches:[331]

- increase the diversity of world supply of strategic metals through exploration and development of promising deposits, both foreign and domestic;
- decrease demand for strategic metals through the implementation of improved manufacturing processes and recycling of strategic materials from scrap and waste;
- identify and test substitute materials for current applications and develop new materials with reduced strategic material content for future applications.

Compared to the EU, the United States have other advantages which bring them in a better position to effectively secure their mineral supply. Those are: less dependency on raw material imports, an economic potential which allows to develop cheaper local resources and create large stocks, the performance of multinational companies which have their decision-making centres in the United States, the existence of a "decision making authority" with full powers centralized (in this case, the Ministry of Defense) in Washington.

The Russian Federation has developed a remarkable resource strategy. With the "Governmental Ordinance No. 494-r of 2003 regarding validation of the **Bases of state policy in the sphere of utilization of mineral raw materials and the use of subsoil**" the basis for strategic planning and use of the mineral raw materials is created[332]. The Ministry for Natural Resources together with further national organs is responsible to establish a fundamental strategy. The main tasks lie in securing raw materials as well as in the efficient raw material management for the sustainable development of the Russian economy, as well as intensified use of domestic deposits to the benefit of present and future generations. An essential element of the strategy is the preservation of the geopolitical interests.

---

that there are 14 rare earth deposits in the US. Industry Newswatch, Mining Engineering, Vol. 63 (2011), 1, p. 26 – [Anonymous] (2011): Mining contributes high wages; New report shows nearly 1.8 million owe their jobs to mining. Industry Newswatch, Mining Engineering, Vol. 63 (2011), 1, p. 10.

331 OECD, (1994), l.c., p 237f.
332 See Annex and http://faolex.fao.org/docs/texts/rus40699.doc. Cp. in this context also Clement, H. (1981): Minerals requirements in the Soviet Union, 1981–1990, The Johns Hopkins Foreign Policy Inst.]. – Cp. also: Morozov, A. F., Lipilin, A. V., Petrov, O. V., Kiselev, E. A., Feoktistov, V. P. (2007): State of predicting resources of mineral raw materials in Russian Federation [in Russian]. Gornyj zurnal, Moscow, Vol. 183 (2007), 10, pp. 47–51.

Countries such as Japan, but now even China, have become importers of raw materials and are dependent on raw material deliveries for the supply of industrial production. In this sense, the raw material strategies of Japan and China are worth mentioning.

Japan, which, because of it unfavourable geologic structure, depends more than other countries on imports of mineral resources, has aimed for years to gain a foothold in developing countries and to develop joint venture structures to become an equal partner in the international mining companies, to develop marine transport and terminal equipment and thus reducing the load on inputs. The responsible Japanese Ministry achieved these goals in an exemplary collaboration with the industry under the motto **"on the domestic market competition and outwards coordination of actions"**.[333] Japan's economic security relies on the stability of its foreign resource supplies. Defining the national goal as comprehensive security including mineral security, the main objectives of the country's minerals policy are the following: secure stable sources of minerals; systematically develop domestic mineral resources; actively promote development of overseas mineral resources through economic co-operation with mineral-rich developing countries; and stockpile rare metals.[334] Japanese authorities administer their overseas exploration and development assistance programmes so as to diversify sources of supply of minerals and metals. The aim is to increase the number of countries supplying a particular mineral to Japan and to diversify sources among the greatest number of countries.[335] The commercial agreement between Japan and Australia from 2005 has to be seen against this background: Japan, with 40% of iron ore imports and 30% of aluminum imports, is a major market for Australia in the mid- and long-term.[336]

Table 21: Japanese import dependence (Crowson, 2008)

	1979/80	1988	1998	2006
Aluminium	70	65	72	69
Arsenic	na	19	95	66
Barytes	36	100	100	100
Bismuth	na	38	83	71

---

333 OECD (1994), l.c., p. 147. See also: www.meti.go.jp/english/press/data/nBackIssue 200803.html.
334 OECD (1994), l.c., p. 135.
335 OECD (1994), l.c., p. 147.
336 Australia-Japan Free Trade Agreement – The Minerals Industry Case, October 2005; Submission to the Department of Foreign Affairs and Trade.

4.4 Approaches towards a minerals policy

	1979/80	1988	1998	2006
Cadmium (refinded)	–	66	61	43
Copper	80	97	88	88
Germanium (refinded)	100	45	na	na
Graphite	na	na	100	100
Indium	na	52	86	na
Industrial diamonds	na	na	67	33
Iodine	na	na	6	–
Kaolin	na	na	94	99
Lead	47	67	52	41
Magnesium metal	9	44	63	100
Mercury	28	17	24	na
Selenium	100	16	2	3
Silicon	43	100	100	100
Silver	57	72	na	na
Soda ash	na	na	33	39
Sulphur	–	10	1	–
Talc	na	na	75	44
Tellurium	na	3	29	na
Tungsten	85	94	100	100
Vanadium	100	100	90	na
Vermiculite	na	na	91	98
Zinc	48	77	59	75

The raw material strategy of **China** focuses on increasing the domestic capability of mineral resources supply. The intention is the promotion of exploration, and also the development of a competitive mining industry. China encourages foreign businesses to invest in prospection and exploitation of mineral resources in the country. Domestic mining enterprises are to cooperate with international mining companies, draw on advanced international experience, import advanced technology and operate in accordance with international practices. Furthermore, China is consequently acquiring external mining rights.

Chapter 4 The concept of a minerals policy

Figure 73: Chinas mineral strategy (Data by Ekdahl 2008)

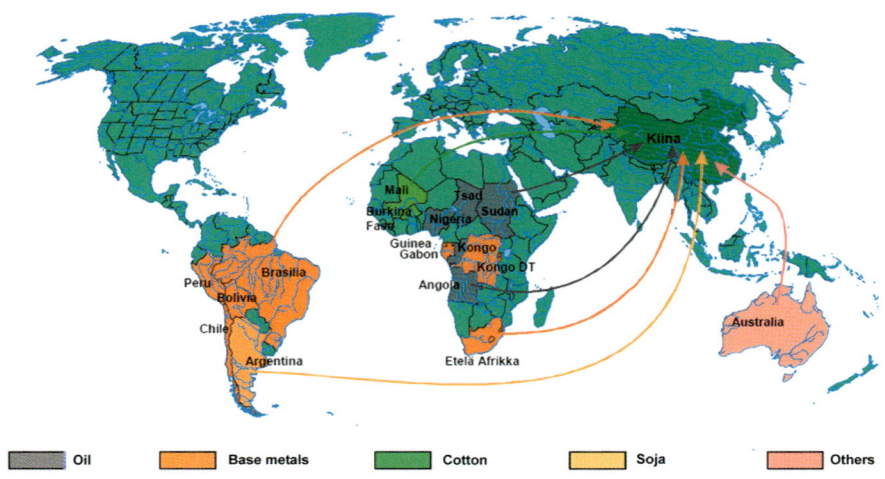

Oil | Base metals | Cotton | Soja | Others

Ad figure 73: Import of China from Africa (2005)

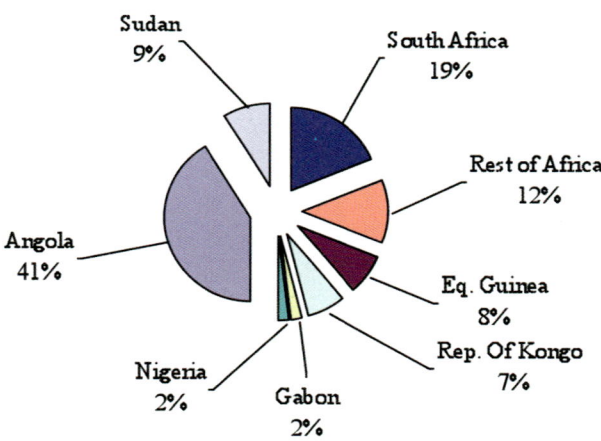

4.4 Approaches towards a minerals policy

Figure 74: "Macro-control" of China by state interventions and commercial restrictions at the example of the added value chain of the copper production (Data by BDI 2007 [Eurométaux])

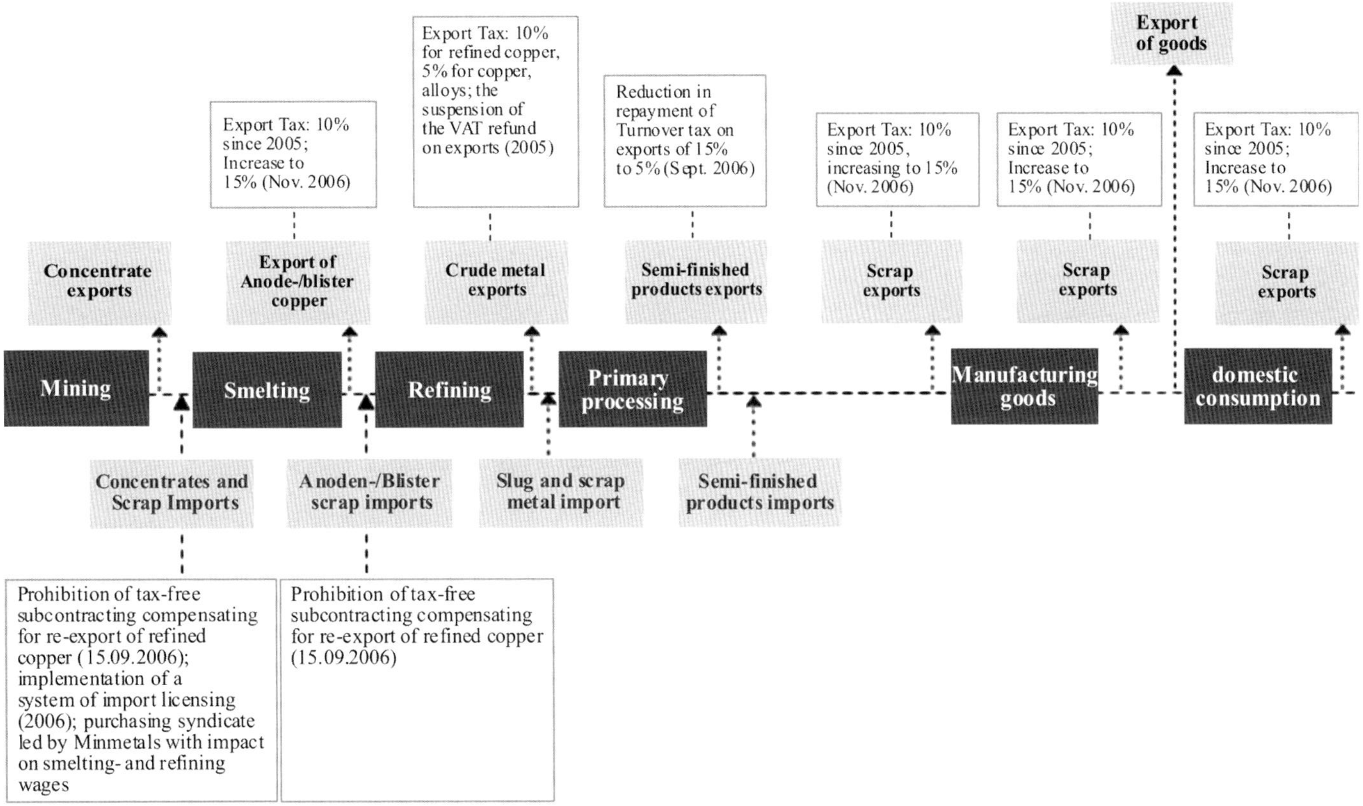

4.4.2 Instruments

The instruments for regulations and **measures** for securing supply with raw materials are extensive. Instruments for the **realization of raw material political objectives** of a state are as diverse as the objectives themselves, ranging from market observation to public enterprises.[337] Below reference will be made to Siebert[338] and Michaelis.

General categorization

Siebert categorises the instruments according to the variables they apply to (table 22):

---

[337] Cp. Gocht, W.R., H. Zantop and R.G.Eggert. International Mineral Economics. Mineral Exploration, Mine Valuation, Mineral Markets, International Mineral policies. Springer 1988. Baron, S., Glismann, H., Stecher, B. (1977): Internationale Rohstoffpolitik: Ziele, Mittel, Kosten. – Tübingen: Mohr, 1977 (Kieler Studien; 150) [International minerals policy: Objectives, means, costs – Tübingen: Mohr, 1977 (Kieler Studien; 150)]. Steger, W. (1980): Ziele und Instrumente internationaler Rohstoffpolitik. – Wien, Wirtschaftsuniversität, Dipl.-Arb., 1980 [Objectives and instruments of international mineral policy. – Vienna University of Economics, diploma thesis, 1980]. Bundesverband der Deutschen Industrie/Arbeitskreis Rohstoffpolitik (1981): Rohstoffversorgungspolitik – Ziele und Instrumente: Fachtagung des Arbeitskreises Rohstoffpolitik des Bundesverbandes der Deutschen Industrie, Köln. [Federation of German Industry/Working Group raw material policy (1981): raw materials supply policy – objectives and instruments: Symposium of Working group raw material policy of the Federation of German Industry, Cologne, Germany.].

[338] Siebert (1983), also additions by Brandstätter 1989, in: Brandstätter (1989)] Cp. also Bilardo, U., Mureddu, G. (1983): Situazione e prospettive del riciclo dei metalli. Energia e materie prime, no 33–34, p 43–53.

4.4 Approaches towards a minerals policy

Table 22: Instruments of minerals policy according to Siebert (with supplements of Brandstätter and Tiess) (Data by Brandstätter, 1989)

Regulative instruments	Instruments regulating sales volume	Specific and ad valorem taxes	Measures regulating profit through taxation	Measures regulating price (prices policy)	Direct interference by the state	Subsidization of alternative technology
1 Mineral right, definition of usage rights for state-owned minerals	1 Licensing strategies	1 Extraction tax as specific tax	1 Tax on profit	1 Price Control	1 Production quota	
2 Protection of deposits and land-use planning (raw materials planning policy)	2 Auctioning strategies	2 Extraction tax as ad valorem tax	2 Taxation of interest income	2 Regulating return	2 Regulations of use for the consumers	
	3 Stockpiling	3 Consumption tax			3 Rationing	
4 Markets (e.g. competition policy, foreign economic aspects)	4 Mineral buffer stock	4 Export tax	4 Depreciation modalities		(4 Public enterprises)	
	5 Production and import quota	5 Import tax	5 Capital profit			
		6 Subsidies	6 Land value tax			

Specific categorisation of instruments

According to the **sort of instruments and the objectives** set, the following categories of interventions can be distinguished:[339]

(a) Domestic economic interventions with intended primary effects on domestic production and processing

Subsidiaries, tax relieves, cut-rate loans, loan guarantees, transport subsidies, absorption of deficits by the tax authorities etc., promotive valorisation, guarantees for domestic disposal, buying up or funding storage of surplus goods;

(b) Foreign economic interventions with intended primary effects on domestic production and processing

(aa) *Import restrictions*
Customs duties and import levies, import quotas, tariff quotas and other quantitative restrictions, other non-tariff barriers;

(bb) *Measures promoting export*
Export facilitation through subsidiaries, particularly export refunds, cut-rate loans, promotive exchange rates, etc;

(b) Domestic economic interventions with the intended primary effect of secured provision or rational use of raw materials
1) Protection of deposits by minerals planning policy
2) Recycling of raw materials
3) Measures aiming at rational use of raw materials in production
4) Measures restricting consumption of raw materials and goods produced thereof
5) Building up strategic reserves

(c) Foreign-trade interventions of the consumer countries to ensure access to certain raw materials. – Interventions of raw material producing countries for encouragement of the domestic resource economy and the promotion of their products

- Royalties, export duties and similar,
- Production and export restrictions, "buffer stocks" and similar,
- Measures for the promotion of vertical diversification,
- Indexing of price,
- Producers' alliances in terms of cartel agreements on prices or quantitative restrictions,

---

[339] Michaelis, H. (1976), l.c., p. 23f.

4.4 Approaches towards a minerals policy

Nationalization of extraction licenses, raw materials occurrences and extraction, processing and transport facilities;

(d) *Interventions by consuming countries to improve access to raw materials and processed products in raw material producing countries* (both as a unilateral action of one or more consuming countries, and because of agreements) with producing countries

- Measures for the liberalization of global commodity markets,
- Autonomous customs duty suspensions and import benefits (subventions, cut-rate loans, etc.),
- Simplifying monetary transactions,
- Creating multilateral financial plans,
- Cooperation with Geological Surveys,
- Promoting storage of raw materials liable to supply problems,
- Promoting initiatives for the development and the extraction of raw materials in producing countries and for the transport of these raw materials,
- Long-term agreements on the supply with raw materials,
- Bilateral agreements on trade and cooperation with developing and state-trading countries,
- "Production sharing" with developing and state-trading countries, international raw materials agreements involving both producing and consuming countries,
- Pooling of co-businesses,
- Possibly a global adjustment with developing countries on the distribution of activities concerning processing and transport of raw materials.

Ad b): **Minerals planning policy**

Minerals planning policy (i. e. raw materials planning policy) is an essential part of minerals policy.[340] **Minerals planning policy is defined as the**

---

[340] From the published literature on minerals planning in Europe follows that minerals planning is commonly seen within the context of land use planning. Cp. Department of the Environment (1995): Minerals planning policy and supply practices in Europe: main report, London. Mineral planning in a European Context – Demand and Supply, Environment and Sustainability. Proceedings of the 1st European Conference on Mineral Planning. Zwolle, Netherlkands, 1997. Geopress 1998. Mineral Planning in Europe. Proceedings of the 2nd European Conference on Minerals Planning, Harrogate, UK, 1999. The Institute of Quarrying 1999. ECMP3, 2002. Sarajevo 2004 und 2006.

protection of mineral deposits through land-use planning. It is particularly responsible for the availability of and access to deposits.

A successful minerals planning policy should create the political, legal and administrative conditions which are necessary to ensure the supply with minerals to society within the framework of sustainable development.[341] All three components are considered to be equally important. Minerals Planning Policies which create an environment of conflict, result in the sterilisation of strategic mineral reserves, lead to an unsustainable minerals industry and result in a shortage in minerals supply, are unlikely to achieve their aim.

Raw materials exploration and national inventarium of the deposit potential including site categorization by the appropriate professional organs of a state are essential elements of a raw material planning policy. A definite goal is locating the deposits and providing with information on quantity, quality and viability of the deposits. At this stage, the role of the State (geologic service, raw material ministry) is vital (aspects of an active raw material planning policy). The State may provide support by funding the exploration, but it can also participate operationally. As already noted above: The financial risk when prospecting for a mineable deposit is relatively high. In Japan, for example, the State has funded exploration programs since 1966 for companies under participation of research institutions. In Finland and Sweden, the Geological Service is fully involved (see Chapter 5).

Importance of land-use planning

Within land-use planning, different types of use are considered. This takes sufficient information land-use the different special fields. The extraction of mineral raw materials is of great importance with respect to land-use and en-

---

341 Cp. Tiess, G., Rossmann, H., Pilgram, R. (2002a): Die Bedeutung des Vorsorgeprinzips bei der Gewinnung mineralischer Baurohstoffe, Teil 1 – in: Recht der Umwelt H 3, S. 84–92 sowie Tiess, G., Rossmann, H., Pilgram, R. (2002b): Die Bedeutung des Vorsorgeprinzips bei der Gewinnung mineralischer Baurohstoffe, Teil 2 – in: Recht der Umwelt H 4, S. 130–136. [The relevance of the precautionary principle in the extraction of mineral construction minerals, Part 1 – in: Right of the Environment H 3, p. 84–92 also Tiess, G., Rossmann, H., Pilgram, R. (2002b): The relevance of the precautionary principle in the extraction of mineral construction minerals, Part 2 – in: Right of the environment H 4, p. 130–136.] Also Wellmer, F. W. (1996): Resource development, land-use planning and sustainability in Germany. Journal of Applied Geology, 42.1996, H.1, p. 62–65. Dingethal, F. J. (2002): Recent Developments in Raw Material Security, Erzmetall p. 247–253]. – See also: Moore, D. J., Tilton, J. E., Shields, D. J. (1996): Economic growth and the demand for construction materials. Resources policy, Vol. 22 (1996), 3, pp. 197–205.

4.4 Approaches towards a minerals policy

vironment.[342] Deposits of mineral raw materials feature three characteristics, namely their location-bound, regional and exhaustible nature (see Chapter 1). Due to these properties the existence of deposits and their extent and constitution need to be determined as soon as possible. Whether or not a deposit can be used later on is a question of land-use planning under consideration of all other aspects. The earlier and more complete information on deposits is collected, the better the solution of possible use conflicts in land-use planning will turn out. Thus, the cooperation of Geological Surveys with public land-use planning is of great importance.

All EU member states develop land-use plans which are in line with national land-use planning principles.[343] According to the principle of planning hierarchy raw materials principles should be considered comprehensively on lower planning levels (regional and local level, operational level).[344]

The characteristics of mining mentioned above cause some implications on land-use and land-use planning. Due to the he exhaustibility of deposits, mining sites are only used temporarily and become available for other use again after the termination of mining activities and recultivation.

Figure 75: Quarry Rotzloch, Alpnachersee/Switzerland (Data by Kündig et al., 1997)

---

342 Tiess, G. (2005): Sustainable supply of the European industry and society with minerals: importance of the non-energy extractive industry, Berg- und Hüttenmännische Monatshefte, S. 415–423 (Journal of Mining. Metallurgical, Material, Geotechnical and Planned Engineering, p. 415–423).
343 A comprehensive presentation of the various land-use planning systems is found in: Department of Mining and Tunnelling, University of Leoben (2004), l.c.
344 Cp. Roberts, P. W., Shaw, T. (1982): Mineral resources in regional and strategic planning, Aldershot: Gower Technical. – The fact that this principle is mostly not observed, is a general problem for the mineral matter: Cp. Department of Mining and Tunnelling, University of Leoben (2004), also in this sense: Chapter 5.

Since mining activities are bound to the location of deposits, potential conflicts with other claims of use of deposits arise. When weighing interests of several parties it should be considered that the possibility for mining to fall back on other sites or areas is very limited, compared to other types of use. From this fact, together with the temporariness of mining due to exhaustion, follows the economic demand for prioritizing mining interests when talking about subsequent land use.[345]

In this connection, extraction of construction raw materials should be mentioned in particular. Construction minerals in general are regionally more evenly distributed and therefore usually closer to the places of consumption. What is special about this group of raw materials is the vast periodic consumption together with an accordingly high volume of transport. In Austria, for example, more than a half of the total traffic volume on the road network as the carrier of traffic is allotted to raw and construction materials. This means that every other ton of goods transported on the road belongs to the group of raw and construction materials.[346]

Moreover, construction minerals are mainly extracted in surface mines, which significantly increases the area of land needed compared to other raw materials. The demand for land together with the transported quantities leads to an increased strain of the environment. For construction minerals needed in large quantities, the availability of deposits near the place of consumption is a significant criterion for the reduction of transport as well as negative effects on the environment.[347]

---

345 Nötstaller (2003a), l.c., p. 27.
346 Cp. Drnek, T. (1995), Die wirtschaftliche Bedeutung der Steine und Erdengewinnung in Österreich, Berg- und Hüttenmännische Monatshefte, p. 447–453. Wagner, H. und R. Nötstaller. The economic significance of industrial minerals extraction, in Journal of Mining. Metallurgical, Material, Geotechnical and Planned Engineering 142. Jg. (1997). P. 339–349.
347 Nötstaller (2003a), l.c., p. 28. Cp. also Tiess, G. (2000): Rohstoffgewinnung im Spannungsfeld, Sand & Kies, H 48. [Raw materials in areas of conflict, Sand & Gravel, H 48]

Figure 76: Construction raw materials mining in Switzerland (Data by Kündig et al., 1997)

The necessity of mining sites near the places of consumption and the increased need of area make this group of raw materials a particular challenge in land-use planning and raw materials planning policy. In order to minimize use conflicts and to avoid possible shortages of construction raw materials due to disadvantageous land-use planning, strategic long-term planning is needed. From this follows that for this group of raw materials the availability of long-term demand prognoses as a basis for macroeconomically advantageous land-use planning is of great importance.[348]

4.4.3 Conceptions

For the realization of raw materials related political goals, strategies (and measures) adequate conception (*"action plan"*) is needed.

A conception is a **comprehensive compilation** of objectives and subsequent strategies and measures for the **realization** of a higher goal. It includes all the information needed as well as **time, measures and resources plans**. Conceptions are usually put into writing and should be **checked for relevance** and topicality regularly.

The conception must be developed including all component policies relevant for the topic of raw materials and the interactions between them. These interactions need to be reasonable, especially for decision-makers. Time and measure plans have to determine at which point which measures have to be taken. Due to the complexity of the topic and various correlations between the policies this is particularly important. The complexity is furthermore relevant regarding resources planning, which can be illustrated using the example of Austrian raw materials planning (see Chapter 5).

---

348 Nötstaller (2003a), l.c., p. 28.

Table 23: Conception of a raw materials policy (the state as a raw materials importer)

Main objective	Overall strategy	Measures	Explanation
Sustainable supply with raw materials			
Subordinate objectives	Singular strategies	Examples	Examples
Supply	Exploration of domestic deposits. Securing access to raw materials in countries outside the EU	Promotion of exploration by the state. Intensified dialogue with international fora	Securing supply with raw materials through raw materials political measures. Securing access to those raw materials needed by the state's industry which are not available inland
Minimization of costs	Increasing competitiveness of the raw materials industry/mineral economy	More efficient licensing procedures	Time- and cost-intensive procedures weigh heavily on the entrepreneurs and lead to a delay of extraction. The raw materials industry is capital-intensive. Every single delay reduces the company's competitiveness.
Protection of mineral resources	Comprehensive raw materials research	Increasing raw materials and materials efficiency	Reduced raw materials and energy consumption result in the protection of primary raw materials and the promotion of environmental protection
Re-use of resources (Recycling)	Improving EU-framework	Unambiguous definition of the term of waste	Guaranteeing less administration effort for businesses and therefore a more cost-efficient recycling of raw materials
Ecology objectives	Efficient cooperation between environmental technology and the raw materials industry	Development of new products and materials	Environmentally damaging processes in mining are to be minimized, not only for the sake of the environment, but also for broader public acceptance.

The phase of implementation of a conception calls for an **effective consultation** process of the different function owners (e. g. geological survey and ministry of economics), where basic conditions such as deposits potential, raw materials criticality or the degree of industrialization of a state are to be considered. In the following some international examples are given, i. e. Phillipines, Jamaica, India, State of Rajasthan and Sierra Leone (further examples are listed in the Appendix):

Phillipines[349]

After the 9-month engagement process, the "National Policy Agenda on Revitalizing Mining in the Philippines" (Executive Order [EO] No 270) was issued in 2004.[350] It contained 12 guiding principles for responsible mining towards sustainable development. It also called for the formulation of a Minerals Action Plan (MAP) which **will detail the strategies and activities for the attainment of the goals** of EO No. 270.

12 Guiding principles of EO 270

Economic principles

1. Recognition of the critical role of investments in the minerals industry in support of national development and poverty alleviation goals;

2. Provision of clear, stable and predictable investment and regulatory policies to facilitate investments;

3. Development of downstream industries or value-adding of minerals;

4. Support to small-scale mining in order to rationalize their activities;

---

349 Department of Environment and Natural Resources (2004): NATIONAL POLICY AGENDA ON REVITALIZING MINING IN THE PHILIPPINES, Issuance of E.O. 270 or the National Policy Agenda on Revitalizing Mining in the Philippines was issued on January 16, 2004 and amended in E.O. 270-A on April 20, 2004 (http://www.mgb6.org/revitalization-of-the-minerals-industry/).
350 In 2003, the government made a policy shift from mere tolerance to promotion for the revitalization of the minerals industry. The government believed on the potential of the minerals sector to attract new investments, generate revenues for the government and provide additional jobs and livelihood opportunities. However, these should be anchored on the basic principles of sustainable development. Thus, the government undertook a social preparation process through the conduct of regional workshops and consultations, local and international mine visits and a national mining conference to thresh out the issues and concerns on mining.

5. Adoption of efficient technologies to ensure judicious extraction and optimum utilization of non-renewable mineral resources;

Environmental principles

6. Protection of the environment in every stage of mining operations;
7. Safeguarding the ecological integrity of areas affected by mining;
8. Pursuing mining within the framework of multiple land use;
9. Rehabilitation of abandoned mines;

Social principles

10. Ensuring the equitable of benefits among direct stakeholders;
11. Sustained information, education and communication ((IEC) programs and respect for the rights of the indigenous people and communities; and
12. Continuous and meaningful consultations with stakeholders.

Minerals Action Plan (MAP) of the Phillipines

The Minerals Action Plan was formulated by the Department of Environment and Natural Resources in *consultation with other government agencies and stakeholders* and approved by the President thru Memorandum Circular No. 67 of 2004. MAP contains 57 strategies and 126 activities to address the problems of mining. It is an *on-going implementation*, with regular review of status and accomplishments with various sectors of civil society.

Jamaica

A successful minerals industry in Jamaica shall be supported by a strategic direction that will lead to greater economic opportunities and higher levels of investment, over the long term.[351] The **broad goals** of this Policy are:

- An industry in which current investments are safeguarded, new investments are attracted and benefits are maximized in the interest of the national economy, local communities and the companies;

---

[351] Government of Jamaica, Ministry of Energy and Mining (2009): The National Minerals Policy, Sustainable Development of the Minerals Industry. (http://209.85.129.132/search?q=cache:JqNqtboC8TsJ:www.men.gov.jm/PDF_Files/Minerals_Policy/NATIONAL%2520MINERALS%2520POLICY%25207TH%2520DRAFT%2520JULY.pdf+Mineral+Policy+Jamaica&cd=11&hl=de&ct=clnk&gl=at).

- A minerals industry that contributes to sustainable national development and integrates the concept and principles of sustainable development in local and national decisions that affect the industry. This includes the effective management of mineral resources and mineral-bearing lands from the pre-mining to post-mining stage;
- A minerals industry that embraces and exemplifies environmental best practices, including the recovery of minerals and other products from mining waste;
- Increased mineral exploitation, production of value-added goods, exportation of minerals and mineral products, and the expansion, diversification and modernization of the Minerals Industry;
- A modern legislative framework and supporting institutions which enable continued development of the minerals industry;
- A strong and profitable industry which includes substantial local interest as outlined in the National Development Plan: 'Vision 2030 Jamaica, the National Industrial Policy and further enunciated in other national policy documents'.

Action plans of Jamaica

The Ministry with responsibility for the minerals portfolio shall develop **appropriate action** plans, which will outline the mechanisms by which the policy goals and **objectives will be achieved, and detailing the strategies**, roles and responsibilities, and *timeframe*. The Ministry shall be responsible for data collection and for evaluation of this policy at the end of the first year after its implementation. *Revision* of the policy and the status of its implementation will be conducted once every three years. A progress and analysis report with respect to its impact and achievements will be presented every year after its implementation.

India, State of Rajasthan

The basic objectives of the new minerals policy are:[352]

(a) To explore mineral wealth of the State expeditiously by adopting modern exploration techniques particularly in the tribal, desert and remote areas.

---

352 Govt of Rajasthan, Department of Information Technology & Communication, Rajasthan (http://www.rajasthan.gov.in/rajgovresources/actnpolicies/MINERAL.html).

(b) To exploit mineral deposits by promoting adoption of mechanised and scientific mining with due regard to the conversation of minerals mines safety and environmental aspects.

(c) Value addition through promotion of processing units and mineral based industries in the State.

(d) To encourage export of minerals having export potential.

(e) To promote development of human resources to meet the requirements of mining and mineral based industries.

(f) To de-mystify procedures and achieve greater transparency in decision making.

(g) To increase employment opportunities in the mining sector particularly, for persons belonging to Scheduled Castes, Scheduled Tribes and other weaker sections.

Strategic action frame work of Rajasthan

(a) To conduct developmental studies in the field of mineral exploration, mineral exploitation and mineral based industries including benefication of low grade minerals.

(b) To take effective measures for checking unauthorized mining and leakages of revenue.

(c) To simplify and adequately modify Minor Mineral Concession Rules to help achieve the objectives of the minerals policy.

(d) To ensure better mineral administration and for adequate delegation of powers to the State Government, reference shall be made to the Government of India.

The above objectives and the action framework of the minerals policy shall be achieved through the following measures:

- Mineral Exploration
- Mineral Administration and Development
- Mineral based Industries
- Export Promotion
- Information Dissemination
- Infrastructural Facilities
- Human Resource Development

Sierra Leone

The main objectives of the *Core Mineral Policy* of Sierra Leone are:[353]

- Review and amend the Mining Law, Regulations and Associated Laws to make them as attractive as possible for investment here rather than in neighbouring countries with similar mineral potential.
- Strengthen the Institutions that administer, regulate and monitor the mineral industry in Sierra Leone to allow the mining industry, especially with respect to the diamond industry to be turned around to become a positive for Sierra Leone;
- Develop and Strengthen Human Resources in the Minerals Sector.
- Attract Private Investments into the Minerals Sector. Encourage private investment to use the implementation of the *Kimberley Process* as a positive at the forefront of selling diamonds for peace and development properly registered by the Kimberley Process;
- Ensure that Sierra Leone's Mineral Wealth supports National Economic and Social Development
- Improve the Regulation and Efficiency of Artisanal and Small-Scale Mines
- Minimise and Mitigate the Adverse Impact of Mining Operations on Health, Communities and the Environment.
- Promote Improved Employment Practices, Encourage Participation of Women in the Mineral Sector and Prevent the Employment of Children in Mines.
- Add Value to Mineral Products and Facilitate Trading Opportunities for Mined Products.
- Improve the Welfare and Benefits of the Individuals and Communities Participating in and Affected by Mining.

Actions of Sierra Leone

The **implementation of the objectives and strategies** considered under the Core Mineral Policy requires *good partnership among* the Government, private sector, civil society and international organizations. The strategy and ob-

---

[353] Ministry of Mineral Resources (2003): Core Mineral Policy, Sierra Leone (http://docs.google.com/viewer?a=v&q=cache:iwastyxj1GAJ:www.daco-sl.org/encyclopedia/4_strat/4_2/mmr_mineralpolicy.pdf+Mineral+Policy&hl=de&gl=at&sig=AHIEtbTKrAHeOJovTjMWhsCPPadWubnHPA).

jectives outlined in the policy will determine the **activities that will be integrated into a comprehensive programme** for the development of the minerals sector in Sierra Leone. Implementation of the Core Mineral Policy will ensure the exploitation of mineral resources in the national interest and will bring improved economic and social benefits for the people of Sierra Leone.

Chapter 5 View of the minerals policies in selected states of Europe

The view of mineral policies is based on the principles discussed in Chapter 4.

5.1 General facts

The term "minerals policy" was already defined in chapter 4. As already mentioned, a **national minerals policy** can be defined as the **entirety of operations of a State for influencing supply of and demand for mineral resources on its territory and beyond that**.

Every nation pursues a fundamental minerals policy. In some states this has been established and published in an official document, whereas in many others investors and undertakers must use and interpret diverse sources of information about the essential aspects of the minerals policy. Examples of minerals policies that take the form of an official document can be found in Canada, India, Malaysia, Pakistan, Sierra Leone, etc.[354] A published, **clearly defined national policy is a very useful regulatory tool that serves two important functions**. Firstly, it provides the mineral industry with a clear statement of the government's expectations and intentions towards the mining activities. Secondly, it provides legislative and regulatory bodies with broad guidance.[355]

It is important that a clearly structured national minerals policy should be written down in a unified document.[356] Ideally, this should be ratified by the responsible minister, as the example of South Africa shows:

Minerals and Mining Policy for South Africa 1998 (Data by http://knowledge. uneca. org/community-of-practice/nepad-regional-integration-and-trade/natural-resources-managment/international-study-group-isg-to-review-africas-mining-codes/mining-codes-in-african-countries/south-africa-mining-code/Minerals_and_Mining_Policy_for_South%20Africa. PDF)

---

354 See Appendix.
355 Otto (1999), l.c., p. 3. See also appendix.
356 For example Canada, South Africa, USA, China and India. Cp. also National Economic and Social Council <Ireland> (1981): National Minerals policy. – Dublin: Stationery Office, 1981 (National Economic and Social Council publications; 60).

Introduction

Chapter One: Business Climate and Mineral
1.1 Investment and Regulatory Climate
1.2 Taxation
1.3 Mineral Rights and Prospecting Information
1.4 Small-scale Mining
1.5 Mineral Beneficiation
1.6 Minerals Marketing
1.7 Research and Development

Chapter Two: Participation in Ownership and Management

Chapter Three: People Issues
3.1 Mine Health and Safety
3.2 Human Resource Development
3.3 Housing and Living Conditions
3.4 Migrant Labour
3.5 Industrial Relations and Employment Conditions
3.6 Downscaling

Chapter Four: Environmental Management

Chapter Five: Regional Co-Operation

Chapter Six: Governance
6.1 Regulation and Promotion
6.2 National and Provincial Governments and Municipalities
6.3 Stakeholder Consultation

Such a document should be signal of a **consensus policy**. Appropriately, it is elaborated together with the relevant stakeholders and thus is based on mutual consent.[357] The result of such a **consultation process** should at best be a completed document that is published and has a certain binding character.

All processes and agreements concerning a minerals policy should have their fundamental base in this document. Thus, a framework of conditions can be created to ensure not only sustainable, but also a regular supply – for all affected stakeholders.

The goals specified in the document should be implemented on the basis of a corresponding raw material strategy and a coherent conception.

---

357 In South Africa the dialog has lasted from 1995 to 1998. In 1998 the ratification of the document "Minerals and Mining Policy for South Africa" was carried out by the Minister for Raw Materials and Energy. See also appendix.

Elements of national minerals policy

The state should create the framework to guarantee a sustainable supply of the economy with mineral raw materials. Substantial elements of national minerals policy are:

General aspects

Apart from providing a secure mineral supply, **measures to reduce consumption of raw materials** are of great importance (i. e. protection of primary mineral reserves). In this connection, **increasing of raw materials efficiency** is also a relevant target. Careful use of raw materials by reducing the intensity of raw materials applies the principle of sustainability.[358]

The aspect of non-renewability should be taken into account by recycling and substitution. The example in Figure 77 illustrates the scale of imports of bauxite and aluminium into the EU, but also the contribution that domestic sources of bauxite and recycling of aluminium make to the total supply of aluminium used by EU fabrication plants. It also illustrates the value chain from mining through refining to metal production and fabrication.

Figure 77: Worldwide supply chain of raw materials and metals for the production of aluminium products in the EU – Sources: European Aluminium Association (EAA) and Organization of European Aluminium Refiners and Remelters (OEA) (Data by EC, DG Enterprise and Industry, Commission Staff Working Document, 2007)

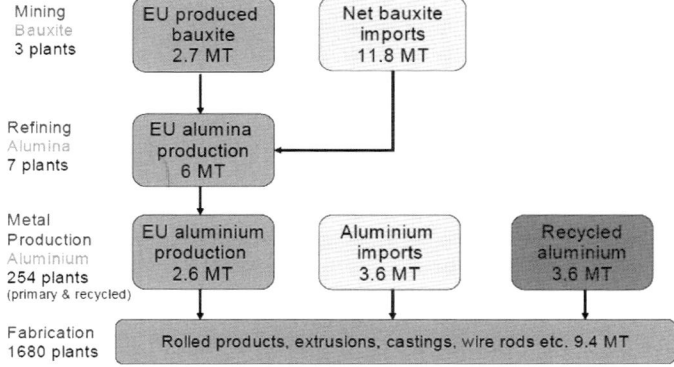

---

[358] Cp. in that regard Weizsäcker, E.U. (2010): Factor Five. Transforming the Global Economy through 80 % Improvements in Resource Productivity, London.

Furthermore the replacement of rare raw materials by general raw materials has to be promoted: e. g. filler materials for the paper and plastics industries, substitution of copper by glass fibres. Using of renewable materials (like timber in construction industry) should be preferred.

Specific aspects for raw materials

A national minerals policy first must provide a "mineral statement" (see figure 78), where two crucial issues have to be included: The first issue is that a national minerals policy first must create the **awareness** of society's needs for minerals.

The second crucial issue is that must it set the supply of minerals, covering all relevant mineral categories (of a country as a force for the benefit of society), and that it sets a **balanced approach** in the assessment of exploration and development of extractive activities (i. e. access to mineral resources) compared to other relevant issues (e. g. environmental issues) in a country *respectively* securing access to mineral resources *outside* a country. The **situation of demand and supply of a nation must be clear**. Knowledge of the deposition potential, critical raw materials, demand forecasts, export and investment intention depending of national mineral economy are of importance. In that regard, the need of a **raw material balance system** is crucial, the responsibility to develop such tool rests with the main responsible mining authority of a state (e. g. ministry of economics). A raw materials balance system shall be based on (amongst others) the analysis of demand and supply situation, *material flow analyses* and focus the *balance between demand and supply* of minerals needed by the state's economy. With regard to the supply offer, it must take into account the primary degree of self-sufficiency, degree of self-sufficiency considering domestic recycling, the technical raw material import dependency, the economical raw materials import dependency and the recycling rate. Based on such raw materials balance system, the state may consequently act in terms of **influencing the (sustainable) secured supply of mineral resources** required from the economy and may initiate **different policy actions**.

Of great importance is the legal base of the raw materials industry, in particular regarding interests of natural and environmental protection, and aspects of health and security. Also fiscal aspects play a role (taxes and fees). The national minerals policy should take into account the **predicted medium to long-term demand for minerals, ensuring that there is a sufficient 'supply offer'** (primary and secondary mineral sources, additionally considering possibilities like substitution etc.) *including* access to minerals outside of a country.

Figure 78: National minerals policy – schematic diagram

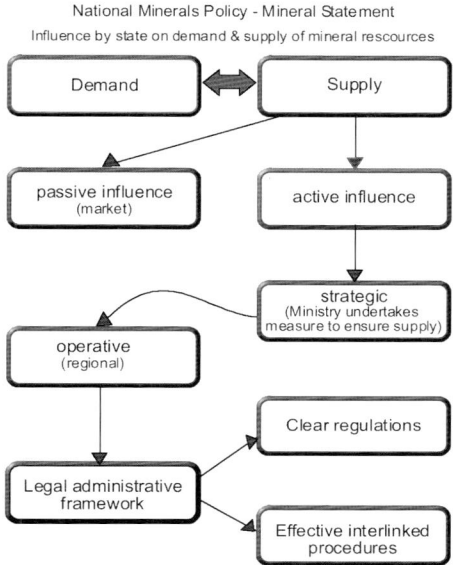

As it was noticed in chapter 4: The state can promote active or passive minerals policy. An **active minerals policy of a state** creates concrete basic conditions for the protection and the use of its raw material potential; and thus, also a cost effective contribution to the Gross Domestic Product.

No country in Europe is able to meet the demand of minerals without relying predominantly on primary resources. Access to primary mineral resources has to be secured for the operator in a long-term, mid-term and short-time perspective based on land use management. Operators need to have investment security as they have to plan their activities in a long-term. Land use management particularly is a responsibility of land use planning and should be considered/included from the national minerals policy frame work. The **minerals planning policy is part of the national minerals policy framework** (see also Chapter 4). In the context of this framework at national level a minerals planning policy must be developed considering strategic issues which then are interrelated to the regional/local (operative) planning level. This is also an important hierarchical planning principle: the planning process starts at (for instance) 1:100 000 and goes to the detailed scale (regional: 1:25 000; local 1:5.000).[359] This is a necessary step in order to consider minerals equally.

---

359 The German web portal www.GisInfoService.de is an example of such a system.

Permitting procedures shall be linked to such plans, to use all existing information (e. g. quarry zones) and – to streamline the permitting process.

Figure 79: National mineral planning policy – schematic diagram

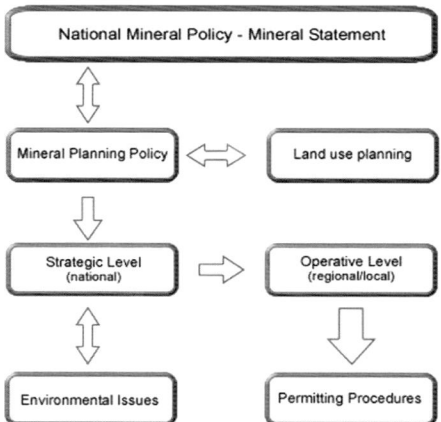

From the above said certain questions related to national minerals policy arise:

- What is the mid and long-term raw materials demand of a country?
- What is the security of minerals supply (i. e. long-term planning)?
  - internal or external supply of a country
  - access to domestic deposits
  - access to external deposits/raw materials,
  - export options
- What is the raw material efficiency, substitution und recycling?
- What are the environmental aspects?
- What are the legal basics (i. e. permitting procedure)?

The record shows that there are **complex** material-economic/political issues. Due to the significance of minerals in the value-added process of a national economy, it is necessary for the state to establish basic conditions for the realization of measures which are of public economic interest. Since a minerals policy is an interdisciplinary subject, an interdisciplinary structured research is also especially important. In that case a minerals policy is necessary to **provide the environment for a balance between demand and supply**. Considering these aspects *and* other conditions, a **national raw materials strategy** including all relevant minerals supply policy issues (e. g. trade policy [external aspect], research policy [internal aspect] **can be determined**.

In the following the minerals policy of different European countries is regarded[360] particularly in view of the goals of supply and minimizing costs. Apart from a short overview of the raw material economy of the respective country[361] four *minerals-political criteria* are used:

1) **National raw materials strategy**: As mentioned in chapter 4 a national raw material strategy defines the objectives of a national minerals policy which then will be implemented by action plans. Such a strategy should (if possible) cover all relevant mineral categories i. e. relevant for the economic of a country *regardless* if the minerals are imported or extracted in the country.

2) **National minerals planning policy:** This is particularly important for the interests of securing (domestic) supply in a long-term way. As mentioned in chapter 4, a minerals planning policy is seen as part of a national minerals policy.

   Securing the supply of minerals includes measures that secure permanent access to economic production and utilization of mineral deposits and ward off claims of third parties, which oppose this goal. It should be distinguished that on one hand raw materials protection is realised on the *national level* (or at least regional level) as planning instrument in the context of mineral planning policy[362] (*if appropriate measures are set*).[363] On the other hand raw materials protection on the private-economic level takes place in the form of the acquisition of exploration and exploitation rights (so-called mineral rights, see point 3).

---

360 Sources used here: Department of Mining and Tunneling (2004), l.c. Besides that: Land Use Consultants (LCU) (2010): consultation process regarding Ad-hoc Working Group on Exchanging Best Practices on Land Use Planning and Geological Knowledge Sharing (for European Commission, DG Enterprise and Industrie). Furthermore: Mining Laws of the respective countries.
361 Main source used here: USGS, BGS, Weber et al. (World Mining Data [note: for metallic minerals the metal content is provided]).
362 As in Section 4 mentioned, minerals planning policy (i.e. securing raw materials) is defined as protection of minerals deposits through land use planning. – The issue of minerals planning policy is addressed in the context of the EU Raw Material Initiative 2008, in particular as criticism of the present EU minerals policy (see section 6).
363 Regarding mineral planning policies in 5.2, only reference is made if planning approaches at national level are existing (for instance Austria: National Mineral Resources Plan).

Chapter 5 View of the minerals policies in selected states of Europe

3) **Mineral rights/licensing procedures**: The procedures for the acquisition of mineral rights are of particular importance.[364] To force an active use of deposits presupposes to make stable and transparent mineral rights available (both for exploration and production) for the entrepreneur (i. e. duration of mineral rights). This provides the entrepreneur with a long-term access to raw materials:[365] Investment security is particularly substantial for the capital-intensive raw materials subject. The principle of the procedure efficiency is also essential. Ineffective procedures mean that costs, complexity and length of the procedures interfere substantially with the competitive ability of the entrepreneur.[366]

4) **Investment policy:** Providing optimal investment conditions for domestic and foreign investors (by the State), also means to focus on active minerals policy.

Figure 80: Countries of Europe (Data by http://www.world-atlas.us)

---

364 Cp. Tiess (2010), l.c.
365 The respective duration of exploration as well as exploitation (i.e. duration of concession) determined in a Mineral law is therefore relevant. It is addressed in the context of the following country discussion. – Cf also Otto (1999), l.c., p.19–20. Cf likewise Tiess (2010), l.c.
366 Addressed in the context of the EU Raw Material Initiative in 2008, there in particular as criticism of the EU Mineral policy (see section 6.5.4).

5.2 EU States

Finland

The Gross Domestic Product increased to $271 billion in 2008 from $129 billion in 1998 (Source Economy watch). The value of the gross output of industry (mining, materials good production, supply of energy) decreased to (2008) 31,6 % from (1998) 33,8 % (Data by OECD). Finland is an industrialized country whose economy is based on trade, primarily exports. The country has a long mining history and a traditional focus on primary resources.[367]

Figure 81: Mining production in Finland between 1998 and 2008 – selected mineral resources (Data by Weber et al. [WMD 1998–2008])

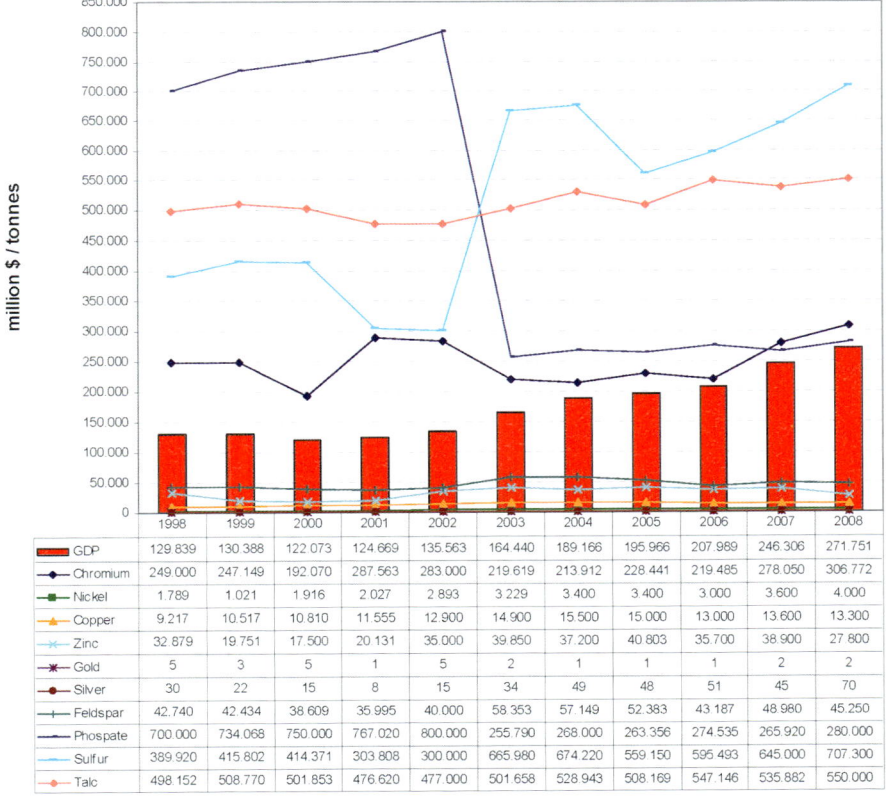

Mining production in Finland between 1998 - 2008 and GDP (current prices)

	1998	1999	2000	2001	2002	2003	2004	2005	2006	2007	2008
GDP	129.839	130.388	122.073	124.669	135.563	164.440	189.166	195.966	207.989	246.306	271.751
Chromium	249.000	247.149	192.070	287.563	283.000	219.619	213.912	228.441	219.485	278.050	306.772
Nickel	1.789	1.021	1.916	2.027	2.893	3.229	3.400	3.400	3.000	3.600	4.000
Copper	9.217	10.517	10.810	11.555	12.900	14.900	15.500	15.000	13.000	13.600	13.300
Zinc	32.879	19.751	17.500	20.131	35.000	39.850	37.200	40.803	35.700	38.900	27.800
Gold	5	3	5	1	5	2	1	1	1	2	2
Silver	30	22	15	8	15	34	49	48	51	45	70
Feldspar	42.740	42.434	38.609	35.995	40.000	58.353	57.149	52.383	43.187	48.980	45.250
Phosphate	700.000	734.068	750.000	767.020	800.000	255.790	268.000	263.356	274.535	265.920	280.000
Sulfur	389.920	415.802	414.371	303.808	300.000	665.980	674.220	559.150	595.493	645.000	707.300
Talc	498.152	508.770	501.853	476.620	477.000	501.658	528.943	508.169	547.146	535.882	550.000

---

367 USGS (2008, 2010): Minerals Yearbook 2008, Volume III (Finland).

Chapter 5 View of the minerals policies in selected states of Europe

Metallic minerals

Between 1998 and 2008 Finland mined chromium, nickel, copper, zinc, gold and silver. The production of chromium (see figure 82) increased to 306.000 t in 2008 from 249.000 t in 1998. Finland is the biggest EU chromium producer, and ranked 7th globally in 2008. Nickel production increased to 4.000 t in 2008 from 1.790 t in 1998, also copper production: to 13.300 t in 2008 from 15.000 t in 2005. Silver production increased to 70 t in 2008 from 30 t in 1998.

The following figure illustrates the production, import and export of chromium in Finland.

Figure 82: Production, export and import of chromium (Data by BGS)

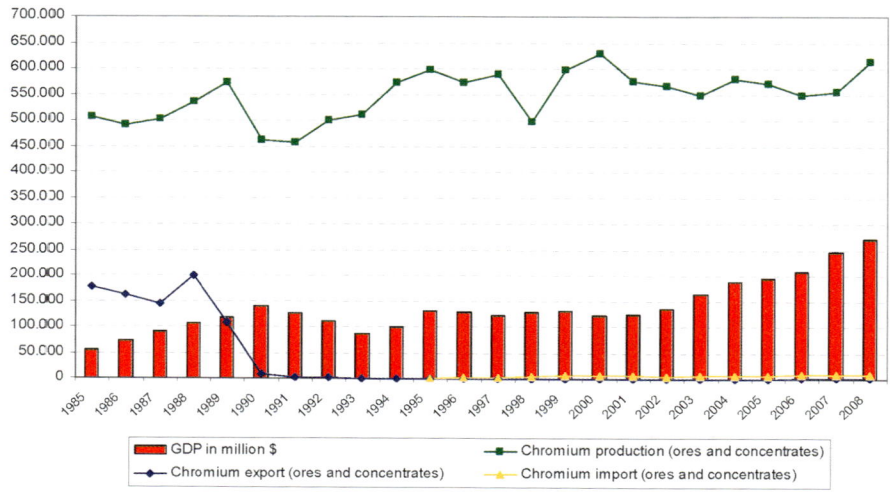

In 2008, Talvivaara's nickel deposit was the largest nickel deposit in Western Europe; it was composed of two polymetallic deposits – the Kolmisoppi and the Kuusilampi – which are located about 30 km from Sotkamo. Based on estimated proven reserves, the deposit was considered to have resources to produce about 2,5 % of the world's nickel during its scheduled 24-year operating life. Talvivaara's bioheap-leach project was planned to produce nickel from an open pit operation and cobalt, copper, and zinc as by-products. The planned nickel production of 50.000 t was anticipated to be reached in 2012. Bioleaching is a process whereby metals are leached from ore as a bacterial action. The bacteria used in the Talvivaara process grow naturally in the ore, and the company reports recovery rates of up to 98% of metal from the leached ore to a solution.[368] The Kevitsa nickel deposit, which is located in northern Finland, was one of the world's major undeveloped nickel sulphide deposits. In April 2008, First Quantum Minerals Ltd. and Scandinavian Minerals Ltd. entered into an agreement in which First Quantum would acquire all outstanding shares of Scandinavian Minerals. In November 2008, Quantum Minerals announced that it was proceeding with the development of the Kevitsa project with the construction of an open pit mine and a 5 million-metric-tons-per-year (Mt/yr)-capacity ore treatment plant. Estimated proven and probable reserves were 107 Mt at a grade of 0,29 % nickel.[369]

---

[368] Mining Technology, 2008, in: USGS (2010), Finland.
[369] USGS (2010), Finland.

Chapter 5 View of the minerals policies in selected states of Europe

Table 24: Metallic minerals: production/mining – import – export of commodities of Finland (Data by BGS, 2010 [European Mineral Statistics])

Commodities related to metallic minerals/products	Production/mining	Import	Export
Silver	mi	m	oc, m
PMG		m, m	m, s
Gold	mi	m, s	s
Zirconium			
Zinc	mi, sz	oc, u	u, u, s
Tin		u	u, s
Rare-earth		cc, orec	rec, m
Mercury	mi	x	x
Lithium		car	
Lead		u	s
Gallium			
Copper	mi, sm, ref	oc, ua, s	u,ur, u, ref, ua, s
Cadmium			
Bismuth			
Bauxite		x	
Arsenic			
Antimony		o	
Aluminium		a, ah, u, ua, s	ua, s
Vanadium			
Tungsten		m	m
Titanium		tm, o	m, o
Tantalum			
Nickel	mi, sm, ref	oc, mat, u, s	oc, u, s, o
Molybdenum			
Manganese		oc, m	
Cobalt	m	oc, m, o	m, o
Chromium	oc, fc	oc, m	oc, fc
Iron	pi, cr, ore, pi, fa, s	or, (bp), pi, fa, s	

a	Alumina	fc	Ferro-cerium
ah	Alumina hydrate	ilm	Ilmenite
bp	Burnt pyrites	m	Metal
c	Concentrate	ma	Metallic arsenic
car	Carbonate	mat	Matte and cement
carb	Carbides	mi	Mine
cc	Cerium compounds	o	Oxide
cr	Crude steel	oc	Ores and concentrates
fa	Ferro alloys	or	Ore

orec	Oher rare earth compounds	sul	Sulfide
otm	Other titanium minerals	sz	Slap zinc
pa	Primary aluminium	tm	Titanium minerals
pen	Pentoxide	ts	Titanium slag
pi	Pig iron	u	uwrought
rec	Rare earth compounds	ua	uwrought alloys
ref	refined	ur	urefined
s	Scrap	vana	Vanadiferous residues
sm	Smelter	wa	White arsenic

Industrial minerals

Between 1998 and 2008, Finland produced (mined) a broad range of industrial minerals that included feldspar, phosphate, sulphur and talc. Feldspar production (see figure 81) increased constant to 45.000 t in 2008 from 42.000 t in 1998, and had its culmination in 2003, 2004 (nearly 60.000 t). Phosphate production increased constant to 800.000 t in 2002 from 700.000 t in 1998, then strongly decreased to 255.000 t in 2003, staying at this level (2008: 280.000 t). Sulphur production increased to 707.000 t in 2008 from 390.000 t in 1998, there was an increase from 300.000 t in 2002 to 665.000 t in 2003. Talc production between 1998 and 2008 was constant, with an average of 500.000 t.

Table 25: Industrial minerals: production/mining – import – export of commodities of Finland (Data by BGS, 2010 [European Mineral Statistics])

Commodities related to industrial minerals/products	Asbestos	Baryte	Bentonite	Diatomite	Diamond	Feldspar	Fluorspar	Gypsum	Graphite	Kaolin	Magnesite	Perlite	Potash	Phosphate rock	Salt	Sulfur	Talc	Zircon	Cement
Production/mining						x								x rec		x	x		
Import	x	x	x		us gcd	x	x	cr ca	x	x	ma ma	x	x		x	x			
Export	x					x					x				sul chl	x	x	x	pc

a	Anhydrite	cr	Crude	i	Industrial	rs	Rock salt
bs	Brinesalt	d	Dust	ma	Magnesia	sib	Salt in brine
ca	Calcined	fs	Fertiliser salts	me	Magnesite	ss	Seasalt
cc	Cement clinker	g	Gypsum	pc	Portland cement	sul	Sulphate
cf	Cement finished	gc	Gem cut	pf	Potassic fertilisers	um	unmanufactured
chl	Chloride	gr	Gem rough	rec	recovered	mi	mine

Construction minerals

Aggregates are economically important mineral raw materials (high need for infrastructure development). The production of aggregates increased constant to 100 Mt in 2006 from 91 Mt in 2002 and then decreased to 86 Mt in 2008. In 2003, Finland produced 92 Mt (i.e. 52 Mt sand and gravel (s&g), 39 Mt crushed rock (cr), 1 Mt recycled aggregates (ra)); in 2004: 98 Mt (i.e. 55 Mt s&g, 43 Mt cr), in 2005: 107 Mt (63 Mt s&g/64 Mt cr / 0,5 Mt ra), in 2006: 100,5 Mt (i.e. 54 Mt s&g, 46 Mt cr and 0,5 Mt ra). In 2008, Finland produced (86 Mt aggregates, i.e.) 25 Mt s&g, 60 Mt cr, 1 Mt ra. The total number of producers (companies) was 400; number of extraction sites (active quarries and pits) was 2.255.[370]

Minerals policy related approaches

Finland published a national minerals strategy at the end of 2010.[371]

The national minerals strategy of Finland was published by the Ministry of Employment and Economy (http://www.mineraalistrategia.fi/materiaalit/fi_FI/materiaalit/). The strategy includes clearly structured objectives and actions. The content of the strategy is the following:

- The significance of minerals
- Global challenges
- Minerals policy in the EU
- The Finnish minerals sector
- The minerals sector as an opportunity for Finland
- Action proposals
- Background material
- The minerals strategy preparation process

The strategy includes three main objectives:

1. Promoting domestic growth and prosperity,
2. solutions for global minerals chain challenges and
3. mitigating environmental impact.

---

[370] UEPG (2010). All data related to aggregates in chapter 5 are provided by UEPG. http://www.uepg.eu/uploads/documents/ (data available up 2002).

[371] The strategy is available at the Finnish Geological Survey. – See also: Ericsson, M. (2010): Global mining towards 2030: food for thought for the Finnish minerals policy process 2010. Tutkimusraporti/Geologian Tutkimuskeskus, 187 (2010), pp. 1–19 – Ericsson, M., Noras, P. (2005): A note on minerals-based sustainable development: one viable alternative. Minerals and energy – raw materials report, 20 (2005), 1, pp. 29–39.

There are four themes of the action proposals:
1. Strengthening minerals policy,
2. securing the supply of raw materials,
3. reducing the environmental impact of the minerals sector and
4. increasing productivity, and strengthening R&D capacities and expertise.

The strategy includes 12 listed actions proposals.

There is no national minerals resource planning system to protect economically attractive deposits.[372]

Regulatory issues

There is a general right for anyone to prospect for mining minerals, regardless of who owns the land. If the applicant can prove that deposits are economically usable, he can apply for the issuing of the exploitation licence (Article 21 ML[373]). Efficient legal mining and environmental structures form the basis of an active exploration. The time used by the Ministry to decide on an application for a pre-claim or prospecting license is about two months. The time needed to process an application for a claim patent or a mining concession can vary between 6 months to almost a year, provided that all the necessary documentation is ready at the time of application.[374] Exploration licences are valid for at least 5 years. If despite systematic exploration during the period mentioned in clause 1, sufficient clarity has not been achieved on the possibilities of exploiting the deposit, the Ministry of Trade and Industry can, upon application made before the stipulated time has run out, grant an extension of up to 3 years (Article 21 (2) ML). The permission is granted for 10 years and can be permanently extended.

Investment policy issues

Finland pursues an active investment policy. The operating environment in Finland is generally favourable for exploration and mining. The country's political and economic stability, the mining and environmental legislation, and the pro-mining attitude are important factors drawing exploration investment. Mining and taxation laws are expected to remain favourable to the industry. The potential of the Fennoscandian Shield to host undiscovered mineral de-

---
372 Nielsen, K. (2004): Country Report Finland, in: Department of Mining and Tunnelling, Minerals Policies and Supply Practices in Europe, University of Leoben.
373 'ML' means 'Mining Law'.
374 For more information, see Tiess (2009): Legal basis for European Raw material policy, Springer, Vienna.

posits has been attracting international mining and exploration companies to Finland.[375] There is significant potential for new discoveries as the region has not been thoroughly examined for minerals. The geologic similarity to the shield areas of Australia and Canada is one reason for the exploration interest. Further positive factors are the good infrastructure and the excellent quality and coverage of geodata created by the Geological Survey of Finland. The country's total expenditure on mineral exploration in 2006, including that by the Geological Survey of Finland, was about € 40 million ($ 56 million),[376] which was the highest in Europe.[377]

Figure 83: Pyhäsalmi, copper-zinc mine in Central Finland

(Data by http://www.inmetmining.com)

The company operating there, Inmet, is a global mining group – headquartered in Canada. The company explores for and produces copper, zinc and gold.

Sweden

The Gross Domestic Product increased to $ 254 billion in 2008 from $ 487 billion in 1998. The value of the gross output of industry (mining, materials good production, supply of energy) decreased to 28 % in 2008. Sweden had Europe's leading mining industry, which contributed 0,3 % of Sweden's GDP. Sweden was a major trading country, and the amount of mineral commodities produced depended mainly on the external demand for these commodities.[378]

---

375 Similar old shields are found on all continents and in some cases contain large deposit potential.
376 USGS (2008), l.c.: Where necessary, values have been converted from EU euros (€) to U.S. dollars ($) at the rate of € 0,71 = $ 1,00.
377 Bergbau Journal 2006, in: USGS 2008 (Länderbericht), p.1 (Mining Journal 2006, in USGS 2008 (country report), p.1).
378 USGS (2008, 2010): Minerals Yearbook 2008, Volume III (Finland).

Figure 84: Mining production in Sweden between 1998 and 2008 – selected mineral resources including aluminium production (Data by Weber et al. [WMD 1998–2008])

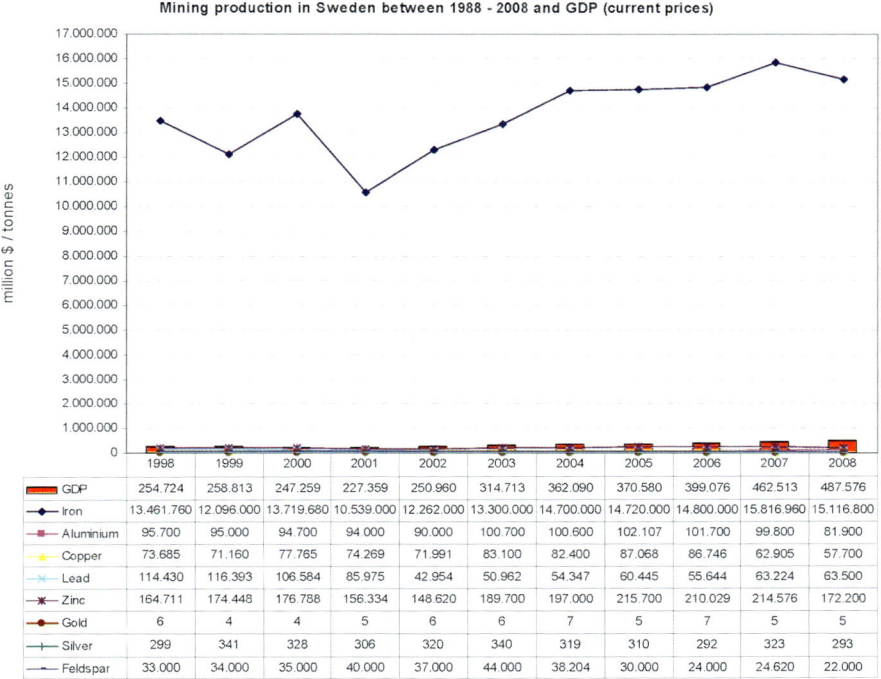

Metallic minerals

Between 1998 and 2008, Sweden mined iron ore, copper, lead, zinc, gold and silver. Iron production increased to 14 Mt in 2008 from 13 Mt in 1998. In 2008 Sweden was the leading producer of iron ore in the EU, and 11th on the global scale. Sweden accounted for about 2% of the world's iron ore production and was responsible for about 90% of the iron ore output of the 27 member countries of the European Union (EU). Copper mine production ranked between 70.000 t and 87.000 t, (see figure 84). Lead mine production reduced continuously to 63.000 t in 2008 from 114.000 t in 1998. Zinc mine production increased to 172.000 t in 2008 from 164.000 t in 1998, while the culmination was in 2005–2007 (215.000 t). Sweden was the second largest EU zinc mine producer. Gold production (4–7 t) and silver production (average 300 t) was moderate.

Chapter 5 View of the minerals policies in selected states of Europe

The following figure illustrates the production, import and export of iron ore in Sweden:

Figure 85: Production, export and import of iron ore (Data by BGS)

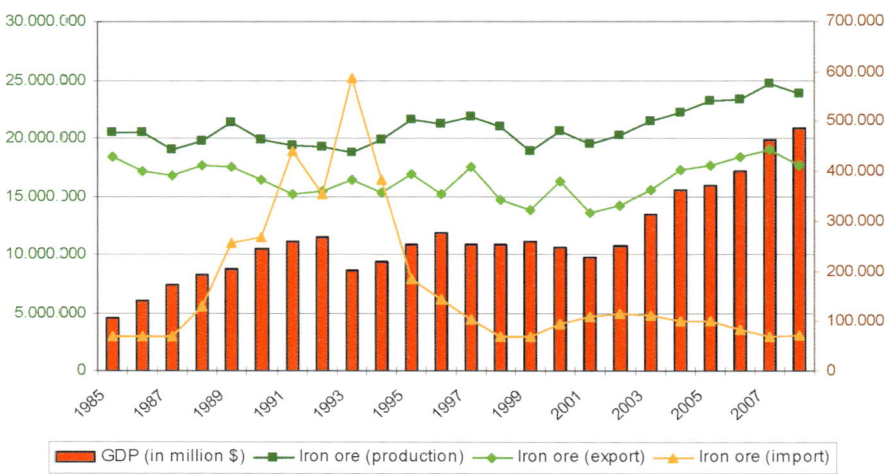

Avalon Resources Ltd. of Australia acquired the Adak and the Viscaria copper projects from Phelps Dodge Exploration Sweden AB in early 2008. Avalon announced that drilling and evaluation work in the first half of 2008 had allowed the company to estimate inferred mineral resources of 26,6 Mt at its Viscaria project. The bulk of the resources were at Viscaria's B Zone, which hosted 24,1 Mt with 0,8 % copper for 182.000 t of contained copper, and the D Zone, which hosted 2,5 Mt with 1,6 % copper for 40.000 t of contained copper. Avalon had undertaken a scoping study and expected to start infield drilling and exploration by year end 2008. Green Leader Equities Research of Australia estimated that it would require a $ 50 million capital investment to construct a throughput mill to process 500.000 t/yr to produce 9.500 t of copper in concentrate.[379]

Between 1998 and 2007 Aluminium production was constant between 90.000 t and 100.000 t. In 2008, primary aluminum production decreased, as did mine production of copper, lead, silver, and zinc and refinery production of ferrochrome. Production of gold, primary and secondary copper increased.

---

[379] Louthean, 2008, in: USGS (2010), Sweden.

5.2 EU States

Table 26: Metallic minerals: production/mining – import – export of commodities of Sweden (Data by BGS, 2010 [European Mineral Statistics])

Commodities related to metallic minerals/products	Production/ mining (x: mine)	Import	Export
Silver	x	oc, m	oc, m
PMG		m, s	m, s
Gold	x	m, s	m, s
Zinc	x	u, ua	oc, u, s
Tin		rec, u	m, o
Rare-earth		rec, m	
Mercury		x	
Lithium		o, car	
Lead	x, ref	oc, u, s	oc, u, s
Gallium			
Copper	x, sm, ref	oc, u,ur, u,ref, ua, s	u,ref, ua, s
Cadmium		m, m	o
Bismuth		m, m	
Bauxite		x	
Arsenic		ma	
Antimony		m, o	
Aluminium		ah, u, ua, s	a, u, ua, s
Vanadium		pen, m	
Tungsten		m, carb	oc, m, carb
Titanium		tm, m, o	
Tantalum			
Nickel		oc, mat, u, ua, s	u, s, o
Molybdenum		oc, m, o	
Manganese		oc, m	m
Cobalt		m	
Chromium		oc, m	m
Iron	ore, pi, cr, fa, s	ore, pi, fa, s	ore, pi, fa, s

a	Alumina	fc	Ferro-cerium	orec	Oher rare earth compounds	sul	Sulfide
ah	Alumina hydrate	ilm	Ilmenite	otm	Other titanium minerals	sz	Slap zinc
bp	Burnt pyrites	m	Metal	pa	Primary aluminium	tm	Titanium minerals
c	Concentrate	ma	Metallic arsenic	pen	Pentoxide	ts	Titanium slag
car	Carbonate	mat	Matte and cement	pi	Pig iron	u	uwrought
carb	Carbides	mi	Mine	rec	Rare earth compounds	ua	uwrought alloys
cc	Cerium compounds	o	Oxide	ref	refined	ur	urefined
cr	Crude steel	oc	Ores and concentrates	s	Scrap	vana	Vanadiferous residues
fa	Ferro alloys	or	Ore	sm	Smelter	wa	White arsenic

Industrial minerals

Between 1998 and 2008 Sweden mined a broad range of industrial minerals that included feldspar and talc. Feldspar production increased to 44.000 t in 2003 from 30.000 t in 1998, since 2004 production decreased to 22.000 t (2008).

Table 27: Industrial minerals: production/mining – import – export of commodities of Sweden (Data by BGS, 2010 [European Mineral Statistics])

Commodities related to industrial minerals/ products	Asbestos	Baryte	Bentonite	Diatomite	Diamond	Feldspar	Fluorspar	Gypsum	Graphite	Kaolin	Magnesite	Perlite	Potash	Phosphate rock	Salt	Sulfur	Talc	Cement
Production/ mining (x: mine)						x									rec		x	cc cf
Import	x	x	x	u	gc i d	x	x	cr ca	x	x	me ma			sul chl	x	x	x	cc pc
Export		x	x			x		cr ca	x	x	ma			sul	x	x	x	pc

a	Anhydrite	cr	Crude	i	Industrial	rs	Rock salt
bs	Brinesalt	d	Dust	ma	Magnesia	sib	Salt in brine
ca	Calcined	fs	Fertiliser salts	me	Magnesite	ss	Seasalt
cc	Cement clinker	g	Gypsum	pc	Portland cement	sul	Sulphate
cf	Cement finished	gc	Gem cut	pf	Potassic fertilisers	um	unmanufactured
chl	Chloride	gr	Gem rough	rec	recovered		

Construction minerals

Aggregates are economically important mineral raw materials (particularly needed for infrastructure development). Figure 86 illustrates the aggregates production between 1984 and 2008; the production increased to about 100 Mt on 2008 from about 85 Mt in 1986. The figure indicates the constant decreasing of sand and gravel production (amongst others due to environmental restrictions) and in turn, the increasing of crushed rock production. Moreover, (according to UEPG data information), the production accounted in 2002: about 74 Mt, in 2003: 75,2 Mt (i.e. 26 Mt sand and gravel (s&g), 41 Mt crushed rock (cr), 8,2 Mt recycled aggregates (ra)), in 2004: 75,2 Mt (26 Mt s&g, 41 Mt cr, 8,2 Mt ra), in: 2005 80,1 Mt (23 Mt s&g, 49 Mt cr, 7,9 Mt ra, 0,2 Mt manufactured aggregates (ma)), in: 2006: 87 Mt (23 Mt s&g, 62 Mt cr, 1,5 Mt ra, 0,2 Mt ma).

In 2008, Sweden produced (93 Mt, i.e.) 19 Mt s&g, 67 Mt cr, and 7 Mt ra. The total number of producers was 985; the total number of extraction sites was 1.802.

Figure 86: Aggregates production in Sweden (Data by Swedish Aggregate Producers Association, 2010)

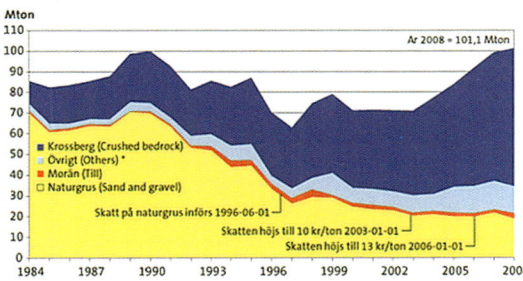

Minerals policy related approaches

The Geological Survey states that Sweden has a 'modern minerals policy'.[380] The Geological Survey has made a classification of deposits of national significance on the basis of the environmental law of 1998 together with the district and local authorities.[381] These are regarded as particularly important for the raw materials supply protection. The competent authorities are responsible for the protection of these deposits in context with spatial planning. This is to ensure that these deposits cannot be sterilised due to building activity and/or over planning.

Regulatory issues

An exploration permit will be granted, if there is reason to assume that exploration in the area can lead to the discovery of a mineral. A concession will be granted if a deposit has been found which can probably be utilized on an economic basis, the location and nature of the deposit does not make it inappropriate that the applicant is granted the concession requested. The production license refers to the area which is determined by position and dimension of a (proven) deposit. Granting an exploration permit takes about 1–2 months, issuance of a mining concession about 1,5 years. An environmental impact assessment takes about 6 months.[382] An exploration permit is valid for (at least) 15 years from the date of the decision. An exploitation concession is valid for

---

380 LCU (2010), table 1 'Minerals Policy' (Sweden), l.c.
381 Nielsen, K. (2004), (Country Report Sweden), in: Department of Mining and Tunnelling, University of Leoben, Minerals Policies and Supply Practices in Europe. About 100 deposits have been declared as deposits of national interest. Mainly: industrial minerals and construction minerals.
382 Ibidem.

twenty-five years (Article 7 ML). The concession period is extended by ten years at a time without special application if regular exploitation is continuing when the period of validity expires.

Investment policy issues

Sweden pursues an active investment policy. The global role of Sweden as an iron ore producer may increase dramatically. Within 5 to 10 years, iron ore production could reach 50 Mt/yr, which is about double that of today.[383] Foreign companies are likely to continue to explore actively in Sweden for base metals, diamond, and gold. The quantity of profitable ores in existing mines is likely to be increased by effective and successful exploration in the vicinity of the mines.

Figure 87: Iron Ore mining Kirunavaara in Sweden (Data by Atzenhofer and Pressler, 2007)

The largest underground iron ore mine in the world.

Large quantities of iron ore are extracted.

Figure 88: Cross section of Kirunavaara Mine (Data by Atzenhofer and Pressler, (2007), Iron ore mining in Kirunavaara)

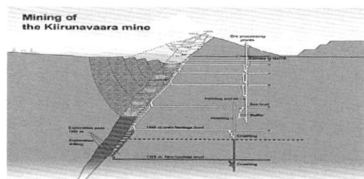

The active group Luossavaara Kiirunavaara AB (LKAB) is an international high-tech company, one of the world's leading iron ore producers. It employs 3.500 employees in 30 companies.

---

383 USGS (2008): 2006 Minerals Yearbook, Volume III – Sweden, p.2.

Denmark

The Gross Domestic Product increased to $ 340 billion in 2008 from 173 $ billion in 1998. The value of the gross output of industry (mining, materials good production, supply of energy) remained on a constant level (25 %). Denmark has an industrialized market economy and the country's economic growth in 2006 depended on imported raw materials and foreign trade.[384] The country has no metallic mineral resources but does have reserves of industrial minerals. Faxe Kalk A/S, which was owned by the Loist Group of Belgium, was a major producer of calcium carbonate from its deposits on the island of Zealand. Faxe was also the leading producer of lime in Denmark.[385]

In 2008, Denmark produced 58 Mt of aggregates, i.e. 43 Mt sand and gravel, 5 Mt marine aggregates and 10 Mt manufactured aggregates. The total number of producers was 350; the total number of extraction sites was 300.

Minerals policy related approaches

Denmark does not have a national raw materials strategy. There is no forward planning for minerals at the national level, with the possible exception that the Ministry of the Environment designates areas for marine mining after comprehensive analyses of the environmental issues and the mapping of resources.[386]

United Kingdom

The United Kingdom is a major European and world economy whose gross domestic product (GDP) in 2006 was ranked second after Germany in the European Union (EU). The Gross Domestic Product increased to $ 2,7 trillion in 2008 $ from 1.456 trillion in 1998. Mining/GDP: 0,2 % share of the total GDP.[387] The output value of the United Kingdom's industry decreased to 23,6 % in 2008 from 28,5 % in 1998 of the GDP. The country is a major regional processor of mineral raw materials and a manufacturer of producer and consumer durables. The country's heavy industries, which included companies that produces automotive and aviation products, chemicals, and machine tools, relies heavily on imported metal ores and concentrates, as well as on some in-

---

384 USGS (2008): Minerals Yearbook 2006, Volume III – Denmark.
385 Industrial Minerals, 2007, in: USGS 2008 (Denmark).
386 Nielsen, K. (2004), Country Report denmark, in: Department of Mining and Tunnelling, Minerals Policies and Supply Practices in Europe, University of Leoben.
387 USGS (2008, 2010): Minerals Yearbook 2008, 2010 Volume III (UK).

dustrial minerals and mineral fuels. In 2006 construction minerals, and industrial minerals constituted about 9% of the total value of mineral production; and metals, an insignificant portion.[388] The London Metal Exchange remaines the dominant world central market for nonferrous metals and the authority on nonferrous metal futures.

Figure 89: Mining production in the United Kingdom between 1998 and 2008 – selected mineral resources including aluminium production (Data by Weber et al. [WMD 1998–2008])

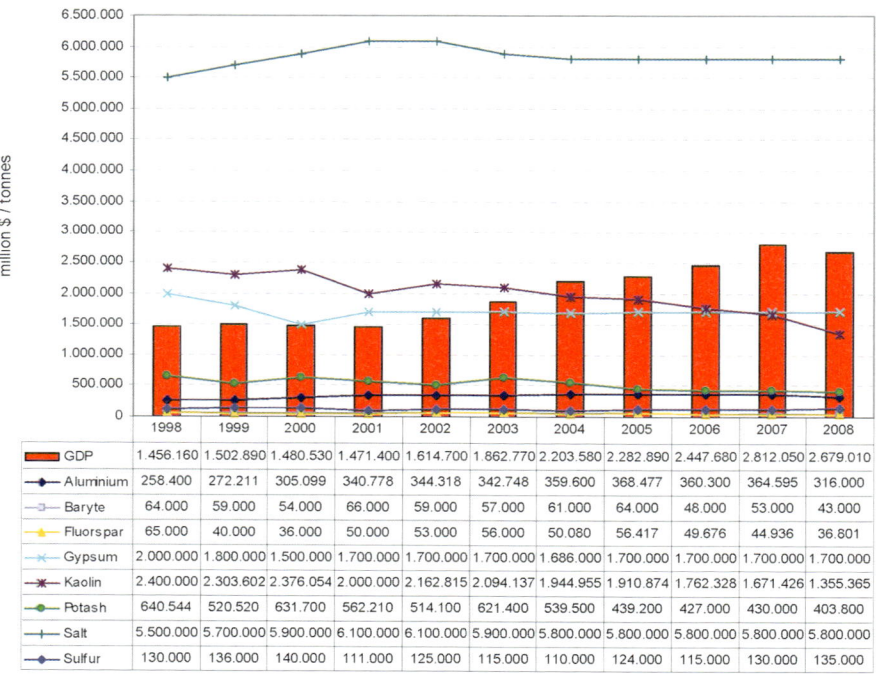

Domestic and internationally owned corporations produced minerals and mineral-based commodities.

---

388 Hetherington and others, 2007, p. 1, in: USGS (2008), p 1.

Metallic minerals

In 2008, the country accounted for about 4% of world refined lead production, about 3% of refined nickel and crude salt output and more than 1% of the world's output of aluminum and crude steel.[389] British Alcan Aluminium Ltd. operated two primary aluminum smelters in the United Kingdom. The nation's third smelter was operated by Anglesey Aluminium Metal Ltd. All the aluminum smelters depended on *imported alumina for feedstock*. In 2006, steel output in the United Kingdom increased by 5,5% compared with that of 2005. Major issues pertaining to the steel industry during the year involved new investments and possible ownership changes at Corus Group plc, which was the dominant producer of steel in the United Kingdom.[390]

---

[389] Bray, 2007, p. 5.17; Ober, 2007, p. 58.9; U.S. Central Intelligence Agency, 2007, p. 598; Carlin, Smith, and Bi, 2008, p. 42.20–42.21; Fenton, 2008, p. 37.13–37.14; Kostick, 2008, p. 63.15–63.17, in: USGS 2008 (UK), p 1. The same is true in 2007 and 2008.
[390] Minerals UK, 2006a, in: USGS 2008 (UK), p 2.

Chapter 5 View of the minerals policies in selected states of Europe

Table 28: Metallic minerals: production/mining – import – export of commodities United Kingdom (Data by BGS, 2010 [European Mineral Statistics])

Commodities related to metallic minerals/products	Production/mining (x: mine)	Import	Export
Silver		oc m s	oc m
PMG		m s	m s
Gold		m s	m s
Zinc		u ua	u ua s
Tin		u ua s	c u ua s
Rare-earth		cc orec fc m	cc orec fc m s
Mercury		x	x
Lithium		o car	o car
Lead	x ref	u s	u ua s
Gallium			
Copper		u,u,r u,ref ua s	mat u ua s
Cadmium		m	m
Bismuth		m	m
Bauxite		x	x
Arsenic		ma	
Antimony		m o	m o
Aluminium	pa	ah u ua s	a ah u ua s
Vanadium		pen m	pen m
Tungsten		m carb	m carb
Titanium		m tm ts o	m o
Tantalum		x	x
Nickel	sm/ref	oc mat u ua s o	mat u ua s
Molybdenum		oc m o	oc m
Manganese		oc m	m
Cobalt		m o	m o
Chromium		oc m	oc m
Iron	ore pi cr	ore pi fa s	pi a s

a	Alumina	otm	Other titanium minerals
ah	Alumina hydrate	pa	Primary aluminium
bp	Burnt pyrites	pen	Pentoxide
c	Concentrate	pi	Pig iron
car	Carbonate	rec	Rare earth compounds
carb	Carbides	ref	refined
cc	Cerium compounds	s	Scrap
cr	Crude steel	sm	Smelter
fa	Ferro alloys	sul	Sulfide
fc	Ferro-cerium	sz	Slap zinc
ilm	Ilmenite	tm	Titanium minerals
m	Metal	ts	Titanium slag
ma	Metallic arsenic	u	unwrought
mat	Matte and cement	ua	unwrought alloys
mi	Mine	ur	unrefined
o	Oxide	vana	Vanadiferous residues
oc	Ores and concentrates	wa	White arsenic
or	Ore		

Figure 90: Anglesey Aluminium Metal Ltd, aluminium Smelter (Data by http://www.angleseyaluminium.co.uk)

Industrial minerals

The United Kingdom remained an important producer of such minerals as baryte, gypsum, kaolin salt, sulphur, calcareous material for cement, clays, and fluorspar. Baryte production (see figure 89) decreased to 43.000 t in 2008 from 64.000 t in 1998. Fluorspar production decreased to 36.000 t in 2008 from 65.000 t in 1998. Gypsum production was at a constant and high level, with an average of 1.700.000 t. Kaolin production was at a constant and high level between 1998 and 2003 (average 2 Mt), slightly decreasing since 2004. Potash production was at a constant level between 1998 and 2003 (average 500.000–600.000 t), slightly decreasing since 2004. In 2006, the country accounted for about 2% of the world's potash production. UK had a significant salt production at a constant level of about 6 Mt. The average of sulphur production amounted to 130.000 t. The major producer of baryte in 2006 was M-I Drilling Fluids (UK) Ltd., which operated the underground Foss Mine near Aberfeldy in Perthshire, Scotland.[391] Cement-manufacturing companies included Lafarge Cement UK, Ltd., Castle Cement Ltd., CEMEX UK Operations, Ltd. (www.cemex.rom), and Buxton Lime Industries Ltd., which had total clinker capacities in 2006 of 6,0 million metric tons per year (Mt/yr), 3,6 Mt/yr, 2,8 Mt/yr, and 800.000 metric tons per year (t/yr), respectively. In 2006, cement production in the United Kingdom amounted to about 11,4 million metric tons (Mt), which was an increase of about 1,6 % compared with that of 2005.[392] In 2006, the Imerys Group, which was a leading producer of kaolin (china clay) in the world continued to produce high-grade clays for ceramics and chemical and rubber products.[393]

---

391 British Geological Survey, 2006a, in: USGS 2008 (UK), p 2.
392 British Geological Survey, 2005, in: USGS 2008 (UK), p 2.
393 Imerys S.A., 2006, in: USGS 2008 (UK), p 2.

Chapter 5 View of the minerals policies in selected states of Europe

Table 29: Industrial minerals: production/mining – import – export of commodities of United Kingdom (Data by BGS, 2010 [European Mineral Statistics])

Commodities related to industrial minerals/products	Baryte	Bentonite	Diatomite	Diamond	Feldspar	Fluorspar	Gypsum	Graphite	Kaolin	Magnesite	Perlite	Potash	Phosphate rock	Salt	Sulfur	Talc	Cement
Production/mining (x: mine)	x	x until 2005				x	x	x		x			chl		rs bs	rec	x	cc cf
Import	x	x	x	u gr gc i d	x	x	cr ca	x	x	me ma	fs chl	x	x	x cr ref	x	x	cc pc
Export	x	x	x	u gr gc i d	x	x	cr ca	x	x	me ma	chl	x	x	x	x	x	pc

a	Anhydrite	d	Dust	ma	Magnesia	ss	Seasalt
bs	Brinesalt	fs	Fertiliser salts	me	Magnesite	sul	Sulphate
ca	Calcined	g	Gypsum	pc	Portland cement	um	unmanufactured
cc	Cement clinker	g	Gem	pf	Potassic fertilisers	us	unsorted
cf	Cement finished	gc	Gem cut	rec	recovered		
chl	Chloride	gr	Gem rough	rs	Rock salt		
cr	Crude	i	Industrial	sib	Salt in brine		

Construction minerals

Aggregates are economically important mineral raw materials (particularly needed for infrastructure development). Figure 91 is illustrating the aggregates extraction between 1955 and 2009. The production increased to 243 Mt in 2008 from about 100 Mt in 1955. The constant increasing of recycled aggregates production is remarkable. Moreover, the production accounted in 2003: 271,8 Mt (i.e. 82,7 Mt s&g, 126,6 Mt crushed rock (cr), 62,5 Mt recycled aggregates (ra)), in 2004: 257 Mt (79 Mt s&g, 124 Mt cr, 54 Mt ra), in 2005: 277 Mt (85 Mt s&g, 124 Mt cr, 56 Mt ra, 12 Mt ma), in 2006: 274 Mt (68 Mt s&g,123 Mt cr, 13 Mt marine aggregates production, 12 Mt manufactured aggregates)

In 2008, the UK produced (243 Mt, i.e.) 55 Mt s&g, 114 Mt cr, 10 Mt marine aggregates, 53 Mt ra and 9 Mt manufactured aggregates. The total number of producers was 450; the total number of extraction sites was 781.

Figure 91: Aggregates Production – Great Britain (Data by Minerals Products Association, 2009)

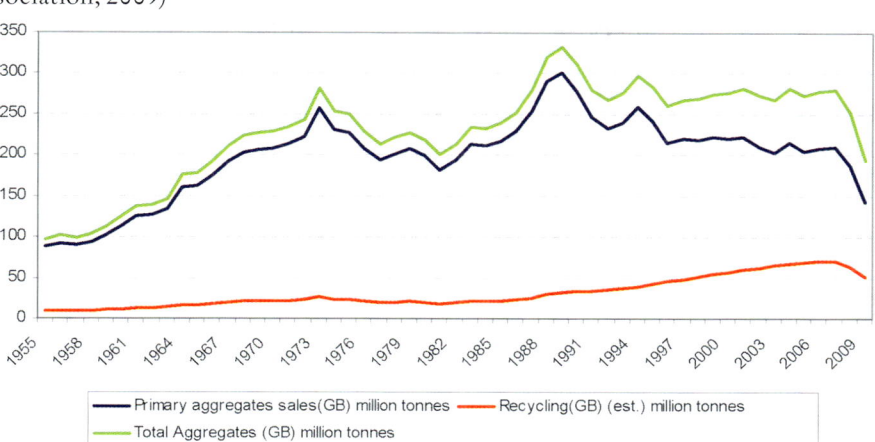

Minerals policy related approaches

The UK does not have a national raw materials strategy, but does have a *National Minerals Policy Statement*: Planning and Minerals. It states that it is essential that there is an adequate and steady supply of minerals to provide the infrastructure, buildings and goods that society, industry and the economy needs, and that this provision is made in accordance with the principles of sustainable development.[394]

Regarding national mineral planning policy the Government in England exercises influence through Mineral Planning Guidance Notes (MPGs) and Marine Mineral Guidance Notes (MMGs). The overall approach to spatial planning at the national level takes the form of general guidance rather than the preparation of spatial specific plans. The guidelines are used for preparation of lower-tier policy instruments, making of decisions on proposed development.[395]

Regulatory issues

Part 22 Class B of Schedule 2 of the General Development Procedure Order (GDPO) 1995 permits exploration operations to be carried out for a period of 6 months. This permission is renewable for the same land by the service of

---

[394] LCU (2010), Table 1 'Minerals Policy' (UK). – See also: UK Minerals Forum (2009): Shaping UK minerals policy, http://www.mauk.org.uk/newsdocs/shaping_uk_policy_final_180809.pdf (accessed: 09.03.2011)
[395] Ike (2004): Department of Mining and Tunneling, Final Report, p 120.

further notice of the Mineral Planning Authority. Through this procedure, flexibility is provided to the mineral operators, which is needed to adjust their exploration programme to the size and complexity of the prospect. The extraction authorization is valid for a limited period: All planning permissions must have a time limit condition, requiring development to cease not later than the expiration of 60 years or such longer or shorter period as the Mineral Planning Authority may specify.[396]

Investment policy issues

The United Kingdom's mineral sector not only serves domestic economic needs but its mining and processing companies continues to play an important role in global mineral prospecting, mineral development, and mineral commodity trade. Falkland Gold and Minerals Ltd. (FGML) continues to explore for gold in the overseas territory of the Falkland Islands. Earlier indications showed subeconomic resources of gold and platinum-group metals.[397] Besides that in Northern Ireland, Tournigan Gold Corp. of Canada continued exploration for gold at Curraghinalt. Galantas was building Northern Ireland's first gold mine.[398]

Regarding copper and Nickel Alba Mineral Resources plc (Alba) continues to study the Arthrath copper-nickel-platinum group project in Aberdeenshire, Scotland. Alba reported that extensive disseminated magmatic nickel-copper sulphide mineralization was present on the Arthrath prospect. Alba also reported having reached a joint-venture and exploration option agreement with Inco Ltd. of Canada. The agreement gave Inco the right to earn a 60% interest in the Arthrath project in return for a capitalization of $ 3.14 million for exploration during a 4-year period.[399]

Ireland

The Gross Domestic Product increased to $ 264 billion in 2008 from $ 88 billion in 1998. The value of the gross output of industry (mining, materials good production, supply of energy) decreased to (2007) 33,6 % from (1998) 40,8 %.[400] Ireland is a trade-dependent country and exports remain a key component of

---

396 Cp. IKE, P. (2004), Country report UK, in: Department of Mining and Tunnelling, l.c.
397 Falkland Gold and Minerals Limited, 2006, in: USGS 2008 (UK).
398 Galantas Gold Corp., 2007, in: USGS 2008 (UK).
399 Alba Mineral Resources plc, 2006, in: USGS 2008 (UK).
400 USGS (2008, 2010): Minerals Yearbook 2008, 2010 Volume III (Ireland).

its gross domestic product. Minerals account for less than 1% of exports. In 2006, mining accounted for about 1% of the GDP and about 1% of the work force (e. g. production of alumina and peat).[401]

Figure 92: Mining production in Ireland between 1998 and 2008 – selected mineral resources (Data by Weber et al. [WMD 1998–2008])

Metallic minerals

In 2006, Ireland was a major European Union (EU) producer of lead and zinc ore also one of the leading exporters of lead and zinc concentrate. Ireland was the leading producer of zinc ore in Europe: There are major producing mines within the Rathdowney Trend in the Midland Orefield. These operations included the world-class ore bodies at Galmoy, Lisheen, and Navan (see figure 85). Arcon International Resources plc had success extending existing reserves in the K and CW zones at the Galmoy Mine.[402] The following figure illustrates the production, import and export of zinc in Ireland.

---

401 USGS (2008): Minerals Yearbook 2006, Volume III – Ireland, p.1.
402 Ibidem.

Chapter 5 View of the minerals policies in selected states of Europe

Figure 93: Production, export and import of zinc (Data by BGS)

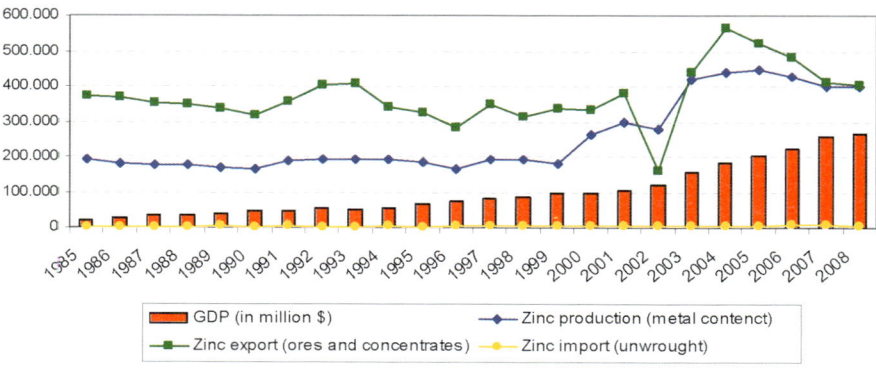

In 2006 the exploration budget at Anglo American plc's Lisheen Mine was $ 2 million. The Government granted AngloAmerican permission to exploit the Bog Zone ore body. At New Boliden AB's Tara Mine, which is located near Navan, exploration within the mine was continuing in the SWEX, SWEX B, and Nevinstown zones.[403] Mineralization was found in several holes, and followup work was continuing.

Ireland is expected to remain an important (EU) supplier of lead and zinc ore as well as aluminiumoxide. Exploration regarding gold, lead and zinc deposits will be an ongoing process.[404]

Between 1998 and 2008 Ireland mined lead, zinc and silver. Lead production increased constantly to 50.000 t in 2008 from 36.000 t in 1998. Zinc production increased to 400.000 t in 2008 from 177.000 t in 1998.

Industrial minerals

Gypsum production increased to 600.000 t in 2008 from 500.000 t in 1998, the culmination was in 2005–2007 (700.000 t).

---

403 Mining Magazine, 2006, in: USGS 2008 (Ireland), p.1.
404 USGS (2008): 2006 Minerals Yearbook, Ireland, p.1.

Construction minerals

Aggregates are economically important mineral raw materials (particularly needed for infrastructure development). The production of aggregates increased to 134 Mt in 2006 from 41 Mt in 2002 and then decreased strongly to 50 Mt in 2008. The production accounted in 2003: 80 Mt (i.e. 35 sand and gravel, 45 Mt crushed rock (cr)), in 2004: 101 Mt (50 Mt s&g, 50 Mt cr, 1 Mt recycled aggregates (ra)), in 2005: 134 Mt (54 Mt s&g, 79 Mt cr, 1 Mt ra), in 2006: 134 Mt (54 Mt s&g, 79 Mt cr, 1 Mt ra). In 2008, Ireland produced (50 Mt, i.e.) 25 Mt s&g and 25 Mt crushed rock. The total number of producers was 150; total number of extraction sites was 355.

Minerals policy related approaches

Northern Ireland's minerals policy is under development.[405] The promotion of exploration and maximisation of the contribution of the mining industry sector to the gross domestic product is a substantial goal of the Irish raw material policy.[406] However, there is no national mineral resource planning system to protect economically attractive deposits. Applications for prospecting and exploration licences can be placed (efficiently) on-line.[407]

---

405 LCU (2010): table 1 'Minerals Policy' (Ireland), lc.
406 The Division for Exploration and Mining (EMD) of the Department for Communication, Marine and Nature Resources (Minister for Transport, Energy and Communications) is the responsible institution for development of the Irish raw material policy.
407 Cp. Ike, P. (2004), Country Report Irland, in: Department of Mining and Tunnelling, University of Leoben, Minerals Policies and Supply Practices in Europe.

Figure 94: Zinc – lead mines in Ireland

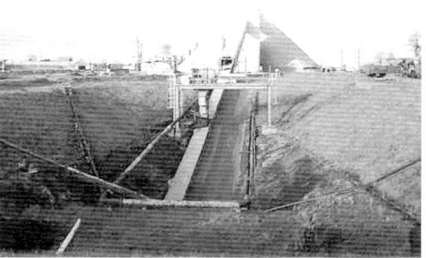

Entrance to underground mining. The mining takes place 100 m below the surface. Lundin Mining Company produces a wide range of base metals (copper, lead and nickel). Mining sites are located in Portugal, Spain, Sweden and Ireland. The company operates presently 6 mines.

Source: http://www.mining-technology.com

Anglo American is a worlwide known company. It is the global headquarters of the Lisheen mine.

Source: http://www.lisheenmine.ie

The Netherlands

The Gross Domestic Product increased to $ 877 billion in 2008 from $ 403 billion in 1998. The value of the gross output of industry (mining, materials good production, supply of energy) decreased to 23,8 % (2005) from 25,3 % (1998) and then increased to 25,5 % (2008). In 2008, the Netherlands was the EU's third ranked exporting country after Germany and France (Port of Rotterdam Authority, 2009). In 2008, the throughput (imports and exports) of the following mineral commodities was the largest in terms of total cargo: 100,4 million metric tons (Mt), ores and scrap. [408]

In terms of world production, in 2006 the Netherlands was a modest producer of metallic and non-metallic minerals and mineral products. Downstream activities included chemical and metallurgical industries, which used mainly imported ores and industrial minerals.[409] The Dutch economy was heavily dependent on international developments. The export, re-export, and import of goods and services together accounted for more than 60 % of the gross do-

---
408 USGS (2008, 2010): Minerals Yearbook 2008, 2010 Volume III (Netherlands).
409 USGS (2008): Minerals Yearbook 2006, Volume III – Netherlands.

mestic product.[410] Rotterdam, which was the world's leading container port and a major European transportation hub, remained extremely important as a shipping and storage centre. In 2006, a record 378 million metric tons (Mt) of cargo passed through the Port of Rotterdam compared with 370 Mt in 2005. Throughput of incoming and outgoing materials in 2006, in thousand metric tons, included mineral products (45.919), ores and scrap (38.524) (Port of Rotterdam Authority, 2006).

In 2006, mining and quarrying accounted for about 3,9% of the value of industrial production.[411]

Figure 95: Salt and aluminium production in the Netherlands between 1998 and 2008 (Data by Weber et al. [WMD 1998–2008])

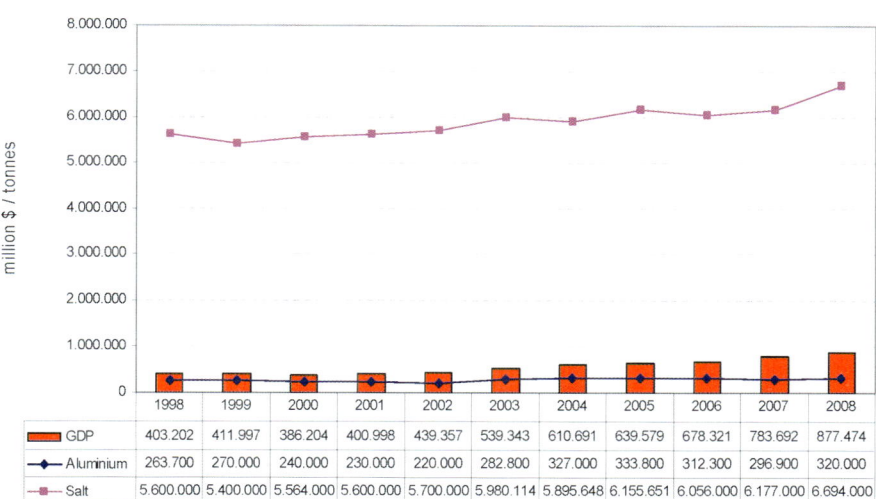

Metallic minerals

Between 1998 and 2008, the Netherlands produced aluminium, lead, and zinc but only in refined metal, as the country does not have resources of these minerals. Aluminium production increased to 320.000 t in 2008 from 263.700 t in 1998.

Pechiney Nederland NV (PLN), a subsidiary of Alcan Inc. of Canada, was a producer of extrusion billets and of rolled aluminum slabs. PLN had a produc-

---
410 Holland Trade, 2006, in: USGS 2008 (Netherlands).
411 Statistics Netherlands, 2006, in: USGS 2008 (Netherlands).

tion capacity of 213.000 metric tons per year (t/yr) of primary aluminum and 230.000 t/yr of aluminum billets and slabs (Pechiney Nederlands CV, 2006). Zinifex Limited's zinc smelter at Budel produced a record 235.913 metric tons (t) of zinc in 2006 exclusively from zinc concentrates from Zinifex's Century Mine in Queensland, Australia. Following an expansion that was completed in 2006, the plant had the capacity to produce 260.000 t/yr.[412]

The proposed acquisition of the Corus Group by Companhia Siderurgica Nacional (CSN) of Brazil could fall within the scope of the European Union's merger regulations according to the preliminary findings of the European Commission (EC). CSN was in competition with Tata Steel of India to acquire Corus. In December 2006, CSN made a $ 9,6 billion bid for Corus. The battle for Corus came at a time of growing consolidation within the global steel industry. A takeover of Corus by either company would create the world's fifth-largest steel group with the capacity to produce 24 million metric tons per year (Mt/yr) of crude steel.[413]

---

412 Zinifex Limited, 2006, in: USGS (Netherlands).
413 Platts, 2006, in: USGS 2008 (Netherlands).

5.2 EU States

Table 30: Metallic minerals: production/mining – import – export of commodities of the Netherlands (Data by BGS, 2010 [European Mineral Statistics])

Commodities related to metallic minerals/products	Production/mining (x: mine)	Import	Export
Silver		m m	m oc
PMG		m m	m m oc s m
Gold		m s	m s
Zinc	sz	oc u ua s	oc u ua s
Tin		u ua s	u ua s
Rare-earth		cc orec fc m	rec fc m
Mercury		x	x
Lithium		o car	o car
Lead	ref	oc u s	oc u s
Gallium			
Copper		u,ur u,ref ua s	u s
Cadmium	x	m	m
Bismuth		m	m
Bauxite		x ma	x m
Arsenic			
Antimony		m o	m o
Aluminium	pa	a ah u ua s	a ah u ua s
Vanadium		pen	pen m
Tungsten		oc m carb	oc m
Titanium		tm m o	tm m o
Tantalum		x	x
Nickel		ua s o	mat u s o
Molybdenum		oc m o	oc m o
Manganese		oc m	oc m
Cobalt		m o	m o
Chromium		oc m	oc m
Iron	pi cr	ore pi fa s	ore pi fa s

a	Alumina	fc	Ferro-cerium	orec	Oher rare earth compounds	sul	Sulfide
ah	Alumina hydrate	ilm	Ilmenite	otm	Other titanium minerals	sz	Slap zinc
bp	Burnt pyrites	m	Metal	pa	Primary aluminium	tm	Titanium minerals
c	Concentrate	ma	Metallic arsenic	pen	Pentoxide	ts	Titanium slag
car	Carbonate	mat	Matte and cement	pi	Pig iron	u	uwrought
carb	Carbides	mi	Mine	rec	Rare earth compounds	ua	uwrought alloys
cc	Cerium compounds	o	Oxide	ref	refined	ur	urefined
cr	Crude steel	oc	Ores and concentrates	s	Scrap	vana	Vanadiferous residues
fa	Ferro alloys	or	Ore	sm	Smelter	wa	White arsenic

Industrial minerals

Between 1998 and 2008 the Netherlands produced a broad range of industrial minerals that included magnesite and salt.

Omya Netherlands BV's ground calcium carbonate (GCC) plant at Moerdijk started production in midyear 2005 and continued in 2006. The initial capacity of the plant was 500.000 t/yr. GCC had became the leading filler in the production of wood-free paper of the resulting paper's pure white color and brightness. The marble used to produce the GCC was imported from Omya's mines in Turkey.[414] Nedmag Industries Mining & Manufacturing BV was Europe's leading producer of high-grade synthetic dead-burned magnesia and other magnesium compounds.[415] Akzo Nobel Salt BV's salt production facilities were located in Delfziji and Hengelo. The raw brine was produced by solution mining in multieffect evaporation plants. With a capacity of more than 2 Mt/yr, the Hengelo plant was the leading vacuum salt plant in the world. Akzo Nobel focused on salt for chemical transformation (electrolysis), road salt for de-icing, and dried salt for salt specialties businesses.[416]

---

414 Industrial Minerals, 2006, in: USGS 2008 (Netherlands).
415 Nedmag Industries Mining & Manufacturing BV, 2006, in: USGS 2008 (Netherlands). – See also: van der Meulen, M.J. (2005): Sustainable mineral development possibilities and pitfalls illustrated by the rise and fall of Dutch mineral planning guiddance. Geological Society Special publication, 250 (2005), pp. 225–232.
416 Akzo Nobel Salt BV, 2006, in: USGS 2008 (Netherlands).

5.2 EU States

Table 31: Industrial minerals: production/mining – import – export of commodities of the Netherlands (Data by BGS, 2010 [European Mineral Statistics])

Commodities related to industrial minerals/products	Asbestos	Baryte	Bentonite	Diatomite	Diamond	Feldspar	Fluorspar	Gypsum	Graphite	Kaolin	Magnesite	Perlite	Potash	Phosphate rock	Salt	Sulfur	Talc	Cement
Production/mining (x: mine)											me				x	rec		cc cf
Import	x	x	x	u gr gc i d	x	x	cr ca	x	x	x	me ma	x	sul chl	x	x	x	x	cc pc
Export	x	x	x	us gc i d	x	x	cr ca	x	x	x	me ma	x	sul chl	x	x	x	x	cc pc

a	Anhydrite	cr	Crude	i	Industrial	rs	Rock salt
bs	Brinesalt	d	Dust	ma	Magnesia	sib	Salt in brine
ca	Calcined	fs	Fertiliser salts	me	Magnesite	ss	Seasalt
cc	Cement clinker	g	Gypsum	pc	Portland cement	sul	Sulphate
cf	Cement finished	gc	Gem cut	pf	Potassic fertilisers	um	unmanufactured
chl	Chloride	gr	Gem rough	rec	recovered		

Construction minerals

Aggregates are economically important mineral raw materials (particularly needed for infrastructure development). The production of aggregates strongly increased to 124 Mt in 2008 from 47 Mt in 2002. The production accounted in 2003: 22 Mt (i.e. 5,2 Mt sand and gravel (s&g), 16,8 Mt crushed rock (cr), 0,2 recycled aggregates (ra)), in 2004: 16,5 Mt (12 Mt s&g, 4 Mt cr, 0,5 Mt ra), in 2005: 48,2 Mt (24 Mt s&g, 4 Mt cr, 20,2 Mt ra), in 2006: 119,5 Mt (44,5 Mt s&g, 50 Mt marine, 25 Mt ra). In 2008, Netherlands produced (124 Mt, i.e.) 46 Mt s&g, 54 Mt marine aggregates and 24 Mt recycled aggregates. In 2008, the total number of producers was 65; the total number of extraction sites was 225.

Minerals policy related approaches

Netherland does not have a national raw materials strategy.

As a consequence of a new market-oriented approach adopted in the Netherlands in 2003, the government has removed the National Structure Plan on

Surface Raw Materials from the Excavation Act. The national policy on surface raw materials was integrated in the National Spatial Plan.[417] In the future minerals planning will be provided by the provincial mineral extraction plans (mainly aggregates).

Regulatory issues

The holder of an exploration licence, who has demonstrated the presence of the relevant minerals, shall be granted a production licence for those minerals in the area to which the exploration licence applies (Article 10 ML). A licence will specify for what period it is valid. This is done such that the period is no longer than necessary for carrying out the activities for which the licence is granted (Article 11 ML). According to the Excavation Act 2002 it is not necessary to have a permit for the exploration of surface minerals.

Belgium

The Gross Domestic Product increased to 506 $ billion in 2008 from 255 $ billion in 1998. The value of the gross output of industry (mining, materials good production, supply of energy) decreased to 23,1 % (2008) from 27,8 % (1998). In 2009, Belgium's industry depended greatly on markets, especially those of other European Union (EU) countries. Its main trading partners were, in order of the percentage of total trade, Germany (which accounted for 19,5 % of Belgium's exports and 17,7 % of its imports), the Netherlands (11,9 % of exports and 17,6 % of imports), France (16,7 % of exports and 11,2 % of imports), the United Kingdom (7,6 % of exports and 6,2 % of imports), the United States (5,7 % of exports and 5,4 % of imports), Italy (5,2 % of exports), Ireland (4,9 % of imports) and China (4,1 % of imports). More than one-half of Belgium's GDP stemmed from foreign sales, which was one of the highest percentages of the industrialized nations.[418]

Belgium, a highly developed market economy, belongs to the Organization for Economic Cooperation and Development (a group of leading industrial countries) and is located at the heart of one of the world's most highly developed industrialized regions.[419] Because it hosts few natural resources of

---

417 Ministeries van VROM, LNV, VenW en EZ (2004), in: Department of Mining and Tunneling, p. 119. – See also: van der Meulen, M. J. (2005): Sustainable mineral development possibilities and pitfalls illustrated by the rise and fall of Duch mineral planning guidance. Geological Society Special publication, 250 (2005), pp. 225–232.
418 USGS (2010): Minerals Yearbook 2008, Volume III (Belgium).
419 USGS (2008): Minerals Yearbook 2006, Volume III – Belgium.

its own, the country must import substantial quantities of raw materials and export a large volume of manufactures, which makes its economy unusually dependent on the state of world markets.[420] With exports equivalent to more than two-thirds of the gross domestic product, Belgium depends heavily on world trade. About three-quarters of Belgium's trade is with other European Union (EU) countries. Germany was Belgium's top customer; the United States was ranked fifth.[421]

Metallic minerals

The country imported substantial quantities of raw materials. The metal processing industries were significant to the Belgian economy.

Belgium's ferrous and nonferrous materials can be divided into the following three categories: base metals, such as aluminum, copper, lead, steel, and zinc; precious metals, such as gold, silver, and platinum; and specialty metals, such as cobalt, germanium, and indium. The refining of copper, minor metals, and zinc, and the production of steel were the leading mineral industries in Belgium. In 2008, the country was also a producer of cadmium, cobalt, germanium, selenium, and tellurium. Umicore Group (Umicore), which was one of Europe's leading metal recyclers and processors, was headquartered in Hoboken, Belgium. It's most profitable business sectors were the zinc processing division; the advanced materials division, which dealt with processing such minerals as cobalt and germanium; and the precious metals processing division.[422] In May 2009, the company acquired the Gordonsville Zinc mine complex in Tennessee, and in November, the company completed its acquisition of an 85 % interest in the Coricancha polymetallic mine in Peru.[423]

Belgium was the 18th ranked producer of steel in the world, producing 10,7 Mt in 2008.[424] Arcelor Mittal's Gent and Liege plants were heavily reliant on demand from the car manufacturing industry; as such, demand for the company's products was reduced by the car industry's slowdown.[425]

---

420 World Group, 2006, in: USGS 2008 (Belgium).
421 Belgium Foreign Trade Agency, 2007, in: USGS 2008 (Belgium).
422 N.V. Umicore S.A., 2009, in: USGS (2010).
423 Nyrstar NV, 2010, in: USGS (2010).
424 World Steel Association, 2009, p. 9, in: USGS (2010).
425 Business Monitor International, 2010, in: USGS (2010).

Chapter 5 View of the minerals policies in selected states of Europe

Table 32: Metallic minerals: production/mining – import – export of commodities of Belgium (Data by BGS, 2010 [European Mineral Statistics])

Commodities related to metallic minerals/products	Production/mining (x: mine)	Import	Export
Silver		oc, m	oc, m
PMG		m	m, m, s
Gold			m, m, s
Zinc	sm, sz	oc, oc, u, ua, s	
Tin	sm	oc, oc, u, ua, s	c, m, u, ua, s
Rare-earth		cc, orec, m	rec, m
Mercury		x	x
Lithium		o, car	o, car
Lead	ref	oc, u, s	oc, u, s
Gallium			
Copper	ref	oc, mat, u,ur, u,ref, ua, s	oc, mat, u,ur, u,ref, ua, s
Cadmium		m, o, sul	m, m, sul
Bismuth		m	m
Bauxite	x	ma, x	
Arsenic	wa	ma	wa
Antimony		m, o	oc, m, o
Aluminium		a, oc, pen, ah, m, u, ua, s	a, oc, u, ua, s
Vanadium			
Tungsten		m, carb	m, carb
Titanium		ilm, otm, m, o	tm, ts, m, o
Tantalum	x		x
Nickel		u, ua, s, o	oc, u, ua, s, o
Molybdenum		oc, m, o	oc, m, o
Manganese		oc, m	oc, m
Cobalt	m	oc, m, o	or, m, o
Chromium		oc, m	oc, m
Iron	pi, cr	or, pi, fa, s	pi, fa, s

a	Alumina	fc	Ferro-cerium
ah	Alumina hydrate	ilm	Ilmenite
bp	Burnt pyrites	m	Metal
c	Concentrate	ma	Metallic arsenic
car	Carbonate	mat	Matte and cement
carb	Carbides	mi	Mine
cc	Cerium compounds	o	Oxide
cr	Crude steel	oc	Ores and concentrates
fa	Ferro alloys	or	Ore

orec	Oher rare earth compounds	sul	Sulfide
otm	Other titanium minerals	sz	Slap zinc
pa	Primary aluminium	tm	Titanium minerals
pen	Pentoxide	ts	Titanium slag
pi	Pig iron	u	uwrought
rec	Rare earth compounds	ua	uwrought alloys
ref	refined	ur	urefined
s	Scrap	vana	Vanadiferous residues
sm	Smelter	wa	White arsenic

Industrial minerals

Between 1998 and 2008 Belgium produced a broad range of industrial minerals that included kaolin, lime and marble.

Belgium is a significant mineral processor and major diamond trader in the world, as well as a globally significant handler of mineral products through its major ports. The diamond district of Antwerp, which comprised four exchanges and about 1.500 diamond companies, is a leading diamond distribution centre. Belgium is the world's leading exporter of diamonds and precious stones. In 2006, 8,7 million carats of polished diamond was exported and 9,3 million carats of polished diamond was imported. The average per carat value of exports in 2006 was $ 1.088, which was up from $ 1.049 per carat in 2005. The United States remained the most important export market for cut diamond. The diamond sector accounted for 8% of Belgium's total exports. Eight in ten rough diamonds in the world are handled in Antwerp.[426]

Belgium's well developed industrial minerals sector included the production of such industrial materials as carbonates and silica sand. Important producers of industrial minerals included SRC-Sibelco S. A., a world leader in silica sand production, and Carmeuse S. A., a world leader in the handling of limestone. Belgium was also a significant producer of cement.[427]

Belgium was also an important producer of marble for more than 2.000 years, was recognized for the diversity and quality of its dimension stone. All the marble quarries are located in the Walloon Region. Most of the marble products were exported in 2008.

---

426 Antwerp World Diamond Center, 2007: in USGS 2008 (Belgium).
427 Mining Journal Online, 2006, in: USGS 2008 (Belgium).

Chapter 5 View of the minerals policies in selected states of Europe

Table 33: Industrial minerals: production/mining – import – export of commodities of Belgium (Data by BGS, 2010 [European Mineral Statistics])

Commodities related to industrial minerals/products	Asbestos	Baryte	Bentonite	Diatomite	Diamond	Feldspar	Fluorspar	Gypsum	Graphite	Kaolin	Magnesite	Perlite	Potash	Phosphate Rock	Salt	Sulfur	Talc	Cement
Production/mining (x: mine)										x						rec		cf
Import	um	x	x		us gr gc i d	x	x	cr ca	x	x	me ma		fs sul chl		x	cr ref	x	cc pc
Export (+ Luxembourg)		x	x		us gr gc i d	x	x	cr ca		x	me ma		fs sul chl	x	x	x	x	cc pc

a	Anhydrite	cr	Crude	i	Industrial	rs	Rock salt
bs	Brinesalt	d	Dust	ma	Magnesia	sib	Salt in brine
ca	Calcined	fs	Fertiliser salts	me	Magnesite	ss	Seasalt
cc	Cement clinker	g	Gypsum	pc	Portland cement	sul	Sulphate
cf	Cement finished	gc	Gem cut	pf	Potassic fertilisers	um	unmanufactured
chl	Chloride	gr	Gem rough	rec	recovered		

Construction minerals

Aggregates are economically important mineral raw materials (particularly needed for infrastructure development). The production of aggregates increased constant to 72 Mt in 2008 from 50 Mt in 2002. The production accounted in 2003: 58 Mt (i.e. 8,7 Mt sand gravel (s&g), 46,6 Mt cr, 3,1 Mt recycled aggregates (ra)), in 2004: 63,6 Mt (9,2 Mt s&g, 47,7 Mt cr, 7 Mt ra), in 2005: 65 Mt (13,9 Mt s&g, 38 Mt cr, 12 Mt ra, 1,2 Mt manufactured aggregates), in 2006: 83 Mt (10,1 Mt s&g, 55,5 Mt cr, 3,5 marine aggregates, 13 Mt ra, 3,0 manufactured aggregates). In 2008, Belgium produced (72 Mt, i.e.) 11 Mt s&g, 42 Mt cr, 4 Mt marine aggregates, 14 Mt recycled aggregates and 2 Mt manufactures aggregates. In 2008, the total number of producers was 280; the total number of extraction sites was 253.

Minerals policy related approaches

Belgium does not have a national raw materials strategy. There is no specific planning for mineral extraction in Belgum at national level. It is mainly the market that determines the nature and quantity of industrial minerals and construction minerals that are to be extracted. However, the Flemish Region was establishing in 2004 a framework for defining plans of surface mineral resources. The plans examine future development over a 25-year period and contain actions for a period of five years. These plans have to be evaluated every five years.[428] There is no regulation, specific for mineral exploration activities. The extraction permit is limited to 20 years.[429]

Investment policy issues

In 2006, Arcelor S. A. and Mittal Steel N. V. announced that they had agreed on an improved bid for Arcelor ($ 33,1 billion) by Mittal Steel that would create the world's first steel producer with a production capacity of more than 100 million metric tons per year (Mt/yr). A combination of the two companies would give Mittal control of an estimated 11 % of the world's annual output. The combination would create a world leader three times larger than its nearest rival, Nippon Steel Corp. of Japan.[430] The proposed merger of Arcelor S. A. and Mittal Steel N. V. was approved by their shareholders in 2007. A combination of the world's number one steel company (Arcelor) and the world's number two steel company (Mittal) would result in control of an estimated 11 % of the world's annual output. The combination, to be named ArcelorMittal and based in Luxembourg, would create a world leader three times larger than its nearest rival, Nippon Steel Corp. of Japan.[431]

In late 2006, Umicore and Ziniflex Ltd. of Australia agreed to a merger that would create the world's leading zinc producer by combining their smelting assets. The proposed deal would result in a smelting and refining entity that would employ 4.500 people and produce 1,2 Mt/yr of metals and materials. That is equal to 10 % of the global demand for zinc in 2006.[432] In 2007, Nyrstar S. A. announced that it had taken ownership of the lead and zinc smelting and alloying assets of N. V. Umicore S. A. and Ziniflex Ltd., thereby formally launching the company and creating the world's leading zinc producer. The Zinifex assets were the Budel (Netherlands), Clarksville (Gordensville, Ten-

---

428 Vervoort, in: Department of Mining and Tunneling (2008), Final Report, p 133, l.c.
429 Ibidem.
430 Forbes, 2006, in: USGS 2008 (Belgium).
431 ArcelorMittal, 2007, in USGS (2009): Minerals Yearbook 2007, Volume III – Belgium.
432 Mining Engineer, 2007, in: USGS 2009 (Belgium).

nessee), and the Hobart and the Port Pirie (Australia) smelting and alloying operations. The Umicore assets were the Auby and the GM Metal (France) and the Balen and the Overpelt (Belgium) smelting and alloying operations. With operations on four continents, Nyrstar produced more than 1 million metric tons of zinc and zinc alloys in 2006, which was equivalent to 10% of the global market. Also, in 2006, Nyrstar was one of the leading primary lead smelting and refining companies in the world.[433]

Poland

The Gross Domestic Product increased to $529 billion in 2008 from $171 billion in 1998. The value of the gross output of industry (mining, materials good production, supply of energy) decreased to 28,7% in 2008 from 32,9% in 1998 and then increased to 32% in 2008. In 2006, the value of mining and quarrying output was 2,5% of the GDP. In 2008, mining and quarrying made up about 2% of the total GDP.[434] In 2006, the value of the gross output of industry represented about 22% of the GDP; the value of mining and quarrying output was 2,5% of the GDP.[435, 436]

---

433 Mining Engineer, 2007, in: USGS 2009 (Belgium).
434 USGS (2008, 2010): Minerals Yearbook 2008, 2010 Volume III (Poland).
435 USGS (2008): Minerals Yearbook 2006, Volume III – Poland.
436 Główny Urząd Statystyczny, 2007a, in: USGS 2008 (Poland).

5.2 EU States

Figure 96: Mining production in Poland of selected mineral resource including aluminium production (Data by Weber et al. [WMD 1998–2008])

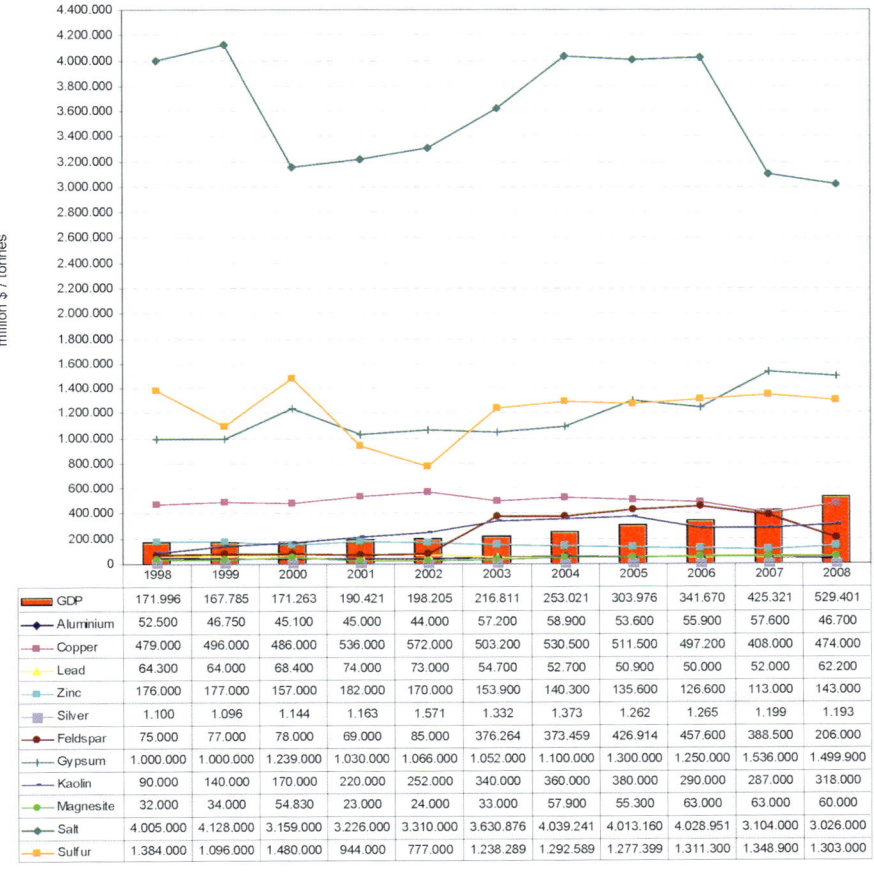

Metallic minerals

Between 1998 and 2008, Poland mined a broad range of metallic minerals that included copper, lead, zinc, gold and silver. Copper mine production was at a constant and high level, average 500.000 t (see figure 96). Lead mine production was also at a constant level (average 60.000 t). Zinc mine production constantly decreased to 143.000 t in 2008 from 176.000 t in 1998. Poland was the largest silver producer of the EU-27, with an average of 1.200 t and a culmination in 2002 with 1.570 t.

In terms of world rankings, Poland was the 6th ranked mine producer of silver, the 11th ranked mine producer of copper. Poland's dependence on imports of such minerals as iron ore and concentrate will likely continue into the foreseeable future and continue to result in a negative trade balance for mineral commodities.[437]

Poland has no bauxite deposits and all alumina for aluminum production was imported. Between 1998 and 2006, aluminium production was on a constant level (average 50.000 t). In 2007, Poland imported 71.600 metric tons (t) of bauxite and 164.000 t of alumina. Bauxite imports were used for the production of aluminous cement, chemicals, refractory materials, and steel. No imported bauxite was used to produce alumina. Germany (88.600 t), Ireland (27.900 t), and Ukraine (14.800 t) were the main sources of Polish alumina imports. The Konin-Impexmetal aluminum smelter at Konin used about 64 % of the alumina imports in the production of primary aluminum; the remaining 36 % of imports was used to produce aluminous cement, chemicals, electroceramics, glass, and high-alumina refractory materials.[438]

Poland has produced no iron ore since 1990 and remained dependent on imported iron ore and concentrates to supply domestic steel producers. In 2007, the Bet total iron ore and concentrates imports were about 8,75 Mt of iron ore, of which 52% came from Russia and 41% from Ukraine. All the imported iron ore and concentrates were used for pig iron production at ArcelorMittal Poland S.A.'s two plants at Dabrowa Gornicza and Krakow.[439]

In 2006, Poland was the leading producer of copper in Europe and Central Eurasia (after Russia) and remained among the top 10 world mine producers of copper.[440] All copper ore in Poland was mined by Kombinat Gorniczo Hutniczy Miedzi (KGHM) Polska Miedz S. A. (KGHM, S. A.), which was a major world copper mining, smelting, and refining complex in the Lubin area (see figure 86). KGHM accounted for about 3,4 % of world mine copper production in 2006. In 2006, the country's reserve base of copper amounted to more than 5 % of the world total. In 2006, Poland was among the major world producers of silver and accounted for more than 6 % of world mine production (Brooks, 2008). The country's copper mining, smelting, and refining complex, which was operated by KGHM in the Lubin area, produced about 98 % of the country's byproduct silver. The top three importers of Polish silver in 2006 were (in

---

437 USGS (2010): Minerals Yearbook 2008, Volume III – Poland.
438 Ney, Smakowski, and Galos, 2009, p. 45–49, in: USGS (2010), Poland.
439 Ney, Smakowski, and Galos, 2009, p. 240–241, in: USGS (2010), Poland.
440 Edelstein, 2008, in: USGS (2008) (Poland), p.4.

descending order of value) the United Kingdom, Germany, and Belgium.[441] Poland's share of global reserves of silver amounted to about 19%.[442]

The following figure illustrates the production, import and export of copper in Poland.

Figure 97: Production, export and import of copper (Data by BGS)

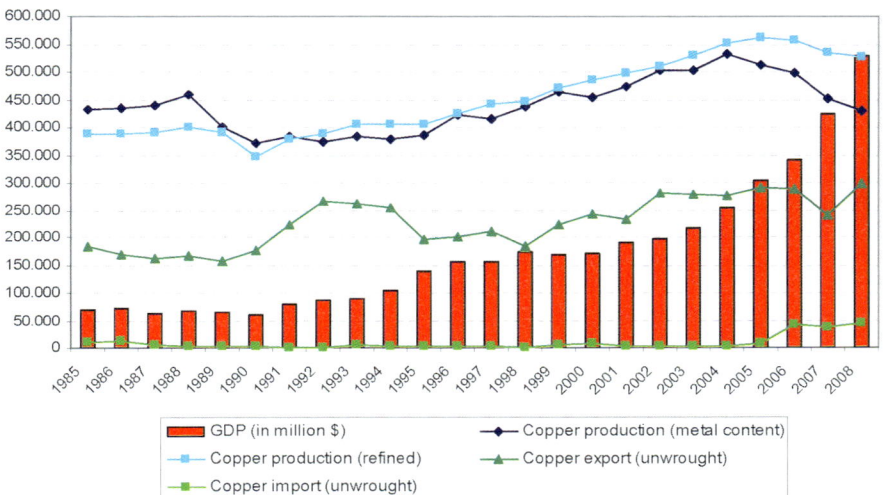

Poland is expected to remain an important world supplier of copper, and a major supplier of lead, and zinc to the European market. The country, however, will continue to rely on mineral imports. According to the Mineral and Energy Economy Research Institute of Poland's Academy of Sciences, of the 121 mineral commodities that were reviewed, 56 (40%) were in the category of total import dependence. Additionally, seven mineral commodities, or about 6% of the total, were in the category of import dependence of more than 50%. The country, especially, will continue to rely on imports of iron ore and concentrate. As domestic sources of nonferrous metals become exhausted (2015 and beyond), Poland will increasingly depend on imports of commodities.

---

441 Główny Urząd Statystyczny, 2007b, p. 261, in: USGS 2008 (Poland), p.4.
442 Brooks, 2008, in: USGS 2008 (Poland), p. 4.

Chapter 5 View of the minerals policies in selected states of Europe

Table 34: Metallic minerals: production/mining – import – export of commodities of Poland (Data by BGS, 2010 [European Mineral Statistics])

Commodities related to metallic minerals/products	Production/mining (x:mine)	Import	Export
Iron	pi, cr, fa	ore, pi, fa, s	ore, pi, fa, s
Chromium	m	oc, m	m
Cobalt		m, o	m
Manganese		oc, m	oc, m
Molybdenum		m	m
Nickel		u, s	u, s, o
Tantalum		x	
Titanium		tm, m, o	m, o
Tungsten		m	m
Vanadium		m	
Aluminium	pa	a, ah, u, ua, s	u, s
Antimony		m, o	
Arsenic		ma	
Bauxite		x	x
Bismuth		m	
Cadmium	m		m
Copper	x, sm, ref	oc, mat, u, s	m, mat, u, ua, s
Gallium			
Lead	x, ref	u, s	oc, u, s
Lithium		o, car	
Mercury		x	x
Rare-earth		cc, orec, m	cc, orec (since 2007), m (since 2008)
Tin		u, s	u, s
Zinc	x	oc, u, ua, s	oc, u, ua, s
Gold	x	x	m
PMG		m, m	m, m, s
Silver	x		m

a	Alumina	fc	Ferro-cerium	orec	Oher rare earth compounds	sul	Sulfide
ah	Alumina hydrate	ilm	Ilmenite	otm	Other titanium minerals	sz	Slap zinc
bp	Burnt pyrites	m	Metal	pa	Primary aluminium	tm	Titanium minerals
c	Concentrate	ma	Metallic arsenic	pen	Pentoxide	ts	Titanium slag
car	Carbonate	mat	Matte and cement	pi	Pig iron	u	unwrought
carb	Carbides	mi	Mine	rec	Rare earth compounds	ua	unwrought alloys
cc	Cerium compounds	o	Oxide	ref	refined	ur	urefined
cr	Crude steel	oc	Ores and concentrates	s	Scrap	vana	Vanadiferous residues
fa	Ferro alloys	or	Ore	sm	Smelter	wa	White arsenic

Industrial minerals

Poland produced a broad range of industrial minerals that included calcareous and silicate rocks, feldspar, gypsum, magnesite, salt, and sulphur, which served the needs of the country's chemical and construction industries. Between 1998 and 2008, Feldspar production (see figure 96) increased to 457.000 t in 2005 from 75.000 t in 1998; between 2002 and 2003 production increased from 85.000 t to 376.000 t, and decreased again to 206.000 t in 2008. Gypsum production between 1998 and 2006 was at a constant level (average) of 1 Mt, in 2008 the production increased to 1,5 Mt. Kaolin production increased to 318.000 t in 2008 from 90.000 t in 1998. Also the magnesite production increased to 60.000 t in 2008 from 32.000 t in 1998. Salt production ranged between 3 Mt and 4 Mt. Sulphur production was at an average value of 1,2 Mt. Poland was among the leading world producers of lime (especially in Europe and Central Eurasia), salt, and sulphur.[443]

Table 35: Industrial minerals: production/mining – import – export of commodities of Poland (Data by BGS, 2010 [European Mineral Statistics])

Commodities related to industrial minerals/products	Asbestos	Baryte	Bentonite	Diatomite	Diamond	Feldspar	Fluorspar	Gypsum	Graphite	Kaolin	Magnesite	Perlite	Potash	Phosphate rock	Salt	Sulfur	Talc	Cement
Production/mining (x: mine)	x	x	x			x		x (+anh)		x	x				rs bs	rec		cc cf
Import	x	x	x	i d	x	x	x	cr ca	x	x	me ma		fs sul chl	x	x	x	x	cc pc
Export	x	x				x		cr ca	x		me ma			pf	x	x	x	cc pc

a	Anhydrite	cr	Crude	i	Industrial	rs	Rock salt
bs	Brinesalt	d	Dust	ma	Magnesia	sib	Salt in brine
ca	Calcined	fs	Fertiliser salts	me	Magnesite	ss	Seasalt
cc	Cement clinker	g	Gypsum	pc	Portland cement	sul	Sulphate
cf	Cement finished	gc	Gem cut	pf	Potassic fertilisers	um	unmanufactured
chl	Chloride	gr	Gem rough	rec	recovered		

Construction minerals

Aggregates are economically important mineral raw materials (particularly needed for infrastructure development). The production of aggregates increased to 203 Mt in 2008 from 147 Mt in 2004. The production was in 2004: 147,5 Mt

---

443 Kostick, 2008; Kramer, 2008; Miller, 2008; Ober, 2008, in: USGS 2008 (Poland), p.6.

(i.e. 105 Mt sand and gravel (s&g), 40 Mt crushed rock (cr), 2,5 recycled aggregates (ra)), in 2005: 150,8 Mt (104,3 Mt s&g, 37,7 Mt crushed rock (cr), 7,2 recycled aggregates (ra), 1,6 Mt manufactured aggregates), in 2006: 169 Mt (115 Mt s&g, 43 Mt cr). In 2008, Poland produced (203 Mt, i.e.) 131 Mt s&g, 49 Mt cr, 22 Mt recycled aggregates and 1 Mt manufactured aggregates. In 2008, the total number of producers was 2.044; the total number of extraction sites was 1.786.

Minerals policy related approaches

The minerals policy for the period 2009–2030 is in preparation. The Ministry of Environment transferred to the Ministry of Economy a draft version of this policy in 2010.[444] There is no national mineral resource planning system to protect economically attractive deposits. According to Article 9 ML the operator may prospect for explore or exploit a designated mineral. The granting of a concession for exploitation of minerals is based on a bidding procedure (Art 10–11 ML).[445] The exploitation license is issued for a maximum of 50 years. The licensing procedures in Poland are long and complex, particularly regarding EIA procedures. In particular, the land use planning procedures are lengthy.[446]

Figure 98: Facility in Rudna, Kombinat Gorniczo Hutniczy Miedzi (KGHM)

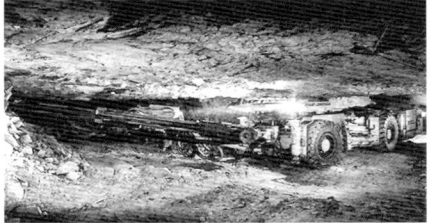

KGHM is an important copper mining industry with processing, smelting and refinery plant in the area of Lubin. **KGHM produces approx. 3.4% of world copper production**. Additionally parallel a silver production takes place in the context of copper production (in 2006 more than **6% of world-wide silver production**).

Source: http://www.mining-technology.com

KGHM – Applied extraction techniques

Source: http://www.mining-technology.com

---

444 LCU (2010): Table 1 'Mineral Policy' (Poland). However, the content of this draft is unkown, it is not clear whether this draft provides a comprehensive raw materials strategy covering all mineral categories.
445 Further information: see Tiess (2010), l.c.
446 Uberman, R., Ostrega, A., in: Department of Mining and Tunnelling (2004), Country Report Poland, l.c.

Germany

The Gross Domestic Product increased to $3,6 trillion in 2008 from $2,2 trillion in 1998. The value of the gross output of industry (mining, materials good production, supply of energy) decreased to 29,8 % (2008) from 30,9 % (1998). Germany was a globally leading exporter of industrial goods and services (including processed and refined mineral products). Germany produced more than 1 % of the total world production of aluminium, barite, bentonite, bromine compounds, cadmium (secondary), cement, diatomite, feldspar, gallium, gypsum, kaolin, crude iron, lime, magnesium compounds (as byproducts of potash mining), nitrogen (ammonia), potash, industrial quartz, salt, selenium (as a by-product of copper refining), silica (industrial sand and gravel), crude steel, and sulphur.[447] In addition, Germany's domestic mineral processing sector accounted for at least 5 % of the world's total production capacity of alumina, fused aluminum oxide, graphite, magnesium metal (secondary), rhenium metal (by-product), strontium compounds, and titanium dioxide pigments.[448] The country's mineral industry, however, depended almost entirely on imported mineral raw materials.[449]

---

447 USGS (2008): Minerals Yearbook 2006, Volume III – Germany.
448 l.c.
449 Bundesanstalt für Geowissenschaften und Rohstoffe, 2006, p. 138, 150, 155, 157, 159, 161, 165; Statistik der Kohlenwirtschaft e.V., 2007; Statistisches Bundesamt, 2007, p. 12; U.S. Library of Congress, Federal Research Division, 2007, p. 10, in: USGS 2008 (Germany), S. 1. [Federal Institute for Geosciences and Natural Resources, 2006, p. 138, 150, 155, 157, 159, 161, 165; Statistics of the Coal Industry inc. soc., 2007; Federal Statistical Office, 2007 p. 12; U.S. Library of Congress, Federal Research Division, 2007, p. 10, in: USGS 2008 (Germany), p. 1.].

Figure 99: Mining production in Germany between 1998 and 2008 – selected mineral resources including iron and aluminium production (Data by Weber et al. [WMD 1998–2008])

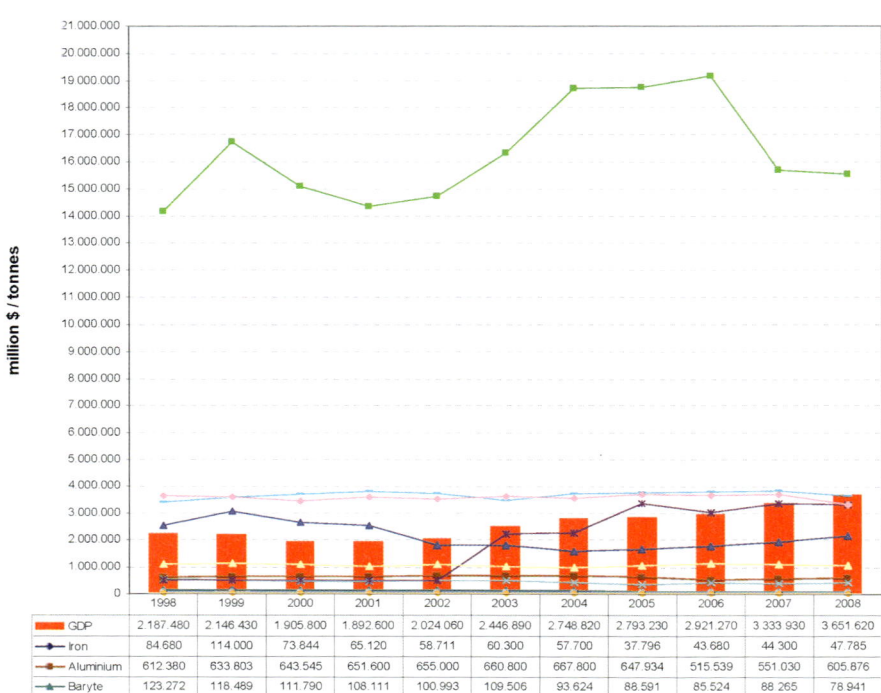

Metallic minerals

Between 1998 and 2008, Germany's metal processing sector relied on imports of metal ores and concentrates and reprocessing of metallic scrap and waste materials (both imported and produced domestically), because no metals were mined in sufficient concentrations for domestic metallurgical use.[450] Iron production decreased to 47.000 t in 2008 from 84.000 t in 1998. Aluminium pro-

---

450 Bundesanstalt für Geowissenschaften und Rohstoffe, 2009, p. 19–20, 25–26, 35–36, 44, 172; Statistik der Kohlenwirtschaft e.V., 2009, in: USGS (2010): Minerals Yearbook 2008, Volume III (Germany).

duction showed an average value of 600.000 t. In 2008, Germany accounted for 1,5 %, 3,8 %, 4,7 %, and 3,4 % of the world's production of each of these metal commodities, respectively. Germany was also the EU's third ranked producer of refined zinc and accounted for about 2,5 % of global production. In addition to any contribution to the GDP, Germany's recycling industry is important to the economy because it reduces the country's reliance on imports of mineral raw materials (especially of metallic minerals for the metal processing sector), helps safeguard the domestic supply, and promotes a sustainable supply of mineral raw materials.

Approximately 1.260 metal-processing plants were operating in Germany in 2006, of which about 70 were focused exclusively on the recovery of metals from secondary sources.

In 2008, the mining industry enterprise KSL Kupferschiefer Lausitz GmbH planned to start with the development of a 700 million € copper ore mine in the Lausitz.[451] Moreover, in 2009 and 2010 KSL wanted to accomplish geophysical field works and drillings, in order to confirm the copper ore reserves known since GDR times and to explore the deposit (due to the financial crisis this did not happen).

---

451 KSL ist am 1. September 2007 als 100%ige Tochter von MINERA gegründet worden. (KSL was founded on 1st September, 2007 as a 100% subsidiary of MINERA.)

Chapter 5 View of the minerals policies in selected states of Europe

Table 36: Metallic minerals: production/mining – import – export of commodities of Germany (Data by BGS, 2010 [European Mineral Statistics])

Commodities related to metallic minerals/products	Production/mining (x: mine)	Import	Export
Silver		oc, m	m
PMG		m, s	m, s
Gold		m, s	m, s
Zinc	sz	oc, u, ua, s	oc, u, ua, s
Tin		u, ua, s	c, u, ua, s
Rare-earth		cc, orec, fc, m	rec, m
Mercury		x	x
Lithium		o, car	o, car
Lead	ref	oc, u, s	oc, u, s
Gallium			
Copper	sm, ref	oc, mat, u,ur, u,ref, ua, s	oc, mat, u,ur, u,ref, ua, s
Cadmium	x	m, o, sul	m, sul
Bismuth		m	m
Bauxite	x		x
Arsenic		ma	
Antimony		m, o	m, o
Aluminium	a, pa	a, ah, u, ua, s	a, ah, u, ua, s
Vanadium		oc, pen, m	oc, pen, m
Tungsten		oc, m, carb	m
Titanium		tm, ts, m, o	tm, m, o
Tantalum	x	x	o
Nickel		oc, mat, u, ua, s	oc, mat, u, ua, s
Molybdenum		oc, m, o	oc, m, o
Manganese		oc, m	
Cobalt	m, o	m, o	ore
Chromium	oc, m	oc, m, o	
Iron	ore, pi, fa	ore, pi, fa, s	ore, pi, fa, s

a	Alumina	fc	Ferro-cerium
ah	Alumina hydrate	ilm	Ilmenite
bp	Burnt pyrites	m	Metal
c	Concentrate	ma	Metallic arsenic
car	Carbonate	mat	Matte and cement
carb	Carbides	mi	Mine
cc	Cerium compounds	o	Oxide
cr	Crude steel	oc	Ores and concentrates
fa	Ferro alloys	or	Ore
otm	Other titanium minerals	sz	Slap zinc
pa	Primary aluminium	tm	Titanium minerals
pen	Pentoxide	ts	Titanium slag
pi	Pig iron	u	unwrought
rec	Rare earth compounds	ua	unwrought alloys
ref	refined	ur	unrefined
s	Scrap	vana	Vanadiferous residues
sm	Smelter	wa	White arsenic
sul	Sulfide		

Industrial minerals

Between 1998 and 2008, production of gypsum, kaolin, potash, salt and sulphur was significant and at high levels (see figure 99): gypsum with average values of 2 Mt, kaolin with an average value of 3,7 Mt, potash with an average value of 3,5 Mt, salt with an average value of 15 Mt and sulphur with an average value of 1 Mt. Feldspar production strongly increased, from 470.000 t (1998) to 3,3 Mt (2008); the increase mainly occurred between 2002 and 2003 (480.000 t to 2,2 Mt). Fluorspar production increased from 30.000 t (1998) to 48.000 t (2008). The graphite production has ceased in 2006. The baryte production decreased from 123.000 t (1998) to 79.000 t (2008). Also the bentonite production decreased from 1998 to 2008 (508.000 t to 414.000 t).

In 2008, Germany was the leading producer of salt (NaCl), kaolin, and potash in the 27-member EU and was the second, third, and fourth ranked producer of these commodities, respectively, in the world. The country was also a leading producer of barite, bentonite, crude gypsum, and feldspar in the EU. The production of industrial minerals is estimated at about 1 billion dollar.

Table 37: Industrial minerals: production/mining – import – export of commodities of Germany (Data by BGS, 2010 [European Mineral Statistics])

Commodities related to industrial minerals/products	Baryte	Bentonite	Diatomite	Diamond	Feldspar	Fluorspar	Gypsum	Graphite	Kaolin	Magnesite	Perlite	Potash	Phosphate rock	Salt	Sulfur	Talc	Cement
Production/mining (x: mine)	x	x			x		x		x			x		rs, bs, sib	rec		cc, cf
Import	x	x	x	g, i, d	x	x	cr, ca	x	x	me, ma		sul, chl	x	x	x	x	cc, pc
Export	x	x	x	us, g, i, d	x	x	cr, ca	x	x	me, ma	x	x	x	x	x	x	pc

a	Anhydrite	d	Dust	ma	Magnesia	ss	Seasalt
bs	Brinesalt	fs	Fertiliser salts	me	Magnesite	sul	Sulphate
ca	Calcined	g	Gypsum	pc	Portland cement	um	unmanufactured
cc	Cement clinker	g	Gem	pf	Potassic fertilisers	us	unsorted
cf	Cement finished	gc	Gem cut	rec	recovered		
chl	Chloride	gr	Gem rough	rs	Rock salt		
cr	Crude	i	Industrial	sib	Salt in brine		

The following figure illustrates the production, import and export of salt in Germany

Figure 100: Production, export and import of salt (Data by BGS)

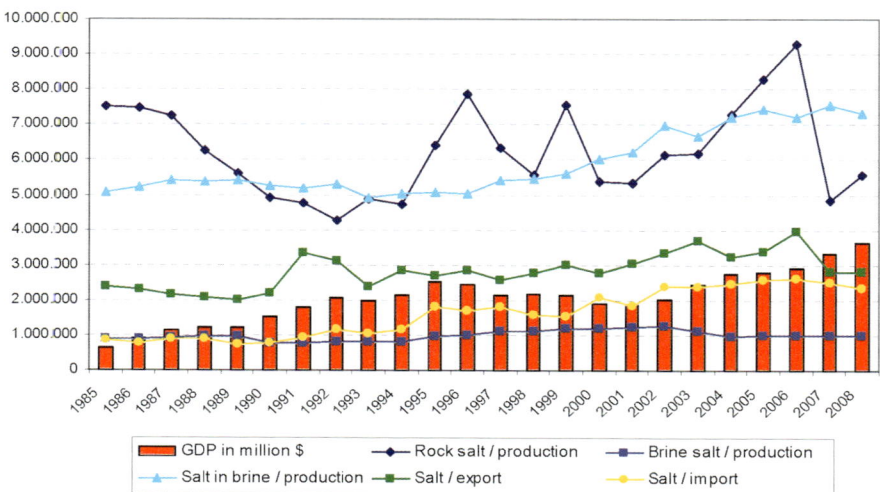

Construction minerals

Aggregates are economically important mineral raw materials (particularly needed for infrastructure development). Germany is one of the largest aggregates producers in Europe. The production of aggregates decreased to 552 Mt in 2008 from 564 Mt in 2002 (see also figure 101). The production accounted in 2003: 502,7 Mt (i.e. 297 sand and gravel (s&g)/165,7 crushed rock (cr), 90 Mt recycled aggregates (ra)), in 2004: 526 Mt (297 Mt s&g /179 Mt cr, 50 Mt ra), in 2005: 513 Mt (263 Mt s&g, 174 Mt cr, 46 Mt ra, 30 Mt manufactured aggregates), in 2006: 625,4 Mt (277 Mt s&g, 270 Mt cr, 0,4 marine aggregates, 48 Mt ra, 30 Mt manufactured aggregates). In 2008, Germany produced (552 Mt aggregates i.e.) 260 Mt s&g, 218 Mt cr, 56 Mt recycled aggregates and 18 Mt manufactured aggregates. In 2008, the total number of producers was 2.300; the total number of extraction sites (active quarries and pits) was 1.510.

Figure 101: Sand & Gravel production in Germany (Data by German sand & Gravel Association)

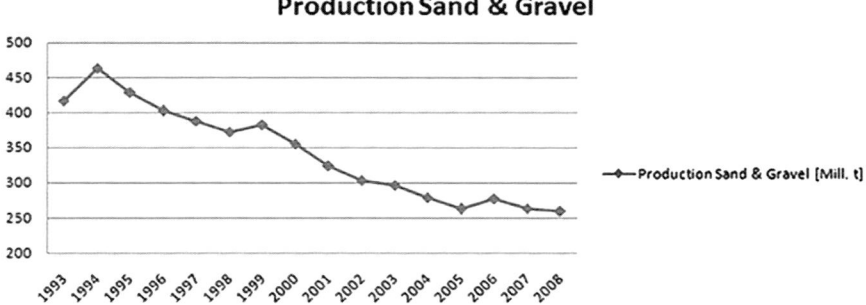

Figure 102: "Material queue" (Data by http://www.bv-miro.org)

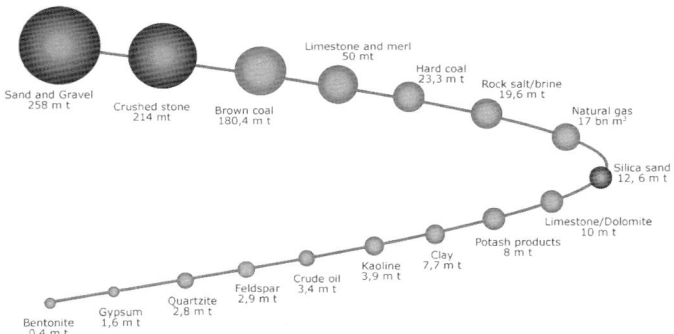

Minerals policy related approaches

The national minerals strategy of Germany was published by the Minister of Economy and Technology (see http://www.bmwi.de/BMWi/Navigation/Service/publikationen,did=365186.html).

The content of the strategy is the following:

1. Introduction
2. Regulatory framework
3. Measures against trade barriers and distortions
4. Measures to diversify sources of supply of raw material
5. Resource efficiency

6. Recycling
7. Mineral resources in the value chain
8. Material Efficiency
9. Promotion of education and training of foreign specialists and managers in the field of raw materials/commodities
10. Importance of derivatives and financial issues in the commodities trade
11. Structural measures
12. Political support
13. Development Cooperation
14. Bilateral commodity partnerships
15. European policy on raw materials
16. Minerals policy in an international context

The strategy includes objectives and actions. Objectives of the strategy are:

- Reduction of trade barriers and distortions
- Support of the German economy to diversify their raw materials sources
- Support of the economy in the development of synergies through sustainable management and increased material efficiency
- Further development of technologies and tools to improve the framework for recycling,
- development of bilateral commodity partnerships with selected countries
- Opening of new options by substitution and materials research
- Focus resource-related research programmes
- establishment of transparency and good governance regarding minerals extraction
- Cooperation among national measures with the European minerals policy.

The Ministry of Economy and Technology has set up the German raw materials agency on October 4th, 2010 at the Federal Institute for Geosciences and Natural Resources. The raw materials agency will be responsible for (amongst others):

- The establishment of a raw materials information system: Based on such a system the transparency in commodity markets shall be improved to provide the German economy a better basis regarding their efforts for minerals supply securing.

- Technical support for Federal Government concerning establishment and implementation of development/promoting programmes in the areas of exploration, extraction as well as material efficiency;
- Research and development projects: New resource potential shall be studied and new commodity-based instruments and methods shall be developed.
- Cooperation with resource-rich countries: In cooperation with developing countries, the raw material agency, will focus the sustainable exploitation of the resource potentials.

National mineral resources planning system

The national mineral resources planning system is based on the Land use planning law at national level (though sometimes not for all minerals)[452], considering minerals and obliging the provinces and districts to carry out land use plans. The national Geological Surveys are involved in mineral exploration, mineral evaluation and mineral protection. The different organization types of the national Geological Surveys in the Federal states, the allocation to the divisions of different national ministries and the resulting different importance of mineral-geological organizational units in the national Geological Surveys, affect perception and estimation of raw material protection in the Federal Countries.

On the basis of a stronger consciousness for mineral raw materials the Federal Ministry for Traffic, Building and Urban Development (BVBS) had planned a first draft for the introduction of a "Federal Development Plan for Raw Materials" for special economically valuable minerals, which should have been obligatory for the regional land use planning of the countries.[453]

---

[452] Department of Mineral Resources and Petroleum Engineering (2010): Planning Policies and Permitting Procedures to Ensure the Sustainable Supply of Aggregates in Europe, commissioned by UEPG, Leoben.

[453] VRB (2008), l.c., p. 47. The recent government draft for the Federal Council and the House of Representatives contains no longer such a provision, as it had not received the necessary approval of all federal ministries. In a joint statement the VRB together with the VKS and the BBS recommended a "Federal Resources Development Plan" and proposed the identification of mineral deposits independent of their need. The concerns of the extraction and mining industries were also discussed in the inter-ministerial working group "Raw Materials". Such an instrument would be appropriate to counteract the insufficient identification of mineral deposits. The identification of mineral deposits and mine reclamation carried out in some federal states should be taken as a legal planning instrument into the ROG.

Regualtory issues

The so-called permission ("Erlaubnis") grants the exclusive right to explore the mineral resources according to the regulations of the German Mining Law (Article 7 ML) in the permitted field (permission field). The approval ("Bewilligung") grants the exclusive right in a certain field (approval field) to explore the mineral resources, to exploit them and other raw materials as well as to acquire the ownership of the resources (Article 8 ML). Moreover, the so-called raw material protection clause of the Federal Mining Law is worth mentioning: Article 1 ML identifies as a goal "to regulate and promote exploration, exploitation and extraction of (state-owned) resources to secure the supply with raw materials". The purpose of the law is elaborated in the "raw material protection clause" (Article 48 paragraph 1 p. 2 Mining Law). Accordingly, "it is to be taken care that the exploration and exploitation are impaired as little as possible".

The permission ("Erlaubnis") is to be limited to a maximum of five years. It can be extended by three years at a time, if the permission field could not yet be sufficiently explored despite of systematic exploration coordinated with the competent authority. (Article 16 (4) Mining Law. The approval ("Bewilligung") or the mining property is granted for the exploitation in each case for an adequate period. Fifty years may only be exceeded if this is necessary in view of the investment normally required for the exploitation. A prolongation up to the prospective exhaustion of the site is allowed in the case of production in accordance with regulations and plan (Article 16 ML). There are several very complex procedures. The time required to obtain a permit, amounts up to eight years in Germany, dependent on Federal states and ecological restrictions.[454]

Figure 103: Amberger Kaolinwerke, Werk Caminau
(Data by http://www.akw-kaolin.com)

Amberger Kaolin: production, processing and refining of quartz, kaolin and feldspar.

---

454 Müller (2004), Country Report Germany, in: Department of Mining and Tunnelling, University of Leoben, Minerals Policies and Supply Practices in Europe.

Investment policy issues

The **export trade and industry** traditionally plays a significant role supporting economic development in Germany. Almost 80 % of the imported raw materials are exported again as refined products and thus are available for recycling only partially.[455] The process for investing in the German mineral industry is unique in some ways. Most importantly, the corporate sector in Germany relies almost exclusively on bank credit (loans) to fund investment instead of issuing securities to the public through market exchanges. This is especially true for small- and medium-scale enterprises (SMEs), which account for the majority of companies invested in the mineral industry of Germany. A foreign company usually has to register a GmbH (private limited-liability company) subsidiary in the country to obtain the same legal benefits as a domestic company. These benefits include Government investment grants, tax benefits, and low-interest loans or bank loans with a state guarantee. These benefits are greater for companies that export.[456]

Czech Republic

The Gross Domestic Product increased to $216 billion in 2008 from $61,8 billion in 1998. The value of the gross output of industry (mining, materials good production, supply of energy) decreased to 37,6 % in 2008 from 39,2 % in 1998.[457] In 2006 the Czech Republic was an important Central European producer of heavy industrial goods manufactured by the country's chemical, machine building, and toolmaking industries. Steelmaking, the mining and processing of industrial minerals and the production of construction materials continued to be of domestic importance.[458] In 2006, mining and quarrying constituted 1,4 % of the Czech economy's net value of output and a 2,5 % share in the value of industrial output.[459]

---

455 BDI (2007), l.c., p. 6.
456 Economist Intelligence Unit Limited, The, 2007; World Tax Inc., 2007, in: USGS 2008 (Germany), p.2.
457 USGS (2008, 2010): Minerals Yearbook 2008, 2010 Volume III (Czech Republic).
458 USGS (2008): Minerals Yearbook 2006, Volume III – Czech Republic.
459 Czech Statistical Office, 2008; GEOFOND, 2007, in: UEPG 2008 (Czech Republik).

Figure 104: Mining production in the Czech Republic between 1998 and 2008 – selected mineral resources (Data by Weber et al. [WMD 1998–2008])

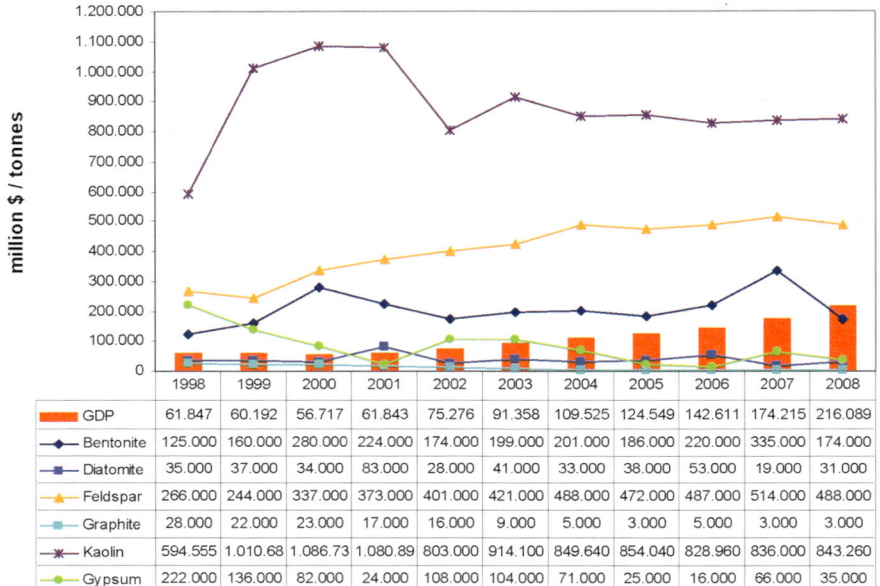

Metallic minerals

The Czech Republic relies on base and precious metals. The Czech Republic's metals sector produced a broad range of base metals and semimanufactures from imported ores and secondary materials (scrap). In 2006, the production increased for such major metals as iron, steel, and steel semimanufactures. Although interest in gold mining continued in some parts of the Czech Republic, other metals were depleted.[460]

---

460 GEOFOND, 2007, in: USGS 2008 (Czech Republic).

Table 38: Metallic minerals: production/mining – import – export of commodities of the Czech Republic (Data by BGS, 2010 [European Mineral Statistics])

Commodities related to metallic minerals/products	Production/mining (x: mine)	Import	Export
Silver		m	m
PMG		m w	m s
Gold		m	m s
Zinc		u ua s	u s
Tin		rec u fc	u s cc orec
Rare-earth			
Mercury		x	x
Lithium		car	
Lead	ref	u s	u s
Gallium			
Copper		u s	u s
Cadmium		m o sul	
Bismuth			
Bauxite		ma x	
Arsenic		m o	
Antimony		m o	o
Aluminium		a ah u ua s	u s
Vanadium		pen	
Tungsten		m m carb	m carb
Titanium		tm m o	tm m o
Tantalum		x	x
Nickel		u o s	s o
Molybdenum		oc m	m
Manganese		oc m	
Cobalt		m o	
Chromium		oc m	oc m
Iron	pi cr fa s	or pi fa s	pi fa s

a	Alumina	orec	Other rare earth compounds
ah	Alumina hydrate	otm	Other titanium minerals
bp	Burnt pyrites	pa	Primary aluminium
c	Concentrate	pen	Pentoxide
car	Carbonate	pi	Pig iron
carb	Carbides	rec	Rare earth compounds
cc	Cerium compounds	ref	refined
cr	Crude steel	s	Scrap
fa	Ferro alloys	sm	Smelter
fc	Ferro-cerium	sul	Sulfide
ilm	Ilmenite	sz	Slap zinc
m	Metal	tm	Titanium minerals
ma	Metallic arsenic	ts	Titanium slag
mat	Matte and cement	u	uwrought
mi	Mine	ua	uwrought alloys
o	Oxide	ur	urefined
oc	Ores and concentrates	vana	Vanadiferous residues
or	Ore	wa	White arsenic

Industrial minerals

Between 1998 and 2008, the Czech Republic mined a broad range of industrial minerals that included bentonite, diatomite, feldspar, graphite, gypsum and kaolin (figure 104). Kaolin production was the highest, it increased to 1 million t (1999) from 594.000t (1998), then the production remained the same until 2001, and then production increased to a constant level of 840.000 t. Feldspar, the second highest production, increased to 488.000 t in 1998 from 266.000 t in 2008 (nearly doubled). Bentonite production increased to 174.000 t in 2008 from 125.000 t in 1998. Diatomite production shows average values of 35.000 t. Graphite production constantly decreased to 3.000 t in 2008 from 28.000 t in 1998. Gypsum production constantly decreased to 35.000 t in 2008 from 222.000 t in 1998.

Table 39: Industrial minerals: production/mining – import – export of commodities of the Czech Republic (Data by BGS, 2010 [European Mineral Statistics])

Commodities related to industrial minerals/products	Asbestos	Baryte	Bentonite	Diatomite	Diamond	Feldspar	Fluorspar	Gypsum	Graphite	Kaolin	Magnesite	Perlite	Potash	Phosphate rock	Salt	Sulfur	Talc	Cement
Production/mining (x: mine)			x	x		x		x	x	x						rec		cc cf
Import	x	x	x	x	gc i d	x	x	cr ca	x	x	me ma	chl	x	x	x	x	x	cc pc
Export		x	x	x	gc	x	x	cr ca	x	x	me ma				x	x	x	cc pc

a	Anhydrite	cr	Crude	i	Industrial	rs	Rock salt
bs	Brinesalt	d	Dust	ma	Magnesia	sib	Salt in brine
ca	Calcined	fs	Fertiliser salts	me	Magnesite	ss	Seasalt
cc	Cement clinker	g	Gypsum	pc	Portland cement	sul	Sulphate
cf	Cement finished	gc	Gem cut	pf	Potassic fertilisers	um	unmanufactured
chl	Chloride	gr	Gem rough	rec	recovered		

Construction minerals

Aggregates are economically important mineral raw materials (particularly needed for infrastructure development). The production of aggregates increased to 76 Mt in 2008 from 67 Mt in 2002. The production accounted in 2003: 67 Mt (i.e. 30 Mt sand and gravel (s&g), 32 crushed rock (cr), 5 recycled aggregates (ra)), in 2004: 52 Mt (24 Mt s&g, 25,5 Mt cr, 2,5 Mt ra), in 2005: 67,2 Mt (25,5 Mt s&g, 38 Mt cr, 3,4 Mt ra, 0,3 manufactured aggregates), in 2006: 72,7 Mt (27,1 Mt s&g, 41,5 Mt cr, 3,8 Mt ra, 0,3 Mt manufactured aggregates). In 2008, the Czech Republic produced (76 Mt aggregates, i.e.) 27 Mt sand and gravel, 44 Mt cr and 4 Mt recycled aggregates. In 2008, the total number of producers was 219; the total number of extraction sites was 489. The following figure is illustrating the moderate development of aggregates consumption per capita/GDP between 1998 and 2008; aggregates consumption increased to 7 t per capita in 2008 from 4 t per capita in 1998.

Figure 105: Aggregates consumption/GDP in Czech Republic (Data by Sitensky, Czech Geological Survey – Geofond, Czech Republic, 2010)

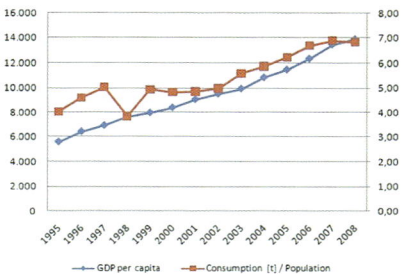

Minerals policy issues

The National Raw Materials Policy was published in 1999 including a comprehensive strategy also covering minerals planning issues for each of the 13 regions. This policy, however, has not been updated since then. An update version of the national raw materials policy is being prepared now.[461]

Regulatory issues

Exploration of mineral deposits can be carried out by organizations in the exploration area that is established pursuant to special regulations (Article 11 ML). Authorization of an operator to extract a *reserved* deposit is established by delimitation of the mining claim. The operator may commence extraction in the delimited mining claim however, only after obtaining a permit from the District Mining Authority. The District Mining Authority can link together administrative procedures for the delimitation of a mining claim and proceedings for the permit of mining activities (Article 24 ML).

---

461 Department of Mining and Tunneling (2004), l.c.; LCU (2010), l.c.

Chapter 5 View of the minerals policies in selected states of Europe

Slovakia

The Gross Domestic Product increased to $95 billion in 2008 from $22 billion in 2008. The value of the gross output of industry (mining, materials good production, supply of energy) increased to 38 % 2008 from 34,6 % (1998).[462] The value of mining and quarrying declined by 9,7 % and constituted 0,5 % of the GDP in 2006.[463]

Figure 106: Mining production in Slovakia between 1998 and 2008 – selected mineral resources including iron and aluminium production (Data by Weber et al. [WMD 1998–2008])

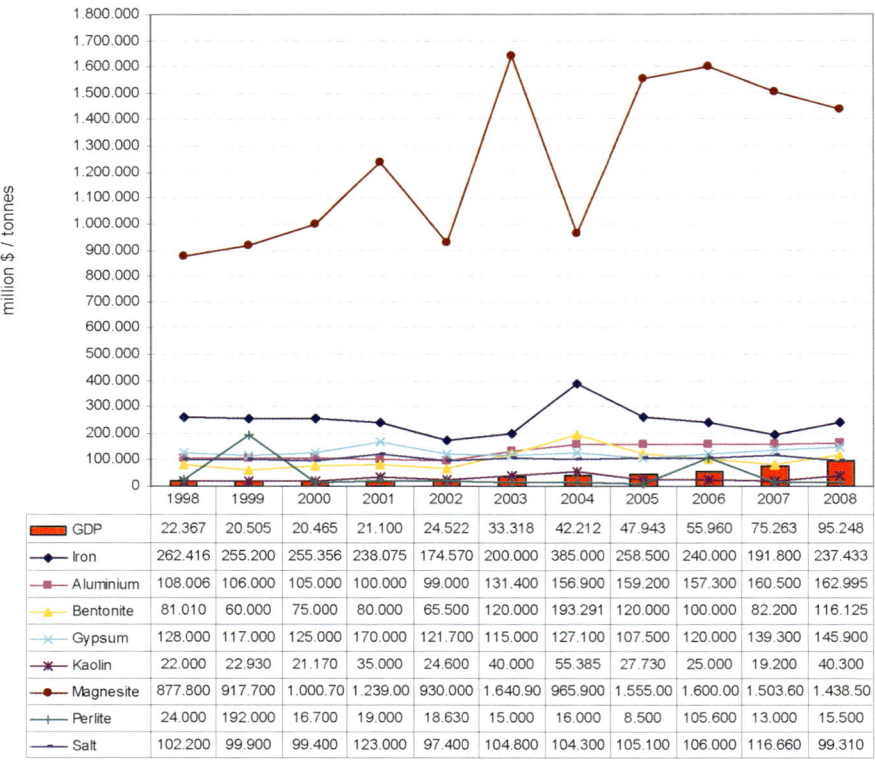

462 USGS (2008, 2010): Minerals Yearbook 2008, 2010 Volume III (Slovakia).
463 Statistical Office of the Slovak Republic, 2007, in: USGS 2008 (Slovakia).

Metallic minerals

Between 1998 and 2008 iron and aluminium production was at a constant level: iron with an average value of 200.000 t (culmination in 2004: 385.000 t), aluminium: average value of 100.000 t. The value of mining and quarrying declined by 9,7 % and constitued 0,5 % of the GDP in 2006.[464]

In 2006, aluminium and steel production formed the dominant elements of the country's metals sector. Steel production was largely based on imported raw materials and that of aluminium was based entirely on imported bauxite. With the possible exception of gold, metal mining has practically ceased owing to depletion of economic reserves. Aluminium and ferrous metals will continue to be produced from imported ores and concentrates. The country will remain dependent on imports metals for its industrial needs. In 2006, Hydro Aluminium AS (Hydro) of Norway acquired a controlling 55,3 % share of Slovalco A.S., which was Slovakia's sole producer of primary aluminium. About 16 % of total sales of aluminium metal was consumed domestically; the balance was exported. In 2006, Italy and Poland were the major importers of Slovakia's aluminium.[465] Activities in the aluminium sector in 2006 included plans announced by Alcan Inc. of Canada to invest $ 35 million to build a new aluminium extrusion plant in Slovakia that would produce products for the construction sector. Completion of the extrusion plant was projected for the first half of 2007.[466]

Figure 107: Slovalco (Data by http://www.slovalco.sk)

---

464 Statistical Office of the Slovak Republic, 2007, in: USGS 2008 (Slovakia).
465 Slovalco A.S., 2006, in: USGS 2008 (Slovakia).
466 Metals Insider, 2006, in: USGS 2008 (Slovakia).

Chapter 5 View of the minerals policies in selected states of Europe

Table 40: Metallic minerals: production/mining – import – export of commodities of Slovakia (Data by BGS, 2010 [European Mineral Statistics])

Commodities related to metallic minerals/products	Production/mining (x: mine)	Import	Export
Silver		m	m
PMG		m	m
Gold	x	m	m u s
Zinc		u	m ua
Tin			m u
Rare-earth		x rec	
Mercury		x	
Lithium			
Lead		u	
Gallium			
Copper	sm since 2007	mat u s	u s
Cadmium			
Bismuth			
Bauxite		ma x	
Arsenic			
Antimony			
Aluminium	pa	a u s	u ua s
Vanadium			
Tungsten		m	
Titanium		tm m	o
Tantalum			
Nickel		u	u, s, since 2006
Molybdenum			
Manganese		oc m	
Cobalt	m		
Chromium	oc m		m
Iron	ore, pi, cr, fa	ore, pi, fa, s	ore, pi, fa, s

a	Alumina	fc	Ferro-cerium	orec	Oher rare earth compounds	sul	Sulfide
ah	Alumina hydrate	ilm	Ilmenite	otm	Other titanium minerals	sz	Slap zinc
bp	Burnt pyrites	m	Metal	pa	Primary aluminium	tm	Titanium minerals
c	Concentrate	ma	Metallic arsenic	pen	Pentoxide	ts	Titanium slag
car	Carbonate	mat	Matte and cement	pi	Pig iron	u	uwrought
carb	Carbides	mi	Mine	rec	Rare earth compounds	ua	uwrought alloys
cc	Cerium compounds	o	Oxide	ref	refined	ur	urefined
cr	Crude steel	oc	Ores and concentrates	s	Scrap	vana	Vanadiferous residues
fa	Ferro alloys	or	Ore	sm	Smelter	wa	White arsenic

Industrial minerals

Between 1998 and 2008, industrial minerals production included that of baryte, bentonite, kaolin, gypsum, perlite, magnesite, salt and talc. Magnesite production (see figure 108) increased to near 1,4 Mt in 2008 from 700.000 t in 1993. Slovakia is the largest magnesite producer of EU-27 and ranked 4th globally in 2008.

The following figure illustrates the production, import and export of magnesite in Slovakia.

Figure 108: Production, export and import of magnesite (Data by BGS)

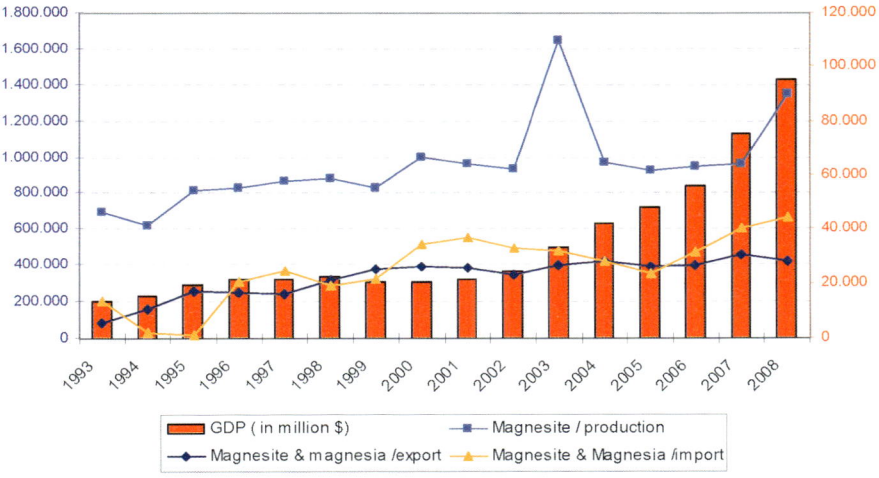

Bentonite production increased to 116.000 t in 2008 from 81.000 t in 1998. The production of kaolin, gypsum, perlite and salt was at a constant level. Kaolin average production was 25.000 t, gypsum average production was 120.000 t, perlite average production was 15.000 t and salt average production was 100.000 t (see figure 106).

Chapter 5 View of the minerals policies in selected states of Europe

Table 41: Industrial minerals: production/mining – import – export of commodities of Slovakia (Data by BGS, 2010 [European Mineral Statistics])

Commodities related to industrial minerals/products	Asbestos	Baryte	Bentonite	Diatomite	Diamond	Feldspar	Fluorspar	Gypsum	Graphite	Kaolin	Magnesite	Perlite	Potash	Phosphate rock	Salt	Sulfur	Talc	Cement
Production/mining (x: mine)	x	x						x		x	x	x			x	rec	x	cf
Import		x	x	x		x	x	cr ca	x	x	me ma	chl	x	x	x	x	x	cc pc
Export		x	x					x					pf	x	x	x	x	cc pc

a	Anhydrite	cr	Crude	i	Industrial	rs	Rock salt		
bs	Brinesalt	d	Dust	ma	Magnesia	sib	Salt in brine		
ca	Calcined	fs	Fertiliser salts	me	Magnesite	ss	Seasalt		
cc	Cement clinker	g	Gypsum	pc	Portland cement	sul	Sulphate		
cf	Cement finished	gc	Gem cut	pf	Potassic fertilisers	um	unmanufactured		
chl	Chloride	gr	Gem rough	rec	recovered				

Construction minerals

Aggregates are economically important mineral raw materials (particularly needed for infrastructure development). The production of aggregates increased to 35 Mt in 2008 from 20 Mt in 2002. The production accounted in 2003: 20 Mt (i.e. 6 Mt sand and gravel (s&g)/14 Mt crushed rock (cr)), in 2004: 20 Mt (4 Mt s&g, 16 Mt cr), in 2005: 26,3 Mt ((8,9 Mt s&g, 16,9 Mt cr, 0,2 Mt recycled aggregates (ra), 0,3 Mt manufactured aggregates)), in 2006: 27 Mt (10 Mt s&g, 16,5 Mt cr, 0,2 Mt ra, 0,3 Mt manufactured aggregates). In 2008, the total number of producers was 219; the total number of extraction sites (active quarries and pits) was 489. The following figure is illustrating the moderate development of aggregates consumption per capita/GDP between 1998 and 2008; aggregates consumption increased to 4 t per capita in 2007 from about 3 t per capita in 1998 and then increased strongly (doubled) to 8 t per capita in 2008.

Figure 109: Aggregates consumption/GDP in Slovakia (Data by Sitensky, 2010)

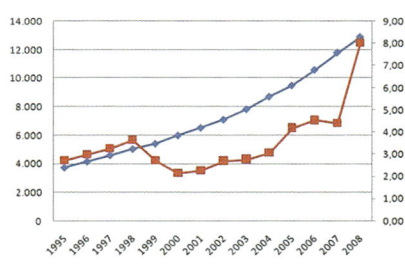

Minerals policy issues

Slovakia does not have a national raw materials strategy. There is no national mineral resource planning system to protect economically attractive deposits.

In 2006, major activities in gold exploration continued by the Tournigan Gold Corp. of Canada at Kremnica. Tournigan's preliminary assessment of the deposit indicated resources at Kremnica to amount to 23,6 Mt of ore at an average grade of 1,37 grams per ton (g/t) gold and 11,36 g/t silver.[467]

---

467 Tournigan Gold Corp., 2008, in: USGS 2008 (Slovakia).

Austria

The Gross Domestic Product increased to $416 billion in 2008 from $212 billion in 1998. The value of the gross output of industry (mining, materials good production, supply of energy) remained at a constant level (30 %).[468]

Figure 110: Erzberg, Steiermark (Data by http://oepg2008.unileoben.ac)

The "Steirische Erzberg" is the largest iron ore open pit in Central Europe. The extracted ore contains 32% iron and 2% manganese. The extracted material is processed in two iron and steel plants of the VOEST Alpine Stahl GmbH, which have a cumulative capacity of 8,5 million tonnes. Austria is self-sufficient in terms of minerals supply from domestic deposits (related to value) for about 23%. However, about 3,9 billion € must be spent annually for the necessary raw materials imports.[469]

---

468 USGS (2008, 2010): Minerals Yearbook 2008, 2010 Volume III (Austria).
469 Weber, L. (2007a): Der Rohstoffplan als Werkzeug einer langfristigen Rohstoffsicherung, Berg- und Hüttenmännische Monatshefte, S. 252–S. 258 [The Raw Material Plan as an instrument for long-term raw material security, Journal of Mining, Metallurgical, Material, Geotechnical and Planned Engineering) p. 252 –258].

5.2 EU States

Figure 111: Mining production in Austria between 1998 and 2008 – selected mineral resources (Data by Weber et al. [WMD 1998–2008])

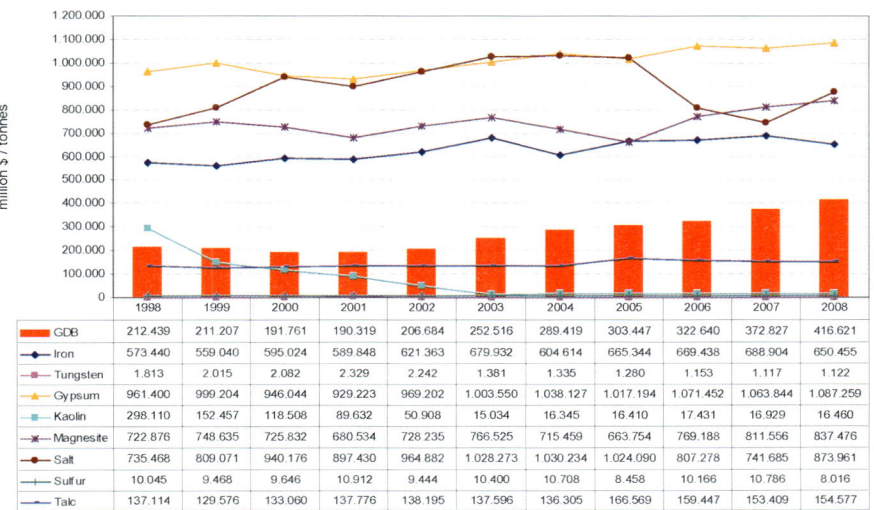

	1998	1999	2000	2001	2002	2003	2004	2005	2006	2007	2008
GDB	212.439	211.207	191.761	190.319	206.684	252.516	289.419	303.447	322.640	372.827	416.621
Iron	573.440	559.040	595.024	589.848	621.363	679.932	604.614	665.344	669.438	688.904	650.455
Tungsten	1.813	2.015	2.082	2.329	2.242	1.381	1.335	1.280	1.153	1.117	1.122
Gypsum	961.400	999.204	946.044	929.223	969.202	1.003.550	1.038.127	1.017.194	1.071.452	1.063.844	1.087.259
Kaolin	298.110	152.457	118.508	89.632	50.908	15.034	16.345	16.410	17.431	16.929	16.460
Magnesite	722.876	748.635	725.832	680.534	728.235	766.525	715.459	663.754	769.188	811.556	837.476
Salt	735.468	809.071	940.176	897.430	964.882	1.028.273	1.030.234	1.024.090	807.278	741.685	873.961
Sulfur	10.045	9.468	9.646	10.912	9.444	10.400	10.708	8.458	10.166	10.786	8.016
Talc	137.114	129.576	133.060	137.776	138.195	137.596	136.305	166.569	159.447	153.409	154.577

Metallic minerals

Although Austria has diverse mineral resources and a long tradition of mining, metal mining activity decreased there during the past few years (through 2008) owing principally to environmental concerns, high mining costs, low ore grades and reserves, and increased foreign competition. All metal mines except the open pit iron ore operation at Erzberg and the underground tungsten operation at Mittersill were closed before 2008. Austria is the biggest tungsten producer of the EU-27 and in 2008 ranked 4th globally.

The following figure illustrates the production, import and export of tungsten in Austria.

Figure 112: Production, export and import of tungsten (Data by BGS)

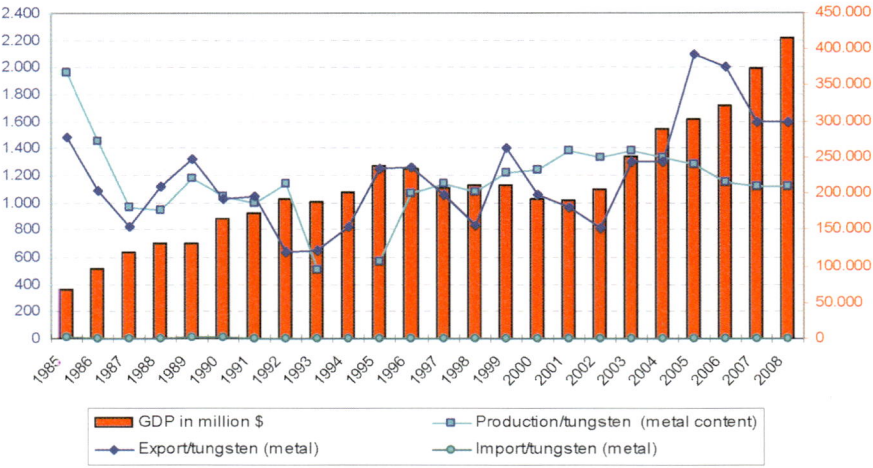

Between 1998 and 2008, iron production (see figure 110) increased to 650.000 t in 2008 from 573.000 t in 1998. Voest-Alpine Erzberg GmbH mined about 2 Mt/yr of iron ore. The Erzberg Mine Rewritten supplied 25% of the iron ore required in the steelmaking process of Voest-Alpine Stahl AG. The rest of the iron ore requirement was imported, mainly from Brazil and South Africa (Voest-Alpine Erzberg GmbH, 2007). Voest-Alpine Stahl AG was an international producer and distributor of steel and steel products. The company's activities were carried out through four divisions: automotive, crude steel, profilform, and railway systems. Voest-Alpine had sales of about $ 14,7 billion in 2007. 88 % of these sales were to the export market. Voest-Alpine was the leading producer in Europe of specialty steel, such as tool steel, and steel strip, steel rails, and high-quality wire.[470]

In 2007, Austria produced secondary aluminum and secondary copper. Montanwerke Brixlegg AG, which was the only copper producer in Austria, specialized in the recycling of copper and other valuable materials to extract pure metals, oxides, and salts from such materials as alloys, blasting grit, residues, and scrap metals using a refining process. Montanwerke Brixlegg processed about 130.000 t/y of secondary material containing copper. Production of secondary copper increased significantly to 81.000 t in 2007 from 72.600 t in 2006.

---

[470] Voest-Alpine Stahl AG, 2007, in USGS (2010), Austria.

Table 42: Metallic minerals: production/mining – import – export of commodities of Austria (Data by BGS, 2010 [European Mineral Statistics])

Commodities related to metallic minerals/products	Production/mining (x: mine)	Import	Export
Silver		m	m
PMG		m m	m s
Gold		u m	u m,s
Zinc		u uo s	u s
Tin		s u	s u
Rare-earth		cc orec m	cc orec m
Mercury		x	x
Lithium		car	
Lead	ref	u,s	u,s
Gallium			
Copper	ref	u,ur u,ref ua	u, ua,s
Cadmium			
Bismuth		x	x
Bauxite		x	x
Arsenic		ma	
Antimony		m o	m
Aluminium		a,ah, u,ua s	u, ua,s
Vanadium		vana oc	
Tungsten	x	m carb	m
Titanium		m o	
Tantalum		x (2004.2007)	x (2007.2008)
Nickel	sm, ref	u, s o	u, s
Molybdenum		m	m m
Manganese		oc m	m
Cobalt		m o	
Chromium		oc m	oc, m
Iron	oc, pi, cr,fa	ore pi fa s	pi, oc, fa, m s

a	Alumina	fc	Ferro-cerium	orec	Oher rare earth compounds	sul	Sulfide
ah	Alumina hydrate	ilm	Ilmenite	otm	Other titanium minerals	sz	Slap zinc
bp	Burnt pyrites	m	Metal	pa	Primary aluminium	tm	Titanium minerals
c	Concentrate	ma	Metallic arsenic	pen	Pentoxide	ts	Titanium slag
car	Carbonate	mat	Matte and cement	pi	Pig iron	u	uwrought
carb	Carbides	mi	Mine	rec	Rare earth compounds	ua	uwrought alloys
cc	Cerium compounds	o	Oxide	ref	refined	ur	urefined
cr	Crude steel	oc	Ores and concentrates	s	Scrap	vana	Vanadiferous residues
fa	Ferro alloys	or	Ore	sm	Smelter	wa	White arsenic

Chapter 5 View of the minerals policies in selected states of Europe

Industrial minerals

Between 1998 and 2008 Austria produced a broad range of industrial minerals that included gypsum, lime, marble, kaolin, magnesite, salt, sulphur and talc. Between 1998 and 2008: gypsum production (see figure 111) was at an average value of 1.000 t. Kaolin production decreased to 16.000 t in 2008 from 298.000 t in 1998. The production of magnesite, salt, sulphur and talc remained at a constant level: the average magnesite production was 700.000 t, the average salt production was 800.000 t, the average sulphur production was 10.000 t and the average talc production was 135.000 t. Austria was the 5th ranked producer of magnesite. The country also accounted for about 1% of the world's production of natural gypsum in 2008.

Table 43: Industrial minerals: production/mining – import – export of commodities of Austria (Data by BGS, 2010 [European Mineral Statistics])

Commodities related to industrial minerals/products	Baryte	Bentonite	Diatomite	Diamond	Feldspar	Fluorspar	Gypsum	Graphite	Kaolin	Magnesite	Perlite	Potash	Phosphate rock	Salt	Sulfur	Talc	Cement
Production/mining (x: mine)							g, a	x	x					rs, sib	rec	x	cc, cf
Import	x	x	x	us, gc, i, d	x	x	cr, ca	x	x	me, ma		fs, sul, chl	x	x	x	x	cc, pc
Export	x	x		us, gc, d	x		cr, ca	x	x	me, ma	x	x	x	x	x	x	cc, pc, oc

a	Anhydrite	cr	Crude	i	Industrial	rs	Rock salt
bs	Brinesalt	d	Dust	ma	Magnesia	sib	Salt in brine
ca	Calcined	fs	Fertiliser salts	me	Magnesite	ss	Seasalt
cc	Cement clinker	g	Gypsum	pc	Portland cement	sul	Sulphate
cf	Cement finished	gc	Gem cut	pf	Potassic fertilisers	um	unmanufactured
chl	Chloride	gr	Gem rough	rec	recovered		

Construction minerals

Aggregates are economically important mineral raw materials (particularly needed for infrastructure development). The production of aggregates ranged between 96 Mt in 2002 and 100 Mt in 2008. The production accounted in 2003: 96 Mt (i.e. 66 Mt sand and gravel (s&g), 27 Mt crushed rock (cr), 3 Mt recycled aggregates (ra)), in 2004: 95 Mt (65 Mt s&g, 27 Mt cr, 3 Mt ra), in 2005: 104,5 Mt

(66 Mt s&g, 32 Mt cr, 3,5 Mt ra), in 2006: 104,5 Mt (66 Mt s&g, 32 Mt cr, 3,5 Mt ra). In 2008, Austria produced (99 Mt aggregates, i.e.) 62 Mt s&g, 32 Mt cr, 4 Mt recycled aggregates and 1 Mt manufactured aggregates. In 2008, the total number of producers (companies) was 960; the total number of extraction sites (active quarries and pits) was 1.290.

Both the construction minerals industry and the processing enterprises secure thousands of jobs and make a large contribution to the Gross Domestic product. Approximately 950 active sand and gravel pits and 250 quarries in Austria secure approx. 6.000 jobs – often in small municipalities and regions. In the building industry there are over 250.000 jobs, which are secured by the raw material economy. Thereby the raw materials extraction industry not only makes a valuable contribution to the national economy, but also to the employment situation.[471]

Minerals policy issues

The Austrian Mineral Resources Plan takes a significant role in the **minerals planning policy**.[472] As it plays a central role for the raw materials protection and likewise for the procedure efficiency[473] in Austria, as well as in the European context[474], this planning instrument will be enlarged upon. **The whole purpose of the Austrian Mineral Resources Plan**[475] is to identify raw materials areas objectively and to protect them after conflict resolution and considerations of nature conservation and environment protection, ground water

---

471 http://www.forumrohstoffe.at/
472 In the course of the Amendment to the Raw Material Act 2001 the National Council agreed to a resolution in which the Federal Minister for Economic Affairs and Labour was appointed to develop an "Austrian Mineral Resources Plan" in reasonable time. This was to document the deposits of minerals needed nationwide and should create the basis of a nationwide mining plan, which should be established according to the specific needs of the federal countries and communities. The project emphasized that Austria has a substantial domestic production of raw materials. Conflicts of aims between resource protection and environmental respectively spatial planning should be harmonized.
473 Maier, A., Weber, L. (2008): Der Österreichische Rohstoffplan, in: Sorger, Veit [Hrsg.] [et al.]: Herausforderung Verwaltungsreform: Best Practice Beispiele für eine effiziente Verwaltung. Wien: Industriellenvereinigung, 2008 (The Austrian Raw Materials Plan, in: Sorger, Veit [Ed] [et al.]: Administrative challenge: best practice examples of efficient management. Vienna: Austrian Industry, 2008).
474 The Austrian Mineral Resources Plan is emphasized as an example of Best Practice in the communication COM (2008) 669.
475 Weber, L. (2007a), l.c. Weber, L. (2007b): The Austrian mineral resources plan, World of mining – surface & underground, H.6, p. 442–452.

protection and other entitled claims to the open space in the spatial planning, so that these can be used without contradiction in the future.[476]

The complexity of the Austrian Minerals Resources Plan has required a division of the works in two phases

The primarily involved working groups estimate a duration of three years for Phase 1. For Phase 2 – planning relevant for the federal countries – two years were assumed.[477]

The aim of Phase 1 of the Austrian Mineral Resources Plan was to determine the resources available in Austria on the basis of the Raw Materials Supply Concept of 1981. For that purpose four working groups were established with task allocations as shown in Figure 91. The major task of the Working Group 1 (geology) was the documentation of raw materials deposits (construction raw materials, ores, industrial minerals, energy resources), as well as the compilation of specific geological base maps ("lithological maps"). The Geological Survey of Austria (GBA) was responsible for the compilation of the maps and the construction materials. The deposits of ores, industrial minerals and energy resources were developed by the Federal Ministry of Economics and Labor (BMWA) together with the Technical Committee for deposit research of the Mining Association in Austria (BVÖ). In Working Group 2 (Mineral Economy) raw materials economic and technical mining issues were treated in individual sections under the auspices of the Institute of Mining Engineering of the University of Leoben (MUL), together with the BMWA.[478]

---

[476] Mining companies must be enabled also in the future to exploit outside of mineral protection areas where this appears necessary and where legal and other relevant conditions allow mining activities.
[477] The results of the Austrian Mineral Resources Plan were published in June 2010.
[478] Weber (2007a), l.c.: e.g. Presentation of the supply situation of Austria and estimated price and demand developments, the presentation of the Austrian Industry, the international situation and trends, the potential supply risks, the improvement of providing the demand with domestic resources, new applications for minerals, environmental impact of rmineral extractive industries, re-use of mining areas and, finally, raw material research in Austria.].

Figure 113: Organigram of the Austrian Mineral Resources Plan (Phase 1) (Data by Weber 2007)

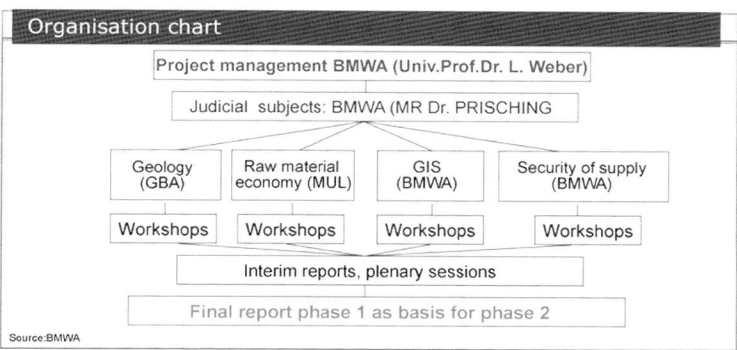

Working Group 3 (GIS application) under leadership of the BMWA had to revise the results of Working Group 1 by use of GIS and to compile a map appropriate for conflict resolution. Working Group 4 (supply security) under the auspices of the BMWA dealt with the main effects of possible supply disruptions.[479] Moreover, the fact should not be ignored that in certain regions of Austria even some surface minerals like sand and gravel might run short, as a consequence of insufficient public awareness and above all because of "overplanning" of the natural space.

In Phase 2 the determined resources from Phase 1 were translated into planning in cooperation with the federal countries. That is, the mineral potential zones identified in Phase 1 were plotted digitally on maps together with such land use plans that oppose or prevent mining activities.[480] Consequently, areas

---

479 Weber (2007a), Cp.: Issues such as self-financing, amount of raw materials imports, sensitivity, multiplier effects, countries of origin, raw materials price and possible substitutes had to be answered. Deposits of critical minerals received a higher priority for security measurements. On account of the investigations the following minerals are classified as potentially critical: iron ore, ores of chrome, manganese, molybdenum, nickel, tantalum, cobalt, copper, magnesium, aluminium, tungsten and lead as well as talc, hard coal, coal coke and the raw materials ferrotitanium, titanoxide, ferroniobium, ferrotantal, ferrovanadium, carbon including soot.

480 Ibidem: Especially with the near-surface minerals it was decided to proceed according to demand. An optimal solution is reached if the distance from producer to consumer is no more than about 30 km, and reserves last for at least 50 years. For this reason in several test areas in Lower Austria an attempt was made to evaluate the demand of the following 50 years considering demographic development, specific regional consumption of gravel sands, possible raw material consuming infrastructure projects, ("Model Lower Austria"). As an example for the district of Melk a demand of about 26,3 million m^3 of gravel sands was assessed.

with restrictive or obstructive land use planning were cut out of the potential mining zones that had been determined by system-analytical methods.[481]

The activities concerning the Austrian Mineral Resources Plan have thus made a fundamental contribution to sustainable resources management for an adequate modern mineral policy.

Regulatory issues

For the exploration of deposits of minerals free for mining an exploration licence is necessary (Article 8 ML).[482] A substantial condition for the issuing of the exploitation licence (Bergwerksberechtigung) forms the payability of raw materials deposits. Raw material deposits are to be called payable if they can presumably be extracted because of their characteristics and location, the quality, quantity and properties of the concerned minerals free for mining with economic benefit (Article 22 ML).[483] The individual licensing procedures can be very complex. The necessary time for the issuing of an extraction permit can reach many years. In one case the time to get a mining license in Lower Austria took 15 years, whereas the extraction duration is limited to 10 years.[484]

Hungary

The Gross Domestic Product increased to $ 155 billion in 2008 from $ 49 billion in 1998. The value of the gross output of industry (mining, materials good production, supply of energy) decreased to 29,1 % (1998) from 31,8 % (2008).[485] In 2006, the mining and quarrying and the production of basic metals, industrial minerals products, and coke and refinery products accounted for 17 % of the value of industrial production.[486]

---

481 These include in particular areas designated as building areas (100 resp. 300 m distance from buildings according to mineral resources law), Water Act-protected areas, nature reserves and Natura 2000 areas.
482 See also Tiess (2010), l.c.
483 See also Tiess (2010), l.c.
484 Verbal communication.
485 USGS (2008, 2010): Minerals Yearbook 2008, 2010 Volume III (UK).
486 USGS (2008): Minerals Yearbook 2006, Volume III – Hungary.

Figure 114: Mining production in Hungary between 1998 and 2008 – selected mineral resources including aluminium production (Data by Weber et al. [WMD 1998–2008])

Metallic minerals

The average bauxite production was about of 400.000–500.000 t. Aluminium production decreased between 1998 and 2007, from 41.000 t to zero. Manganese production increased to 13.400 t in 1998 from 8.700 t in 2008. In 2008, Hungary was the 2nd ranked producer of gallium in the world (5 t).

To meet its economic requirements, Hungary continued to depend on imports of most metals (ores and concentrates). In 2006, the value of iron and steel imports exceeded exports by about 97%. Bauxite mining and refining to alumina and manganese mining were the only major metal mining and processing operations in Hungary.

Chapter 5 View of the minerals policies in selected states of Europe

Table 44: Metallic minerals: production/mining – import – export of commodities of Hungary (Data by BGS, 2010 [European Mineral Statistics])

Commodities related to metallic minerals/products	Production/mining (x: mine)	Import	Export
Silver		m	m
Pmg		m	m s
Gold		m	m
Zinc		u	u u s
Tin		u	s
Rare-earth		rec	orec
Mercury		x	
Lithium			
Lead		u	s
Gallium			
Copper		u s	s
Cadmium			
Bismuth			
Bauxite	x	x	
Arsenic			
Antimony			o
Aluminium	a pa (until 2006)	a ah u ua s	a ah u s
Vanadium			
Tungsten			m
Titanium		tm m o	o m
Tantalum			
Nickel			
Molybdenum		m	
Manganese	x	oc	
Cobalt			
Chromium		oc pi m	
Iron	pi cr	ore pi fa s	fa s

a	Alumina	fc	Ferro-cerium	orec	Other rare earth compounds	sul	Sulfide
ah	Alumina hydrate	ilm	Ilmenite	otm	Other titanium minerals	sz	Slap zinc
bp	Burnt pyrites	m	Metal	pa	Primary aluminium	tm	Titanium minerals
c	Concentrate	ma	Metallic arsenic	pen	Pentoxide	ts	Titanium slag
car	Carbonate	mat	Matte and cement	pi	Pig iron	u	uwrought
carb	Carbides	mi	Mine	rec	Rare earth compounds	ua	uwrought alloys
cc	Cerium compounds	o	Oxide	ref	refined	ur	urefined
cr	Crude steel	oc	Ores and concentrates	s	Scrap	vana	Vanadiferous residues
fa	Ferro alloys	or	Ore	sm	Smelter	wa	White arsenic

Industrial minerals

Between 1998 and 2008, Hungary produced a broad range of industrial minerals that included bentonite, diatomite, gypsum, kaolin, and perlite. Such industrial minerals (and also cement) continued to play an important role in Hungary's economy, especially in the modernization of the country's infrastructure. The average production of bentonite (see figure 114) was about 10.000 t. Gypsum production increased to 251.000 t in 2000 from 185.000 t in 1998, and decreased to 231.000 t in 2008 t from 16.000 t in 2001. Also perlite production decreased to 68.000 t in 2007 from 148.000 t in 1998, and increased between 2007 and 2008 (132.000 t).

The industrial minerals continued to play an important role in Hungary's economy, especially in the modernization of the country's infrastructure. Highway construction planned for the next years continue to be an important element in the development of the country's infrastructure.

Table 45: Industrial minerals: production/mining – import – export of commodities of Hungary (Data by BGS, 2010 [European Mineral Statistics])

Commodities related to industrial minerals/products	Asbestos	Baryte	Bentonite	Diatomite	Diamond	Feldspar	Fluorspar	Gypsum	Graphite	Kaolin	Magnesite	Perlite	Potash	Phosphate rock	Salt	Sulfur	Talc	Cement
Production/mining (x: mine)			x	x				x		x		x			rec			cc cf
Import	x	x	x	x	gc	x	x	cr ca	x	x	ma me		chl		x	x	x	cc pc
Export		x	x					cr ca		x		pf			x	x	x	cc pc

a	Anhydrite	cr	Crude	i	Industrial	rs	Rock salt
bs	Brinesalt	d	Dust	ma	Magnesia	sib	Salt in brine
ca	Calcined	fs	Fertiliser salts	me	Magnesite	ss	Seasalt
cc	Cement clinker	g	Gypsum	pc	Portland cement	sul	Sulphate
cf	Cement finished	gc	Gem cut	pf	Potassic fertilisers	um	unmanufactured
chl	Chloride	gr	Gem rough	rec	recovered		

Construction minerals

Highway construction planned for the next years continue to be an important element in the development of the country's infrastructure. The following figure is illustrating the aggregates production between 1998 and 2008; production increased strongly to 8 t per capita in 2006 from 1,5 t per capita but then decreased (also strongly) to about 4,0 t in 2007.

Figure 115: Per capita aggregates production in Hungary between 1998 and 2008 (Data by Kovács, Hungarian Office for Mining and Geology, 2010)

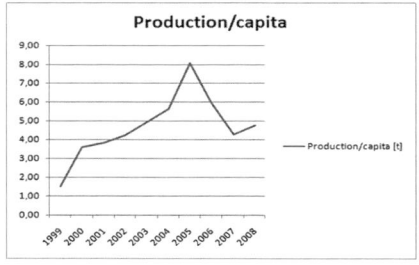

Minerals policy issues

Hungary does not have a national raw materials strategy. There is no national mineral resources planning system to protect economically attractive deposits. However, Minerals Management Zones are designated in accordance with the Spatial Development Act 2003. These cover all mineral reserves in the national Minerals Inventory, but do not imply that extraction will be permitted.[487]

Figure 116: Bauxite mining in Hungary (Data by http.//deutsch.mal.hu/)

In the course of privatisation the Bakonyer Bauxitbergbau, Ajkaer Tonerdefabrik, Inotaer aluminium smelter, closely linked with the aluminum industry, passed into the ownership of MAL Hungarian Aluminium Production and Trading Ltd (MAL AG)

Mineral rights may be issued according to Articles 5–7 and Articles 22–23 of the Mining Law or based on a bidding procedure according to Articles 8–19 of the Mining Law. The planned period of prospecting within the period of the concession may not be longer than 4 years, and may, on one occasion, be extended by half thereof, at the most Article 14 ML). Concession contracts may be concluded

---

[487] LCU (2010): table 2 'land use planning' (Hungary).

for a maximum period of 35 years, which may be extended on one occasion, at the most, by half the period of the concession contract (Article 12 ML).

Slovenia

The Gross Domestic Product development increased to $155 billion in 2008 from $48 billion in 1998. The value of the gross output of industry (mining, materials good production, supply of energy) decreased to 33,9 % (2008) from 36,4 % (1998).[488] In terms of value, imports of mine and quarry products exceeded exports by more than 11 times, demonstrating Slovenia's dependence on imports of mineral products.[489]

Metallic minerals

In 2007, Slovenia produced aluminium, secondary refined lead, and steel as well as a modest amount of industrial minerals.

The following figure illustrates the production, import and export of alumina in Slovenia.

Figure 117: Production, export and import of alumina (Data by BGS)

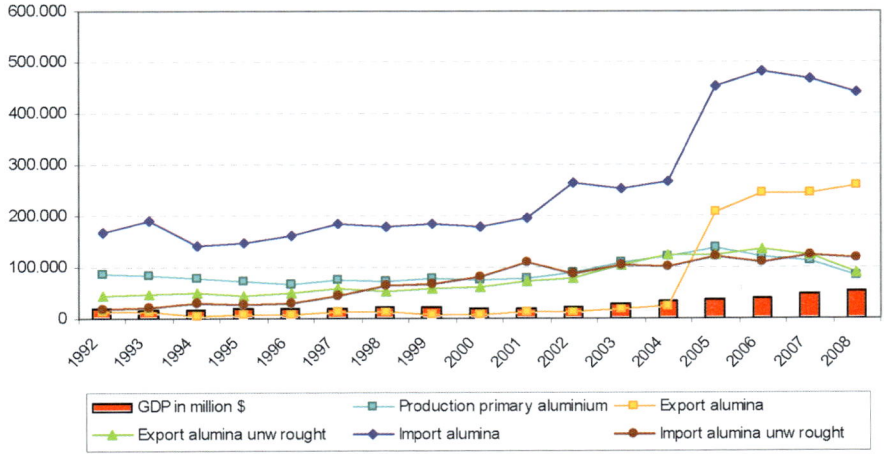

---

488 USGS (2008, 2010): Minerals Yearbook 2008, 2010 Volume III (Slovenia).
489 Statistical Office of the Republic of Slovenia, 2008, in: USGS 2008 (Slovenia).

Chapter 5 View of the minerals policies in selected states of Europe

Table 46: Metallic minerals: production/mining – import – export of commodities of Slovenia (Data by BGS, 2010 [European Mineral Statistics])

Commodities related to metallic minerals/products	Production/mining (x: mine)	Import	Export
Silver		m	m
PMG		m	m m
Gold		m s	m
Zinc		u	u s
Tin		u	
Rare-earth		rec	m
Mercury			x
Lithium		o	car
Lead	ref	u s	u s
Gallium			
Copper		u s	s
Cadmium			
Bismuth		m	
Bauxite		x	x
Arsenic		ma	
Antimony		m	
Aluminium	pa	ah u ua s	a ua s
Vanadium		m	
Tungsten			
Titanium		tm ts m	m o
Tantalum			
Nickel		u	u
Molybdenum		oc	
Manganese		oc m	m
Cobalt		m	m
Chromium		oc m	
Iron	cr fa	pi fa s	pi fa s

a	Alumina	fc	Ferro-cerium
ah	Alumina hydrate	ilm	Ilmenite
bp	Burnt pyrites	m	Metal
c	Concentrate	ma	Metallic arsenic
car	Carbonate	mat	Matte and cement
carb	Carbides	mi	Mine
cc	Cerium compounds	o	Oxide
cr	Crude steel	oc	Ores and concentrates
fa	Ferro alloys	or	Ore

orec	Oher rare earth compounds	sul	Sulfide
otm	Other titanium minerals	sz	Slap zinc
pa	Primary aluminium	tm	Titanium minerals
pen	Pentoxide	ts	Titanium slag
pi	Pig iron	u	uwrought
rec	Rare earth compounds	ua	uwrought alloys
ref	refined	ur	urefined
s	Scrap	vana	Vanadiferous residues
sm	Smelter	wa	White arsenic

Industrial minerals

Between 1998 and 2008, Slovenia produced some industrial minerals that included salt.

Table 47: Industrial minerals: production/mining – import – export of commodities of Slovenia (Data by BGS, 2010 [European Mineral Statistics])

Commodities related to industrial minerals/products	Asbestos	Baryte	Bentonite	Diatomite	Diamond	Feldspar	Fluorspar	Gypsum	Graphite	Kaolin	Magnesite	Perlite	Potash	Phosphate	Salt	Sulfur	Talc	Cement
Production/ mining (x: mine)															x			cf
Import		x	x	d	x	x	cr ca	x	x	me ma		pf	x	x	x	x	cc pc
Export							cr ca							x	x			cc pc

a	Anhydrite	cr	Crude	i	Industrial	rs	Rock salt
bs	Brinesalt	d	Dust	ma	Magnesia	sib	Salt in brine
ca	Calcined	fs	Fertiliser salts	me	Magnesite	ss	Seasalt
cc	Cement clinker	g	Gypsum	pc	Portland cement	sul	Sulphate
cf	Cement finished	gc	Gem cut	pf	Potassic fertilisers	um	unmanufactured
chl	Chloride	gr	Gem rough	rec	recovered		

Construction minerals

Aggregates are economically important mineral raw materials (particularly needed for infrastructure development). The following figure is illustrating the constant development of aggregates consumption per capita/GDP between 1998 and 2008; aggregates consumption increased to 10 t per capita in 2008 from 5 t per capita in 1998.

Figure 118: Aggregates consumption/GDP in Slovenia (Data by Solar, Slovenian Geological survey, 2010)

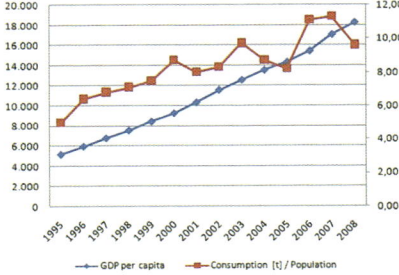

Minerals policy issues

Slovenia does not have a national raw materials strategy.

However, according to Article 5 Mining Law a National Mineral Resources Management Programme should be developed.[490] It provides goals, policies and conditions for the co-ordinated exploration and exploitation of mineral resources. The National Programme shall consist of a general plan and individual mineral resource management plans. Since the adoption of the policy (National Mineral Resource Programme) in April 2009 only "Statement on accordance" was introduced.[491]

An exploration concession shall be granted for a period of not more than five years, and can be extended for a maximum of three years. An exploitation concession is granted for a definite period of time, normally required for cost-effective exploitation of a particular mineral resource in a particular area. This period of time cannot be more than fifty years, except in the case that, because of substantial investment in the exploitation of a certain mineral resource in a certain area, and subject to regular and well performed exploitation, the reserves in the exploitation area cannot be extracted in full (Article 13 ML).

Italy

The Gross Domestic Product increased to $ 2,3 trillion in 2008 from $ 1,2 trillion in 1998. Contribution of industry (mining, materials good production, supply of energy) to GDP decreased to 27 % (2008) from (1998) 29,4 %. Italy, which is one of the largest EU members in terms of its population and the size of its industrial sector, is expected to continue to be a major consumer and producer of durable goods and to continue to rely on imported and recycled mineral raw materials.[492]

In 2006, Italy's gross domestic product (estimated at $ 1,8 trillion) was the fourth largest economy in the European Union (EU), after Germany, the United Kingdom, and France. The country's economy was complex and highly developed with industry accounting for about 28 % of the GDP.[493] Heavy industry comprised facilities for the production of chemicals and iron and steel, for machine building and metal working, and for automotive assembly. These

---

490 Cp. Department of Mining and Tunneling (2004), Final Report, p 121.
491 Land Use Consultans (2010): table 1 'Minerals Policy' (Slovenia).
492 USGS (2008, 2010): Minerals Yearbook 2008, 2010 Volume III (Italy).
493 USGS (2008): Minerals Yearbook 2006, Volume III – Italy.

sectors were heavily dependent on imported nonfuel and fuel mineral inputs.[494] Italy's trade was mainly with other member-countries of the EU.

Figure 119: Mining production in Italy between 1998 and 2008 – selected mineral resources including aluminium production (Data by Weber et al. [WMD 1998–2008])

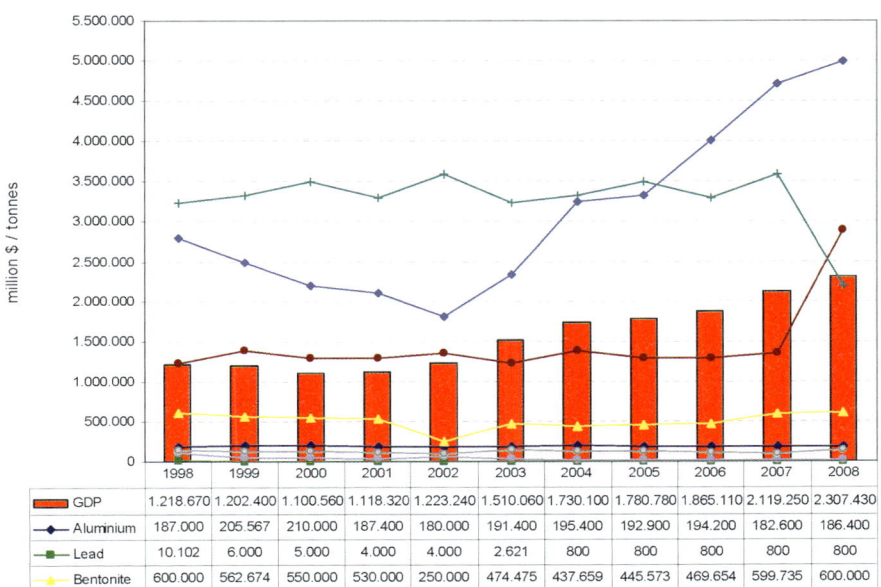

Metallic minerals

The mine output of lead decreased to 800 t in 2008 from 10.000 t in 1998. The production of aluminium remained at a constant level of about 190.000 t.

Italy's minerals industries produced such metals as copper, iron and steel, lead, and zinc, all important materials for the country's manufacturing sector. The raw materials used to produce these and other metals stemmed mostly from imported ores and concentrates and from secondary scrap recovery.

In 2006, the iron and steel sector reported production increases for pig iron and crude steel of about 1% and 9%, respectively, compared with output levels

494 U.S. Central Intelligence Agency, 2006, in: USGS 2008 (Italy).

in 2005. During the same period, the nonferrous sector showed mixed results. Aluminium and copper production increased by about 2% and 13%, respectively, compared with that of 2005; the output of lead and zinc metal, however, declined by about 10% each. In the industrial minerals sector, cement output rose by about 7,3% compared with that of 2005.

In 2006, Italy's chief producers of alumina and primary aluminium were Alcoa Italia S. p. A. and Eurallumina S. p. A. In August, Eurallumina was acquired by UC Rusal, which was a major Russian producer of aluminium. Rusal purchased 56,2% of Eurallumina's stock from Rio Tinto plc of the United Kingdom.[495]

Regarding copper the KME Group S. p. A. (a major European refiner and fabricator of copper headquartered in Florence) conducted its operations within Italy under its subsidiary Europa Matalli S. p. A. at Barga and at Scrivia. Copper and copper semimanufactures were the main products at these locations. Italy imported small amounts of copper concentrate and relied mainly on scrap recovery and imports of copper metal. In 2008, the refined primary and secondary copper production increased.

Italy is a major European producer and consumer of pig iron and crude and finished steels. In 2006, Italy produced more than 11,4 million metric tons (Mt) of pig iron and more than 29 Mt of crude steel, which placed the country as the third ranked pig iron producer in Europe [excluding the Commonwealth of Independent States (CIS)] after Germany and France, and the second ranked producer of crude steel after Germany. Italy's apparent consumption of crude steel ranked the country second in Europe (excluding the CIS) after Germany. Italy's apparent consumption of finished steel amounted to more than 31 Mt, which was the second largest consumption level in Europe and Central Eurasia, after Germany.[496]

Moreover, Italy depended on imports of lead and zinc ores and concentrates. Italy imported most of its requirements for lead and zinc concentrates; a minor amount of lead and zinc concentrate, however, was produced in Sardinia. Glencore International AG of Switzerland remained the country's principal processor (smelter and refiner) of lead and zinc.

---

495 Rio Tinto plc, 2006; UC Rusal, 2006a, b, in: USGS 2008 (Italy).
496 International Iron and Steel Institute, 2006, in: USGS 2008 (Italy).

5.2 EU States

Table 48: Metallic minerals: production/mining – import – export of commodities of Italy (Data by BGS, 2010 [European Mineral Statistics])

Commodities related to metallic minerals/products	Production/mining (x: mine)	Import	Export
Silver	x	oc, m	m
Pmg		m, s	m, s
Gold		m, s	m, s
Zinc	sz	oc, u, ua, s	oc, u, ua, s
Tin		oc, u, ua	u, s
Rare-earth		cc, orec, fc, m	rec, m
Mercury		x	x
Lithium		o, car	car
Lead	x, ref	oc, u, s	u, s
Gallium			
Copper	ref	u,ur, u,ref, ua, s	mat, u, ua, s
Cadmium		mo, m	m
Bismuth		m	
Bauxite		x	x
Arsenic		ma	ma
Antimony		oc, m, o	oc, o
Aluminium	a, pa	a, ah, u, ua, s	a, ah, u, ua, s
Vanadium		oc, pen	oc
Tungsten		m, carb	m
Titanium		tm, m, o	ts, m, o
Tantalum			
Nickel		mat, u, ua, s	u, s, o
Molybdenum		oc, m, o	oc, m, s
Manganese	ore	oc, m	m
Cobalt		oc, m, o	m, o
Chromium		oc, m	oc, m, s
Iron	pi, cr, fa	ore, pi, fa, s	ore, pi, fa, s

a	Alumina	fc	Ferro-cerium	orec	Oher rare earth compounds	sul	Sulfide
ah	Alumina hydrate	ilm	Ilmenite	otm	Other titanium minerals	sz	Slap zinc
bp	Burnt pyrites	m	Metal	pa	Primary aluminium	tm	Titanium minerals
c	Concentrate	ma	Metallic arsenic	pen	Pentoxide	ts	Titanium slag
car	Carbonate	mat	Matte and cement	pi	Pig iron	u	uwrought
carb	Carbides	mi	Mine	rec	Rare earth compounds	ua	uwrought alloys
cc	Cerium compounds	o	Oxide	ref	refined	ur	urefined
cr	Crude steel	oc	Ores and concentrates	s	Scrap	vana	Vanadiferous residues
fa	Ferro alloys	or	Ore	sm	Smelter	wa	White arsenic

Industrial minerals

Between 1998 and 2008, Italy was a significant EU- and world producer of a variety of industrial minerals, which included bentonite, feldspar, fluorspar, gypsum, salt and sulphur. Feldspar production (world top producer, in 2008 1st) increased significantly to 5 Mt in 2008 from more than 1 Mt in 1985 (see figure 120). Between 1998 and 2008 the production of bentonite remained at a constant average level of 500.000 t (see figure 119). Gypsum production shows an average value of 1,3 Mt between 1998–2007, and increased between 2007–2008: from 1,3 Mt to 2,9 Mt. The average salt production was about 3,2 Mt between 1998–2007, and decreased between 2007–2008: from 3,6 Mt to 2,2 Mt. Fluorspar production decreased to 15.000 t in 2008 from 104.000 t in 1998. In 2008, Italy remained a leading European and global producer of such industrial minerals as pumice (25% of world output), feldspar (25%), bentonite (4%), lime (2%), cement (1.7%), and gypsum (1%).[497]

The following figure illustrates the production, import and export of feldspar in Italy.

Figure 120: Production, export and import of feldspar (Data by BGS)

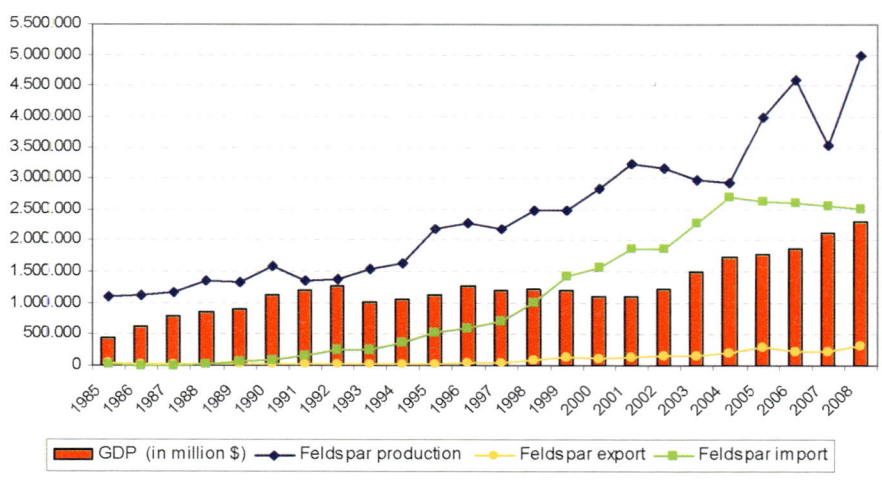

___

[497] Crangle, 2009a, b; Miller, 2009; Potter, 2009; van Oss, 2009; Virta, 2009, in: USGS (2010), Italy.

5.2 EU States

In 2006, Italy remained a leading European and global producer of such industrial minerals as bromine (among top 13 world producers), pumice (27% of world output), feldspar (19% of world output), bentonite (4% of world output), cement (2% of world output), lime (2% of world output), and gypsum (1% of world output).[498] Industrial minerals for which there were insufficient domestic resources and had to be imported included ball clay, barite (Ukraine), chamotte (Germany), fluorspar (China), kaolin (United States), magnesite (Turkey), mica (China), and talc (China).[499]

Table 49: Industrial minerals: production/mining – import – export of commodities of Italy (Data by BGS, 2010 [European Mineral Statistics])

Commodities related to industrial minerals/products	Asbestos	Baryte	Bentonite	Diatomite	Diamond	Feldspar	Fluorspar	Gypsum	Graphite	Kaolin	Magnesite	Perlite	Potash	Phosphate rock	Salt	Sulfur	Talc	Cement
Production/mining (x: mine)	x	x				x	x until 2006	x		x		x			x	rec	x	cc cf
Import	x	x	x	u gr gc i d	x	x	cr ca	x	x	x	me ma		sul chl	x	x	x cr ref	x	cc pc
Export	x	x	x	gc d	x	x	cr ca	x	x	x	me ma		pf		x	x	x	cc pc

a	Anhydrite	cr	Crude	i	Industrial	rs	Rock salt
bs	Brinesalt	d	Dust	ma	Magnesia	sib	Salt in brine
ca	Calcined	fs	Fertiliser salts	me	Magnesite	ss	Seasalt
cc	Cement clinker	g	Gypsum	pc	Portland cement	sul	Sulphate
cf	Cement finished	gc	Gem cut	pf	Potassic fertilisers	um	unmanufactured
chl	Chloride	gr	Gem rough	rec	recovered		

Construction minerals

Aggregates are economically important mineral raw materials (particularly needed for infrastructure development). Italy is one of the largest aggregates producers in Europe. The production of aggregates increased to 368 Mt in 2008 from 350 Mt in 2002. The production accounted in 2003: 359,6 Mt (i.e. 210 Mt sand and gravel (s&g), 140 Mt crushed rock (cr), 4,5 Mt recycled aggregates

---

498 Founie, 2007a, b; Lyday, 2007; Miller, 2007; Potter, 2007; van Oss, 2007; Virta, 2007, in: USGS 2008 (Italy).
499 Wilson, 2007, in: USGS 2008 (Italy).

(ra)), in 2004: 358 Mt (220 Mt s&g, 135 Mt cr, 3 Mt ra), in 2005: 377,5 Mt (125 Mt s&g, 145 Mt cr, 4,5 Mt ra, 3,0 Mt manufactured aggregated), in 2006: 354 Mt (210 Mt s&g, 135 Mt cr, 5,5 Mt ra, 3,0 Mt manufactured aggregates). In 2008, Italy produced (368 Mt, i.e.) 225 Mt s&g, 135 Mt cr and 5 Mt recycled aggregates. The total number of producers was 1.796; total number of extraction sites was 2.360.

Minerals policy issues

Italy does not have a national raw materials strategy. There is no national mineral resources planning system to protect economically attractive deposits.

Regulatory issues

Permission to exploration can be arranged for three years and may be extended upon confirmation, which should be done at the expense of the explorer (Article 6 ML). The explorer (of a deposit) is preferred to any other applicant, provided that the competent authority finds that he possesses the technical and economic suitability. The granting of the concession is temporary (Article 21). The concession and its appliances are subject to the provisions of law governing the property (Article 22 ML).

Investment policy issues

If the explorer/operator receives the concession, he has the right to obtain a premium (in relation to the importance of the discovery) and compensation on account of the operating system. The premium and the benefits are provisionally established in the concession act (Article 16 ML). The acquisition of the Lucchini Steel Group by OAO Severstal of Russia was one of the highlights in the country's mineral industry in 2006.

Midyear 2006, Sargold Resource Corp. of Canada finalized the acquisition of Sardinia Gold Mines S. p. A. (SGM) from Medoro Resources Ltd. Sargold received 90% of SGM's shares for about € 1 million ($ 1.32 million). Sargold also acquired a 75% interest in SGM Ricerche S. p. A. (also for about $ 1.32 million), which held major interests in the Monte Ollastedu gold exploration project.[500] By yearend, Sargold held the largest share of gold exploration and development assets in Sardinia.[501]

---

500 Sargold Resource Corp., 2006c, in: USGS 2008 (Italy).
501 Sargold Resource Corp., 2006b, in: USGS 2008 (Italy).

The Furtei deposit was worked by the Furtei Gold Mine until 2003; the company beneficiated about 400.000 metric tons per year (t/yr) of ore. Sargold indicated that the mine and beneficiation plant were well maintained; plans to restart operations were under review during the year. The deposit holds strata-bound mineralization of Tertiary age and was estimated to contain a resource of 584.000 troy ounces of gold. Additional gold deposits had been delineated by SGM in the Furtei region, which Sargold reportedly planned to explore further.[502]

Figure 121: Furtei mine (Data by http://www.mindat.org)

France

France's gross GDP remained the third largest GDP in the European Union (EU) after Germany and the United Kingdom. France is expected to continue to rely on imported mineral raw materials, to produce consumer and producer durables and such intermediate products as ferrous and nonferrous metals and semi manufactures, construction materials, and chemicals.[503] Its GDP was the third largest in the European Union (EU) after Germany and the United Kingdom. The output value of France's industry of the GDP decreased from 23,4 % 1998 to 20,4 % in 2008. The country was a major processor of mineral raw materials

---

502 Sargold Resource Corp., 2006a, in: USGS 2008 (Italy).
503 USGS (2008, 2010): Minerals Yearbook 2008, 2010 Volume III (France).

Chapter 5 View of the minerals policies in selected states of Europe

and a manufacturer of producer and consumer durable goods.[504] France's heavy industries, which, among other product categories, produced machine tools, chemicals, and automotive and aviation products for domestic consumption and export, relied mainly on imported metal ores and concentrates, and on imported industrial minerals and mineral fuels.[505] Owing to the size of France's economy, the upstream input of minerals was the key to continued maintenance and growth of the country's heavy industries. In 2006, the value of crude material imports alone, which included ores and concentrates of metals and industrial minerals, base metals, and mineral fuels, amounted to almost 3% of the GDP.[506]

Figure 122: Mining production in France between 1998 and 2008 – selected mineral resources including aluminium production (Data by Weber et al. [WMD 1998–2008])

---

504 USGS (2008): Minerals Yearbook 2006, Volume III – France (Steblez, W.G.).
505 U.S. Central Intelligence Agency, 2007, in: USGS (France).
506 Eurostat, 2007, in: USGS, 2008 (France).

Metallic minerals

Between 1998 and 2008, France depended on imports. Bauxite production increased/doubled to 160.000 t in 2008 from 80.000 t in 1998. Aluminium production increased to 714.000 t in 2002 from 423.000 t in 1998, and decreased between 2002 and 2003 (444.000 t), and then remained at a constant level of (average) 400.000t. France's Aluminum Pechiney SA, which was owned by Rio Tinto Alcan Inc. (Rio Tinto), was the country's sole producer of primary aluminum. Rio Tinto also operated facilities for the production of alumina and aluminum semi manufactures.

In 2008, France's production of secondary lead declined by 7% compared to that of 2007 whereas primary lead production for 2008 seems to have completely stopped with the decommissioning of Metaleurop's plant in Noyelles-Godault. Total zinc metal production in 2008 amounted to 118.900 t, which was a decrease of 9,2 % compared with output in 2007. France's output of crude steel consumption during the 2001 to 2005 period (the latest years for which data were available) averaged slightly more than 18 million metric tons per year (Mt/yr); the consumption of finished steel averaged about 16,7 Mt/yr.[507]

---

[507] Metaleurop, 2007; World Bureau of Metal Statistics, 2007, p. 90, 137; International Lead and Zinc Study Group, 2009, in: USGS (2010).

Chapter 5 View of the minerals policies in selected states of Europe

Table 50: Metallic minerals: production/mining – import – export of commodities of France (Data by BGS, 2010 [European Mineral Statistics])

Commodities related to metallic minerals/products	production/mining (x: mine)	import	export
Silver		oc m	oc m
PMG		m s	m s
Gold		oc m s	oc m s
Zinc	sz	oc u ua s	oc u ua s
Tin		oc u ua s	u ua s
Rare-earth		cc orec fc m	rec fc m
Mercury		x	
Lithium		o car	car
Lead	ref	u s	u s
Gallium			
Copper		u,ur ua s	u ua
Cadmium	x	m o sul	x
Bismuth		x x	x
Bauxite	x	x a ah u ua s	x
Arsenic		m ma	m ma
Antimony		m o	m o
Aluminium	a	a ah u ua s	
Vanadium		oc pen m	
Tungsten		m carb	m carb
Titanium		m o	tm m o
Tantalum	x		x
Nickel	sm/ref	mat u ua s o	mat u ua s o
Molybdenum		oc m o	oc m
Manganese		oc m	oc m
Cobalt		m o	m o
Chromium		oc m	oc m
Iron	pi cr fa	ore pi fa	ore pi fa s

a	Alumina	fc	Ferro-cerium	orec	Oher rare earth compounds	sul	Sulfide
ah	Alumina hydrate	ilm	Ilmenite	otm	Other titanium minerals	sz	Slap zinc
bp	Burnt pyrites	m	Metal	pa	Primary aluminium	tm	Titanium minerals
c	Concentrate	ma	Metallic arsenic	pen	Pentoxide	ts	Titanium slag
car	Carbonate	mat	Matte and cement	pi	Pig iron	u	uwrought
carb	Carbides	mi	Mine	rec	Rare earth compounds	ua	uwrought alloys
cc	Cerium compounds	o	Oxide	ref	refined	ur	urefined
cr	Crude steel	oc	Ores and concentrates	s	Scrap	vana	Vanadiferous residues
fa	Ferro alloys	or	Ore	sm	Smelter	wa	White arsenic

Industrial minerals

Between 1998 and 2008, France mined a broad variety of industrial minerals that included baryte, diatomite, feldspar, fluorspar, gypsum, kaolin, salt, sulphur and talc. Production of gypsum and salt (see figure 122) increased considerable (gypsum) to 4,8 Mt from 3,5 Mt and (salt) to 6 Mt from 4,5 Mt. Diatomite production remained at a constant level of (average) 75.000 t, the same is true for the production of feldspar (650.000 t), fluorspar (50.000–70.000 t), kaolin (300.000 t), sulphur (700.000 t) and talc (300.000 t – 400.000 t). Baryte production decreased from 75.000 t (1998) to 20.000 t (2008).

In 2006, Imerys, which was a major French producer of industrial minerals, mined and processed ball clays, carbonates, feldspar, and red clays not only domestically but from deposits in such countries as China, Germany, Spain, the United States, and Vietnam for *domestic use and export*.[508] France's principal cement manufacturers were Lafarge S. A. (www.lafarge.com) and Société des Ciment Français (Ciments Français); apart from cement-producing facilities in France, both companies had major capital assets abroad. Ciments Français was a subsidiary of Italcementi S. p. A of Italy. The other important producers of cement in France were the Vicat Group, which had five plants with a total cement production capacity of 6 Mt/yr, and Ciments d'Origny, which had six plants and a total cement production capacity of 4,2 Mt/yr.[509]

Figure 123: Vicat Group, The La Courbaisse quarry and facilities in the Alpes Mari times, France (Data by http://www.vicat.com/)

---

508 Imerys, 2007, in: USGS 2008 (France).
509 PR Newswire Europe Ltd., 2006; International Cement Review, 2003, in: USGS 2008 (France).

Table 51: Industrial minerals: production/mining – import – export of commodities of France (Data by BGS, 2010 [European Mineral Statistics])

Commodities related to industrial minerals/products	Asbestos	Baryte	Bentonite	Diatomite	Diamond	Feldspar	Fluorspar	Gypsum	Graphite	Kaolin	Magnesite	Perlite	Potash	Phosphate	Salt	Sulfur	Talc	Cement
Production/mining (x: mine)	x	x				x	x			x					x	rec	x	cc cf
Import	x	x	x	us gr gc i d	x	x	cr ca	x	x	x	ma me	x	fs sul chl	x	x	x cr ref	x	cc pc
Export	x	x	x	us g	x	x	cr ca	x	x	x	me ma	x	sul chl	x	x		x	cc pc

a	Anhydrite	cr	Crude	i	Industrial	rs	Rock salt
bs	Brinesalt	d	Dust	ma	Magnesia	sib	Salt in brine
ca	Calcined	fs	Fertiliser salts	me	Magnesite	ss	Seasalt
cc	Cement clinker	g	Gypsum	pc	Portland cement	sul	Sulphate
cf	Cement finished	gc	Gem cut	pf	Potassic fertilisers	um	unmanufactured
chl	Chloride	gr	Gem rough	rec	recovered		

Construction minerals

Aggregates are economically important mineral raw materials (particularly needed for infrastructure development). France is one of the largest aggregates producers in Europe. The production of aggregates increased to 432 Mt in 2008 from 406 Mt in 2002. The production accounted in 2003: 401 Mt (i.e. 168 Mt sand gravel (s&g), 215 Mt crushed rock (cr), 18 Mt recycled aggregates (ra)), in 2004: 402 Mt (168 Mt s&g, 225 Mt cr, 9 Mt ra), in 2005: 410 Mt (170 Mt s&g, 223 Mt cr, 10 Mt ra, 7 Mt manufactured aggregates), in 2006: 430 Mt (167 Mt s&g, 233 Mt cr, 7 Mt ra, 14 Mt manufactured aggregates). In 2008, France produced (432 Mt, i.e.) 165 Mt s&g, 237 Mt cr, 7 Mt marine aggregates, 15 Mt recycled aggregates and 8 Mt manufactured aggregates. The total number of producers was 1.640; the total number of extraction sites: 3.050. Figure 124 is illustrating the development of aggregates consumption per capita/GDP between 1998 and 2008; production ranged between 6 and 7 t per capita.

Figure 124: Aggregates consumption/GDP in France (Data by Rodriguez Chavez/Schleifer, 2010)

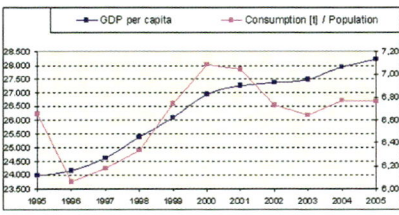

Minerals policy issues

France has a national raw materials strategy. The strategy defers in function of the kind of mineral. For industrial minerals and construction materials access to the resources is privileged because they are present in the French soil and base rock. The environmental factor is completely incorporated into the feasibility study of an extraction project. For ore minerals the absence of economically profitable deposits within the French soil and the sub-soil involve a different policy. Strategic metals are listed and particular attention is devoted to them. Thus, according to the problems which are specific to each one of them provisions are taken, on a case by case basis, in order to guarantee the security of supply. For the metals of current use with its participation in the international groups (INSG, ICSG and IZLSG), France ensures regularly updated knowledge and a transparency of the markets. Three specialised actors, which in fact are Public institutions of an industrial and commercial nature advice the government. ADEME: Agence de l'Environnement et de la Maîtrise de l'Energie (environment and energy management agency) IFREMER: Institut Français de Recherche pour l'Exploitation de la Mer (research and sea exploitation French office) and BRGM: Bureau de recherches géologiques et minières (geology and mining research office).[510]

Recently the Minister of State, Minister for Ecology, Energy, Sustainable Development and Sea published a communication related to strategic metallic minerals (http://www.gouvernement.fr/gouvernement/les-metaux-strategiques). Access to these metals is needed to ensure the development of the French industry and to enable the development of products more competitive. An action plan was adopted, which focuses on the following issues:

- Knowledge of strategic metals shall be improved. The Office of Geological and Mining Research (BRGM) will invest specifically in this field to iden-

---

510 LCU (2010): table 1 'Minerals Policy' (France).

tify a possible vulnerability in different sectors, considering the identification of appropriate ways to address that.

- The extension of the geological knowledge through exploration campaigns shall be targeted.
- The development of new exploration tools facilitating the extraction and processing of strategic metals shall be done (BRGM and the French Research Institute for Exploitation of Sea (l'Institut français de recherche pour l'exploitation de la mer [IFREMER]).
- A recycling policy shall be developed, the Agency for Environment and Energy Management (ADEME) will provide coordination.
- The strengthening of governmental action shall be realized. A senior official in charge of the case will soon be appointed and a dialogue shall be held between the government and industry on matters relating to the security of metallic minerals supply.

Regulatory issues

The exclusive license is granted by the competent authority for five years. This license gives the holder the exclusive right to all research work within the scope of that license (Article 9 ML). At the request of its holder, the validity of a permit may be extended twice, each time for five years and under the same conditions as those laid down for its issue (Article 10 ML). Article 29 ML specifies that a permit is valid for a maximum of 50 years. It can be prolonged by successive periods of a maximum of 25 years. Article 109 of the Code Minier allows the public authorities to define areas in which authorisations for exploration and extraction permits may be issued without the consent of the owner of the land. Those powers may also be used when the utilisation of a quarried material is sufficiently compromised to threaten regional and national balance.

Investment policy issues

The major event in the steel industry was the agreement reached in midyear of 2006 to merge Acelor S. A. and Mittal Steel Company N. V. to form the world's leading steel producer. The initial value of Mittal's acquisition of Arcelor was set at $ 33,1 billion. Arcelor was created in 2002 through a merger of steel companies in France (Usinor Group), Luxembourg (Arbed S. A.), and Spain (Ac-

5.2 EU States

eralia Corporacion Siderurgica S. A). Virtually France's entire steel sector was encompassed by the merger.[511]

Spain

The Gross Domestic Product increased to $ 1,6 trillion in 2008 from $ 601 billion in 1998. The value of the gross output of industry (mining, materials good production, supply of energy) decreased to 28,4 % (2008) from 29,1 % (1998). Spain has one of Europe's most important and diversified mining sectors. However, Spanish mining is not sufficient to satisfy domestic demand, and continues to be a large-scale importer of minerals.[512]

Spain was the fifth ranked economy in the European Union (EU) and ome of the world's highest ranked exporting country. Spain's international economic profile has grown appreciably in recent years.[513] The share of foreign trade in Spain's gross domestic product was about 55 % in 2006. The mining and mineral processing industries represented almost 1 % of the GDP in 2006.[514]

Figure 125: Mining production in Spain between 1998 and 2008 – selected mineral resources including aluminium production (Data by Weber et al. [WMD 1998–2008])

---

[511] Kanter and others, 2006, in: USGS 2008 (France).
[512] USGS (2008, 2010): Minerals Yearbook 2008, 2010 Volume III (Spain).
[513] USGS (2008): Minerals Yearbook 2006, Volume III – (Spain).
[514] Sociedad Estatal de Participaciones Industriales, 2006; U.S. Central Intelligence Agency, 2007; U.S. Department of State, 2007, in: USGS 2008 (Spain).

Chapter 5 View of the minerals policies in selected states of Europe

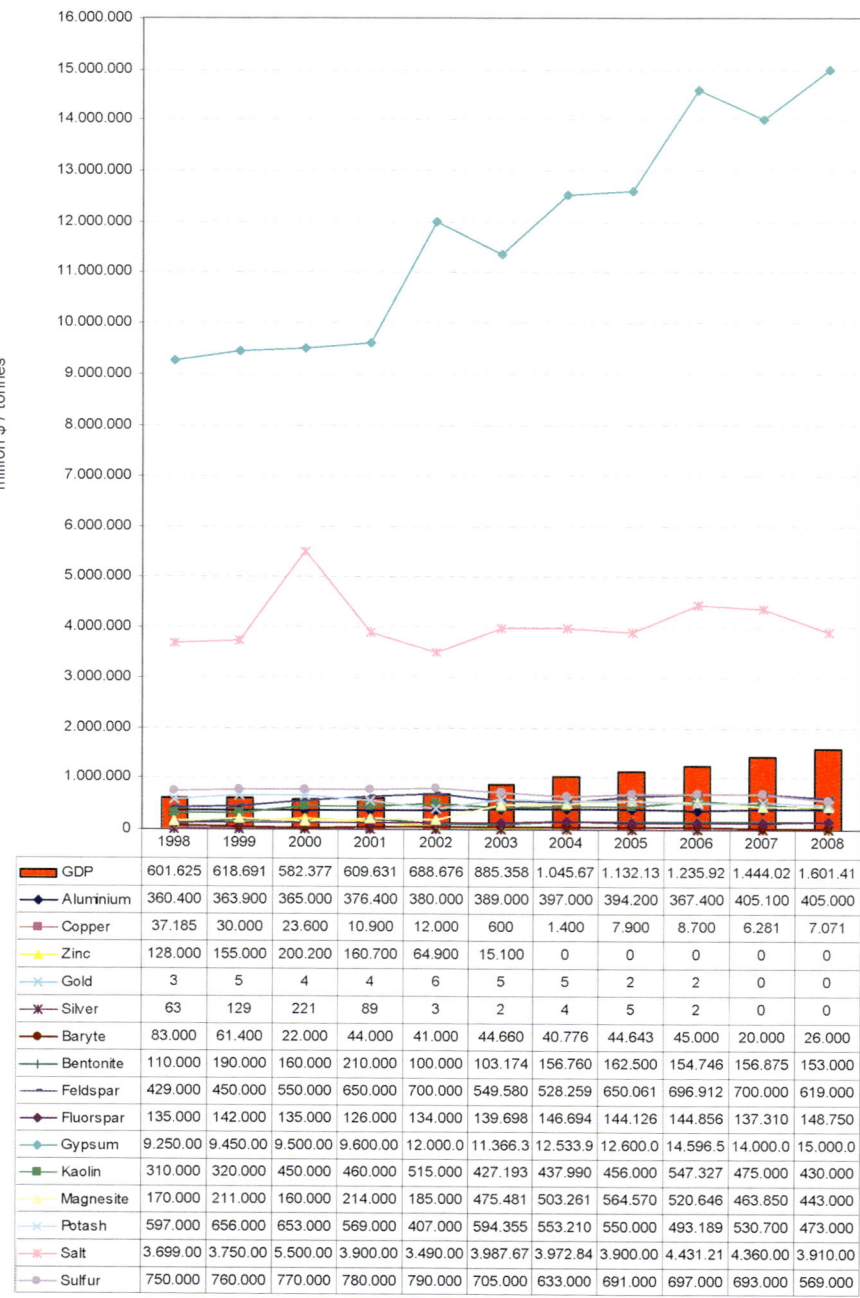

Metallic minerals

Spain was an important European producer of mineral commodities such as, copper, gold, silver, and zinc. Spain occupies about 85% of the Iberian Peninsula and has some of the most mineralized territory in Western Europe, including the volcanic-hosted massive sulphide (VMS) deposits of the Iberian Pyrite Belt (IPB) of southern Spain. The IPB stretches from Seville in southern Spain to south of Lisbon in Portugal. At least 80 VMS deposits are thought to exist. The IPB alone was estimated to have yielded 1,2 billion metric tons of sulphides (mined ore and reserves) from which, in order of value, more than 880 metric tons (t) of gold, 35 million metric tons (Mt) of zinc, 13 Mt of lead, and 15 Mt of copper have so far been produced.[515] In Spain, Cambridge Mineral Resources plc (CMR) owns 100% of its local subsidiary, Recursos Metallicos SA, which holds mining licenses over the Lomero-Poyatos auriferous polymetallic massive sulphide deposit with the right to production. The licenses are valid for 45 years.[516]

Figure 126: CMR, Lomero – Poyatos (Data by http://www.cambmin.co.uk)

Primary aluminium metal production increased to 405.000 t in 2008 from 360.400 t in 1998 (see figure 125). Production of mined copper content increased to 8.700 t in 2006 from 7.900 t in 2005, and refined copper production decreased to 263.700 t from 284.200 t in 2005. Gold mine production remained about the same level of 2005. Mine production of nickel increased to 6.400 t from 5.380 t in 2005.

In 2008 production of refined copper remained similar to that of 2007 (290.000 t); Spain's zinc metal output was 456.050 t in 2008 compared with 494.090 t in 2007. Asturiana de Zinc S.A. continued production at its San Juan de Nieva Castillon plant. Asturiana's core business was the refining and production of zinc metal, mainly zinc ingots. According to Asturiana, the San Juan de Nieva plant, which had a capacity of 500.000 t/yr of zinc metal, was

---
515 USGS, 2010.
516 Cambridge Mineral Resources plc, 2007; Encyclopedia of the Nations, 2007, in: USGS 2008 (Spain).

the leading single zinc smelter in the world and also one of the world's lowest cost operations.[517]

Also in 2006, CMR's Lomero-Poyatos gold deposit was reported to contain estimated reserves of 3,71 Mt at grades of 3,26 grams per metric ton (g/t) gold, 27,9 g/t silver, 0,87 % copper, 1,57 % lead, and 1,16 % zinc. Ormonde sought to build a mid-size suite of projects in Spain that would be focused primarily on gold. The Salomon project, which was located in northwest Spain, was the most advanced project and contained an estimated 20.000 kg of inferred resources of gold. Ormonde's objective was to establish mining operations that could produce copper, gold, and silver. Projects included the Salamanca, the Salamon, the Tracia, and the Trives gold projects and the La Zarza gold-copper project which was situated within a mining concession in the IPB in southwest Spain. Ormonde was earning a 70 % interest in the La Zarza gold-copper project by providing € 1,8 million ($ 2,2 million) during a 3-year period under an option agreement with the property owner, Nueva Tharsis S. A. L.[518]

Figure 127: La Zarza Copper-Gold Project (Data by http://www.ormondemining.com)

In 2006, Spain produced 17,8 Mt of crude steel, which was about the same level as that of 2005 and 19 Mt in 2008. Compañia Española de Laminación S. L. (Celsa) produced about 2,2 million metric tons per year (Mt/yr) of steel 6. Sidenor Industrial, S. L. (Sidenor), which was a leading producer of special steels in Spain, was planning to start producing stainless steel by expanding

---
517 USGS, 2010.
518 Cambridge Mineral Resources plc, 2007; Ormonde Mining plc, 2007a, b, in: USGS 2008 (Spain).

its existing electric arc furnace at its works in Basauri, northern Spain. A joint venture of Gerdau Group of Brazil, Santander Group of Spain, and executives of Sidenor signed a € 443,8 million ($ 556 million) agreement to acquire the entire capital stock of Sidenor. The investment would allow Gerdau to enter the strategic EU market and provide it with access to large international automobile makers.[519]

Nickel Rio Narcea Gold Mines Ltd.'s (RNG's) Aguablanca Mine consistes of an open pit, an onsite processing mill, and a potential underground mine. The initial open pit mine life was estimated to be 10,5 years based on the mineral reserves of 15,7 Mt of ore containing 0,66% Ni, 0,46% Cu, 0,47 g/t platinum-group metals (PGM), and 0,13 g/t Au. The assumptions of the feasibility study contemplated an annual production of approximately 8.165 t (18 million pounds) of nickel; 6.350 t (14 million pounds) of copper; and 620 kg (20.000 ounces) of PGM.[520]

Figure 128: The Aguablanca nickel-copper sulfide deposit (Data by http://www.lundinmining.com)

---

519 Sidenor Industrial, S.L., 2007, in: USGS 2008 (Spain).
520 Rio Narcea Gold Mines Ltd., 2007, in: USGS 2008 (Spain).

Table 52: Metallic minerals: production/mining – import – export of commodities of Spain (Data by BGS, 2010 [European Mineral Statistics])

Commodities related to metallic minerals/products	Production/Mining (x: mine)	Import	Export
Silver	x until 2006	oc	m
PMG		m s	m s
Gold	x until 2006	m s	m s
Zinc		oc u ua s	oc u ua s
Tin		c u ua s	u s
Rare-earth		rec fc m	cc orec fc
Mercury		x	x
Lithium	x	o car	car
Lead	ref	oc u s	oc u s
Gallium			
Copper	x sm ref	oc u,ur u,r ua s	oc u,ur u,ref ua s
Cadmium		m o	
Bismuth		m	m
Bauxite		x	x
Arsenic		ma	
Antimony		oc m o	m o
Aluminium	a pa	ah u ua s	a ah u ua s
Vanadium			
Tungsten		oc pen m carb	oc m
Titanium		tm ts m o	tm m o
Tantalum		x	
Nickel	x	u ua o	oc u s
Molybdenum		oc m	
Manganese		oc m	oc m
Cobalt		m o	m o
Chromium		oc m	oc m
Iron	x until 2004 pi cr fa	ore pi fa s	ore pi fa s

a	Alumina	fc	Ferro-cerium	orec	Oher rare earth compounds	sul	Sulfide
ah	Alumina hydrate	ilm	Ilmenite	otm	Other titanium minerals	sz	Slap zinc
bp	Burnt pyrites	m	Metal	pa	Primary aluminium	tm	Titanium minerals
c	Concentrate	ma	Metallic arsenic	pen	Pentoxide	ts	Titanium slag
car	Carbonate	mat	Matte and cement	pi	Pig iron	u	uwrought
carb	Carbides	mi	Mine	rec	Rare earth compounds	ua	uwrought alloys
cc	Cerium compounds	o	Oxide	ref	refined	ur	urefined
cr	Crude steel	oc	Ores and concentrates	s	Scrap	vana	Vanadiferous residues
fa	Ferro alloys	or	Ore	sm	Smelter	wa	White arsenic

Industrial Minerals

In 2006, Spain was a significant European producer of such mineral commodities as cement (seventh after China, India, the United States, and others), fluorspar (sixth after China, Mexico, Mongolia, and others), gypsum (second after the United States), and industrial sand and gravel (sixth after the United States, Slovenia, Germany, Austria, and France).[521]

Between 1998 and 2008 Spain mined a broad variety of industrial minerals, which included baryte, bentonite, feldspar, fluorspar, gypsum, kaolin, magnesite, potash, salt and sulphur. Baryte production (see figure 125) decreased to 26.000 t in 2008 from 83.000 t in 1998. Production of bentonite (average 150.000 t), feldspar (average 550.000 t), kaolin (average 400.000 t), potash (average 500.000 t), sulphur (average 700.000 t) and salt (average 3,5 Mt – 4, 0 Mt) remained at a constant level. Fluorspar production increased to 149.000 t in 2008 from about 135.000 t in 1998. This is also true for magnesite production, which increased to 443.000 t from 170.000 t in the same period. Production of gypsum increased to 15 Mt in 2008 from 9,2 Mt in 1998.

The following figure illustrates the production, import and export of fluorspar in Spain.

Figure 129: Production, export and import of fluorspar (Data by BGS)

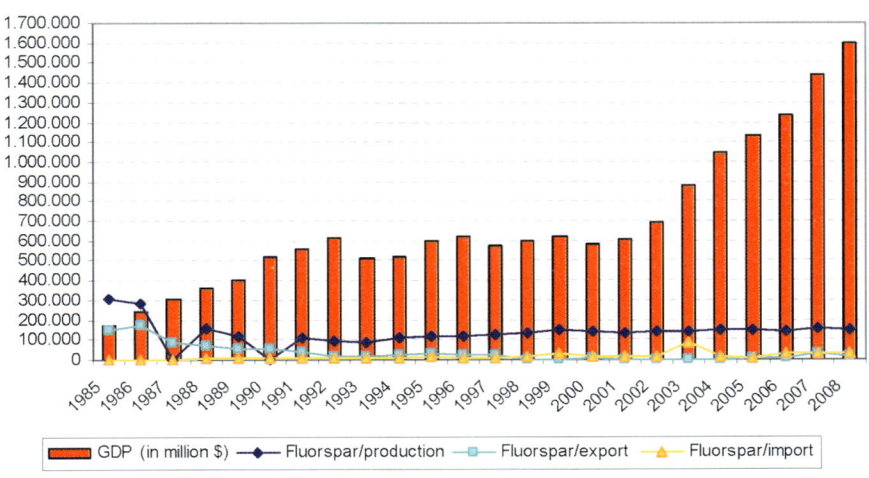

---

521 Dolley, 2007; Founie, 2007; Miller, 2007; van Oss, 2007, in: USGS (2008.2010), Spain.

Spain's barite output remained at about the same level as that of 2005. Minerales y Productos Derivados S. A. (Minersa) was a main supplier of drilling grade material. Minersa continued to operate a surface mine and plant at Vera.

Spain's cement output decreased slightly to 50 Mt in 2006 from 50,3 Mt in 2005. Cementos Portland Valderrivas planned to boost its white cement production rate from 700 metric tons per day (t/d) to 900 t/d while achieving a significant emission reduction by converting its El Alto plant near Madrid. The conversion would include the integration of a calciner into the existing preheater, the installation of a new rotary kiln drive, and replacement of the clinker cooler.[522]

Regarding clays, Spain, whose reserves of sepiolite in the Tagus Basin represent a 70% of the world's reserves, maintained in 2006 its world leadership in sepiolite production. The largest deposit was thought to be in excess of 15 Mt.[523] Minersa was Europe's leading fluorspar producer owing to its three deposits in the Province of Asturias in northern Spain. The Emilio, the Jaimina, and the Moscona underground mines produced a combined 420.000 t/yr of crude fluorspar. Iberpotash S. A. was a 100% owned subsidiary of Dead Sea Works Ltd., which was a leading producer of potash and an important potash resource in Western Europe. Iberpotash mined sylvinite and sylvite ore from the Cataluña deposit in the Suria area.

---

522 Polysius AG, 2007, in: USGS 2008 (Spain).
523 Grupo Tolsa, 2007, in: USGS 2008 (Spain).

Table 53: Industrial minerals: production/mining – import – export of commodities of Spain (Data by BGS, 2010 [European Mineral Statistics])

Commodities related to industrial minerals/products	Asbestos	Baryte	Bentonite	Diatomite	Diamond	Feldspar	Fluorspar	Gypsum	Graphite	Kaolin	Magnesite	Perlite	Potash	Phosphate rock	Salt	Sulfur	Talc	Cement
Production/mining (x: mine)	x	x	x			x	x	x		x	me		chl		rs ss		x	cf
Import	x	x	x	u gc i d		x	x	cr ca	x	x	me ma	sul chl	x		x	x cr ref	x	cc pc
Export	x	x	x	u gc i d		x	x	cr ca	x	x	me ma	sul chl	x		x	x	x	cc pc

a	Anhydrite	cr	Crude	i	Industrial	rs	Rock salt
bs	Brinesalt	d	Dust	ma	Magnesia	sib	Salt in brine
ca	Calcined	fs	Fertiliser salts	me	Magnesite	ss	Seasalt
cc	Cement clinker	g	Gypsum	pc	Portland cement	sul	Sulphate
cf	Cement finished	gc	Gem cut	pf	Potassic fertilisers	um	unmanufactured
chl	Chloride	gr	Gem rough	rec	recovered		

Construction minerals

Aggregates are economically important mineral raw materials (particularly needed for infrastructure development). Spain is one of the largest aggregates producers in Europe. The production of aggregates increased to 485 Mt in 2006 from 396 Mt in 2002, and then decreased to 383 Mt. The production was in 2003: 421 Mt, in 2004: 438 Mt (i.e. 155 Mt sand gravel (s&g), 282 Mt crushed rock (cr), 1 Mt recycled aggregates (ra)), in 2005: 460 Mt (159 Mt s&g, 300 Mt cr, 1 Mt ra, in 2006: 485,5 Mt (170 Mt s&g, 314 Mt cr, 1,5 Mt ra). In 2008, Spain produced (383 Mt aggregates, i.e.) 134 Mt s&g, 244 Mt cr production, 5 Mt recycled aggregates, and 1 Mt manufactured aggregates. In 2008, the total number of producers was 1.600; the total number of extraction sites was 2.060.

Minerals policy issues

Spain does not have a national raw materials strategy.

No national plan exists, since regions are responsible for mineral resources management. Planning practices are very different between regions. Many of the Autonomous Communities have their own nature conservation laws and have already established their own network of nature areas.[524] Some Autonomous Communities have also developed a mining plan. In few of them, a specific approach to aggregates quarrying was developed. Others have a special mining and quarrying plan, which, however, is not integrated with land planning. Castilla La Mancha states that the Region is developing a territory strategy that takes mining into account in their regional land uses planning.[525] However, most of the 17 regions do not have integrated extraction planning.[526]

Exploration permits (metallic ores) are granted for one year, renewable (Article 40 ML). The exploitation concession is granted for 30 years renewable for periods equal to a maximum of 90 years (Article 62 ML).

Investment policy issues

Several gold, nickel, tungsten, and base metal projects are undergoing feasibility studies and most of them are focused on the Iberian Pyrite Belt. According to the Encyclopedia of the Nations (2009), the IPB is a focus of interest for mining companies and it is a prime target for exploration activities because of past successes in discovering large deposits and possible increased production of copper, gold, nickel, silver, and sepiolite in Spain. The Government is expected to continue with its privatization and liberalization efforts in the mineral industry.

International minerals investment is encouraged by several important factors, including the highly prospective geology of the IPB in the south and the gold discoveries at the Boinas, the Carles, and the El Valle deposits in the Rio Narcea Belt in the north. International minerals investment interest is encouraged by the country's transparent legislative framework, positive fiscal environment for the extraction of natural resources, well-developed infrastructure and skilled workforce, long mining tradition, track record of successful exploration and mine development, and the availability of non-refundable Government grants for both exploration and mine development.[527]

In 2006, Inmet Mining Corp. of Canada announced that it had completed the acquisition of a 70% interest in Las Cruces copper project, and that a wholly owned subsidiary, Leucadia National Corp., would retain the remaining 30% interest. Las Cruces is a high-grade volcanic massive sulfide copper deposit located on the eastern edge of the IPB about 15 km northwest of Seville. The

---

524 Department of Mining and Tunneling (2004), Final Report, p 121, p 140.
525 Land Use Consultans (2010): table 1 'Minerals Policy' (Spain).
526 Department of Mineral Resources and Petroleum Engineering (2010), l.c.
527 Cambridge Mineral Resources plc, 2007; Federation of International Trade Associations, 2007, in: USGS 2008 (Spain).

project had estimated proven and probable copper reserves of 17,6 Mt grading 6,2 % copper. The mine had a production capacity of 72.000 metric tons per year (t/yr) of copper cathode, and the projected life of the mine was 15 years (2008 to 2022).[528]

Ormonde Mining plc of Ireland is a mineral exploration and development company focused on Spain, with the objective of developing mining projects and taking them into production. Ormonde's main project is La Zarza copper-gold project in southern Spain where a prefeasibility study for the development of an underground mining operation was completed in September 2006 and a follow-on bankable feasibility study was being carried out.[529]

Portugal

The Gross Domestic Product increased to $ 253 billion in 2008 from $ 122,7 billion in 1998. The value of the gross output of industry (mining, materials good production, supply of energy) decreased to 23,9 % (2008) from 28,8 (1998). Portugal's mining and mineral processing industries represented about 1% of the gross domestic product (GDP) in 2008.[530] Portugal was a significant European minerals producer and one of Europe's leading copper producers and an important producer of tungsten concentrates.[531]

---

[528] Inmet Mining Corp., 2007, in: USGS 2008 (Spain).
[529] Ormonde Mining plc, 2007a, in: USGS 2008 (Spain).
[530] USGS (2010): Minerals Yearbook 2008, Volume III (Portugal).
[531] MBendi Information Services (Pty) Ltd., 2007a; Ober, 2007; Potter, 2007; Shedd, 2007, in: USGS 2008 (Portugal).

Figure 130: Mining production in Portugal between 1998 and 2008 – selected mineral resources (Data by Weber et al. [WMD 1998–2008])

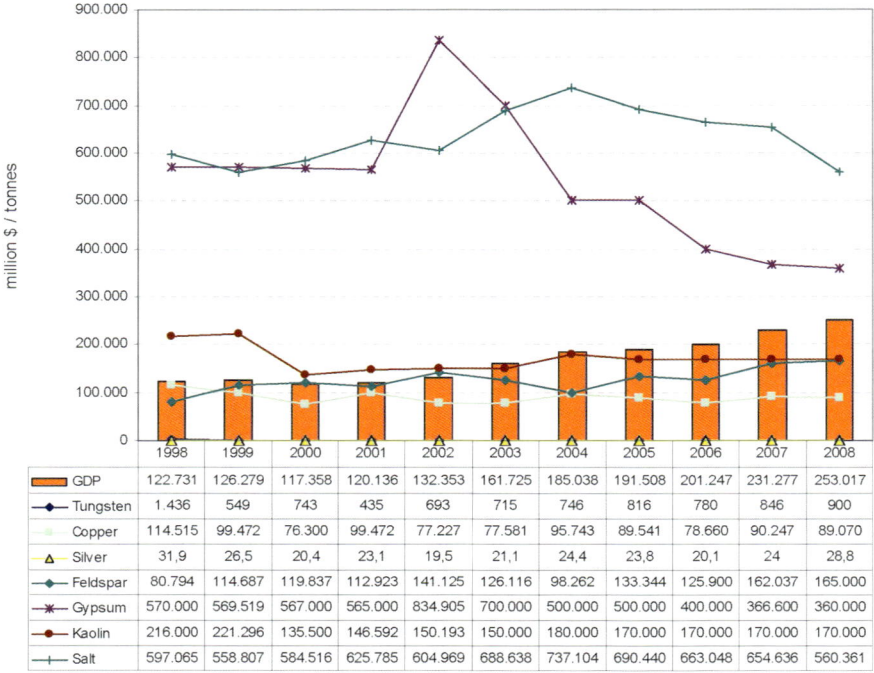

Metallic minerals

Between 1998 and 2008, Portugal mined a broad range of metallic minerals including copper, lead, zinc, lithium, tin and silver. The average tungsten mine production (see figure 130) is between 700 and 800 t. Copper mine production decreased to 89.000 t in 2008 from 114.000 t in 1998, however still remained at a high (European) level. Silver production remained in general at a constant level (average) of 20 t.

In 2008, Portugal was a significant world producer of lithium (sixth after Chile, Australia, China, Argentina, and Canada), tin (tenth after China, Indonesia, Peru, Bolivia, Brazil, Russia, Vietnam, Malaysia, and Australia), and tungsten (sixth after China, Russia, Canada and others).[532]

---

532 Carlin, 2009; Jaskula, 2009; Shedd, 2009, in: USGS (2010), Portugal.

Portugal's Iberian Pyrite Belt (IPB) is one of the most mineralized geological provinces of Western Europe and is geologically complex. Massive sulphides linked to synorogenic volcanism were deposited in the southwestern part of the Iberian Peninsula where the IPB's volcanogenic massive sulphide (VMS) deposits are located. The IPB, which has 85 known VMS deposits, was an important source of base metals in the EU.[533]

In October 2006, the Canadian corporations Eurozinc Mining Corp. and Lundin Mining Corp. merged into a joint venture, which was consolidated before yearend. The new company, Lundin Mining Corp. (LMC), became a part of the European copper, lead, and zinc industries. In Portugal, LMC acquired the Aljustrel zinc-lead-silver project and the Neves Corvo copper-zinc mine and was set to conduct greenfields exploration for base and precious metals near the Neves Corvo Mine as well.

Figure 131: The Neves-Corvo Copper–Zinc mine (Data by http://www.nafinance.com)

Copper production from the Neves Corvo Mine was 78.660 metric tons (t) in concentrate compared with 89.541 t in 2005, which was a decrease of 12,2% compared to 2006 production. The country's copper output was valued at $ 385 million, which was 26,5% higher than that of 2005; the increase in value was a result of the increase in the price of copper to $ 2.829 per pound in 2006 from $ 1.549 per pound in 2005. The Neves Corvo Mine was one of the highest-grade copper mines in the world. It consists of five ore bodies that contain copper, tin, and zinc. The mine has proven copper reserves of 6.835 Mt at an average grade of 5,73%, probable copper reserves of 9.975 Mt at an average grade of 5,29%, and probable zinc reserves of 10.626 Mt at an average grade of 7,96%.[534]

Also in 2006, tungsten production from the Panasqueira tungsten mine in Beira Baixa Province of central Portugal was 780 t in concentrate (W content)

---
533 MBendi Information Services (Pty) Ltd., 2007a, in: USGS 2008 (Portugal).
534 Instituto Nacional de Estatística, 2007; Lundin Mining Corp., 2007a, b, in: USGS 2008 (Portugal).

compared with 816 t in 2005, which was a decrease of 4,4%, owing to low tungsten prices. The Panasqueira Mine continued to be one of the world's leading producers of tungsten concentrates and produced a 75% tungsten oxide (WO_3) concentrate. In June 2006, PMI announced that the mine has proven and probable reserves of 1,4 million metric tons (Mt) at a grade of 0.233% WO_3 and additionally indicated (3,3 Mt at a grade of 0,263% WO_3) and inferred (1,6 Mt at a grade of 0,224% WO_3) resources. The main end-use application for tungsten is in the manufacture of cemented carbides (60%), steel and alloy (21%), electrical and electronics (11%), and catalysts and pigments (8%).[535] In 2006, Primary Metals Inc. (PMI) of Canada, through its subsidiary Beralt Tin & Wolfram S. A., mined tungsten at its Panasqueira Mine, which is located in central Portugal.[536]

---

[535] Primary Metals Inc., 2007, in: USGS 2008 (Portugal).
[536] Primary Metals Inc., 2007, in: USGS 2008 (Portugal).

5.2 EU States

Table 54: Metallic minerals: production/mining – import – export of commodities of Portugal (Data by BGS, 2010 [European Mineral Statistics])

Commodities related to metallic minerals/products	Production/mining (x: mine)	Import	Export
Silver	x	x	oc (since 2007) m
PMG			s
Gold		m	s
Zinc	x	u ua	ua oc s
Tin	x	u ua	ua s
Rare-earth		rec fc m	
Mercury		x	
Lithium	x		
Lead	x ref	u	u s
Gallium			
Copper	x	u s	oc u s
Cadmium			
Bismuth			
Bauxite			
Arsenic		wa	
Antimony		o	m
Aluminium		a ah u ua s	ua s
Vanadium			
Tungsten	x	m carb	oc m
Titanium		o	o
Tantalum		x	x
Nickel		u	
Molybdenum			
Manganese		oc	
Cobalt			
Chromium		oc m	oc
Iron	cr	pi fa s	s

a	Alumina	fc	Ferro-cerium	orec	Oher rare earth compounds	sul	Sulfide
ah	Alumina hydrate	ilm	Ilmenite	otm	Other titanium minerals	sz	Slap zinc
bp	Burnt pyrites	m	Metal	pa	Primary aluminium	tm	Titanium minerals
c	Concentrate	ma	Metallic arsenic	pen	Pentoxide	ts	Titanium slag
car	Carbonate	mat	Matte and cement	pi	Pig iron	u	uwrought
carb	Carbides	mi	Mine	rec	Rare earth compounds	ua	uwrought alloys
cc	Cerium compounds	o	Oxide	ref	refined	ur	urefined
cr	Crude steel	oc	Ores and concentrates	s	Scrap	vana	Vanadiferous residues
fa	Ferro alloys	or	Ore	sm	Smelter	wa	White arsenic

Industrial minerals

Between 1998 and 2008 Portugal mined a broad variety of industrial minerals that included feldspar, gypsum, kaolin, marble, pyrites, and salt.[537] Feldspar production (see figure 130) increased to 165.000 t in 2008 from 81.000 t in 1998, while kaolin production decreased to 170.000 t in 2008 from 216.000 t in 1998. Gypsum production (average 500.000 t) and salt production (average 600.000 t) remained at a constant level. The country was one of the leading producers of salt rock, talc in the EU. Cimpor was Portugal's leading cement producer and was the second ranked cement company on the Iberian Peninsula after Cemex SA. The development of Portugal's infrastructure was expected to create a substantial demand for Cimpor's products in the coming years.[538]

Table 55: Industrial minerals: production/mining – import – export of commodities of Portugal (Data by BGS, 2010 [European Mineral Statistics])

Commodities related to industrial minerals/products	Asbestos	Baryte	Bentonite	Diatomite	Diamond	Feldspar	Fluorspar	Gypsum	Graphite	Kaolin	Magnesite	Perlite	Potash	Phosphate rock	Salt	Sulfur	Talc	Cement
Production/mining (x: mine)						x		x		x					rs ss	rec	x	cc cf
Import	x	x	x		gr gc i d	x	x	cr ca	x	x	me ma		sul chl	x	x	x	x	cc pc
Export		x	x		gc	x		cr ca		x			pf		x	x	x	cc pc

a	Anhydrite	cr	Crude	i	Industrial	rs	Rock salt
bs	Brinesalt	d	Dust	ma	Magnesia	sib	Salt in brine
ca	Calcined	fs	Fertiliser salts	me	Magnesite	ss	Seasalt
cc	Cement clinker	g	Gypsum	pc	Portland cement	sul	Sulphate
cf	Cement finished	gc	Gem cut	pf	Potassic fertilisers	um	unmanufactured
chl	Chloride	gr	Gem rough	rec	recovered		

Construction minerals

Aggregates are economically important mineral raw materials (particularly needed for infrastructure development). The production of aggregates decreased to 88 Mt in 2005 from 125 Mt in 2002 and then increased to 93 Mt in

---

537 MBendi Information Services (PTY) LTD., 2007a,b, in: USGS 2008 (Portugal).
538 Hoover's, Inc., 2006, in: USGS 2008 (Portugal).

2008. The production accounted in 2003: 101 Mt, in 2004: 88,3 Mt (i.e. 6,3 Mt sand and gravel (s&g), 82 Mt crushed rock (cr)), in 2005: 88,3 Mt (6,3 Mt s&g, 82 Mt cr), in 2006: 97,5 Mt. In 2008, Portugal produced (93 Mt, i.e.) 61 Mt s&g, 15 Mt cr, and 17 Mt manufactured aggregates. In 2008, the total number of producers was 350; the total number of extraction sites was 200.

Moreover Portugal's construction minerals sector was a modern and efficient producer of a variety of materials, most notably dimension stone and minerals for the manufacture of ceramics. The dimension stone industry continued to be an important segment of the mining industry in terms of value and trade.

Minerals policy issues

Portugal has an Energy Raw Materials Strategy, however does not have a comprehensive national raw materials strategy including also non-energy mineral resources. New laws (1999 as amended in 2003) on spatial planning in Portugal created at the national level the provision for a Sectorial Plan for minerals which must be produced by the national government. The legal nature of this plan is considered as policy guidance to assist in the decision making process, both for the preparation and approval of the low level plans (e. g. municipal) and the whole administration.[539] However, presently, minerals are accorded a low priority.[540]

Investment policy issues

As a target for foreign direct investment, Portugal has been overshadowed by lower-cost producers in Central Europe and Asia.[541] Several gold and base metal projects were in 2008 undergoing feasibility studies and most of them are focused on the Portuguese Zone of the IPB. The IPB is a focus of interest for mining companies and a prime target for exploration activities because it appears to have a good potential for success on the basis of the large deposits discovered so far.

---

539 Department of Mining and Tunneling (2008): Final Report, p 121, l.c.
540 Land Use Consultants (2010): table 2 'land use planning' (Portugal), l.c.
541 Instituto Nacional de Estatística, 2009; U.S. Central Intelligence Agency, 2009; USGS 2010 (Portugal).

Chapter 5 View of the minerals policies in selected states of Europe

Romania

The Gross Domestic Product increased to $ 204,3 billion in 2008 from $ 42,1 billion in 1998.[542] Industrial production increased by about 5,7% and accounted in 2007 for about 24% of the GDP.[543] The Government of Romania continued to develop policies that were aimed at reforming the industrial sector both to increase its competitiveness in preparation for privatization and entry into the European Union (in 2007) and to abate pollution from mine-based point sources.[544]

Figure 132: Mining production in Romania between 1998 and 2008 – selected mineral resources (Data by Weber et al. [WMD 1998–2008])

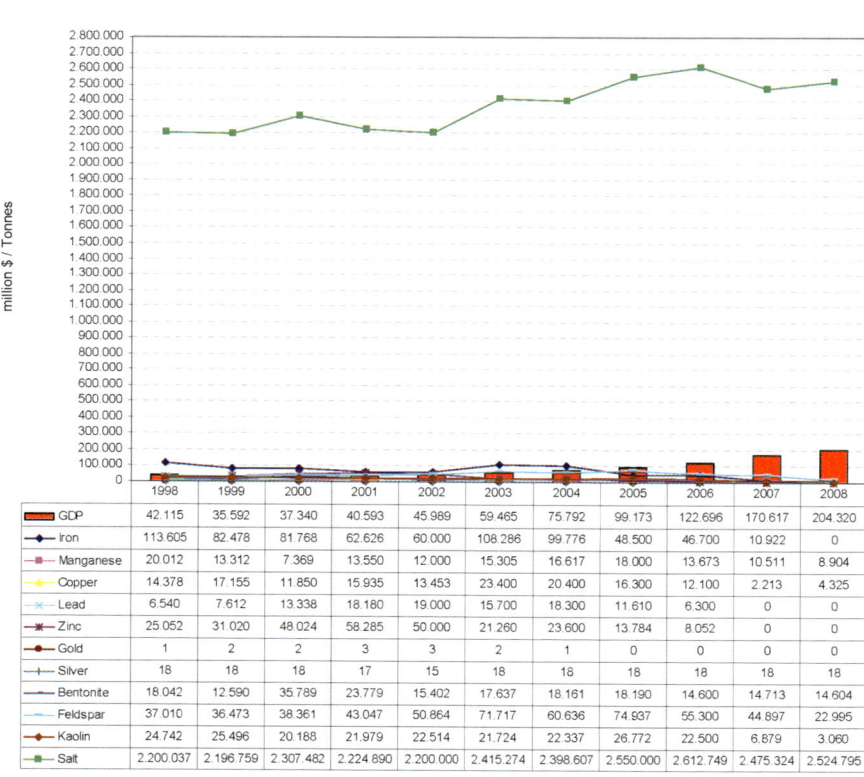

---

542 USGS (2008): Minerals Yearbook 2006, Volume III – Romania.
543 National Institute of Statistics, 2007, p. 1; U.S. Central Intelligence Agency, 2007, in: USGS 2008 (Romania).
544 USGS (2008): 2006 Minerals Yearbook, (Romania).

Metallic minerals

Between 1998 and 2008, Romania mined a broad range of metallic minerals that included iron ore, manganese, lead, zinc, gold and silver Iron production (see figure x) decreased to 11.000 t in 2007 from 113.000 t in 1998. Lead mine production decreased to 58.000 t in 2001 from 25.000 t in 1998, and furthermore between 2001 and 2008 to zero. The same is true for zinc mine production: it decreased to 58.000 t in 2001 from 25.000 t in 1998, and furthermore constant decreasing near to zero. Production of manganese also decreased to 9.000 t in 2008 from 20.000 t in 1998. AT the same time, copper mine production decreased between 1998 and 2003, from 14.000 t to 32.000 t, and decreased between 2003 and 2008: from 32.000 t to 4.000 t. A small production of gold and silver can be observed. Aluminium was produced between 2004 and 2008 (222.000 t to 289.000 t) and bauxite was produced between 1998 and 2002 (134.000 t.).

In 2006, the total aluminium production increased by almost 6% compared with that of 2005. At yearend 2005, Romania's primary aluminium producer, SC Alro S. A. (Alro), announced the acquisition of a majority stake in Alum S. A. Alro indicated that investment for modernizing Alum would continue.[545] The integration of the two companies proceeded in 2006 and included investments by Alro in alumina loading and storing equipment and pollution control technologies at the Alum alumina operation.[546]

Figure 133: Gold mining in Romania (Rosia Montana) (Data by http://www.gabrielresources.com)

Copper was mined at the northeastern part of the country (mainly at the Baia Sprie, the Cavnic, and the Lesul Ursului Mines), and at the southwestern part of the country (the major mines were the Moldova Noua, the Rosia Poieni, and the Rosia Montana Mines). Generally, such major producing mines as Moldova Noua and Rosia Poieni were hoisting ore that graded about 0,35%

---
545 International Bank for Reconstruction and Development, 2006, in: USGS 2008 (Romania).
546 Marco Group, 2006, in: USGS 2008 (Romania).

copper or less. Concentrates from these areas were smelted and refined at Baia Mare and Zlatna. At Baia Mare, SC Allied Deals Phoenix SA operated an Outokumpu flash smelter, an electrolytic copper refinery, and a continuous caster. At Zlatna, SC Ampelum SA processed copper concentrates and operated a smelter and an electrolytic refinery.[547]

Domestic production of iron ore ceased in 2008, which necessitated total reliance on imported raw materials. Romania's net imports of iron ore for the 2002 to 2006 period were about 6,2 Mt, 7 Mt, 6,6 Mt, 6,6 Mt, and 5,2 Mt, respectively. The country's consumption of crude steel during the same period was 3,4 Mt, 3,7 Mt, 3,9 Mt, 4,0 Mt, and 5,1 Mt, respectively.[548]

---

[547] Moreno, 2000, in: USGS 2008 (Romania).
[548] International Iron and Steel Institute, 2007, in: USGS 2010 (Romania).

Table 56: Metallic minerals: production/mining – import – export of commodities Romania (Data by BGS, 2010 [European Mineral Statistics])

Commodities related to metallic minerals/products	Production/mining (x: mine)	Import	Export
Silver	x	m	
PMG			
Gold	x x		
Zinc		x sz	oc u
Tin			u
Rare-earth			rec
Mercury			x
Lithium			
Lead		x (until 2007) ref	oc u
Gallium			
Copper		x sm ref	u s
Cadmium			
Bismuth		x	
Bauxite			x
Arsenic			
Antimony		m o	
Aluminium	a pa	a ah u s	a o ua s
Vanadium			
Tungsten		m	
Titanium		tm m o	
Tantalum			
Nickel		u	
Molybdenum			
Manganese	ore	oc m	since 2006
Cobalt			
Chromium		oc m	
Iron	ore (until 2007) pi cr fa	ore pi fa s	pi fa s

a	Alumina	fc	Ferro-cerium	orec	Oher rare earth compounds	sul	Sulfide
ah	Alumina hydrate	ilm	Ilmenite	otm	Other titanium minerals	sz	Slap zinc
bp	Burnt pyrites	m	Metal	pa	Primary aluminium	tm	Titanium minerals
c	Concentrate	ma	Metallic arsenic	pen	Pentoxide	ts	Titanium slag
car	Carbonate	mat	Matte and cement	pi	Pig iron	u	uwrought
carb	Carbides	mi	Mine	rec	Rare earth compounds	ua	uwrought alloys
cc	Cerium compounds	o	Oxide	ref	refined	ur	urefined
cr	Crude steel	oc	Ores and concentrates	s	Scrap	vana	Vanadiferous residues
fa	Ferro alloys	or	Ore	sm	Smelter	wa	White arsenic

Industrial minerals

Between 1998 and 2008, Romania produced a broad range of industrial minerals that included baryte, feldspar, graphite, gypsum, salt and silica-group minerals. The modernization of the country's economy and infrastructure was expected to increase the domestic demand for industrial minerals materials. The production of bentonite (average 40.000–50.000 t), feldspar (average 50.000 t), kaolin (average 20.000 t) and salt (average 2,5 Mt) remained at a constant level.

Table 57: Industrial minerals: production/mining – import – export of commodities of Romania (Data by BGS, 2010 [European Mineral Statistics])

Commodities related to industrial minerals/ products	Asbestos	Baryte	Bentonite	Diatomite	Diamond	Feldspar	Fluorspar	Gypsum	Graphite	Kaolin	Magnesite	Perlite	Potash	Phosphate rock	Salt	Sulfur	Talc	Cement
Production/ mining (x: mine)		x	x	x until 2007		x	x until 2004	x	x until 2005	x					x	rec	x	cf
Import	x	x	x		x	x		cr ca	x	x	me ma			sul chl	x	x	x	cc pc
Export		x				x		cr ca			ma					x		cc pc

a	Anhydrite	cr	Crude	i	Industrial	rs	Rock salt
bs	Brinesalt	d	Dust	ma	Magnesia	sib	Salt in brine
ca	Calcined	fs	Fertiliser salts	me	Magnesite	ss	Seasalt
cc	Cement clinker	g	Gypsum	pc	Portland cement	sul	Sulphate
cf	Cement finished	gc	Gem cut	pf	Potassic fertilisers	um	unmanufactured
chl	Chloride	gr	Gem rough	rec	recovered		

Construction minerals

In 2008, the modernization of the country's economy and infrastructure was expected to increase the domestic demand for construction materials. In 2008, Romania produced 26 Mt of aggregates, i.e. 18 Mt sand and gravel, 7 Mt crushed rock and 1 Mt recycled aggregates. In 2008, the total number of producers was 500 and the total number of extraction sites was 730.

The modernization of the country's economy and infrastructure was expected to increase the domestic demand for construction materials.

Minerals policy issues

The sustainable development strategy of Romania refers to the strategic reserves of raw materials.[549] The strategy considers different metallic raw materials, besides (partly low-quality reserves) gold and silver. Additionally a categorization of deposits exists according to the following criteria

- Deposits which ensure a long-term supply of raw materials for the industry
- Deposits which ensure a medium-term supply of raw materials for the industry,
- Deposits which ensure a slight supply of raw materials for the industry,
- Important deposits which cannot be developed for various reasons

Finally the question arises whether a protection of these deposits occurs in the context of land use planning.

Regulatory issues

The exploration license is granted to the winner of a public offering, organized by the Competent Authority (Article 15 ML). The exploration license is granted for a maximum period of 5 years, with a renewal right of no more than 3 years (Article 16 ML). The issuing of an exploitation license can be done in different ways (Article 18–37 ML): a) Directly to the title holder of the exploration license, on his request for any of the mineral resources discovered, within a maximum of 90 days from the submission of final exploration report, given the satisfaction of the Competent authority; b) to the winner of a public offering, organized by the Competent Authority (Article 18 ML). The exploitation license is granted for maximum 20 years, with the right of continuation for successive periods of 5 years each.[550]

Finally the investment policy of Romania should be considered and the Strategia Industriei Miniere pentru perioada (2004–2010) should be noted. By this strategy, the Romanian mining industry is to be made competitive. Romania is expected to continue to develop its industrial minerals sector, including quarries and processing facilities for the production of construction materials. Gold exploration is expected to continue to be an important aspect

---

549 National Sustainable Development Strategy, www.sdnp.ro/ncdpublications/nssd.pdf. In 1999 a government decree established a working group for their development. However, the question arises, whether this strategy is including all mineral categories and wether there is an action plan to implement the overall objectives of a national minerals policy.
550 For further information see Tiess (2011), l.c.

of foreign investment in the country's mineral industry. The effort of the copper mining industry enterprises Compania nationala REMIN S. A. as well as Compania Nationala Minvest which are the property of the government is considerable. Both enterprises force the integration of investors based on regulation 590/2006, namely the creation of Joint ventures and the establishment of subsidiaries.

Bulgaria

The Gross Domestic Product increased to $49.904 billion in 2008 from $12,8 billion in 1998. In 2006, the overall output value of the mining sector increased by 1,8% compared with that of 2005.[551] The mining industry's branches reported a 2,7% decrease in the output value of metal ores, and an 11% increase in the output value of industrial mineral mining and quarrying.[552]

---

[551] USGS (2008): Minerals Yearbook 2006, Volume III – Bulgaria.
[552] National Statistical Institute, 2007,in: USGS (2008): 2006 Minerals Yearbook, (Bulgaria).

Figure 134: Mining production in Bulgaria between 1998 and 2008 – selected mineral resources (Data by Weber et al. [WMD 1998–2008])

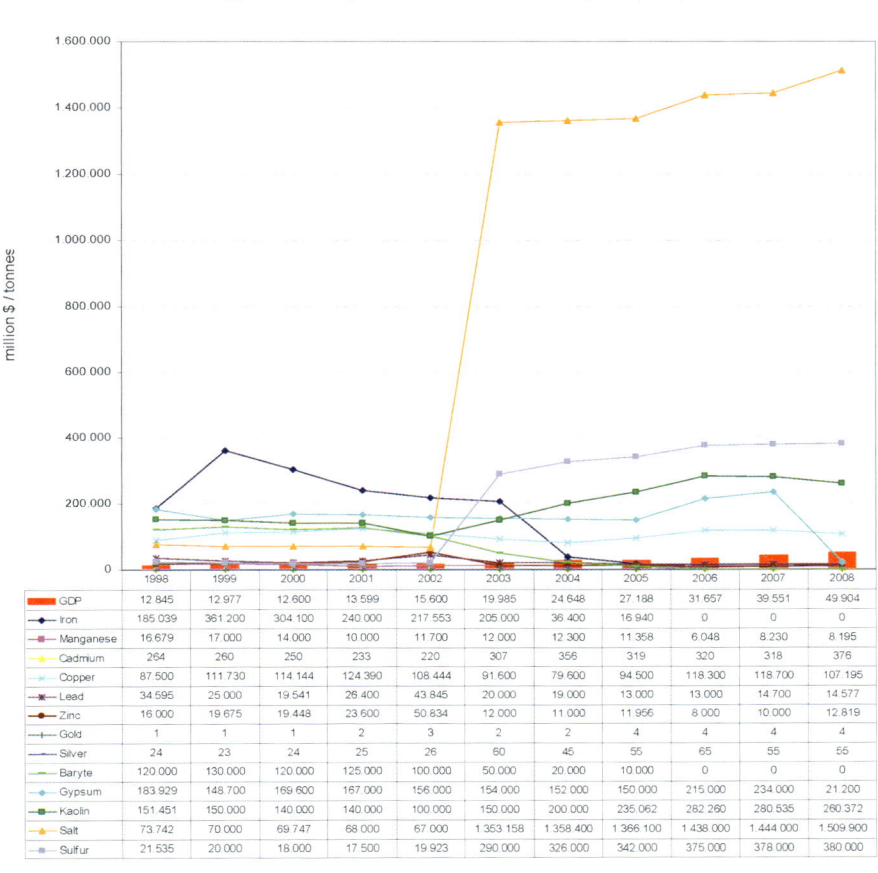

Metallic minerals

Between 1998 and 2008, Bulgaria mined a broad range of metallic minerals that included bismuth, cooper, gold, iron ore, lead, manganese ore, silver and zinc. Iron production (see figure 134) decreased to 17.000 t in 2005 from 185.039 t in 1998. Manganese production: decreased to 8.195 t in 2008 from 16.679 t in 1998. Copper production increased to 107.195 t in 2008 from 87.500 t in 1998. Lead production decreased to 14.577 t in 2008 from 34.595 t in 1998. Zinc production decreased to 14.577 t in 2008 from 16.000 t in 1998. Gold production increased to 4 t in 2008 from 1 t in 1998 and silver production increased to 55 t in 2008 from 24 t in 1998.

Chapter 5 View of the minerals policies in selected states of Europe

The following figure illustrates the production, import and export of copper in Bulgaria.

Figure 135: Production, export and import of copper (Data by BGS)

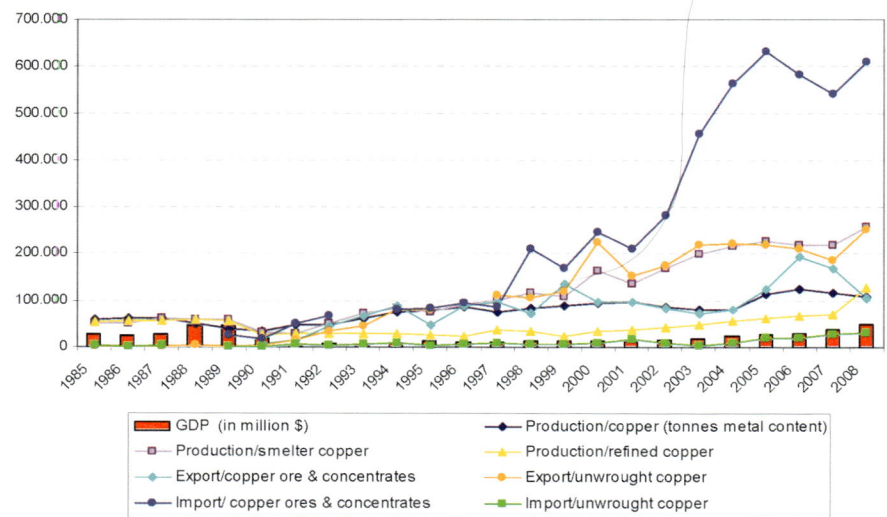

The lead and zinc industry in Bulgaria was based on mining and processing operations near Plovdiv in the Ossogovo Mountains of western Bulgaria, near the Thundza River in southeastern Bulgaria, and in the Madan area near the Greek border. Lead and zinc smelting and refining operations were based in Kurdjali in the Madan area and in Plovdiv. The underground lead and zinc mining complex at Gorubso continued to be the country's major producer of lead and zinc ore.[553]

In 2006, the total mine production of lead in concentrate declined by almost 18% compared to that of 2005. The production of zinc in concentrate declined by almost 23% compared to that of 2005. Primary and secondary lead production amounted to about 84.300 t, which was almost a 10% decrease compared with that of 2005. Lead & Zinc Complex PLC Kurdzhali (OTZK) anticipated significant production increases as a result of equipment modernization and the acquisition of several mines in the Kurdzhali area. OTZK indicated that lead production would increase by 122% to 30.000 t, and that zinc output would increase by about 6% to 27.500 t.[554]

---

553 USGS 2008 (Bulgaria).
554 Metals Insider, 2006, in: USGS 2008 (Bulgaria).

Table 58: Metallic minerals: production/mining – import – export of commodities of Bulgaria (Data by BGS, 2010 [European Mineral Statistics])

Commodities related to metallic minerals/products	Production/mining	Import	Export
Silver	mi		
PMG		m	s
Gold	mi	mi	
Zinc		mi, sz	oc, u	oc, u, s
Tin		u	
Rare-earth			
Mercury			
Lithium			
Lead		mi, ref	oc, u, s	oc, u
Gallium			
Copper		mi, sm, ref	oc, u, s	
Cadmium		mi	x
Bismuth		mi	
Bauxite			
Arsenic		ma	
Antimony		m	
Aluminium		a, ah, u, ua, s	ua, s
Vanadium			
Tungsten			
Titanium		o	m
Tantalum			
Nickel		u	
Molybdenum		m	oc, m
Manganese	mi	oc	oc
Cobalt			
Chromium			
Iron	mi, pi, cs, fa	pi, fa, s	pi, fa, s

a	Alumina	fc	Ferro-cerium	orec	Oher rare earth compounds	sul	Sulfide
ah	Alumina hydrate	ilm	Ilmenite	otm	Other titanium minerals	sz	Slap zinc
bp	Burnt pyrites	m	Metal	pa	Primary aluminium	tm	Titanium minerals
c	Concentrate	ma	Metallic arsenic	pen	Pentoxide	ts	Titanium slag
car	Carbonate	mat	Matte and cement	pi	Pig iron	u	uwrought
carb	Carbides	mi	Mine	rec	Rare earth compounds	ua	uwrought alloys
cc	Cerium compounds	o	Oxide	ref	refined	ur	urefined
cr	Crude steel	oc	Ores and concentrates	s	Scrap	vana	Vanadiferous residues
fa	Ferro alloys	or	Ore	sm	Smelter	wa	White arsenic

Industrial minerals

Between 1998 and 2008, Bulgaria mined a broad range of industrial minerals that included baryte, feldspar, gypsum, kaolin, perlite and salt. Baryte production decreased from 120.000 t to 10.000 t in 2005. Gypsum production remained at a constant level until 2007. Kaolin production increased from 151.451 t (1998) to 250.372 t (2008). Salt production increased from 73.742 t (1998) to 1,5 Mt (2008).

Figure 136: Kaolin and Bentonite industry in Bulgaria (Data by http://www.kaolin.bg/)

Kaolin AD is the 4-largest producer in Europe.

Bentonit AD and Kaolin AD were major producers of industrial minerals in Bulgaria. Kaolin AD operated quarries and processing facilities at Kaolinovo and Vetovo (kaolin), Shoumen and Varna (quartz-feldspathic sands), and Konarata and Ustrem (potassium and sodium feldspars). Final output included kaolin, glass sand, and dry and wet silica sands. Bentonit AD was a major Bulgarian producer of bentonite, perlite, and zeolite. With bentonite, zeolite, and perlite resources amounting to about 7,3 Mt, 2,8 Mt, and 0,8 Mt, respectively, Bentonit AD's corresponding processing capacities were reported to be 200.000 t/yr, 50.000 t/yr, and 150.000 cubic meters per year.[555]

---

555 USGS 2008 (Bulgaria).

5.2 EU States

Table 59: Industrial minerals: production/mining – import – export of commodities of Bulgaria (Data by BGS, 2010 [European Mineral Statistics])

Commodities related to industrial minerals/ products	Baryte	Bentonite	Diatomite	Diamond	Feldspar	Fluorspar	Gypsum	Graphite	Kaolin	Magnesite	Perlite	Potash	Phosphate rock	Salt	Sulfur	Talc	Cement
Production/ mining (x: mine)	x	x			x		x		x		x			x	rec		cc cf
Import		x	x		x	x	cr ca	x	x	me ma				x	x	x	x cc pc
Export	x	x			x		cr ca	x						x	x		cc pc

a	Anhydrite	cr	Crude	i	Industrial	rs	Rock salt
bs	Brinesalt	d	Dust	ma	Magnesia	sib	Salt in brine
ca	Calcined	fs	Fertiliser salts	me	Magnesite	ss	Seasalt
cc	Cement clinker	g	Gypsum	pc	Portland cement	sul	Sulphate
cf	Cement finished	gc	Gem cut	pf	Potassic fertilisers	um	unmanufactured
chl	Chloride	gr	Gem rough	rec	recovered		

Construction minerals

In 2008, Bulgaria produced 40 Mt aggregates, i.e. 18 Mt sand and gravel and 22 Mt crushed rock. The total number of producers (companies) was 200 and the total number of extraction sites (active quarries and pits) was 100.

Minerals policy issues

Bulgaria does not have a national raw materials strategy, however, is in preparation. There is no national mineral resource planning system to protect economically attractive deposits.

Regulatory issues

Article 39 ML regulates the conditions concerning licenses for prospecting for and/or exploration of mineral resources. Mineral resources covered by Article 2 ML can be granted through: a) competitive bidding or tender; b) direct selection of the license holder.[556] According to Article 39 ML concessions for extraction of mineral resources under Article 2 can be granted either a) through competitive bidding or tender; or b) to a holder of a license for prospecting and

---

556 Further informations see Tiess) (2010), l.c.

exploration or a license for exploration by right in pursuance of Article 29 ML. According to Article 31 the validity term of a license for prospecting and/or exploration is up to three years and may be renewed twice by up to two years. The validity term of the extraction concession is up to 35 years and can be renewed by up to 15 years (Article 36 ML).

Investment policy issues

The National Program for Sustainable Development of Mining in Bulgaria was drafted and approved in 1998. The Government continued to work to improve the country's environmental condition and began enforcement of environmental regulations that would meet the EU standards in all new mineral industry projects. In 2006, Bulgaria's Chamber of Mining and Geology (BCMG) indicated that the country's mining and mineral processing sector would need about 1,2 billion Euros (€) ($ 1,6 billion)[557] to meet the EU's ecological standards.[558]

Bulgaria is expected to continue to develop its metallic minerals, industrial minerals and construction minerals sector, especially quarries and processing facilities for the production of construction materials. Exploration for precious metals also is expected to increase. The Law on Transformation and Privatization of State and Municipal-Owned Enterprises was adopted by Parliament in 1992, and the Mining Law 1998 was adopted to promote private enterprise and foreign investment. The principle of equal treatment of foreign and domestic investors was promulgated in law in the Constitution and in the Law on Encouragement of Investments.[559]

Greece

The Gross Domestic Product increased to $ 351,9 billion in 2008 from $ 133,8 billion in 1998. The value of the gross output of industry (mining, materials good production, supply of energy) decreased to 19 % (2008) from 20,6 % (1998).

In 2006, the mineral industry, which consisted of the sectors that mine and process metallic and non-metallic minerals and mineral fuels, was a small but important segment of the Greek economy. The high metal prices in world markets supported the Greek mineral industry in 2006.[560] In terms of the value of

---

557 USGS 2008 (Bulgaria): Where necessary, values have been converted from European euros (€) to U.S. dollars ($) at an average rate of € 0,758 = $ 1,00.
558 Sofia Echo, 2006e, in: USGS 2008 (Bulgaria).
559 Krastanova, 2005, in: USGS 2008 (Bulgaria).
560 USGS (2008): Minerals Yearbook 2006, Volume III – Greece.

5.2 EU States

production, bauxite was the most important of Greece's mineral commodities. Production of mineral commodities in Greece was closely tied to the export market since about 50% of the country's mineral production was exported.[561]

Figure 137: Mining production in Greece between 1998 and 2008 – selected mineral resources including aluminium production (Data by Weber et al. [WMD 1998–2008])

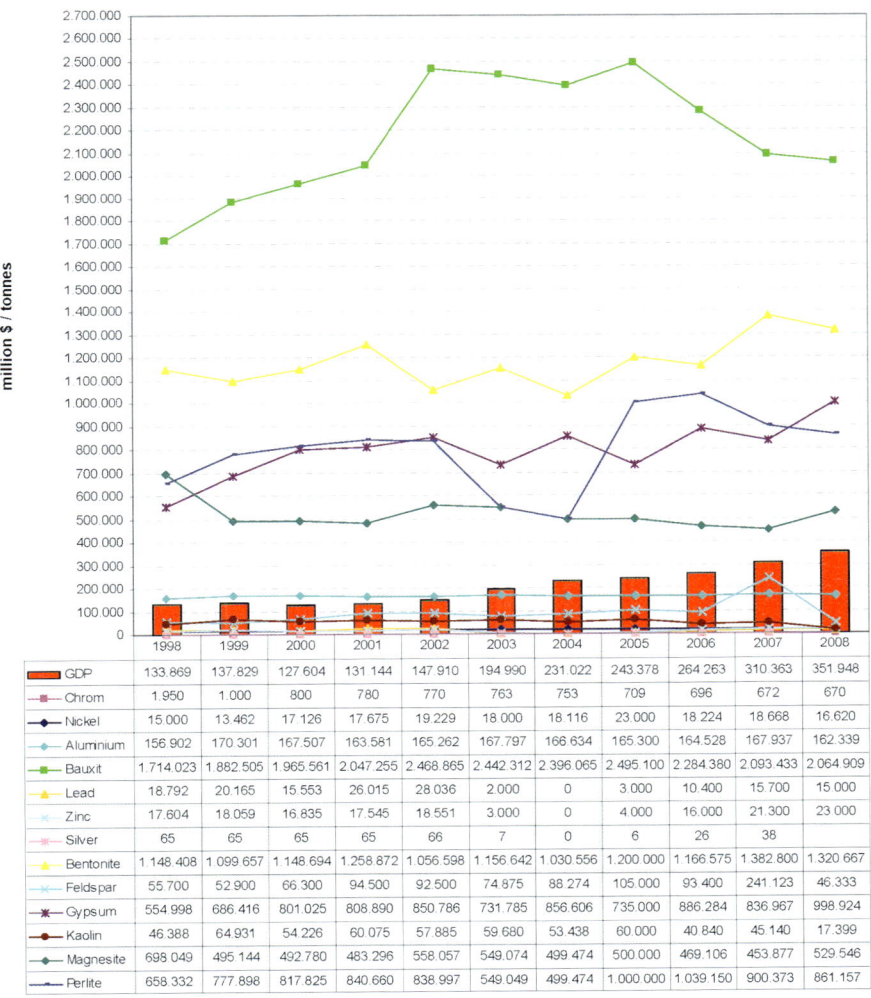

---

561 USGS 2008 (Greece).

Metallic minerals

Greece mined a broad range of metallic minerals that included nickel, bauxite, lead, zinc and silver. Nickel mine production increased between 1998 and 2008, from 15.000 t to 16.600 t. Greece ranked 15th globally in 2008. Zinc production increased from 17.600 t to 23.000 t. Lead production increased between 1998 (18.790 t) and 2002 (18.550 t) and decreased between 2002 and 2008 (15.000 t). Production of chromium decreased from 1.950 t (1998) to 670 t (2008).

Bauxite is the principal raw material used in the production of alumina, which is the major source of aluminum. Aluminium production increased from 156.900 t to 162.339 t (1998–2008), bauxite production increased (1998–2008) from 1,7 Mt to 2,0 Mt. Greece is the largest EU bauxite producer. Greece's estimated 100 Mt of bauxite reserves were of boehmitic and diasporic type. Although the bauxite ore had an average aluminum oxide content of 53 %, it also had a high silica content that made it hard to process.

The following figure illustrates the production, import and export of Greece.

Figure 138: Production, export and import of bauxite (Data by BGS)

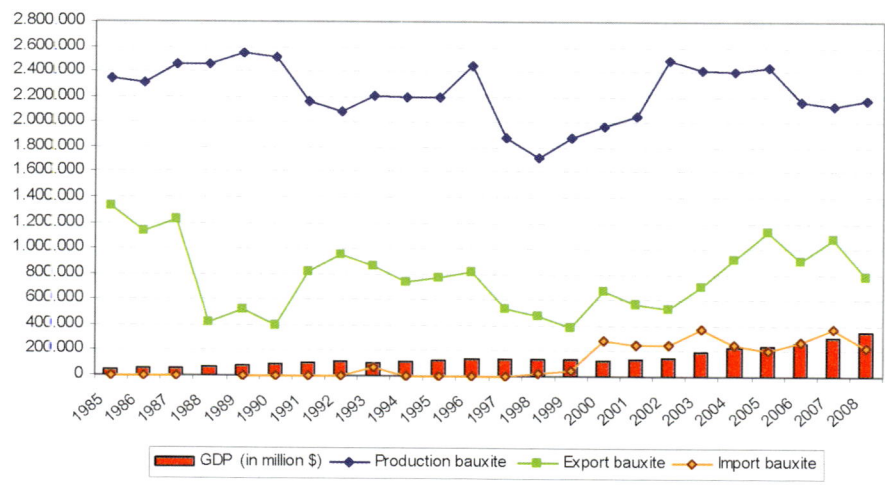

The metallic minerals sector involved a relatively small number of large, capital-intensive companies vertically integrated and engaged in both mining and metallurgical processing. Typical examples were Aluminium de Grèce S. A., (AdG) a producer of bauxite, alumina and aluminium.[562]

---

562 USGS 2008 (Greece).

The major bauxite deposits are located in central Greece within the Parnassos-Ghiona geotectonic zone and on Evoia Island. S&B industrial minerals S.A. which leases and owns mines in these regions controlled the most significant bauxite reserves, in terms of size, in Europe. In 2006, the recovery rate for all the bauxite mined was about 85 %.[563] AdG and S&B announced that a new agreement was reached to supply the AdG alumina plant with bauxite from S&B's mines. The new 10-year agreement covered the period from January 1, 2007, to December 31, 2016. According to the agreement, AdG commissioned S&B to supply its plant with 700.000 t/yr of bauxite, a quantity significantly larger than the 400.000 t/yr provided under the existing contract. Moreover, the new contract included an option for an additional 70.000 t/yr of bauxite, depending on the needs of AdG.[564]

In 2006, Larco was among the leading ferronickel producers in the world and the only producer of nickel in Europe that used domestic nickel ores. Larco had three main mining areas—Evia (open pit), with an annual production of about 1,5 Mt of ore; Agios Ioannis (underground), with an annual production of about 700.000 t of ore; and Kastoria (open pit), with an annual production of about 300.000 t. The Larymna metallurgical plant consisted of four rotary kilns, five electric arc furnaces, and two converters with a metal production capacity of 50 t each. The 17.736 t of ferronickel produced in 2006 was sold to Acerinox S. A. of Spain, ArcelorMittal Group of Luxembourg, Outokumpu oyj of Finland, and ThyssenKrupp AG of Germany, based on long-term contracts. In 2006, Glencore S. A. was added to Larco's customer portfolio.[565]

Figure 139: Larco G. M. M. S. A.

Agios Ioannis Mines (http://www.larco.gr)

Servia Lignite Mine (http://www.larco.gr)

In 2006, European Goldfields Ltd. held a 95 % interest in Hellas Gold S. A., which owned three gold and base-metal deposits in northern Greece. These were the Olympias deposit, which contains gold, lead, silver, and zinc; the Skouries copper/gold deposit; and the polymetallic Stratoni deposit. Hellas Gold started production at Stratoni in late 2005 and continued in 2006, which

---

563 S&B Industrial Minerals S.A., 2006a, in: USGS 2008 (Greece).
564 S&B Industrial Minerals S.A., 2006c, in: USGS 2008 (Greece).
565 Larco G.M.M S.A., 2006, in: USGS 2008 (Greece).

resulted in higher production of lead and silver than in 2005. The total estimated proven and probable reserves at Stratoni in 2006 were 1,9 Mt at grades of 8,1% lead, 190 g/t silver, and 10,8% zinc. Production of ore was expected to reach 170.000 t by the end of 2006 and to increase steadily to 400.000 t/yr by 2010. Based on historical production levels, the Stratoni Mine was expected to produce at grades of between 8% and 10% lead, 200 g/t silver and 8% to 10% zinc, with concentrator metals recovery of about 90%.[566]

---

566 European Goldfields Ltd., 2007, in: USGS (Greece).

5.2 EU States

Table 60: Metallic minerals: production/mining – import – export of commodities of Greece (Data by BGS, 2010 [European Mineral Statistics])

Commodities related to metallic minerals/products	Production/mining (x: mine)	Import	Export
Silver	x	m	m
PMG		m	
Gold		m	m
Zinc	x	u, oc	oc, u, s
Tin		u	
Rare-earth		cc, fc	
Mercury			
Lithium			
Lead	x, ref	u, s	u, s
Gallium			
Copper		u, ua, s	u, ref, ua, s
Cadmium			
Bismuth		x	
Bauxite	x	x	x
Arsenic		ma	
Antimony		oc, m	m
Aluminium	a, pa	a, ah, u, ua, s	a, ah, u, ua, s
Vanadium			
Tungsten			
Titanium		tm, o	o
Tantalum			
Nickel	x, sm/ref	oc, u	oc
Molybdenum		m	
Manganese		oc, m	
Cobalt			
Chromium	oc	oc	
Iron	cr, fa	ore, pi, fa, s	fa, s

a	Alumina	fc	Ferro-cerium	orec	Oher rare earth compounds	sul	Sulfide
ah	Alumina hydrate	ilm	Ilmenite	otm	Other titanium minerals	sz	Slap zinc
bp	Burnt pyrites	m	Metal	pa	Primary aluminium	tm	Titanium minerals
c	Concentrate	ma	Metallic arsenic	pen	Pentoxide	ts	Titanium slag
car	Carbonate	mat	Matte and cement	pi	Pig iron	u	uwrought
carb	Carbides	mi	Mine	rec	Rare earth compounds	ua	uwrought alloys
cc	Cerium compounds	o	Oxide	ref	refined	ur	urefined
cr	Crude steel	oc	Ores and concentrates	s	Scrap	vana	Vanadiferous residues
fa	Ferro alloys	or	Ore	sm	Smelter	wa	White arsenic

Industrial minerals

Between 1998 and 2008, Greece mined a broad range of industrials that included bentonite, feldspar, kaolin, magnesite, perlite, salt and talc. Bentonite production (see figure 137) increased from 1,15 Mt to 1,3 Mt. In 2008, Greece ranked 1st in the EU and 3rd globally. S&B was the leading bentonite producer in Europe with sales of more than 1 Mt/yr; it was also the world's third ranked bentonite producer after the United States. S&B remained focused on supplying bentonite to foundries and the drilling industry. S&B mined bentonite from its reserves on the island of Milos and produced about 85 % of total Greek bentonite output.

Feldspar production (see figure 137) increased to 241.120 t in 2007 from 55.700 t in 1998. Gypsum production increased from about 555.000 t (1998) to 998.920 t (2008). Kaolin production increased from 46.388 t in 1998 to 60.000 t in 2005 and decreased to about 17.400 t between 2005 and 2008. The average magnesite production was 500.000 t. Perlite production between 1998 and 2006 increased from 658.300 t to 1,0 Mt, and then decreased between 2006 and 2008. Moreover, the marble industry was active in the quarrying, processing, and sale of blocks and finished products. More than 60 different marble types with many colours were available. The Greek marble industry continued to play a leading role in the international market and its exports included blocks, slabs, and tiles.

S&B Industrial Minerals S. A. was Greece's – and one of the world's – leading producers of industrial minerals. Composed of five divisions (the Bauxite, Bentonite, Fluxes, Perlite, and Specialty Minerals Divisions), the company had facilities around the globe; its main mining and processing operations, however, were located in Greece.[567] S&B Industrial Minerals S. A. was the leading bentonite producer in Europe, the world's second ranked producer after the United States, and the leading bentonite supplier worldwide. S&B's reserves were located on the island of Milos.[568] Heracles General Cement S. A., a member of the Lafarge Group, had three cement plants. One of these plants, located in Volos, was the largest-capacity cement plant in Europe [4,6 Mt/yr]; the other plants were located in Halkis Evia and Milaki Evia. The total combined production capacity of the three plants was 9,6 Mt/yr. Heracles was also active in the production and sale of ready-mix and aggregate products.[569] With a production capacity of 650.000 t/yr, S&B was the leading producer of raw perlite worldwide and the leader in the European market for perlite used in building materials, cyrogenics, formed products, and horticulture, and as filter aids. S&B mined perlite at Provatas, Trachylas, and Tsigrado on Milos Island.[570]

---

567 Industrial Minerals, 2006, in: USGS 2008 (Greece).
568 S&B Industrial Minerals S.A., 2006a, in: USGS 2008 (Greece).
569 Google Finance, 2006, in: USGS 2008 (Greece).
570 S&B Industrial Minerals S.A., 2006d, in: USGS 2008 (Greece).

Figure 140: S&B Industrial Minerals Company, Voudia bay, Milos Island (Data by http://www.s.andb.gr)

Grecian Magnesite S. A. (GM) was the leading magnesite exporter in Europe and the only active magnesite operation in Greece. GM's open pit mine and plant are located southeast of Thessaloniki in the Chalkidiki region. In 2006, GM extracted about 2,5 Mt of crude run-of-mine material to produce 200.000 t of magnesium oxide, about 97 % of which was exported.[571] Also, the Greek marble industry continued to play a leading role in the international market as a result of continued marble production in almost all areas of the country.

Table 61: Industrial minerals: production/mining – import – export of commodities of Greece (Data by BGS, 2010 [European Mineral Statistics])

Commodities related to industrial minerals/ products	Asbestos	Baryte	Bentonite	Diatomite	Diamond	Feldspar	Fluorspar	Gypsum	Graphite	Kaolin	Magnesite	Perlite	Potash	Phosphate rock	Salt	Sulfur	Talc	Cement
Production/mining (x: mine)		x		x		x		x	x	x					x	rec	x	
Import		x	x	x	u gr gc i d	x	x	cr ca	x	x	me ma	sul chl	x	x	x	x		cc pc
Export		x		x				cr ca		x	ma	p			x	x	x	cc pc

a	Anhydrite	cr	Crude	i	Industrial	rs	Rock salt
bs	Brinesalt	d	Dust	ma	Magnesia	sib	Salt in brine
ca	Calcined	fs	Fertiliser salts	me	Magnesite	ss	Seasalt
cc	Cement clinker	g	Gypsum	pc	Portland cement	sul	Sulphate
cf	Cement finished	gc	Gem cut	pf	Potassic fertilisers	um	unmanufactured
chl	Chloride	gr	Gem rough	rec	recovered		

---

571 Grecian Magnesite S.A., 2006, in: USGS 2008 (Greece).

Construction minerals

In 2008, Greece produced 40 Mt aggregates, i.e. 20 Mt sand and gravel and 20 Mt crushed rock. The total number of producers was 300 and the total number of extraction sites was 200.

Minerals policy issues

Greece does not have a national raw materials strategy. There is no national mineral resource planning system to protect economically attractive deposits. Minerals are included to a certain grade in the National Spatial Plan and sectoral Land Plan.[572]

There are two different procedures to obtain permission for exploration activities. One pertains to metallic minerals and one to industrial minerals and aggregates. For the metallic minerals one has to obtain a Preliminary Exploration Permit (Articles 20–43 Mining Code) or exploration concession. This is valid for two years. If successful, it should be followed by an exploitation concession (Articles 44–64 Mining Code), which is valid for 50 years (renewable).

Investment policy issues

Because northern Greece was thought to contain a significant amount of exploitable mineral resources, it received the most attention for exploration activity of anywhere else in the country. Mining companies renewed exploration funding, which indicated their optimism about future production. For instance, activities continued to be directed toward the search for gold. Greece is expected to remain a major supplier of industrial minerals in the international market. Mineral exploration activities in Greece will be intensified to secure additional high-quality reserves. The Government could be involved in planning investment programs to improve the existing installations and lower operating costs.

---

572 LCU (2010): table 2 'land use planning' (Greece).

5.3 Non-EU States

Russian Federation

The GDP-development increased to $1.6 trillion in 2008 from $271 billion in 1998. The value of the gross output of industry (mining, materials good production, supply of energy): decreased to 36.1 % in 2008 from 37.4 % in 1998. The minerals industry contributes more than a half to industrial production. The raw materials extraction plays a substantial role for the Russian Federation, which is one of the **largest mineral producers of the world**. The Russian Federation holds amongst others the following top rankings: Iron, cobalt, platinum, nickel, palladium.[573]

Figure 141: Production of selected metallic and industrial minerals in Russia (European part) (Data by Weber et al. [WMD 1998–2008])

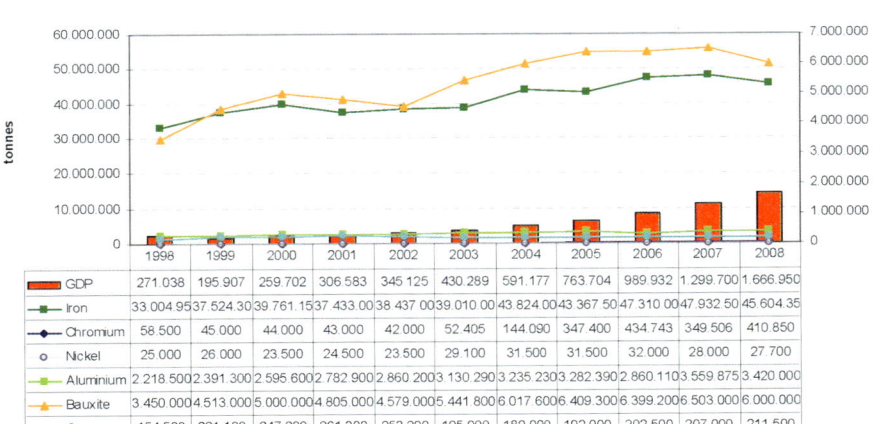

---

[573] Russia, though, holds a rank among the Top 10 of the World for a variety of other commodities: cp. Weber/Zsak, World Mining Data (2008).

Chapter 5 View of the minerals policies in selected states of Europe

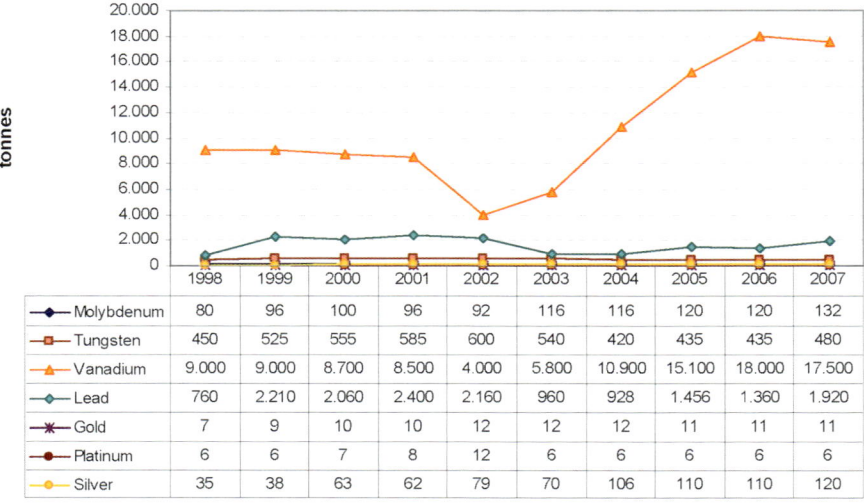

Production of selected metallic minerals in Russia/European Part between 1998 - 2008 and GDP (current prices) - Section II

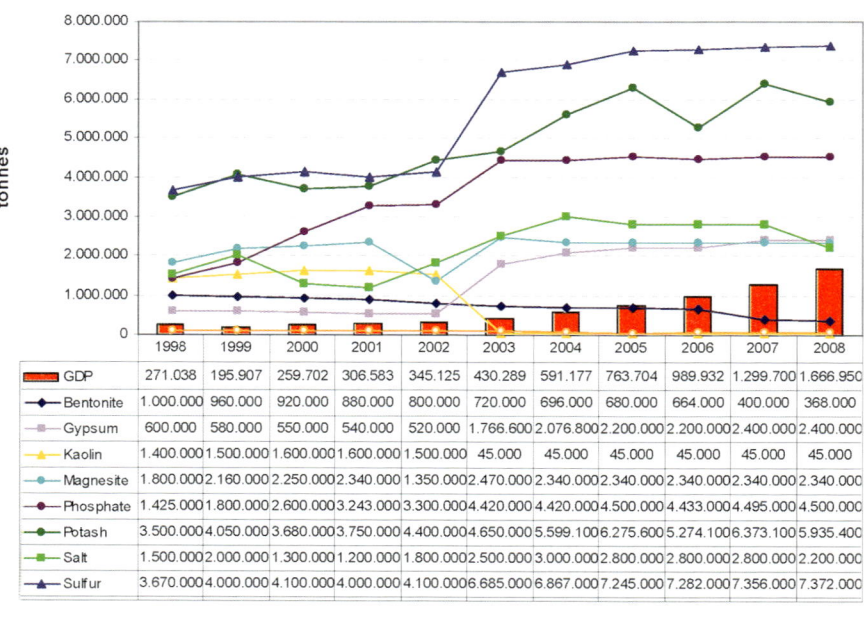

Production of selected industrial minerals in Russia/European Part between 1998 - 2008 and GDP (current prices) - Section I

5.3 Non-EU States

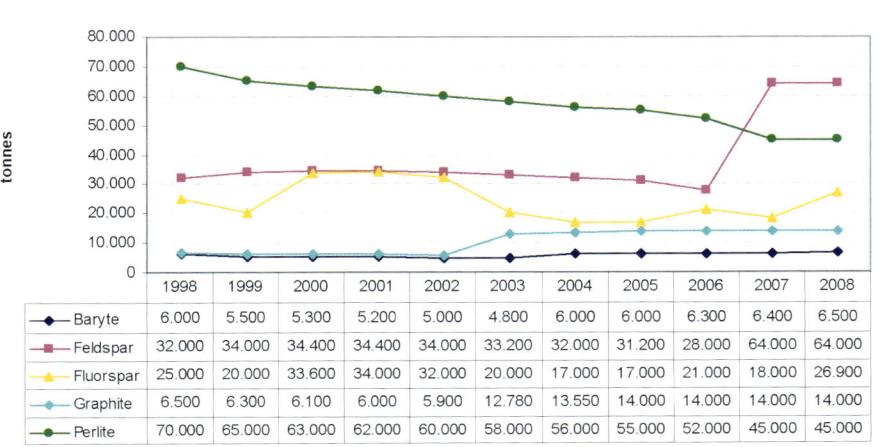

The following figure illustrates the production, import and export of nickel in Russia.

Figure 142: Production, export and import of nickel (Data by BGS)

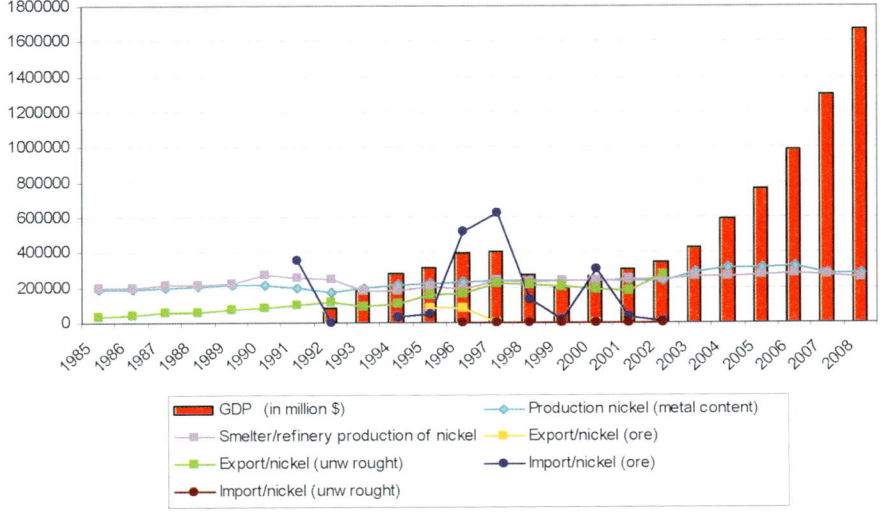

Figure 143: Deposit Stoilensky GOK http://www.nlmksteel.com

NOVOLIPETSK STEEL (NLMK) is one of the largest steel producers of the world. Among other things pig iron, plates, transformer and dynamo steel are produced. Products in 2007 were supplied to Europe, America, Asia, Africa and in the Middle East (over 60 countries). NLMK produces altogether 13 % of Russia steel. http://www.nlmksteel.com/

Minerals policy issues

Russian has a national raw materials strategy.[574]

A precise categorisation of deposits as well as an accurate knowledge about the domestic mineral potential is crucial for the raw materials economy of Russia.[575] Specific conditions are determined by the state; especially a minimum mineral content of the deposit is prescribed. This is to ensure economic benefit of mining for all deposits, which are listed in the Russian resources register. Article 31 Subsoil Law includes (the establishment of) a balance system of raw materials. Thus the actual condition of the mineral reserves shall be pointed out. It contains information about quantity, quality, and degree of exploration of different raw materials as well as production data. The competent authority (Federal Agency on Subsoil Management) is responsible for that.

The rights of disposal of strategic raw materials can be substantially limited. Concerning national strategic mineral reserves there are restrictions for foreign investors as follows:[576] If, for example, the purchase of shares by foreign investors of more than 10 % of the total votes of shareholders of the Company is intended, a resolution of a particular national commission is required.

---

574 Already mentioned in Chapter 4; see also Annex.
575 Pticyn, A. M., Ljudin, J. K., Polonskij, G. V. (2004): State and measures for strenghtening of mineral and raw material base of Russian metallurgy. Gornyj zurnal, Moscow, Vol 180 (2004), 3, pp. 45–53.
576 Based on the law: "On Procedures for Foreign Investments in Companies of Strategic Significance for National Defense and Security".

Restrictions for foreign investors which are regarded by the Russian state as strategic, can be widely interpreted. They also concern mining companies for strategic minerals, such as uranium, diamond, high-quality quartz, the group of elements yttrium, nickel, cobalt, tantalum niobium, beryllium, lithium and platinum group elements. According to Article 10 Subsoil Law the right to use subsoil plots will be granted either for a fixed or unlimited period of time. The right to use subsoil plots might be granted for a fixed period of time in the following cases:

- For the geological study – for a period of up to five years or for a period of up to 10 years during the works of the geological exploration of the subsoil section in the inland sea waters, the territorial sea and the continental shelf of the Russian Federation.
- For the period of extraction of the mineral deposit which is calculated on the basis of the feasibility study for the extraction of the mineral deposit ensuring the rational use and protection of subsoil (Article 10 Subsoil Law).

Norway

The Gross Domestic Product increased to $446 billion in 2008 from $151 billion in 1998. The value of the gross output of industry (mining, materials good production, supply of energy) increased to 46,2% (2008) from 32,7% (1998). Norway has a varied geology and a broad spectrum of mineral resources for exploration and exploitation.[577]

The country's long shoreline and close proximity to the large European Union market are major competitive advantages for some raw materials, particularly aggregates, natural stone, and certain other industrial minerals.

---

577 USGS (2008, 2010): Minerals Yearbook 2008, 2010 Volume III (Norway).

Chapter 5 View of the minerals policies in selected states of Europe

Figure 144: Mining production in Norway between 1998 and 2008 – selected mineral resources including aluminium production (Data by Weber et al. [WMD 1998–2008])

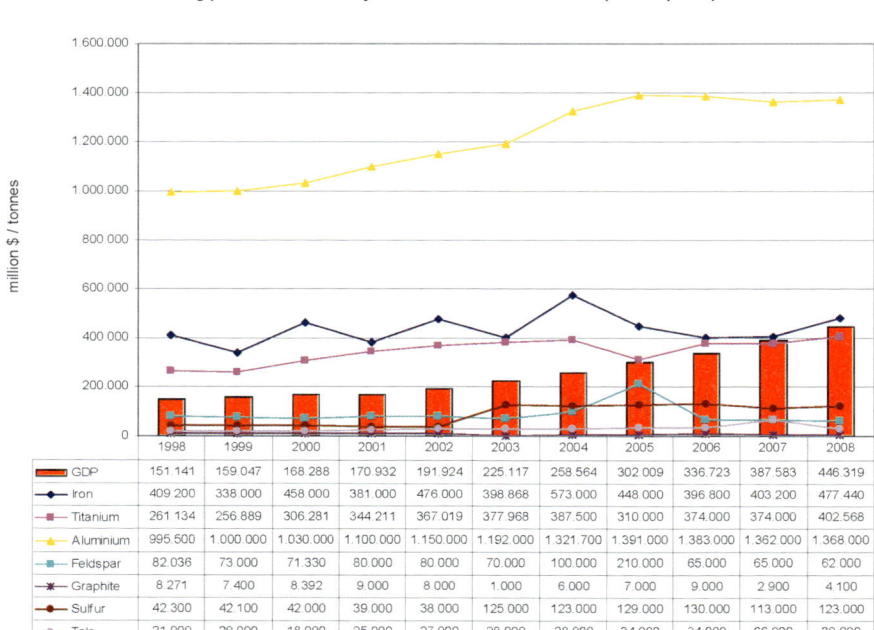

Metallic minerals

Between 1998 and 2008, Norway mined iron ore, cadmium, titanium, nickel and gold. Iron production (see figure 144) increased to about 446.000 t in 2008 from 409.200 t in 1998. In 2006, production of titanium accounted for 7,2 % of world production, and increased to 402.000 t in 2008 from 261.000 t in 1998. In 2008, the country's production of titanium accounted for 6,7 % of world production. Titania A/S was one of the world's leading producers of ilmenite with about 6% of world production. It was the leading producer of ilmenite in Europe. Production from the Tellnes open pit mine resulted in about 915.000 metric tons (t) of ilmenite concentrate containing 44,7 % titanium dioxide.[578] Finally, aluminum production increased to 1.368.000 t in 2008 from 995.000 t in 1998.

---

578 Norwegian Geological Survey, 2009b, p. 20, in: USGS (2010), Norway.

5.3 Non-EU States

The following figure illustrates the production, import and export of titanium in Norway.

Figure 145: Production, export and import of titanium (Data by BGS)

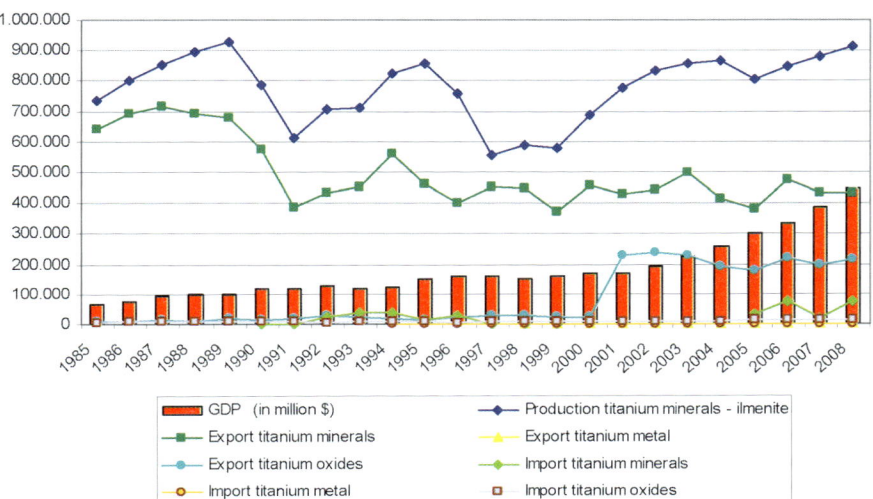

Table 62: Metallic minerals: production/mining – import – export of commodities of Norway (Data by BGS, 2010 [European Mineral Statistics])

Commodities related to metallic minerals/products	Production/mining (x: mine)	Import	Export
Silver		oc, m	m
PMG		m, s	
Gold		m, s	m, s
Zinc		oc, u	u, ua
Tin		u	s
Rare-earth		m	rec
Mercury			
Lithium			
Lead		u	s
Gallium			
Copper	sm, ref	u, s	u, ua, s
Cadmium		m	m
Bismuth			
Bauxite		x	
Arsenic			
Antimony		m, o	orec
Aluminium	pa	a, ah, u, ua, s	u, ua, s
Vanadium			
Tungsten		m	
Titanium	x	tm, m, o	tm, m, o
Tantalum			
Nickel	x, sm/ref	mat, u	oc, u, s
Molybdenum			
Manganese		oc, m	oc, m
Cobalt	m	m	m
Chromium		oc, m	m
Iron	ore, pi, cr, fa	ore, pi, fa, s	ore, pi, fa, s

a	Alumina	fc	Ferro-cerium
ah	Alumina hydrate	ilm	Ilmenite
bp	Burnt pyrites	m	Metal
c	Concentrate	ma	Metallic arsenic
car	Carbonate	mat	Matte and cement
carb	Carbides	mi	Mine
cc	Cerium compounds	o	Oxide
cr	Crude steel	oc	Ores and concentrates
fa	Ferro alloys	or	Ore
orec	Oher rare earth compounds	sul	Sulfide
otm	Other titanium minerals	sz	Slap zinc
pa	Primary aluminium	tm	Titanium minerals
pen	Pentoxide	ts	Titanium slag
pi	Pig iron	u	uwrought
rec	Rare earth compounds	ua	uwrought alloys
ref	refined	ur	urefined
s	Scrap	vana	Vanadiferous residues
sm	Smelter	wa	White arsenic

5.3 Non-EU States

Industrial minerals

Between 1998 and 2008 mine production included feldspar, graphite, ilmenite, and limestone. Feldspar production (see figure 144) increased from 82.000 t (1998) to 210.000 (2005), and then decreased to a constant level of 65.000 t. Graphite production decreased from 8.270 t to 4.100 t (1998–2008). Sulphur production increased from 42.300 t–123.000 t (1998–2008). Talc production was at a constant level of (average) 25.000 t.

In 2007, Huustadmarmor AS was the world's leading producer of calcium carbonate slurry for the paper industry.[579]

Table 63: Industrial minerals: production/mining – import – export of commodities of Norway (Data by BGS, 2010 [European Mineral Statistics])

Commodities related to industrial minerals/products	Asbestos	Baryte	Bentonite	Diatomite	Diamond	Feldspar	Fluorspar	Gypsum	Graphite	Kaolin	Magnesite	Perlite	Potash	Phosphate rock	Salt	Sulfur	Talc	Cement
Production/mining (x: mine)						x			x								x	cf
Import	x	x	x			x		cr ca	x	x	me ma			sul chl	x	x	x	cc pc
Export	x		x	x		x		cr ca			chl				x		x	cc pc

a	Anhydrite	cr	Crude	i	Industrial	rs	Rock salt
bs	Brinesalt	d	Dust	ma	Magnesia	sib	Salt in brine
ca	Calcined	fs	Fertiliser salts	me	Magnesite	ss	Seasalt
cc	Cement clinker	g	Gypsum	pc	Portland cement	sul	Sulphate
cf	Cement finished	gc	Gem cut	pf	Potassic fertilisers	um	unmanufactured
chl	Chloride	gr	Gem rough	rec	recovered		

Construction minerals

Aggregates are economically important mineral raw materials (particularly needed for infrastructure development). The country's long shoreline and close proximity to the large European Union market are major competitive advantages particularly for aggregates. Production ranged between 53 Mt (1998) and 58 Mt (2006). It accounted in 2002: 53,4 Mt, in 2003: 53,3 Mt (i.e. 14,7 Mt sand and gravel (s&g), 35,7 Mt crushed rock (cr), 0,9 Mt recycled aggregates (ra)), in 2004: 51,3 Mt (14,7 Mt s&g, 35,7 Mt cr, 0,9 Mt ra), in 2005: 53,2 Mt (15 Mt s&g,

---

579 Norwegian Geological Survey, 2009c, p. 15, in USGS (2010), Norway.

38 Mt cr, 0,2 Mt ra), in 2006: 58,2 Mt (13,4 Mt s&g, 45 Mt cr). In 2008, Norway produced 68 Mt, i.e. 15 Mt s&g, 52 Mt cr. The total number of producers was 690; the total number of extraction sites was 713.

Figure 146: Mining on the coast of Norway Tinfos Titan & Iron KS – Tyssedal (Data by http://www.tinfos.no)

Minerals policy issues

Norway does not have a national raw materials strategy. Norway does not have an active national planning policy; there are virtually no guidelines for forward planning of minerals supply. The Planning and Building Act does not give any specific guidelines with regard to mineral resources at the national level.[580] The Geological Survey identified and classified in the last 10–15 years deposits of national interest. The criteria used for the classification are deposits for export, deposits for domestic market and deposits to be mined in the next 50 years. According to this, the Geological Survey developed data bases of economic resources, but without a legal status in spatial planning. Thus, no protection is possible against sterilisation/over planning.[581] In connection with county planning it is stated that: "The county plan shall have guidelines for the use of areas and natural resources in cases where such issues involve several local municipalities or cannot be satisfactorily handled by a local municipality inside its own borders."[582]

The Mining Law states that mineral occurrences – regardless who the property owner is – can be explored and extracted if necessary.[583]

---

580 Nielsen, K. (2004), Country Report Norway, in: Department of Mining and Tunnelling, Minerals Policies and Supply Practices in Europe, University of Leoben.
581 Nielsen, K. (2004), Country Report Norway, in: Department of Mining and Tunnelling, Minerals Policies and Supply Practices in Europe, University of Leoben.
582 Department of Mining and Tunneling (2004), Final Report, p 119, lc.
583 See also Tiess "Legal Basics", 2011.

Switzerland

The Gross Domestic Product increased to $502 billion in 2008 to $272 billion in 1998. The value of the gross output of industry (mining, materials good production, supply of energy) increased to 28,2 % (2008) from 27,9 % (1998).

Trade is the key to prosperity in Switzerland. The country is dependent upon export markets and upon imports of raw materials. In 2006, the value of 1 metric tonne of exported goods was two and a quarter times more than that of the same amount of imports.[584] Switzerland's main trading partner is Germany, followed by France, Italy, and the United Kingdom.[585]

The country serves as a major diamond exchange; it is actively involved in the cutting and polishing of diamonds and plays a significant role in international diamond trade activities. Lucerne was the world's third most important diamond trading centre after London and Antwerp. In 2006, Switzerland's exports of polished diamonds were valued at $ 661 million and imports of polished diamonds were valued at $ 592 million.[586]

The reserves of the small deposits of metalliferous ores that once existed in Switzerland have been depleted. Consequently, metals were not mined in 2006. In 2006, metal processing was confined mainly to the production of primary and secondary aluminium, copper, secondary lead, pig iron, and steel. Mining was predominantly carried out for building materials. The following industrial minerals were produced: gypsum, lime and salt.[587]

---

584 USGS (2008): Minerals Yearbook 2006, Volume III – Switzerland.
585 Swissworld, 2006, in: USGS 2008 (Country report Switzerland), p.1.
586 HRD – Antwerp World Diamond Centre, 2006, in: USGS 2008 (Country report), p.1.
587 USGS (2008): 2006 Minerals Yearbook, Switzerland, p.1.

Table 64: Metallic minerals: production/mining – import – export of commodities of Switzerland (Data by BGS, 2010 [European Mineral Statistics])

Commodities related to metallic minerals/products	Production/mining (x: mine)	Import	Export
Silver		m	m
PMG		m s	m s
Gold		m s	m s
Zinc		u	u s
Tin		u	s
Rare-earth		cc	orec
Mercury			
Lithium		o	o
Lead	ref	u	u s
Gallium			
Copper		u	ua s
Cadmium		m	
Bismuth		m	
Bauxite		x	
Arsenic		oc ma	orec
Antimony		oc m	o
Aluminium	pa	ah u ua s	u s
Vanadium			
Tungsten		m	m
Titanium		m	o
Tantalum		x	
Nickel		u	u s
Molybdenum			
Manganese		oc m	
Cobalt	m	m	
Chromium	oc m	m	
Iron	pi cr	ore pi fa s	fa s

a	Alumina	fc	Ferro-cerium	orec	Other rare earth compounds	sul	Sulfide
ah	Alumina hydrate	ilm	Ilmenite	otm	Other titanium minerals	sz	Slap zinc
bp	Burnt pyrites	m	Metal	pa	Primary aluminium	tm	Titanium minerals
c	Concentrate	ma	Metallic arsenic	pen	Pentoxide	ts	Titanium slag
car	Carbonate	mat	Matte and cement	pi	Pig iron	u	uwrought
carb	Carbides	mi	Mine	rec	Rare earth compounds	ua	uwrought alloys
cc	Cerium compounds	o	Oxide	ref	refined	ur	urefined
cr	Crude steel	oc	Ores and concentrates	s	Scrap	vana	Vanadiferous residues
fa	Ferro alloys	or	Ore	sm	Smelter	wa	White arsenic

5.3 Non-EU States

Table 65: Industrial minerals: production/mining – import – export of commodities of Switzerland (Data by BGS, 2010 [European Mineral Statistics])

Commodities related to industrial minerals/products	Asbestos	Baryte	Bentonite	Diatomite	Diamond	Feldspar	Fluorspar	Gypsum	Graphite	Kaolin	Magnesite	Perlite	Potash	Phosphate rock	Salt	Sulfur	Talc	Cement
Production/mining (x: mine)								x							x			cf
Import	x	x	x		gr gc i d	x	x	cr ca	x	x	me ma		sul chl		x			cc pc
Export																		cc pc

a	Anhydrite	cr	Crude	i	Industrial	rs	Rock salt
bs	Brinesalt	d	Dust	ma	Magnesia	sib	Salt in brine
ca	Calcined	fs	Fertiliser salts	me	Magnesite	ss	Seasalt
cc	Cement clinker	g	Gypsum	pc	Portland cement	sul	Sulphate
cf	Cement finished	gc	Gem cut	pf	Potassic fertilisers	um	unmanufactured
chl	Chloride	gr	Gem rough	rec	recovered		

Construction minerals

Aggregates are economically important mineral raw materials (particularly needed for infrastructure development). The production of aggregates increased to 61 Mt in 2006 from 49 Mt in 2002. The production accounted in 2003: 49,1 Mt, in 2004: 32 Mt (i.e. 26 Mt sand and gravel (s&g), 3 Mt crushed rock (cr), 3 recycled aggregates (ra)), in 2005: 57,1 Mt (46,5 Mt s&g, 5,3 Mt cr, 5,3 Mt ra), in 2006: 61,4 Mt (50 Mt s&g, 5,7 Mt cr, 5,7 Mt ra). In 2008, Switzerland produced (47 Mt aggregates, i.e.) 37 Mt s&g, 5 Mt cr and 5 Mt recycled aggregates. The total number of producers was 350 and the total number of extraction sites was 505. For instance, Holcim is a global supplier of (cement and) aggregates, founded already in 1912 (www.holcim.com).

The Swiss sand and gravel industry plays an important role. It is mostly medium-structured. There are about 100 companies with 200 gravel plants and 600 sites. In Switzerland annually approximately 50 million tonnes of sand and gravel are needed, about 60 % of them for cement production. Further approximately 10 % are used for bituminous coatings; the remainder is needed for a great variety of free applications in the building industry. A portion of approx. 10 % of these mineral raw materials originates from the deconstruction of old buildings and plants. To reach this percentage, about 75 % of construction waste is recycled.[588]

Figure 147: Extraction of sand and gravel in Switzerland (Data by Kündig et al, 1997)

Gravel transport from the face to processing plant (Data by Kündig et al, 1997)

Gravel mining with high pressure water jet and Pneutrax (Data by Kündig et al, 1997)

Minerals policy issues

Switzerland does not have a national raw materials strategy.

In Switzerland all claims related to area and ground depend on the land use planning right of the federation (land use planning law). It sets the principles for the separation of building and not- building areas, the selection of protected areas, the establishment of supply and disposal sites etc. Based on this law the state is responsible (i. e. has a planning obligation) for all supply and disposal tasks of public interest, including mineral raw materials.

The implementation for planning rests with the cantons, which usually provide concepts for mineral supply in cooperation with the industry. Such concepts shall identify the need of minerals, current supply situation and possible future locations.[589] A particularly successful approach is the minerals supply concept of the Aargau canton.[590]

---

588 Grob, J. (2005), Die schweizerische Sand- und Kiesindustrie (The Swiss sand and gravel industry), BHM Berg- und Hüttenmännische Zeitschriften [Journal of Mining, Metallurgical, Material, Geotechnical and Planned Engineering], p. 42.
589 Grob (2005), l.c., p. 43.
590 Cp. Tiess, G., Pilgram, R. (2003): Zentrale Aufgabe der Raumplanung: die nachhaltige Sicherung mineralischer Rohstoffe (Important task of land use planning: to ensure sustainable supply with mineral raw materials), BHM Berg- und Hüttenmännische Zeitschriften Zeitschriften [Journal of Mining, Metallurgical, Material, Geotechnical and Planned Engineering], pp. 408–410.

As an example for mining law, the Berne canton is mentioned. Article 30 ML regulates the privileges of the exploration-entitled. If the exploration-entitled can prove a deposit worthy of exploitation of the mineral raw material designated in the digging grant, he has a claim on issuance of the exploitation concession. Article 31 regulates the exploitation concession. Due to the exploitation concession the concessionaire receives the right to extract the state-owned minerals within certain properties.

The sand and the gravel industry faces a complex situation (large mass density, structural problems). The licensing procedures are complex. For instance, it takes one to two years for investigation and planning of a EIA until an application for mining can be submitted and, depending upon canton, again up to two years, until the permit is approved. The duration of the procedure can be extended for many years, if appeals take place. The industry must cope with today's situation – particularly regarding the year-long procedures with unpredictable outcome. As a result the supply of raw materials has become very capital-intensive in the last decades.[591]

Serbia

The Gross Domestic Product increased to $ 48 billion in 2008 to $ 8 billion in 2000. The overall value of mining and quarrying increased by 4,9 % compared with that of 2006.[592] Serbia was a modest producer of bauxite, copper, and lead and zinc ores, and such industrial minerals as clays, feldspar, gypsum, magnesite, pumice, and salt. In 2006, the value of non-energy mineral output increased by more than 112 %.[593] In 2006, the export volumes of coal and metal ores declined by about 19 % and 27 %, respectively, compared to those of 2005. The export value of metal ores increased by about 29 % compared to 2005.[594]

In 2008, within the Southern Balkans, the country was an important producer of such metals as copper, iron and steel. The lithium project of Rio Tinto plc in Jadar had 125,3 Mt of inferred resources of ore containing 1,8 % lithium oxide and 16,2 Mt of contained borates. The project was still in the early stages of development, but Rio Tinto estimated that the project could begin production within 6 years.[595]

---

591 Grob (2005), l.c., p. 45.
592 USGS (2008): Minerals Yearbook 2006, Volume III – Serbia.
593 Republicki Zavod za Statistiku Srbije, 2007, p. 126–129, in: USGS (2008).
594 Republicki Zavod za Statistiku Srbije, 2007, p. 126–129, in USGS (2008).
595 Rio Tinto plc, 2010, p. 72, 74; Tony Shaffer, Principal Adviser, Media Relations, Rio Tinto plc, written commun., October 15, 2010, in: USGS (2011): Minerals Yearbook 2009 Volume III (Serbia).

Figure 148: Copper mine near the city of Bor (Veliki Krivelj Mine) (Data by http://www.euromaxresources.com)

Minerals policy issues

In 2007 the Ministry for Mining and Energy started with the execution of the "**Study on a Master Plan** for the Promotion of the Mining Industry in the Republic of Serbia". The Japan International Cooperation Agency (JICA) is intensely involved in the implementation of this study. The master plan shall cover the resource potential which is not yet developed. Similarly a strategy shall be established for restructuring the mining industry and the development of efficient technologies. In the framework of the study a first evaluation of the Serbian raw materials economy was accomplished. The second phase is focussing on the establishment of the master plan. Beyond that, the **National Sustainable Development Strategy of Serbia** includes strategies for the exploration of new deposits as well as approaches to sustainable resource use.

The issuing of exploration permits takes place on basis of the Exploration Law (1995), the issuing of production permits on basis of the Concession Law and the Mining Law (1995).[596] The master plan mentioned above includes a reconstruction and a renewal strategy for the Serbian mining industry sector as well as an **investment plan** for domestic and foreign companies.

The Government continued to reduce restrictions on foreign investment in the country's minerals industry. In 2006, foreign investment in the country's minerals industry encompassed such targets as the Rudarsko Topionicki Bazen Bor (RTB Bor) (a copper mining and processing complex in Serbia) and magnesite and nickel interests in Serbia's Kosovo Province.

---

596 For further information see Tiess (2010), l.c.

Albania

The Gross Domestic Product increased to $ 12,9 billion in 2008 from $ 2,7 billion in 1998.

Figure 149: Mining production in Albania between 1998 and 2008 – selected mineral resources (Data by Weber et al. [WMD 1998–2008])

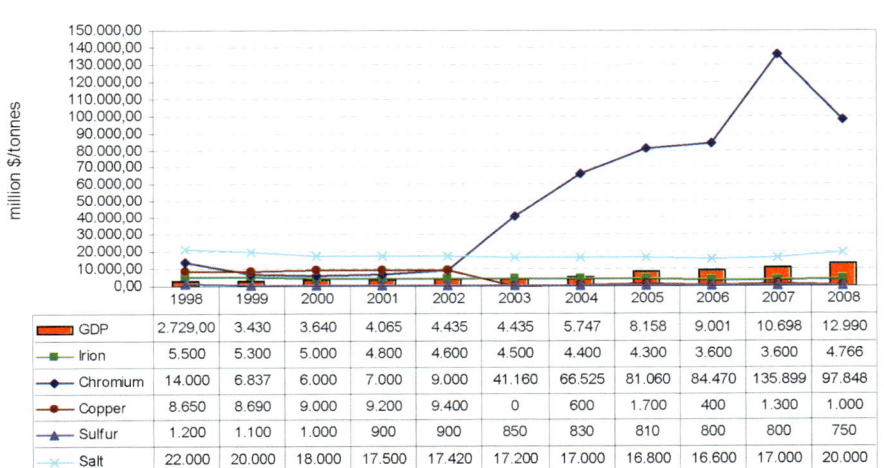

Metallic minerals

Between 1998 and 2008, mine production included iron ore, chromium and copper. Iron production (see figure 149) decreased to 4.700 t in 2008 from 5.500 t in 1998, copper production decreased to 1.000 t in 2008 from 8.600 t in 1998. Chromium mine production is important, production increased to 98.000 t in 2008 from 14.000 t in 1998. Albania's chromite output remained insubstantial compared with routine production levels reached during the 1960s through the late 1980s.[597] Reported chromite ore reserves were 32,8 million metric tons (Mt), including 12,8 Mt at a grade of more than 30% Cr_2O_3.[598]

The following figure illustrates the production, import and export of chromium in Albania.

---

[597] USGS (2008, 2010): Minerals Yearbook 2008, 2010 Volume III (Albania).
[598] Albinvest, 2009, p. 15, in: USGS (2010), Albania.

Chapter 5 View of the minerals policies in selected states of Europe

Figure 150: Production, export and import of chromium (Data by BGS)

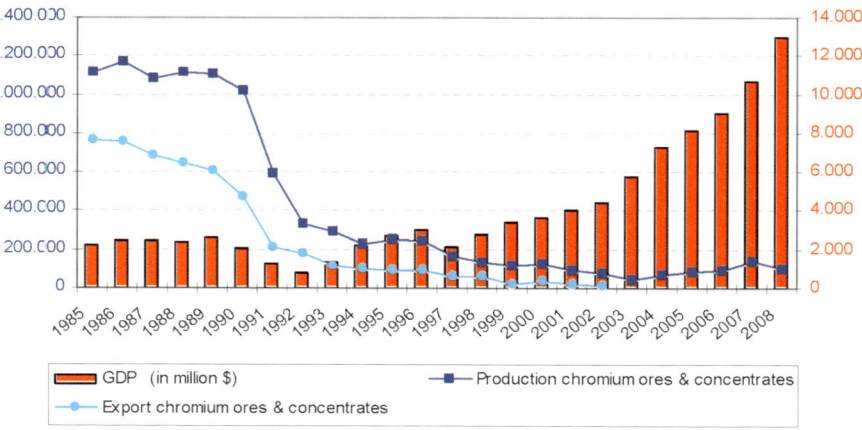

The rights to chromite mines near Bulquize and the Burrel and the Elbasan ferrochromium plants were acquired in 2007 by a partnership of the Austrian firm DCM DECOmetal GmbH and the Russian firm Terwingo Ltd. The partnership announced plans to increase investment significantly at its new facilities to increase annual production of marketable chromite ore and concentrate to 100.000 t from approximately 82.000 t and to achieve annual production of 33.000 t of low carbon ferrochromium and 15.000 t of high carbon ferrochromium. Production of ferrochromium was expected to be restarted at the end of 2007 or the beginning of 2008.[599]

European Nickel PLC (ENickel) continued to make progress on exploratory drilling at its Devolli nickel project in southeastern Albania. ENickel reported inferred resources of 35,6 Mt of nickel ore with an average grade of 1,20% in November 2007 and planned to complete its initial project study in 2008.[600]

Industrial Minerals

Albania mined a broad variety of industrial minerals. Important developments took also place in the cement industry. The Greek company Antea Cement Sh. A. (a subsidiary of the Titan Group) and Spanish company Cementos Aquila sh. p. k. (a subsidiary of Grupo Empresarial Aricam) each were granted

---

[599] DCM DECOmetal GmbH, 2007; 2008a, b; Metal-Pages, 2007, in: USGS 2009 (Albania), p 1.
[600] European Nickel PLC, 2008, in: USGS 2009 (Albania), p 1.

permits to construct cement manufacturing plants in Albania. Antea intended to invest about $ 125 million[601] into the construction of a 1,5 Mt/yr-capacity cement plant in the Kruje region with an expected completion date of late 2009.[602] Cementos Aquila planned to invest about $ 145 million to construct a plant in Mamurras with an initial capacity of 1,3 Mt/yr and an eventual capacity of 2,6 Mt/yr. The Mamurras plant was expected to begin operations in 2010.[603]

Minerals policy issues

Albania has a national raw materials strategy, which was published in 2006. The strategy and its action plan shall be updated every four years based on the strategic objectives of the four years program of the government. After a period of 15 years a new strategy will be drafted or the strategy itself will be updated taking in consideration mid and long term objectives. However, the strategy until now has not been updated.[604]

There is no national mineral resource planning system to protect economically attractive deposits.

Regulatory issues

The initial time limit of the exploration permit is two years and object to three extensions of one year each if requested by the permit possessor (Article 34 ML). The possessor of the exploration permit has within the validity of his permit the right of conversion of parts or of the whole permitted area in exploitation permits, excluding those parts from the obligations of the exploration permit (Article 43 ML). If the applicant of the exploitation permit does not possess the exploration permit, the application must include complete details of the financial and technical capacity of the applicant and information regarding his experience in the mining field (Article 46 ML). The competent authority grants an exploitation permit to natural persons or legal entities by considering financial and technical resources and mining experience. This permit guarantees the exclusive right to use one or more minerals specified in the permitted area (Article 44). The time limit of a mining exploitation permit is up to twenty years, and is subject of four extensions up to five years each, if the permit possessor requests these extensions in a time no later than one year prior to the expiration data of the previous time limit (Article 49 ML).

---

601 Where necessary, values have been converted from European Union euros (€) to U.S. dollars at the rate of 0,73€ = US$ 1,00.
602 Titan Group, 2008a, p. 52; b, in: USGS 2009 (Albania), p 1.
603 Grupo Empresarial Aricam, 2008, in: USGS 2009 (Albania), p 1.
604 LCU (2010): table 1 'mineral policy' (Albania).

Figure 151: Steel works in Elbasan/Albanien (Data by http://www.travelbilder.de)

The Albanian iron and steel industry is dominated by the Metallurgical Combine Elbasan. Today the work is in possession of the Turkish Kurum Steel. 15 million US Dollar are to be invested in the 20-year duration of the concession.

Investment policy issues

The development of additional mineral exploration and development projects in Albania will likely depend on the country's ability to attract foreign investment and further its privatization program. Improvements to the business environment in Albania could increase foreign investment and increase domestic economic activity, which could result in increases in mineral production. The investment agency Albinvest ranked the mining industry under the **key sectors** for investments.[605] According to Article 100 Mining Law the competent authority can support private investors to carry out mineral exports.[606]

Kosovo

Kosovo has deposits of construction minerals, bauxite, chromium, lead and zinc, lignite, magnesite, nickel, and silver.

In 2008, the real gross domestic product (GDP) growth was estimated to be 5,4% compared to 3,9% in 2007. Unemployment was about 40%, but because a large amount of economic activity in Kosovo takes place in the informal sector, this number probably overestimates the actual unemployment rate.[607]

---

605 Near clothing industry, shoe and leather production and agribusiness.
606 E.g. Account management of the enterprise in US dollar or any other currency accepted by the Albanian National Bank.
607 Independent Commission for Mines and Minerals, 2005, in: USGS (2010): Minerals Yearbook 2008, Volume III – Kosovo.

Based on trade data, it can be assumed that this category was a significant contributor to the composition of the GDP.[608]

By 2008, exports of base metals and articles of base metals, which were valued at $ 180 million (62.9 % of total exports), were the leading exports in terms of value and exports of mineral products, which were valued at $ 28 million (9.4 % of total exports), were the second most important export category in terms of value. The leading import category in terms of value from 2004 to 2008 was mineral products, largely because this category includes minerals fuels. In 2008, imports of mineral products were valued at $ 557 million (19.6 % of total imports), and of this value, mineral fuels accounted for about $ 505 million (17.8 % of total imports).[609]

Kosovo's lead and zinc industry was based on five mines, two concentration plants, and two smelters that made up the Trepca Complex. At the time that the Trepca Complex was fully operational, the Stan Terg Mine was producing about 600.000 t/yr of lead-zinc ore, the Artana Mine's designed capacity was about 250.000 t/yr, the Hajvalija Mine was producing about 100.000 t/yr, the combined capacity of the Belo Brdo and the Crnac Mines was about 100.000 t/yr, and the production capacities of the lead smelter at Zvecan and the zinc smelter at Mitrovica were each about 80.000 t/yr. Production capacities in 2008 were unknown, but were significantly lower than the figures listed above mainly owing to damage sustained during fighting in 1999 and a lack of investment.[610]

Ferronikeli, which was Kosovo's ferronickel plant at Gllogovac, was one of the most important components of Kosovo's minerals industry in terms of the current and potential value of production and employment. In 2006, the plant and Kosovo's three nickel mines were sold together and became the first important privatization in Kosovo's minerals industry.

---

608 Central Bank of the Republic of Kosovo, 2009a, p. 20, 22; Statistical Office of Kosovo, 2009, p. 13, in: USGS.
609 Central Bank of the Republic of Kosovo, 2009b; Republic of Kosovo Ministry of Trade and Industry, 2009, p. 15, in: USGS (2010), Kosovo 2008.
610 Nelles, 2003, p. 8–9; Palairet, 2003, p. 6, in: USGS (2010), Kosovo 2008.

Figure 152: Belo Brdo Mining facilities on the surface (Data by www.kosovomining.org/)

In 2005, the Kosovo Trust Agency (KTA) named the United Kingdom-based company Alferon Management Ltd. the winner of the auction for Ferronikeli with a bid of $ 49 million. The deal also required that the company invest a minimum of $ 29 million on capital improvements within the first 3 years of the sale and employ a minimum of 1.000 workers by the end of the first year and for a minimum of 2 years thereafter. Alferon Management Ltd. was connected with Eurasian Natural Resources Corp. (ENRC) of Kazakhstan. Ownership of Ferronikeli was transferred to Cunico Resources N. V., which was registered in Amsterdam and was the holding company for a joint venture by International Mineral Resources BV (IMR) (IMR is owned by the three founding shareholders of ENRC) and BSG Resources Ltd., which had its head office in Guernsey, Channel Islands [United Kingdom]. Ferronikeli restarted production in September 2007, but the company reduced production significantly at the end of 2008 because of the low market price of ferronickel; the company expected to halve production in 2009.[611]

Minerals policy issues

According to Article 61 Mining Regulation (MR) the Independent Commission for Mines and Minerals (ICMM) which was established in 2005, is assigned to provide a **raw materials plan** (Mineral Resources Management Plan). This plan has to be submitted to the Ministry for Energy and Mining (MEM).

The plan shall cover the goals and measures suggested by the commission for the implementation of a coordinated mine planning. The MEM will adapt

---

611 Kosovar Report, 2005; Kosovo Trust Agency, 2005, p. 5, 20; Mining Journal, 2007; Eurasian Natural Resources Corp., 2008, p. 6, 106; Cunico Resources N.V., 2009; SeeNews, 2009, in: USGS (2010), Kosovo 2008.

the submitted mineral resources plan if necessary. The local land use planning authorities have to consider the mineral resources plan, which is superordinated to any land-use plan.

For the purpose of promoting the development of and securing large-scale investment in the mining sector in Kosovo, the Commission can arrange a Mine Development Agreement with a Licensee or a proposed Licensee (Article 16 MR).

Regulatory issues

An Exploration License for Construction Minerals is valid for no more than two years and can be extended once only for a maximum of two additional years (Article 20 (1) MR). An Exploration License for all other minerals is valid for no more than two years and can be extended a maximum of three times, each such extension to be for a period of no more than two years.[612] According to Article 29 MR a Mining License for Construction Minerals is valid of no more than 25 years; and is extendable for further terms of up to 25 years. It applies to such area as required for the concerned Mineral Resource. A Mining License for all other minerals is valid for no more than 40 years and is not extendable.

Investment policy issues

In the Investor Guide 2008 the mining industry and energy sector is emphasized in the list of **sectors interesting for investment.** Kosovo has a liberal trading system, which is focussing on a good investment climate with the neighbour states, and on export options. Based on the EU Autonomous Trade Preference for Kosovo there are duty-free commercial flows of trade. Regarding the imported goods mineral raw materials products are of paramount importance, while base metals and their refined products account for more than half of the exports.

In the export sector the other mineral products take second place, which makes the **mining industry the driving force of Kosovo**. The investment climate in Kosovo is promoted particularly by incentives of the government (incentives in the tax and tariff range). In addition, the Multilateral Investment Guarantee Agency of the World Banks guarantees investments in Kosovo up to a height of 20 million Euros.[613]

---

612 For further information see Tiess, (2010), l.c.
613 PRESSE of 27 March 2008: Mineral wealth counts as the biggest potential of the Kosovo. Lignite, zinc, nickel, copper and magnesite are stored in the underground of the youngest state of Europe. Half of its export earnings are recently achieved by the exportation of auto scrap.

5.4 Concluding remarks

The concluding remarks are based on the **analysis of the minerals economy and minerals policy** as treated in section 5.2 and 5.3 and on the **summarized** data/information in this section. More than 20 European countries were selected to illustrate the European minerals economy: EU Member States (Austria, Belgium, Bulgaria, Czech Republic, Finland, France, Germany, Greece, Hungary, Ireland, Italy, Netherlands, Poland, Portugal, Romania, Slovakia, Slovenia, Spain, Sweden and UK) and Non-EU Member States (Albania, Norway, Switzerland). The tables listed below refer to selected minerals/commodities and concern selected European countries. Lacking a relieable homogenous data base, different **data sources** had to be used, which implicates a certain inhomogeneity:

- **GDP 1998 – 2008** (including contribution of industry to GDP)

- Production between **1998 and 2008** (WMD, see section 5.3 and 5.4); to indicate trends. Production data of metallic minerals and industrial minerals from 10 years have been considered (EU-countries and Non-EU-countries including Russia [European part]). **Aggregates production data are used from UEPG**, data are referring to 2008; for several countries production data for 10 years are available.

- **Production, import and export of minerals/commodities** from BGS (the only source to provide these data). In section 5.3 and 5.4 some examples/diagrams of different raw materials are given which indicate production, import and export between **1985 and 2008**: such diagrams facilitate a better overview of the development of a certain commodity. The examples mostly refer to a top raw material produced in a certain country, for instance feldspar in Italy (in 2008 the largest producer world wide). Besides that, for each observed country tables are provided which illustrate the production, imports and exports of metallic minerals and industrial minerals, respectively minerals related commodities data based on the **commodity** (e.g. iron: ore, concentrates, crude steel, pig iron, ferro alloys, scrap) **itself**. A summary of this is included (here) in section 5.4. It is not meant to enable comparison of data, but gives an indication of the relation **between** production, import and export of commodities, what is crucial: it indicates the high import dependency of all observed countries versus their domestic production of the respectively needed metallic minerals and also industrial minerals.

- **USGS data** indicating important mineral industries in Europe

5.4 Concluding remarks

Also considered (in terms of the analysis) is the EU-report on critical minerals which was published by DG Enterprise in 2010 (http://ec.europa.eu/enterprise/policies/raw-materials/critical/index_en.htm, also mentioned in Chapter 3 and 7). This report determines critical minerals, economically very important minerals and economically important minerals regarding the European minerals economy. These indicated mineral resources will be compared with the used data sources in terms of production, imports and exports of minerals and mineral related commodities.

Summary of tables

Concluding remarks are based on the following tables:

Table 66:
- Production of metallic raw materials in selected European countries, based on collected information from the listed tables in section 5.2 and 5.3; however, only the production data of the years 1998 and 2008 are provided for all countries to yield a better overview and summary. The highlighted data indicate the maximum production data of each country between 1998 and 2008.

Table 67:
- Production ranking of selected metallic raw materials – comparison between Europe and the world in 2008.

Table 68:
- Production of industrial minerals in selected European countries, based on collected information from the listed tables in section 5.2 and 5.3; however, only the production data of the years 1998 and 2008 are provided for all countries to yield a better overview and summary. The highlighted data indicate the maximum production data of each country between 1998 and 2008.

Table 69:
- Production ranking of selected industrial minerals – comparison between selected European countries and the world in 2008.

Table 70:
- Production of aggregates in selected European countries – comparison based on different features in 2008.

Table 71:
- European minerals economy (including mining and minerals related industries) – overview.

Chapter 5 View of the minerals policies in selected states of Europe

Table 72 and 73:
- Production, import and export of mineral commodities in selected European countries (1998–2008).
- Summary of different features (national raw material strategy, national minerals plan, aspects of legal framework, investment policy) of the minerals policy framework in selected European countries.

Table 66: Production of metallic raw materials (t) in selected European countries between 1998 and 2008 (Data by Weber et al., 2004, 2009, 2010)

Production (t)	GDP Billion $		Iron		Chromium		Manganese		Nickel		Titanium		Tungsten		Copper		Lead		Zinc		Aluminium		Bauxit		Gold		Silver	
	1998	2008	1998	2008	1998	2008	1998	2008	1998	2008	1998	2008	1998	2008	1998	2008	1998	2008	1998	2008	1998	2008	1998	2008	1998	2008	1998	2008
Albania	2,7	12,9	5.500	4.766	14.000	97.848									8.650	1.000												
Austria	212	416	573.440	650.455									1.813	1.122														
Bulgaria	12,8	49					16.679	8.195							87.500	107.195	34.595-	14.577	16.000	12.819					1	4	24	55
Czech R.	61	216																										
Finland	129	271	249.000	306.772					1.789	4.000					9.217	13.300			32.879	27.800					5	2	30	70
France	1.474	2.865																			423.600	431.600	80.000- 160.000					
Germany	2.187	3.651	84.680	47.785																	612.380	605.876						
Greece	133	351	1.950	670					15.000	16.620					18.792	15.000	17.604	23.000			156.902	162.339	1,7 Mt	2.0Mt			65	38
Hungary	48	155					8.700	13.386							40.700- 300 (2006)						460.600	511.300						

5.4 Concluding remarks

Production (t)	GDP Billion $ 1998	GDP Billion $ 2008	Iron 1998	Iron 2008	Chromium 1998	Chromium 2008	Manganese 1998	Manganese 2008	Nickel 1998	Nickel 2008	Titanium 1998	Titanium 2008	Tungsten 1998	Tungsten 2008	Copper 1998	Copper 2008	Lead 1998	Lead 2008	Zinc 1998	Zinc 2008	Aluminium 1998	Aluminium 2008	Bauxit 1998	Bauxit 2008	Gold 1998	Gold 2008	Silver 1998	Silver 2008
Ireland	88	264															35.900	50.300	177.200	398.200							11	8,5
Italy	1.218	2.307															10.102	800			187.000	186.400						
Norway	151	446	409.200	477.440							261.134	402.568									995.500	1.3Mt						
Netherl.	171	529																										
Poland	171	529													479.000	474.000	64.300	62.200	176.000	143.000	52.500–46.700				1.100	1.193		
Portugal	122	253											1.436	900	114.515	89.070									31,9	28,8		
Romania	42	204	113.605– 10.922 (2007)				20.012	8.904							14.378	4.325	6.540	6.300 (2006)	25.052	8.052 (2006)			1		18	18		
Slovakia	22	95	262.416	237.433																	108.006	162.995						
Sweden	601	1.601	13.4Mt	15.1Mt											73.685	57.700	114.430	63.500	164.711	172.200					6	5	299	293
Spain	254	487													37.185	7.071			128.000		360.400	405.000	3		2 (2006)	63	15.100 (2003)	2 (2006)
UK	1.456	2.679																			258.400	316.000						

White background: country with the highest production

349

Table 67: Production ranking of selected metallic raw materials – comparison between Europe and the world in 2008 (Data by Weber et al., 2010)

	Iron		Chromium		Manganese		Nickel		Titan		Tungsten	
	EU	Global	EU	Global	EU	Global	EU	Global	EU	Global	EU	Global
1th	Sweden 15.1 Mt	11th 1,35%	Finland 306,772 t	8th 2,91%	Hungary 13,386 t	19th 0,09%	Greece 16,620 t	15th 1,11%	Norway 402,568 t	6th 5,90%	Austria 1,122 t	6th 1,95%
2th	Austria 650,455 t	27th 0,06%	Albania 97,848 t	13th 0,93%	Romania 8,904 t	21th 0,06%	Spain 8,136 t	18th 0,54%			Portugal 900 t	7th 1,56%
3th	Norway 477,440 t	29th 0,04%	Greece 670 t	21th 0,01%	Bulgaria 8,195 t	22th 0,06%	Finland 4,000 t	21th 0,27%			Spain 245 t	15th 0,43%
4th	Slovakia (imported) 237.433	30th 0,02					Poland 530 t	23th 0,04%				
5th	Germany (imported) 47–785 t	38th					Norway 400 t	24th 0,03%				
Σ		1,47%		3,85%		0,21%		1,99%		5,90%		3,4%

5.4 Concluding remarks

	Bauxite		Copper		Lead		Zinc		Silver	
	EU	Global	EU	Global	EU	Global	EU	Global	EU	Global
1th	Greece 2,0 Mt	12th 1,04%	Poland 474,000 t	10th 3,07%	Sweden 63,500 t	9th 1,64%	Ireland 398,200 t	8th 3,43%	Poland 1,193 t	7th 5,59%
2th	Hungary 511.300 t	20th 0,26%	Bulgaria 107,195 t	21th 0,70%	Poland 62,200 t	10th 1,60%	Sweden 172,200 t	14th 1,48%	Sweden 293 t	12th 1,37%
3th	France 160.000 t	22th 0,08%	Portugal 89,070 t	22th 0,58%	Ireland 50,300 t	12th 1,30%	Poland 143,000 t	15th 1,23%	Finland 69,9 t	21th 0,33%
4th			Sweden 57,700 t	26th 0,37%	Greece 15,000 t	21th 0,39%	Finland 27,800 t	27th 0,24%	Portugal 28,8 t	32th 0,13%
5th			Finland 13,300 t	34th 0,09%	Bulgaria 14,577 t	22th 0,38%	Portugal 22,566 t	30th 0,19%	Romania 18 t	35th 0,08%
Σ		1,38%		4,81%		5,31%		6,57%		7,5 %

351

Chapter 5 View of the minerals policies in selected states of Europe

Table 68: Production of industrial minerals in selected European countries 1998 and 2008 (Data by Weber et al., 2004, 2009, 2010)

Production	GDP – Billion $		Baryte		Bentonite		Feldspar		Graphite		Gypsum		Kaolin		Potash	
	1998	2008	1998	2008	1998	2008	1998	2008	1998	2008	1998	2008	1998	2008	1998	2008
Albania	2.7	12,9														
Austria	212	416									961.400	1Mt	298.110	16.460		
Bulgaria	12.8	49	120.000	10.000 (2005)							183.929	21.200	151.451	260.372		
Czech R.	61	216			125.000	174.000	266.000	488.000	28.000	3.000	222.000	35.000	594.555	843.260		
Finland	129	271					42.740	45.250								
France	1.474	2.865	75.000	20.000			670.000	650.000			3,5M	2.3Mt	350.000	300.000		
Germany	2.187	3.651	123.272	78.941	508.738	414.336	469.928	3.3Mt	1.200	2.638 (2005)	2.5Mt	2.1Mt	3.3Mt	3.6Mt	3.5Mt	3.2Mt
Greece	133	351			1.1Mt	1.3Mt	55.700	46.333			554.998	998.924	46.388	17.399		
Hungary	48	155			26.700	7.464					185.000	15.940				
Ireland	88	264									500.000	600.000				

5.4 Concluding remarks

Production	GDP – Billion $		Baryte		Bentonite		Feldspar		Graphite		Gypsum		Kaolin		Potash	
	1998	2008	1998	2008	1998	2008	1998	2008	1998	2008	1998	2008	1998	2008	1998	2008
Italy	1.218	2.307			600.000	600.000	2.8Mt	5Mt			1.2Mt	2.9Mt				
Norway	151	446					82.036	62.000	8.271	4.100						
Netherl.	403	877														
Poland	171	529					75.000	206.000			1Mt	1.5Mt	90.000	318.000		
Portugal	122	253					80.794	165.000			570.000	360.000	216.000	170.000		
Romania	42	204			18.042	14.604	37.010	22.995					24.742	3.060		
Slovakia	22	95			81.010	116.125					128.000	145.900				
Spain	601	1.601	83.000	26.000	110.000	153.000	429.000	619			9,2 Mt	15 Mt	310.000	430.000	597.000	473.000
Sweden	254	487					33.000	22.000								
UK	1.456	2.679	64.000	43.000							2Mt	1.7Mt	2.4Mt	1.3Mt		

Chapter 5 View of the minerals policies in selected states of Europe

Production	Salt 1998	Salt 2008	Sulphur 1998	Sulphur 2008	Magnesite 1998	Magnesite 2008	Talc 1998	Talc 2008	Perlite 1998	Perlite 2008	Diatomite 1998	Diatomite 2008	Phosphat 1998	Phosphat 2008	Fluorspar 1998	Fluorspar 2008
Albania	22.000	20.000	1.200	750												
Austria	735.468	873.961	10.045	8.016	722.876	837.476	137.114	154.577								
Bulgaria	73.742	1.5Mt	21.535	380.000												
Czech R.											35.000	31.000				
Finland			389.920	707.300			498.152	550.000					700.000	280.000		
France	4.5Mt	6Mt	848.400	654.000			340.000	420.000			75.000	75.000			105.000	40.000
Germany	14.1Mt	15.5Mt	1.1Mt	1Mt											30.641	48.519
Greece					698.049	529.546			658.332	861.157						
Hungary									148.500	132.000						
Ireland																
Italy	3.2Mt	2.2Mt			134.000	140.000									104.188	15.000
Norway			42.300	123.000			21.000	30.000								
Netherl.																

354

5.4 Concluding remarks

Production	Salt 1998	Salt 2008	Sulphur 1998	Sulphur 2008	Magnesite 1998	Magnesite 2008	Talc 1998	Talc 2008	Perlite 1998	Perlite 2008	Diatomite 1998	Diatomite 2008	Phosphat 1998	Phosphat 2008	Fluorspar 1998	Fluorspar 2008
Poland	4Mt	3Mt	1.4Mt	1.3Mt	32.000	60.000										
Portugal	597.065	560.361														
Romania	2.2Mt	2.5Mt														
Slovakia	102.200	99.310			877.800	1.4Mt					24.000	15.500				
Spain	3.7Mt	3.9Mt	750.000	569.000	170.000	443.000									135.000	148.750
Sweden																
UK	5.5Mt	5.8Mt	130.000	135.000											65.000	36.801

White background: country with the highest production

Chapter 5 View of the minerals policies in selected states of Europe

Table 69: Production ranking of selected industrial minerals – comparison between Europe and the world in 2008 (Data by Weber et al, 2010)

	Baryte		Bentonite		Diatomite		Feldspar		Fluorspar		Graphite		Gypsum		Kaolin	
	EU	global	EU	global	EU	global	EU	global	EU	global	EU	global	EU	global	EU	global
1st	Germany 78.941 t	9th 0.90%	Greece 1,32 Mt	3rd 8.73%	France 75.000 t	4th 4.65%	Italy 5,0 Mt	1st 22.24%	Spain 148.750 t	6th 2.45%	Norway 4.100 t	12th 0.36%	Spain 15,0 Mt	3rd 10.33%	Germany 3,6 Mt	3rd 13.65%
2nd	Spain 26.000 t	17th 0.30%	Italy 600.000 t	6th 3.96%	Spain 46.192 t	5th 2.86%	Germany 3,3 Mt	3rd 14.68%	Germany 48.519 t	13th 0.80%	Czech R. 3.000 t	13th 0.27%	Italy 2,9 Mt	9th 2.00%	UK 1,35 Mt	5th 5.12%
3rd	Slovakia 7.300 t	24th 0.08%	Germany 414.336 t	8th 2.74%	Czech R. 31.000 t	7th 1.92%	France 650.000 t	7th 2.89%	France 40.000 t	14th 0.66%			France 2,3 Mt	13th 1.61%	Czech R. 843.260 t	8th 3.19%
4th	Italy 5.000 t	29th 0.06%	Czech R. 174.00 t	14th 1.15%	Romania 50 t	20th %	Spain 619.000 t	9th 2.75%	UK 36.801 t	15th 0.61%			Germany 2,1 Mt	15th 1.45%	Spain 430.000 t	12th 1.63%
5th	Portugal 25 t	35th %	Spain 153.000 t	16th 1.01%			Czech R. 488.000 t	10th 2.17%	Italy 15.000 t	19th 0.25%			UK 1,7 Mt	17th 1.17%	Poland 318.000 t	14th 1.20%
Σ		2,68%		17,59%		18,86%		44,73%		4,77%		0,63%		16,56%		24,79%

5.4 Concluding remarks

	Magnesite		Perlite		Phosphate		Potash		Salt		Sulfur		Talc	
	EU	global	EU	global	EU	global	EU	global	EU	global	EU	global	EU	global
1th	Slovakia 1,43 Mt	4th 7.05%	Greece 861.157 t	1st 42,20%	Finalnd 280.800 t	16th 0.56%	Germany 3,28 Mt	4th 9.66%	Germany 15,5 Mt	4th 6.27%	Poland 1,3 Mt	10th 2.22%	Finland 550.000 t	5th 7.05%
2th	Austria 837.476 t	6th 4.10%	Hungary 132.000 t	5th 6.47%			Spain 473.000 t	10th 1.39%	NL 6,69 Mt	9th 2.70%	Germany 1,02 Mt	12th 1.76%	France 420.000 t	6th 5.39%
3th	Greece 529.546 t	7th 2.59%	Slovakia 15.500 t	11th 0.76%			UK 403.800 t	11th 1.19%	France 6,0 Mt	10th 2.42%	Italy 740.000 t	16th 1.26%	Austria 154.577 t	9th 1.98%
4th	Spain 443.000 t	8 2.17%					UK 5,8 Mt	11th 2.34%			Finland 707.300 t	17th 1.21%	Italy 140.000 t	11th 1.80%
5th	Poland 60.000 t	15 0.29%					Spain 3,9 Mt	14th 1.58%			France 654.000 t	18th 1.12%	Spain 60.000 t	17th 0.77%
Σ	16,2%		49,43%		0,56%		12,24%		15,31%		7,57%		16,99%	

Table 70: Production of aggregates Mt in selected European countries– comparison based on different features (UEPG 2010)

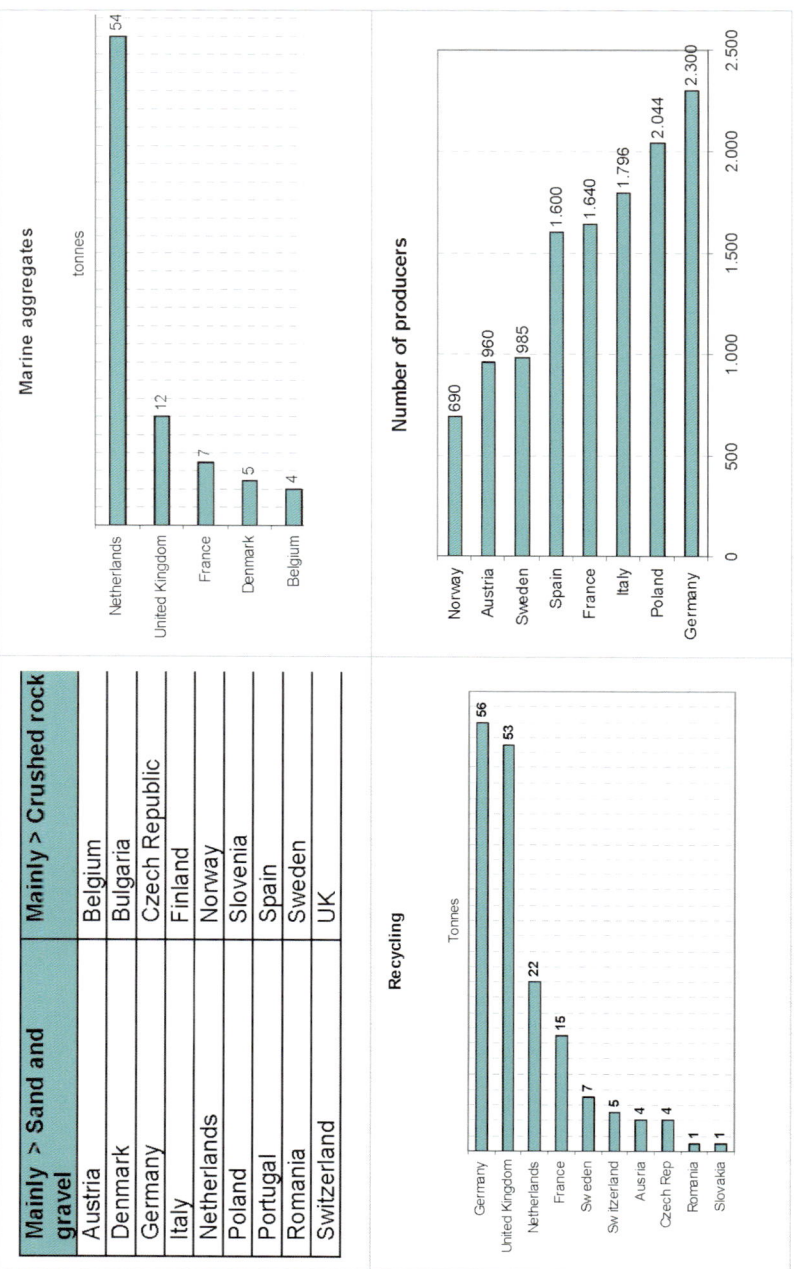

5.4 Concluding remarks

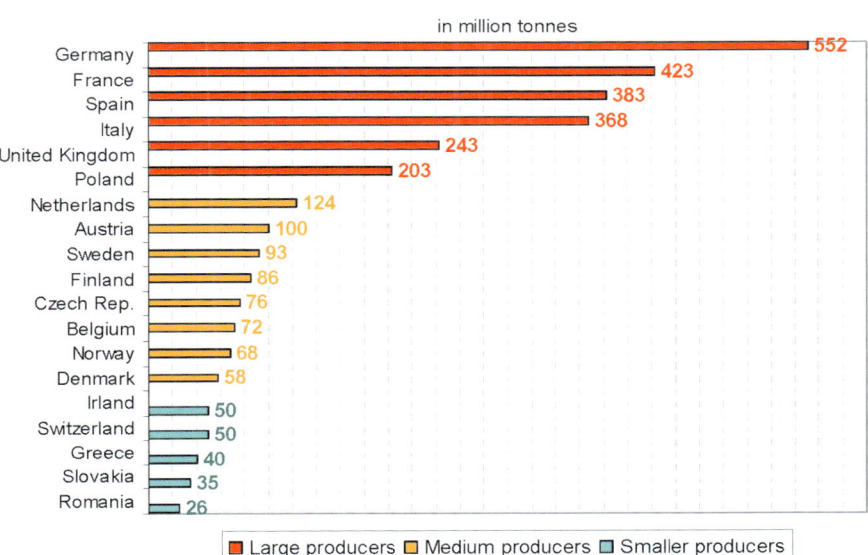

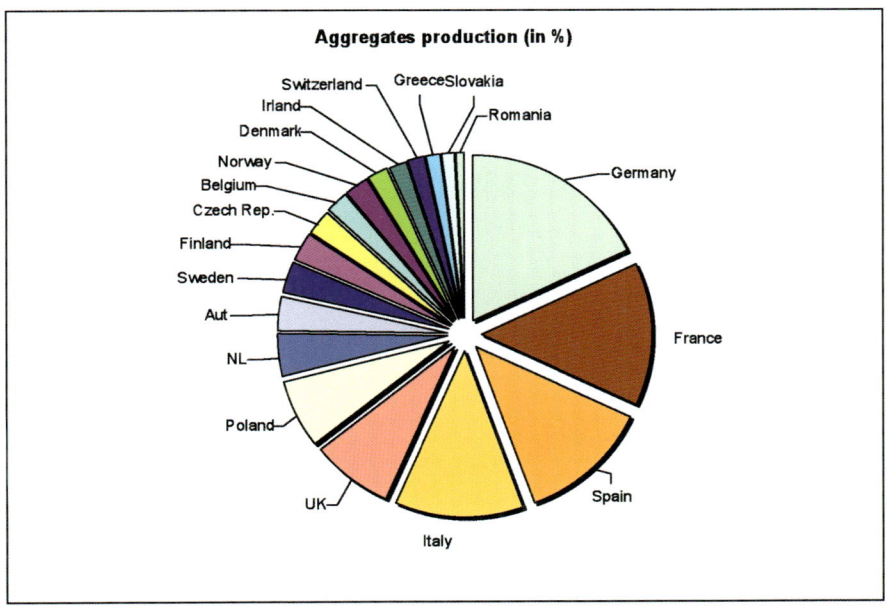

Chapter 5 View of the minerals policies in selected states of Europe

Table 71: European minerals economy (including mining and minerals related industries) – overview (Economy Watch, OECD, USGS, WMD and UEPG)

	GDP Billion $		Contribution of Industry to GDP 1998–2008		Contribution of mining to GDP (according to available information)	Metallic minerals	Industrial minerals	Construction minerals (aggregates production in 2008)
	1998	2008	1998	2008				
Albania	2,7	12,9				European top producer in terms of chromite. The rights to chromite mines near Bulquize and the Burrel and the Elbasan ferrochromium plants were acquired in 2007 by a partnership of the Austrian firm DCM DECOmetal GmbH and the Russian firm Terwingo Ltd.		
Austria	212	416	30,8%	30,7%	0,41% (2007)	European top producer in terms of iron and tungsten. Austria is the fifth-largest producer of iron ore in Europe after Russia, Ukraine, Sweden and Turkey. The "Steirische Erzberg" is the largest iron ore open pit in Central Europe. The extracted material is processed in two iron and steel plants of the VOEST Alpine Stahl GmbH. The Austrian tungsten mining industry is one of the worldwide important producers.	European top producer in terms of magnesite and talc.	Total number of producers (companies): 960 Total number of extraction sites (active quarries and pits): 1.290 Sand and gravel production: 62Mt Crushed rock production: 32Mt Recycled aggregates: 4Mt
Belgium	255	506	27,8%	23,1%		The refining of copper, zinc, and the production of steel are the leading mineral industries in Belgium. Zinc N. V. Umicore S. A. is an international metals and materials producer. The country is also a producer of cadmium, cobalt, germanium, selenium, and tellurium. And a significant mineral processor and major diamond trader in the world.	SRC-Sibelco S. A. is a world leader in silica sand production. Belgium is an important producer of marble; all the marble quarries are located in the Walloon Region.	Total number of producers (companies): 180 Total number of extraction sites (active quarries and pits): 253 Sand and gravel production: 11Mt Crushed rock production: 42Mt Recycled aggregates: 14Mt

5.4 Concluding remarks

	GDP Billion $		Contribution of Industry to GDP 1998–2008		Contribution of mining to GDP (according to available information)	Metallic minerals	Industrial minerals	Construction minerals (aggregates production in 2008)
	1998	2008	1998	2008				
Bulgaria	12,8	49		26% (4% related to mining)	In 2008, about 29.600 people were employed in mining and quarrying	European top producer in terms of copper and lead.	Bentonit AD and Kaolin AD are major producers in Europe.	Total number of producers (companies): 200 Total number of extraction sites (active quarries and pits): 100 Sand and gravel production: 18Mt Crushed rock production: 22Mt Recycled aggregates: 0
Czech R.	61	216	39,2%	37,6%	1,4%	Production of major metals as iron, steel, and steel semi manufactures	European top producer in terms of bentonite, diatomite, feldspar, graphite and kaolin.	Total number of producers: 219 Total number of extraction sites (active quarries and pits): 489 Sand and gravel production: 27Mt Crushed rock production: 44Mt Recycled aggregates: 4Mt
Finland	129	271	33,8%	31,6%		European top producer in terms of chromium, nickel, copper, zinc and silver	European top producer in terms of phosphate rock.	Total number of producers: 400 Total number of extraction sites: 2.255 Sand and gravel production: 25Mt Crushed rock production: 60Mt Recycled aggregates: 1Mt

Chapter 5 View of the minerals policies in selected states of Europe

	GDP Billion $		Contribution of Industry to GDP 1998–2008		Contribution of mining to GDP (according to available information)	Metallic minerals	Industrial minerals	Construction minerals (aggregates production in 2008)
	1998	2008	1998	2008				
France	1,474	2,865	23,4%	20,4%		Nickel and aluminium production (based primarily on imported raw material). Iron and steel production: France's output of crude steel consumption during the 2001 to 2005 period averaged more than 18Mt/yr; the consumption of finished steel averaged about 16,7Mt/yr.	European top producer in terms of diatomite, feldspar, gypsum, salt, sulphur and talc.	Total number of producers: 1.640 Total number of extraction sites: 3.050 Sand and gravel production: 165Mt Crushed rock production: 237Mt Recycled aggregates: 15
Germany	2.187	3.651	30,9%	29,8%		Domestic mineral processing sector accounts for at least 5% of the world's total production capacity of alumina, fused aluminum oxide, (graphite) magnesium metal (secondary), rhenium metal (by-product).	European top producer in terms of barite bentonite, feldspar, fluor-spar, gypsum, kaolin, potash, salt and sulphur.	Total number of producers: 2.300 Total number of extraction sites (active quarries and pits): 1.510 Sand and gravel production: 260Mt Crushed rock production: 218Mt Recycled aggregates: 56Mt
Greece	133	351	20,6%	19%		European top producer in terms of bauxite and nickel. Aluminium de Grèce S. A., (AdG) is an important producer of bauxite, alumina and aluminium. Greece (Larco) is also among the leading ferronickel producers in the world and one of the producers of nickel in Europe that uses domestic nickel ores.	Major global supplier of several key industrial minerals, perlite, bentonite, and magnesite. Also the marble industry plays a leading role in the international market.	Total number of producers: 300 Total number of extraction sites: 200 Sand and gravel production: 20Mt Crushed rock production: 20Mt Recycled aggregates: 0
Hungary	48	155	31,8%	29,1%		European top producer in terms of bauxite (refined to alumina) and manganese. Moreover, in 2008, Hungary is the second world widest producer of gallium, after China.	European top producer in terms of perlite.	

5.4 Concluding remarks

	GDP Billion $		Contribution of Industry to GDP 1998–2008		Contribution of mining to GDP (according to available information)	Metallic minerals	Industrial minerals	Construction minerals (aggregates production in 2008)
	1998	2008	1998	2008				
Ireland	88	264	40,8%	33,6% (2007)	1%	Major producer of lead and zinc ore and also one of the leading exporters of lead and zinc ore. There are three major producing mines within the Rathdowney Trend in the Midland Orefield.		Total number of producers: 150 Total number of extraction sites: 355 Sand and gravel production: 25Mt Crushed rock production: 25Mt Recycled aggregates: 0
Italy	1.218	2.307	29,4%	27%		Italy's mineral industries produces such metals as copper, iron and steel, lead and zinc, all important material for the country's manufacturing sector (the required raw materials are mostly imported).	European and global producer of such industrial minerals as feldspar, bentonite, gypsum, fluorspar. Moreover: bromine (among top 13 world producers), pumice (27% of world output), cement (2% of world output) and lime (2% of world output).	Total number of producers: 1.796 Total number of extraction sites: 2.360 Sand and gravel production: 225Mt Crushed rock production: 135Mt Recycled aggregates: 5Mt
Kosovo	151	446				Kosovo's lead and zinc industry is based on several mines, two concentration plants, and two smelters that made up the Trepca Complex. Ferronikeli, Kosovo's ferronickel plant at Gllogovac is one of the most important components of Kosovo's mineral industry in terms of the current and potential value of production and employment.		

363

Chapter 5 View of the minerals policies in selected states of Europe

	GDP Billion $		Contribution of Industry to GDP 1998–2008		Contribution of mining to GDP (according to available information)	Metallic minerals	Industrial minerals	Construction minerals (aggregates production in 2008)
	1998	2008	1998	2008				
Nether-lands	171	529	25.3%	25.5%		Important aluminium and zinc industry: Pechiney Nederland NV (PLN) is a producer of extrusion billets and of rolled aluminum slabs. PLN has a production capacity in 2006 of 213.000 metric tons per year (t/yr) of primary aluminum and 230.000 t/yr of aluminum billets and slabs. Zinifex Limited's zinc smelter at Budel produced 235.913 metric tons (t) of zinc in 2006 exclusively from zinc concentrates from Zinifex's Century Mine in Queensland, Australia.	European top producer in terms of salt. Akzo Nobel Salt BV's salt production facilities are located in Delfzijl and Hengelo. In 2008, with a capacity of more than 2Mt/yr, the Hengelo plant is the leading vacuum salt plant in the world. GCC had become the leading filler in the production of wood-free paper of the resulting paper's pure white colour and brightness. The marble used to produce the GCC is imported from Omya's mines in Turkey. The initial capacity of the Omya Netherlands BV's ground calcium carbonate plant is about 500.000 t/yr. Moreover, Nedmag Industries Mining & Manufacturing BV is Europe's leading producer of high-grade synthetic dead-burned magnesia and other magnesium compounds.	Total number of producers: 65 Total number of extraction sites: 225 Sand and gravel production: 46Mt Crushed rock production: 0 Recycled aggregates: 24Mt

5.4 Concluding remarks

	GDP Billion $		Contribution of Industry to GDP 1998–2008		Contribution of mining to GDP (according to available information)	Metallic minerals	Industrial minerals	Construction minerals (aggregates production in 2008)
	1998	2008	1998	2008				
Norway	171	529	32,7%	46,2%		European top producer in terms of titanium, nickel and olivine. Norway's production of olivine accounts for more than 60% of world production; production of titanium accounts for more than 6% of world production.	European top producer in terms of graphite, ilmenite,	Total number of producers: 690 Total number of extraction sites: 713 Sand and gravel production: 15Mt Crushed rock production: 52Mt Recycled aggregates: 0
Poland	122	253	32,9	32%	2,5%	Poland is (after Russia) the leading producer of copper in Europe and Central Eurasia and ranks among the top 10 world mine producers of copper. All copper ore in Poland is mined by Kombinat Gorniczo Hutniczy Miedzi (KGHM). In Europe and Central Eurasia, Poland is also a significant producer of lead and zinc. Additionally, a major world producer of silver, more than 6% of world mine production.	European top producer in terms of kaolin, magnesite and sulphur.	Total number of producers: 2.044 Total number of extraction sites: 1.786 Sand and gravel production: 131Mt Crushed rock production: 49Mt Recycled aggregates: 22Mt
Portugal	42	204	28,8%	23,9%	1%	European top producer in terms of copper, tin, tungsten, and silver. Portugal is a significant world producer of lithium (seventh after Chile, China, Australia, and others), and tungsten.	An important producer of feldspar, high-quality marble, pyrites, and rock salt.	Total number of producers: 350 Total number of extraction sites: 200 Sand and gravel production: 61Mt Crushed rock production: 15Mt Recycled aggregates: 0

Chapter 5 View of the minerals policies in selected states of Europe

	GDP Billion $ 1998	GDP Billion $ 2008	Contribution of Industry to GDP 1998	Contribution of Industry to GDP 2008	Contribution of mining to GDP (according to available information)	Metallic minerals	Industrial minerals	Construction minerals (aggregates production in 2008)
Romania	22	95		24%		European top producer in terms of manganese and silver. Gabriel Resources Ltd. of Canada estimated (2009) that the Rosia Montana gold project in north-western Romania could produce an average of 511.000 troy ounces per year of gold during a 16-year mine life.	Produces a broad range of industrial minerals that includes barite, various calcareous rocks, clays, feldspar and mica of granitic/pegmatitic sources, graphite, gypsum, salt, and silica-group minerals.	Total number of producers: 500 Total number of extraction sites: 730 Sand and gravel production: 18Mt Crushed rock production: 7Mt Recycled aggregates: 1Mt
Russian Federation	271	1.666		38,80% from industry (2006)	More than half to industrial production	Russian federation is one of the largest mineral producers of the world, amongst others iron, cobalt, platinum, nickel, palladium.	Top producer of many industrial minerals.	Infrastructure development requires significant aggregates production
Serbia	601	1.601				Important producer of bauxite, copper, lead and zinc ores.	Produces clays, feldspar, gypsum, magnesite, pumice, and salt	
Slovakia	254	487	34,6%	38%	0,5%	Aluminium and steel production formed the dominant elements of the country's metals sector.	European top producer in terms of magnesite, perlite and barite.	Total number of producers: 170 Total number of extraction sites: 92 Sand and gravel production: 13Mt Crushed rock production: 21Mt Recycled aggregates: 1Mt
Slovenia	1.456	2.679	36,4%	33,9%	0,4% (2007)	Produces aluminium, secondary refined lead, and steel.		

5.4 Concluding remarks

	GDP Billion $		Contribution of Industry to GDP 1998–2008		Contribution of mining to GDP (according to available information)	Metallic minerals	Industrial minerals	Construction minerals (aggregates production in 2008)
	1998	2008	1998	2008				
Spain	601	1.601	29,1%	28,4%	1%	Important European producer of copper, tungsten, zinc, gold and silver. Spain occupies about 85% of the Iberian Peninsula and has some of the most mineralized territory in Western Europe.	Significant European producer of such mineral commodities as cement, fluorspar, gypsum, kaolin, magnesite and industrial sand and gravel (in 2007 sixth after the United States, Slovenia, Germany, Austria, and France).	Total number of producers: 1.600 Total number of extraction sites: 2.060 Sand and gravel production: 134Mt Crushed rock production: 244Mt Recycled aggregates: 5Mt
Sweden	254	487	29,1%	28%		European top producer in terms of iron, copper, lead, zinc and silver. Leading producer of iron ore in the EU, producing more than 2% of the world's total.		Total number of producers: 985 Total number of extraction sites: 1.802 Sand and gravel production: 19Mt Crushed rock production: 67Mt Recycled aggregates: 7Mt
United Kingdom	1.456	2.679	28,5%	23,6%		4% of world refined lead production, 4% of world refined lead production, about 3% of refined nickel and more than 1% of the world's output of aluminum and crude steel. Most of the needed material must be imported.	European top producer in terms of potash, kaolin, fluorspar, gypsum, and salt. Additionally an important cement producer.	Total number of producers: 450 Total number of extraction sites: 781 Sand and gravel production: 55Mt Crushed rock production: 114Mt Recycled aggregates: 53Mt

Chapter 5 View of the minerals policies in selected states of Europe

Table 72: Production, import and export of metallic minerals related commodities in selected countries of Europe between 1998 and 2008 (Data by BGS [European minerals statistics]) Legend: see end of the table

Commodities related to metallic minerals/products	Iron	Chromium	Cobalt	Manganese	Molybdenum	Nickel	Tantalum	Titanium	Tungsten	Vanadium	Aluminium	Antimony	Arsenic	Bauxite	Bismuth	Cadmium	Copper	Gallium	Lead	Lithium	Mercury	Rare-earth	Tin	Zinc	Gold	PGM	Silver
Austria																											
Production/mining (mi: mine)	oc, pi, cr, fa	m	oc	oc	m	sm, ref		m	mi			m					ref		ref								
Import	ore pi fa s	oc m	m o	m o	m	u, s o	x (2004.2007)	m o	m carb oc	vana	a, ah, m u, ua s	m o	ma		x		u, ur u, ref ua		u, s	car	x	cc orec m	u s	u uo s	m m, s	m	m
Export	pi, fa, s	oc, m		m	m	u, s	x(2007.2008)		m		u, ua, s	m		x	x		u, ua, s		u, s		x	cc orec m	u s	u s	m, s	m s	m
Belgium																											
Production/mining (mi: mine)	pi cr		m										wa	mi			ref		ref				sm s	sz			

5.4 Concluding remarks

Commodities related to metallic minerals/products	Import	Export	Bulgaria Production/mining (mi: mine)	Import	Export
Silver	oc, m	oc, m	mi		
PGM	m	m, s	m	s	
Gold		m, s	mi		
Zinc	oc, u, ua, s		mi, sz	oc, u	oc, u, s
Tin	oc, u, ua, s	c, u, ua, s		u	
Rare-earth	cc, orec, m	rec, m			
Mercury	x	x			
Lithium	o, car	o, car			
Lead	oc, u, s	oc, u, s	mi, ref	oc, u, s	oc, u
Gallium					
Copper	oc, mat, u, ur, ref, ua, s	oc, mat, u, ur, ref, ua, s	mi, sm, ref	oc, u, s	
Cadmium	m, o, sul	m, sul	m		x
Bismuth	m	m	m		
Bauxite	x				
Arsenic	ma	wa	ma		
Antimony	m, o	oc, m, o	m		
Aluminium	a, ah, u, ua, s	a, u, ua, s		a, ah, u, ua, s	ua, s
Vanadium	oc, m, pen	oc			
Tungsten	m, carb	m, carb			
Titanium	ilm, otm, m, o	tm, ts, m, o		o	m
Tantalum	x	x			
Nickel	u, ua, s, o	oc, u, ua, s, o		u	
Molybdenum	oc, m, o	oc, m, o		m	m
Manganese	oc, oc, m	oc, m	mi	oc	oc
Cobalt	oc, or, m, o	or, m, o			
Chromium	oc, m	oc, m			
Iron	or, pi, fa, s	pi, fa, s	mi, pi, cs, fa	pi, fa, s	pi, fa, s

Chapter 5 View of the minerals policies in selected states of Europe

Commodities related to metallic minerals/products	Czech Rep. Production/mining (mi:mine)	Czech Rep. Import	Czech Rep. Export	Finland Production/mining	Finland Import	Finland Export
Iron	pi cr fa	or pi fa s	pi fa s	pi cr	ore pi fa s	or (bp) pi fa s
Chromium		oc m	oc m	oc fc	oc m m o	oc fc o
Cobalt		oc m o	oc m	oc m	oc oc o	m o
Manganese		oc m			oc oc	
Molybdenum		oc m	m			
Nickel		u o	m u s o	x sm ref	oc mat u s	oc u s o
Tantalum		x	x			
Titanium		tm m o	tm m o		tm o	m o
Tungsten		m	m carb		m	m
Vanadium		pen				
Aluminium		a ah u ua s	u s		a ah u ua s	ua s
Antimony		m o	o		o	
Arsenic		ma				
Bauxite		x			x	
Bismuth		m				
Cadmium		m o sul				
Copper		u s	u s	x sm ref	oc ua s	u, ur u, ref, ua s
Gallium						
Lead	ref	u s	u s		u	s
Lithium		car			car	
Mercury		x	x	x	x	x
Rare-earth		rec fc	cc orec		cc orec	rec m
Tin		u	u s		u	u s
Zinc		u ua s	u s	x sz	oc u	u s
Gold		m s	m s	x	m s	s
PGM		m w	m s		m	m s
Silver		m	m	x	m	oc m

5.4 Concluding remarks

Commodities related to metallic minerals/products	France Production/mining (mi: mine)	France Import	France Export	Germany Production/mining (mi: mine)	Germany Import	Germany Export
Silver		oc, m	oc, m		oc, m	m
PGM		m, s	m, s		m, s	m, s
Gold		m, s	m, s		m, s	m, s
Zinc	sz	oc, u, ua, s	oc, u, ua, s	sz	oc, u, ua, s	oc, u, ua, s
Tin		oc, u, ua, s	u, ua, s		u, ua, s	c, u, ua, s
Rare-earth		cc, orec, fc, m	rec, fc		cc, orec, fc, m	rec, m
Mercury		x	x		x	x
Lithium		o, car	car		o, car	o, car
Lead	ref	u, s	u, s	ref	oc, u, s	oc, u, s
Gallium						
Copper		u, ur, ua, s	u, ua	sm, ref	oc, mat, u, ur, ua, s	oc, mat, u, ref, ua
Cadmium	mi	m, o, sul	x	mi	m, o, sul	sul
Bismuth		x	x		m	m
Bauxite	mi	x	x		x	x
Arsenic		ma	ma		ma	
Antimony		m, o	m, o		m, o	m, o
Aluminium	a	a, ah, u, ua, s	a, ah, u, ua, s	a, pa	a, ah, u, ua, s	a, ah, u, ua, s
Vanadium		oc, m	pen, m		oc, pen, m	oc, m
Tungsten		m, carb	m, carb		oc, m, carb	m
Titanium		m, o	tm, m, o		tm, ts, m, o	tm, m
Tantalum		x	x		x	
Nickel	sm/ref	oc, mat, u, ua, s, o	oc, mat, m, u, ua, s, o		oc, mat, u, ua, s, o	oc, mat, u, ua
Molybdenum		oc, m, o	oc, m		oc, m, o	oc, m
Manganese		oc, m	oc, m		oc, m	oc, m
Cobalt		oc, m, o	oc, m, o		oc, m, o	ore, ore
Chromium		oc, m	oc, m		oc, m	oc, m
Iron	ore, pi, cr, fa	ore, pi, fa	ore, pi, fa, s	ore, pi, fa	ore, pi, fa	ore, pi, fa, s

Chapter 5 View of the minerals policies in selected states of Europe

Commodities related to metallic minerals/products	Greece Production/mining (mi: mine)	Greece Import	Greece Export	Hungary Production/mining (mi: mine)	Hungary Import	Hungary Export
Iron	cr, fa	ore, pi, fa, s	fa, s	pi, cr	ore, pi, fa, s	fa, s
Chromium		oc	oc		oc, m	
Cobalt						
Manganese		oc, m		mi	oc	
Molybdenum		m			m	
Nickel	mi, sm/ref	oc, u	oc			
Tantalum						
Titanium		tm, o	o		tm, m, o	o
Tungsten						m
Vanadium						
Aluminium	a, pa	a, ah, u, ua, s	a, ah, u, ua, s	a, pa (until 2006)	a, ah, u, ua, s	a, ah, u, s
Antimony		oc, m	m		o	
Arsenic		ma				
Bauxite	mi	x	x	mi	x	
Bismuth		x				
Cadmium						
Copper		u, ua, s	u, ref, ua, s		u, s	s
Gallium				mi		
Lead	mi, ref	u, s	u, s		u	s
Lithium						
Mercury					x	
Rare-earth		cc, fc			rec	orec
Tin		u			u	u, s
Zinc	mi	u, oc	oc, u, s		u	u, s
Gold		m	m		m	m
PGM		m			m	s
Silver	mi	m	m		m	m

5.4 Concluding remarks

Commodities related to metallic minerals/products	Italy Production/mining (x: mine)	Italy Import	Italy Export	Netherlands Production/mining (mi: mine)	Netherlands Import	Netherlands Export
Silver	x	oc, m	m		m	oc, m
PGM		m, s	m, s		m, s	m, s
Gold		m, s	m, s		m, s	m, s
Zinc	sz	oc, u, ua, s	oc, u, ua, s	sz	oc, u, ua, s	oc, u, ua, s
Tin		oc, u, ua, s	u, ua, s		u, ua, s	u, ua, s
Rare-earth		cc, orec, fc, m	rec, m		cc, orec, fc, m	rec, fc
Mercury		x	x		x	x
Lithium		o, car	car		o, car	o, car
Lead	x, ref	oc, u, s	u, s	ref	oc, u, s	oc, u, s
Gallium						
Copper	ref	u, ur, u, ref, ua, s	mat, u, ua, s		u, ur, u, ref, ua, s	u, s
Cadmium		mo	m	mi	m	m
Bismuth		m			m	m
Bauxite		x	x		x	x
Arsenic		ma	ma		ma	
Antimony		oc, m, o	oc, o		m, o	m, o
Aluminium	a, pa	a, ah, u, ua, s	a, ah, u, ua, s	pa	a, ah, u, ua, s	a, ah, u, ua, s
Vanadium		oc, pen	oc		pen	pen, m
Tungsten		m, carb	m		oc, m, carb	oc, m
Titanium		tm, m, o	ts, m, o		tm, m, o	tm, m, o
Tantalum					x	x
Nickel		oc, mat, u, ua, s, o	oc, u, s, o		ua, s, o	oc, mat, u, s, o
Molybdenum		oc, m, o	oc, m, o		oc, m, o	oc, m, o
Manganese	ore	oc, m	m		oc, m	oc, m
Cobalt		oc, m, o	oc, m, o		m, o	m, o
Chromium		oc, m	oc, m		oc, m	oc, m
Iron	pi, cr	ore, pi, fa, s	ore, pi, fa, s	pi, cr	ore, pi, fa, s	ore, pi, fa, s

Chapter 5 View of the minerals policies in selected states of Europe

Commodities related to metallic minerals/products	Norway Production/mining (mi: mine)	Norway Import	Norway Export	Poland Production/mining (mi: mine)	Poland Import	Poland Export
Iron	ore, pi, cr, fa	ore, pi, fa, s	ore, pi, fa, s	pi, cr, fa	ore, pi, fa, s	ore, pi, fa, s
Chromium	oc, m	m	m	oc, m	m	m
Cobalt	m	oc, m	m, m, oc	oc, m, o	oc, m	oc, m
Manganese		oc, m	oc, m		oc, m	oc, m
Molybdenum					m	m
Nickel	mi, sm/ref	mat, u	oc, u, s		u, s	u, s, o
Tantalum					x	
Titanium	mi	tm, m, o	tm, m, o		tm, m, o	m, o
Tungsten		m			m	m
Vanadium					m	
Aluminium	pa	a, ah, u, ua, s	u, ua, s	pa	a, ah, u, ua, s	u, s
Antimony		m, o			m, o	
Arsenic				ma		
Bauxite		x			x	x
Bismuth					m	
Cadmium		m	m		m	m
Copper		sm, ref	u, s		sm, ref	oc, mat, u, s
Gallium						
Lead			u		mi, ref	u, s
Lithium						o, car
Mercury					x	x
Rare-earth			m	rec	cc, orec, m	cc, orec (since 2007), m (since 2008)
Tin			u	s	u, s	u, s
Zinc			oc, u	u, ua	mi	oc, u, ua, s
Gold			m, s	m, s	mi	m	m, x
PGM			m, s			m	m, s
Silver			oc, m	m	mi	x	m

5.4 Concluding remarks

Commodities related to metallic minerals/products	Portugal Production/mining (mi: mine)	Portugal Import			Portugal Export	Romania Production/mining (mi: mine)	Romania Import		Romania Export
Silver		mi	x	oc (since 2007) m		mi		m	
PGM				s					
Gold			m	s		mi			
Zinc		mi	u	ua	oc s	mi sz		oc u	oc u s
Tin		mi	u	ua	ua s			u	
Rare-earth			rec fc	m				rec	
Mercury			x					x	
Lithium		mi							
Lead		mi ref	u		u s	mi (until 2007) ref		oc u	oc u s
Gallium									
Copper		mi	u	s	oc u s	mi sm ref		u s	oc mat u s
Cadmium									
Bismuth						mi			
Bauxite								x	x
Arsenic		wa							
Antimony			o		m			m o	
Aluminium			a ah u	ua s		a pa		a ah u s	a o ua s
Vanadium									
Tungsten		mi		m carb	oc m			m	
Titanium			x	o	o			tm m	o
Tantalum					x				
Nickel			u					u	
Molybdenum									
Manganese			oc			ore		oc m	since 2006
Cobalt									
Chromium			oc m		oc			oc m	
Iron		cr	pi fa s		s	ore (until 2007) pi cr fa		ore pi fa s	pi fa s

Chapter 5 View of the minerals policies in selected states of Europe

Commodities related to metallic minerals/products	Slovakia Production/mining (mi: mine)	Slovakia Import	Slovakia Export	Slovenia Production/mining (mi: mine)	Slovenia Import	Slovenia Export
Silver		m	m		m	m
PGM		m			m	m
Gold	mi	m	m s		m s	m
Zinc		u	u ua		u	u s
Tin		u	m		u	
Rare-earth		rec			rec m	
Mercury		x				x
Lithium					o	car
Lead			u	ref	u s	u s
Gallium						
Copper	sm since 2007	mat u s	u s		u s	s
Cadmium						
Bismuth					m	
Bauxite		x			x	x
Arsenic		ma			ma	
Antimony		o			m	
Aluminium	pa	a u s	u ua s	pa	ah u ua s	a ua s
Vanadium					m	
Tungsten			m			
Titanium		tm m o			tm ts m o	m o
Tantalum						
Nickel		u	u, s, since 2006		u	u
Molybdenum					oc	
Manganese		oc m			oc m	m m
Cobalt		m			m	
Chromium		oc m	m			
Iron	ore pi cr fa	ore pi fa s	ore pi fa s	cr fa	pi fa s	pi fa s

5.4 Concluding remarks

Commodities related to metallic minerals/products	Spain Production/Mining (mi: mine)	Spain Import	Spain Export	Sweden Production/mining (mi: mine)	Sweden Import
Silver	mi until 2006	oc m	m	mi	oc m
PGM		m s	m s		m s
Gold	mi until 2006	m s	m s	mi	m s
Zinc		oc u ua s	oc u ua s	mi	u ua s
Tin		c u ua s	u s		u
Rare-earth		rec fc m	cc orec fc		rec m
Mercury		x	x		x
Lithium	mi	o car	car		o car
Lead	ref	oc u s	oc u s	mi ref	oc u s
Gallium					
Copper	mi sm ref	oc u,ur u,r ua s	oc u,ur u,ref ua s	mi sm ref	oc u,ur u,ref ua s
Cadmium		m o			m
Bismuth		m	m		m
Bauxite		x	x		x
Arsenic		ma	ma		ma
Antimony		oc m o	m o		m o
Aluminium	a pa	ah u ua s	ah u ua s		ah u ua s
Vanadium		tm m oc carb pen	oc m		tm m carb pen m
Tungsten		tm m ts m o	tm m m o		tm m o
Titanium		x			
Tantalum					
Nickel	mi	oc u ua o	oc u s		oc mat u ua s
Molybdenum		oc m	oc m		oc m o
Manganese		oc m	oc m		oc m
Cobalt		oc m o	oc m o		oc m o
Chromium		oc m	oc m		oc m
Iron	mi until 2004 pi cr fa	ore pi fa s	ore pi fa s	ore pi cr fa	ore pi fa s

Chapter 5 View of the minerals policies in selected states of Europe

Commodities related to metallic minerals/products	Switzerland Export	Switzerland Production/mining (mi: mine)	Switzerland Import	Switzerland Export	UK Production/mining (mi: mine)	UK Import	UK Export
Silver	oc, m			m		oc, m, s	oc, m
PGM	m, s		m, s	m, s		m, s	m, s
Gold	m, s		m, s	m, s		m, s	m, s
Zinc	oc, u, s			u, s		u, ua	u, ua, s
Tin	m, o			u, s		u, ua, s	c, u, ua, s
Rare-earth			cc, orec			cc, orec, fc, m	cc, orec, fc, m
Mercury						x	x
Lithium			o	o		o, car	o, car
Lead	oc, u, s	ref	u	u, s	mi, ref	u, s	u, ua, s
Gallium							
Copper	u, ref, ua, s		u, ua, s		u, ur, u, ref, ua, s	mat, u, ua, s	
Cadmium	o		m	m		m	m
Bismuth			m				
Bauxite		x			x		x
Arsenic		ma			ma		
Antimony		oc, m, o				m, o	m, o
Aluminium	a, u, ua, s	pa	ah, u, ua, s	u, s	pa	ah, u, ua, s	a, ah, u, ua, s
Vanadium					pen m		pen m
Tungsten	oc, m, carb		m		m, carb m		m, carb m
Titanium					tm, ts, o	m, o	m, o
Tantalum			x		x		x
Nickel	u, s, o		u	u, s	sm/ref	mat, u, ua, s, o	mat, u, ua, s
Molybdenum					oc, m, o	oc, m	
Manganese	m		oc, m		oc, m	m	
Cobalt		oc, m	m		oc, m, o	oc, m, o	
Chromium	m	oc, m	m		oc, m	oc, m	
Iron	ore, pi, fa, s	pi, cr	ore, pi, fa, s	fa, s	ore, pi, cr	ore, pi, fa, s	pi, a, s

378

a	Alumina	orec	Oher rare earth compounds
ah	Alumina hydrate	otm	Other titanium minerals
bp	Burnt pyrites	pa	Primary aluminium
c	Concentrate	pen	Pentoxide
car	Carbonate	pi	Pig iron
carb	Carbides	rec	Rare earth compounds
cc	Cerium compounds	ref	refined
cr	Crude steel	s	Scrap
fa	Ferro alloys	sm	Smelter
fc	Ferro-cerium	sul	Sulfide
ilm	Ilmenite	sz	Slap zinc
m	Metal	tm	Titanium minerals
ma	Metallic arsenic	ts	Titanium slag
mat	Matte and cement	u	uwrought
mi	Mine	ua	uwrought alloys
o	Oxide	ur	urefined
oc	Ores and concentrates	vana	Vanadiferous residues
or	Ore	wa	White arsenic

Chapter 5 View of the minerals policies in selected states of Europe

Table 73: Production, import and export of industrial minerals related commodities in selected countries of Europe between 1998 and 2008 (Data by BGS [European minerals statistics]) legend: see end of the table

Commodities related to industrial minerals/products	baryte	bentonite	diatomite	diamond	feldspar	fluorspar	gypsum	graphite	kaolin	magnesite	perlite	potash	Phosphate rock	salt	sulfur	talc	cement
Austria																	
Production/mining (mi: mine)							g,a		mi	mi				rs, sib	rec	mi	cc, cf
Import	x	x	x	us, gc, i, d	x	x	cr, ca	x	x	me, ma		fs sul chl	x	x	x	x	cc pc
Export	x	x		us, gc, d	x	x	cr, ca	x	x	me, ma		x	x	x	x	x	cc, pc, oc
Belgium																	
Production/mining (mi: mine)									mi						rec		cf
Import		x	x	us gr gc i d	x	x	cr ca	x	x	me ma		fs sul chl		x	cr ref	x	cc pc
Export (+ Luxembourg)	x	x	x	us gr gc i d	x	x		cr ca		x	me ma		fs sul chl	x	x	x	x	cc pc
Bulgaria																	
Production/mining (mi: mine)	mi	mi			mi		mi		mi	mi				mi	rec		cc cf
Import		x	x		x	x	cr ca	x	x	me ma			x	x	x	x	cc pc
Export	x	x				x	cr ca		x					x	x		cc pc
Czech Rep.																	
Production/mining (mi: mine)		mi	mi		mi		mi	mi	mi							rec	cc cf
Import	x	x	x	gc i d	x	x	cr ca	x	x	me ma	chl		x	x	x	x	cc pc
Export	x	x	x	gc	x	x	cr ca	x	x	me ma			x	x	x		cc pc

5.4 Concluding remarks

Commodities related to industrial minerals/products	baryte	bentonite	diatomite	diamond	feldspar	fluorspar	gypsum	graphite	kaolin	magnesite	perlite	potash	Phosphate rock	salt	sulfur	talc	cement
Finland																	
Production/ mining (mi: mine)				mi									mi	rec	mi	mi	
Import	x	x	x	us gc d	x	x	cr ca	x	x	ma ma	x	x			x	x	
Export	x			x			x			sul chl	x	x				x	pc
France																	
Production/ mining (mi: mine)	mi		mi		mi	mi			mi				mi	rec	mi		cc cf
Import	x	x	x	us gr gc i d	x	x	cr ca	x	x	ma me	fs sul chl	x	x	x cr ref	x	x	cc pc
Export	x	x	x	us g	x	x	cr ca	x	x	me ma	sul chl	x	x		x	cc pc	
Germany																	
Production/ mining (mi: mine)	mi	mi				mi	mi		mi			mi	rs bs sib	rec			cc cf
Import	x	x	x	g i d	x	x	cr ca	x	x	me ma	sul chl	x	x	x	x	cc pc	
Export	x	x	x	us g i d	x	x	cr ca	x	x	me ma		x	x	x	x	pc	
Greece																	
Production/ mining (mi: mine)		mi			mi		mi		mi	mi	mi		mi	rec	mi		
Import	x	x	x	u gr gc i d	x	x	cr ca	x	x	me ma	sul chl	x	x	x	x	cc pc	
Export		x			x		cr ca		x	ma		p		x	x	x	cc pc

Chapter 5 View of the minerals policies in selected states of Europe

Commodities related to industrial minerals/products	baryte	bentonite	diatomite	diamond	feldspar	fluorspar	gypsum	graphite	kaolin	magnesite	perlite	potash	Phosphate rock	salt	sulfur	talc	cement
Hungary																	
Production/mining (mi: mine)		mi	mi				mi		mi		mi					rec	cc cf
Import	x	x	x	gc	x	x	cr ca	x	x	ma me	chl			x	x	x	cc pc
Export		x	x				cr ca		x		pf			x	x	x	cc pc
Italy																	
Production/mining (mi: mine)	mi	mi			mi	mi until 2006	mi		mi		mi			x	rec	x	cc cf
Import	x	x	x	u gr gc i d	x	x	cr ca	x	x	me ma	sul chl	x	x	x cr ref	x	x	cc pc
Export	x	x	x	gc d	x	x	cr ca	x	x	me ma	pf	x	x	x	x	x	cc pc
Netherlands																	
Production/mining (mi: mine)										me				mi	rec		cc cf
Import	x	x	x	u gr gc i d	x	x	cr ca	x	x	me ma	sul chl	x	x	x	x	cc pc
Export	x	x	x	us gc i d	x	x	cr ca	x	x	me ma	sul chl	x	x	x	x	cc pc
Norway																	
Production/mining (mi: mine)					mi			mi								mi	cf
Import	x	x	x		x		cr ca	x	x	me ma	sul chl	x	x	x	x	cc pc
Export	x		x		x		cr ca				chl			x		x	cc pc

5.4 Concluding remarks

Commodities related to industrial minerals/products	baryte	bentonite	diatomite	diamond	feldspar	fluorspar	gypsum	graphite	kaolin	magnesite	perlite	potash	Phosphate rock	salt	sulfur	talc	cement
Poland																	
Production/ mining (mi: mine)	mi	mi	mi		mi		mi (+anh)		mi	mi				rs bs		rec	cc cf
Import	x	x	x	i d	x	x	cr ca	x	x	me ma		fs sul chl	x	x	x	x	cc pc
Export	x		x		x		cr ca		x	me ma		pf	x	x	x		cc pc
Portugal																	
Production/ mining (mi: mine)					mi		mi		mi					rs ss		rec mi	cc cf
Import	x	x	x	gr gc i d	x	x	cr ca	x	x	me ma		sul chl	x	x	x	x	cc pc
Export		x	x	gc	x		cr ca		x			pf	x	x	x		cc pc
Romania																	
Production/ mining (mi: mine)		mi	mi until 2007		mi	mi until 2004	mi	mi until 2005	mi					mi		rec mi	cf
Import	x	x	x		x	x	cr ca	x	x	me ma		sul chl	x	x	x	x	cc pc
Export		x			x		cr ca			ma						x	cc pc
Slovakia																	
Production/ mining (mi: mine)	mi	mi					mi		mi	mi	mi			mi		rec mi	cf
Import	x	x	x		x	x	cr ca	x	x	me ma		chl	x	x	x	x	cc pc
Export	x	x			x							pf	x	x	x	x	cc pc
Slovenia																	
Production/ mining (mi: mine)														mi			cf
Import		x	x	d	x	x	cr ca	x	x	me ma		pf	x	x	x	x	cc pc
Export							cr ca						x	x			cc pc

Chapter 5 View of the minerals policies in selected states of Europe

Commodities related to industrial minerals/products	baryte	bentonite	diatomite	diamond	feldspar	fluorspar	gypsum	graphite	kaolin	magnesite	perlite	potash	Phosphate rock	salt	sulfur	talc	cement
Spain																	
Production/mining (mi: mine)	mi	mi	mi		mi	mi	mi		mi	me		chl		rs ss		mi	cf
Import	x	x	x	u gc id	x	x	cr, ca	x	x	me ma	x	sul chl	x	x	x cr ref	x	cc pc
Export	x	x	x	u gc id	x	x	cr, ca	x	x	me ma	x	sul chl	x	x	x	x	cc pc
Sweden																	
Production/mining (mi: mine)							mi								rec	mi	cc cf
Import	x	x	x	u gc id	x	x	cr, ca	x	x	me ma	x	sul chl	x	x	x	x	cc pc
Export		x	x			x	cr, ca	x	x	ma		sul	x	x	x	x	pc
Switzerland																	
Production/mining (mi: mine)							mi					mi					cf
Import	x	x	x	gr gc id	x	x	cr ca	x	x	me ma		sul chl	x				cc pc
Export																	cc pc

5.4 Concluding remarks

Commodities related to industrial minerals/products	baryte	bentonite	diatomite	diamond	feldspar	fluorspar	gypsum	graphite	kaolin	magnesite	perlite	potash	Phosphate rock	salt	sulfur	talc	cement
UK																	
Production/mining (mi: mine)	mi	mi until 2005			mi	mi	mi		mi			chl		rs bs		rec mi	cc cf
Import	x	x	x	u gr gc i d	x	x	cr ca	x	x	me ma	fs chl	x	x	x cr ref	x		cc pc
Export	x	x	x	u gr gc i d	x	x	cr ca	x	x	me ma	chl	x	x	x	x		pc

a	Anhydrite	cr	Crude	i	Industrial	rs	Rock salt
bs	Brinesalt	d	Dust	ma	Magnesia	sib	Salt in brine
ca	Calcined	fs	Fertiliser salts	me	Magnesite	ss	Seasalt
cc	Cement clinker	g	Gypsum	pc	Portland cement	sul	Sulphate
cf	Cement finished	gc	Gem cut	pf	Potassic fertilisers	um	unmanufactured
chl	Chloride	gr	Gem rough	rec	recovered		

Chapter 5 View of the minerals policies in selected states of Europe

Table 74: Summary of different issues of the Minerals Policy Framework in selected European Countries (Department of Mining and Tunnelling 2004, Land Use Consultants 2010, European mining laws, USGS)

	National Mineral Strategy	National Minerals Plan	Mineral rights – Duration of exploration permit (particularly metallic minerals, industrial minerals)	Mineral rights/ Mining concession securing long–term access – (particularly metallic minerals, industrial minerals)	Investment Policy
Albania	✓	–	2 years, extension is possible (3 × 1 year)	20 years, extension is possible (4 × 5 years)	The development of mineral exploration and development projects will depend on the country's ability to attract foreign investment and further its privatization program. Improvements to the business environment could increase foreign investment and increase domestic economic activity, which could result in increases in mineral production. Recently, the investment agency Albinvest ranked the mining industry under the key sectors for investments. According to the mining law the competent authority can support private investors to carry out mineral exports.
Austria	–	✓	5 years, extension up to 5 years.		
Belgium	–	–	Depending on regional laws	Depending on regional laws	
Bulgaria	In preparation	–	3 years, may be renewed twice by up to 2 years	35 years, can be renewed by up to 15 years	Bulgaria continues to develop its whole minerals sector, including quarries. Exploration for precious metals also is expected to increase. The Law on Transformation and Privatization of State and Municipal–Owned Enterprises, and the Mining Law was adopted to promote private enterprise and foreign investment. The principle of equal treatment of foreign and domestic investors was promulgated in law in the Constitution and in the Law on Encouragement of Investments.
Czech R.	In preparation	–		Unlimited periode	

	National Mineral Strategy	National Minerals Plan	Mineral rights – Duration of exploration permit (particularly metallic minerals, industrial minerals)	Mineral rights/ Mining concession securing long-term access – (particularly metallic minerals, industrial minerals)	Investment Policy
Finland	Published in 2010.	–	5 years, extension up to 3 years	10 years, can be permanently extended.	Active investment policy related to exploration and mining. The operating environment in Finland is generally favourable for exploration and mining. The country's political and economic stability, the mining and environmental legislation, and the promining attitude are important factors drawing exploration investment. Mining and taxation laws are expected to remain favourable to the industry. The potential of the Fennoscandian Shield to host undiscovered mineral deposits has been attracting international mining and exploration companies to Finland.
France	✓	–	5 years, may be extended twice	50 years, can be prolonged by 25 years.	
Germany	Published in (October) 2010.	–	5 years, extension up to 3 years.	50 years, extension is possible depending on investment justification	Supporting investment policy actions by useful bank credits; applicable for domestic and foreign investors
Greece	–	–	2 years	50 years (renewable)	Northern Greece contains significant amount of exploitable mineral resources, and receives considerable attention for exploration activity. Mining companies renewed exploration funding in 2006, which indicated their optimism about future production. Greece is expected to remain a major supplier of industrial minerals in the international market. Exploration activities will be intensified to secure additional high-quality reserves. The Government could be involved in planning investment programs to improve the existing installations and lower operating costs.
Hungary	–	–	4 years, extension up to 2 years	35 years, extension by half of the concession period contract	

Chapter 5 View of the minerals policies in selected states of Europe

	National Mineral Strategy	National Minerals Plan	Mineral rights – Duration of exploration permit (particularly metallic minerals, industrial minerals)	Mineral rights/ Mining concession securing long-term access – (particularly metallic minerals, industrial minerals)	Investment Policy
Ireland	–	–	3 years, extension is possible		According to the Mining Law: If the explorer/operator receives the concession, he has the right to obtain a premium (in relation to the importance of the discovery).
Italy	–	–	Depending on regional laws	Depending on regional laws	
Kosovo		✓	2 years, extension three times (up to 2 years)	40 years	In the Investor Guide 2008 the mining industry and energy sector is emphasized in the list of sectors interesting for investment. Kosovo has a liberal trading system, which is focussing on a good investment climate with the neighbour states, and on export options. Based on the EU Autonomous Trade Preference for Kosovo there are duty-free commercial flows of trade. Regarding the imported goods mineral raw material products are of paramount importance, while base metals and their refined products account for more than half of the exports. In the export sector the other mineral products take second place, which makes the mining industry the driving force of Kosovo. The investment climate in Kosovo is promoted particularly by incentives of the government (incentives in the tax and tariff range).
Netherland	–	–			
Norway	–	–	7 years	Operating licence for 10 years, renewable	
Poland	In preparation	–	5 years	50 years	
Portugal	–	–	initial period of 1 to 2 years followed by 3, or 3 one-year extensions	initial period (usually up to 30 years) and one or two possible extensions	Feasibility studies on Portguese zone of the IPB (gold and base metal projects)

5.4 Concluding remarks

	National Mineral Strategy	National Minerals Plan	Mineral rights – Duration of exploration permit (particularly metallic minerals, industrial minerals)	Mineral rights/ Mining concession securing long-term access – (particularly metallic minerals, industrial minerals)	Investment Policy
Romania	–	–	5 years, extension up to 3 years.	20 years, continuously (periods) extension of 5 years	The Strategia Industriei Miniere pentru perioada (2004 – 2010) has to be noted. By this strategy, the Romanian mining industry is to be made competitive. Romania is expected to continue to develop its industrial minerals sector, including quarries and processing facilities for the production of construction materials. Gold exploration is expected to continue to be an important aspect of foreign investment in the country's mineral industry. The effort of the copper mining industry enterprises Compania nationala REMIN S. A. as well as Compania Nationala Minvest which are the property of the government force the integration of investors based on regulation 590/2006, namely creation of joint ventures and the establishment of subsidiaries.
Russian Federation	✓	–	5 years	Depending on mineral category	The rights of disposal of strategic raw materials can be substantially limited. Concerning national strategic mineral reserves there are restrictions for foreign investors as follows: If for example the purchase of shares by foreign investors of more than 10 % of the total votes of shareholders of the company is intended, a resolution of a particular national commission is required.
Serbia	✓	–	3 years		In 2007 the Ministry for Mining and Energy started with the execution of the "Study on a Master Plan for the Promotion of the Mining Industry in the Republic of Serbia". The Japan International Cooperation Agency is involved in this study. The master plan shall cover the resource potential which is not yet developed. Similarly a strategy shall be established for restructuring the mining industry and the development of efficient technologies. The master plan mentioned includes a reconstruction and a renewal strategy for the Serbian mining industry sector as well as an investment plan for domestic and foreign companies.
Slovakia	–	–	period of maximum 10 years	unlimited period	

Chapter 5 View of the minerals policies in selected states of Europe

	National Mineral Strategy	National Minerals Plan	Mineral rights – Duration of exploration permit (particularly metallic minerals, industrial minerals)	Mineral rights/ Mining concession securing long-term access – (particularly metallic minerals, industrial minerals)	Investment Policy
Slovenia	–	–	Five years, extension up to 3 years.	50 years, extension is possible depending on investment justification	
Spain	–	–	One year, renewable	30 years, extendable up to 90 years	Strong investment actions i.e. exploration and feasibility study projects on the Iberian Pyrite Belt regarding for instance gold, nickel, tungsten.
Sweden	–	–	15 years	25 years, extendable by 10 years continuously.	Sweden pursues an active investment policy. Foreign companies are likely to continue to explore actively in Sweden for base metals, diamond, and gold. The quantity of profitable ores in existing mines is likely to be increased by effective and successful exploration in the vicinity of the mines. The global role as an iron producer may increase considerably.
Switzerland	–	–	Depending on regional laws	Depending on regional laws	The sand and gravel industry faces a complex situation (large mass density, structural problems). The industry must cope with today's situation – particularly regarding the year–long procedures with unpredictable outcome. Not least therefore investment has become very capital – intensive in the last decades.
United Kingdom	–	✓	6 months, renewable	60 years	

5.4 Concluding remarks

The following remarks of section 5,4 are divided in general and specific remarks. The latter again are divided in remarks referring to the minerals economic part and minerals policy part.

General remarks

ONE SIDE:

Minerals play an essential role in our lives with 70% of the EU manufacturing production dependent on raw materials. The minerals economy is a **relevant pillar of the total economy of each of the observed** countries: The indicator is particularly the **high portion of imports** (for all observed countries), but also for many countries the **remarkable portion of exports** of mineral related commodities.

Any national minerals policy/strategy must be based on the existing respectively *intended* minerals economy of a country. In other words: usually, the **minerals economy determines the objectives and actions of a national minerals policy/strategy.** In turn, it is also possible to influence the minerals economy by certain policy actions (as mentioned above, to reduce the production of primary raw materials and to increase the use of secondary raw materials, as this is the case for recycled aggregates). The minerals economy of a country can be divided in the **production, import and export parts**. It is important to consider all three parts similarly (interrelated); to analyse one part isolated from the other two parts would not be correct. In fact, the minerals consumption of a country equals production + imports – exports. Moreover, the national minerals policy/strategy can influence whether a country should, for instance, increase its exploration activities, or influence the export of produced minerals. At least, it is important to know the detailed issues of the minerals economy.

THE OTHER SIDE:

Only **three countries of the EU** – Finland, Germany and France - provide a national minerals strategy (Finland and Germany published their strategies only last year). These documents reflect the minerals economic activities of the country *and* develop/provide appropriate objectives and action proposals/plans. However, all in all, this is a **sign of a passive minerals policy pursued by many European countries** – particularly *against the background of the developing countries.* China, India, Brazil, Indonesia and others are very actively envolved in the world raw materials and commodity markets, taking advantage of increased prices, reduced access to minerals like rare earths etc., which lead to serious trade distortions (more information is given in chapters 6 and 7).

Specific remarks on mineral economic issues

Issue: GDP and minerals economy

Regarding the different tables which illustrate the European minerals economy and its development, the analysis shows that **minerals economy is a substantial element of the added value of a state**. It clearly shows that the European economy needs mineral resources, i.e. metallic minerals, industrial minerals and construction minerals for the manufacture of different *products*.

GDP development (1998–2008) of all observed countries has increased constantly, some have doubled, or even more. Indicator regarding mining: The **contribution of the industry (including mining) to GDP in European countries is still high** (see table 71: European minerals economy); for several countries information of the GDP in terms of mining is available (compare chapter 1: the contribution is moderate, however, mining is interconnected with other mining based processes).

It is obvious that all basic and high-tech productive industries, construction and infrastructure projects are dependent on sufficient supply of raw materials. Only if this is secured, they can make their important contribution to the GDP (compare also chapter 1.)

Another fact which underlines that minerals economy is a substantial element of the added value of a state is the **considerable amount** of imports and exports of minerals related commodities (see tables 72 and 73: production, imports and exports of commodities related to metallic minerals and industrial minerals).

Equally, the **high need of aggregates** in all of the observed countries demonstrates the important contribution of mining to GDP development, particularly concerning residential, social and commercial **infrastructure development** (see tables referring to aggregates production): This issue is clearly visible in the figures illustrating the development of GDP per capita *and* aggregates production of 10 years and more (i.e. Czech Republic, etc.). Moreover, this issue is also visible by the commodity cement, required in all of the observed countries for different construction applications ([for instance concrete production requires a mixtur of aggregates, cement and water] see table 73: production, imports and exports of commodities related to industrial minerals).

5.4 Concluding remarks

Issue: National minerals economy – a mix of certain mineral resources/commodities is required

This issue is crucial: the different tables (BGS data on *production, import and export*) indicate that European countries require – usually based on their GDP-level - a **certain mineral resources/commodities mix** to secure the development of their (minerals) economy. The maximum raw materials commodity mix diversity can be generally found in the larger countries like France, Germany, Italy, Spain and UK but also surprisingly in Belgium and the Netherlands (for logistic reasons).

Moreover, it is obvious that at one side certain *basic* raw materials and at the other side more *specific* raw materials are required, depending on the respective intended application in a mineral industry (application possibilities of raw materials see chapter 2).

Basic metallic minerals in terms of various commodities are for example: **iron** (ore, oc, pi, cr, fa, s), **nickel** (u, ua, s, o), **aluminium** (a, ah, u, ua, s) and **copper** (oc, mat, u, ur, ref, ua, s). These metallic minerals (particularly aluminium) are imported in different commodities into each of the observed countries, not only being required for the internal market of a country but also needed to export newly generated products to external markets (i.e. EU countries and non-EU countries).

Specific (high tech) metallic minerals are amongst others: **rare earth minerals** (cc, orec, m), **lithium** (car), **cobalt**, **vanadium** (vana, oc) etc., usually required in countries with a high GDP. These mineral resources are not only required for the internal market of a country but also exported to external markets within and outside Europe: The European market *remains* very important for the economy of a number of exporting third countries, including many developing ones.

Basic industrial minerals (*amongst others*) are: bentonite, diatomite, feldspar, gypsum etc. They are used for manifold purposes in production industries, e.g. chemical, glass and ceramic, polymer, paper, metallurgical industries, as well as for agricultural and pharmaceutical uses and many more.

Specific industrial minerals (*amongst others*) are: baryte, graphite, fluorspar, diamond (us, gr, gc, i, d). They are required in a variety of production industries, similar to the basic industrial minerals. All industrial minerals are fundamental for industrial production and thus are prerequisite for a prospering economy and growing GDP.

Construction minerals: Aggregates are required for different applications, e.g. natural stone; sand and gravel for concrete and construction fill, road base and coverings, bituminous mixtures. Because of their wide range of applications they are essential for every economy.

Issue: Import and export of mineral commodities

Import and export, particularly of metallic minerals and industrial minerals, are an important pillar part of the economy. Remarkable is the fact, that *regardless* if certain mineral resources are mined/produced in a country, **all kinds of required raw materials also have to be imported**. Possibly, this is often not visible enough (for stakeholders). The required amount of mineral resources and mineral commodities in terms of import and export in fact are crucial elements of not only the mineral, but the whole economy of a country.

The possibilities of domestic production of the required mineral raw materials mix of a country are **rather small, primarily due to geological reasons** but for sure also due to limitations of land access. However, the first mentioned aspect (i.e. the mineral resources potential of a country) is particularly relevant. Reflecting these aspects, the question arises from which sources and in which manner a European country is able to import its needed raw materials. A good example in that regard is figure 13 (chapter 1), which shows the producing countries of German metallic minerals imports and indicates Germany's high import dependency.

Regarding production, import and export of metallic and industrial minerals, also the following has to be considered (see BGS data in section 5,2 and 5.2): Finland is the largest chromium producer of EU-27 but uses nearly all produced chromium for domestic applications; only a small part is exported. The same is true for Italy regarding feldspar production. In turn, Ireland is the largest producer of zinc of EU-27 but exports most of the produced zinc. Thus, to give a correct description of the economic situation of a country, it is essential to make detailed analyses of its production, import and export of minerals and minerals related commodities.[614] Therefore it has to be differentiated between:

- European countries which are top producers of certain raw materials but mainly use them for their domestic production and export small amounts, like: Finland (chromium), Italy (feldspar), Albania (chromium), Spain (fluorspar).

---

614 At least, data of 10 years are needed for such analyses to indicate trends, priorities of the present but also for future demand forecasting.

5.4 Concluding remarks

- European countries which are top producers of certain raw materials, use them for their domestic production, but also record considerable exports, like: Austria (tungsten), Bulgaria (copper), Germany (salt), Greece (bauxite), Ireland (zinc), Norway (titanium), Poland (copper), Slovakia (magnesite), Sweden (iron ore).

Issue: Contribution of European mining production to the world market
Data comparison for 2008 (Weber et al, 2010) indicates a *low* European contribution of metallic minerals and *moderate* contribution of industrial minerals to the world market. Figures 67 and 69 are illustrating the global contribution (in %) of European metallic production and industrial production (based on table regarding the five European top producers in comparison with the global rankings).

Figure 153: Global contribution of metallic and industrial minerals production in Europe (%)

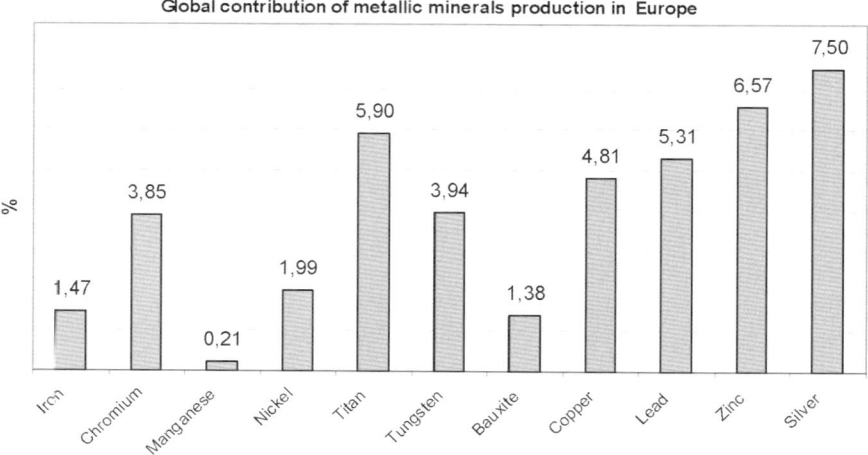

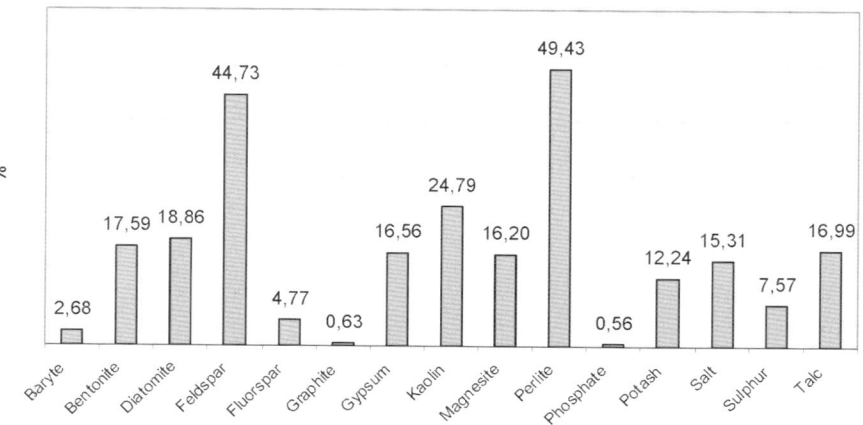

Remarkable is the global contribution of some industrial minerals, particularly feldspar production (about 45 % in 2008 [Italy 22 %] and Germany [about 15 %]) and perlite production (nearly 50 % in 2008 [42 % from Greece]).

As already mentioned, the analysis (of the BGS data sources) indicates the **high import dependency** on many metallic minerals (in different commodities) **and** industrial minerals. On the contrary, the production of these minerals/metals/commodities in the European countries – in comparison with the import – is rather **unbalanced**. The reasons can be multiple: geological reasons (i.e. no minerals deposit potential), access to minerals is not permitted, etc.

5.4 Concluding remarks

Concerning the **European domestic deposit potential**, it is believed that certain regions in Europe, namely the Baltic states, still include a high minerals potential (e.g. development of nickel deposits in Finland, but also the Iberian deposit potential is worth to mention [Spain and Portugal]). In that context, also the investment policy as part of a national minerals policy plays an import role (more information is provided in the section below). However, within Europe there are still many unexplored and unexploited resources (that contain also a significant recycling potential) but due to environmental, societal or economic reasons these resources are not fully explored and used (see also chapter 6).

Issue: Economic importance and supply risk of mineral resources for the European raw materials economy

An analysis of the economic importance and supply risk of 41 minerals was performed based on the EU report on critical minerals (EC, 2010). The supply risk may be accentuated by the low political-economic stability of the main supplier(s), as well as by the low substitutability and low recycling rates of raw materials. There are a number of reasons for this heightened supply risk, one of which is the high concentration of the production of specific raw materials in a given non-EU country (see figure 154). This issue can be qualified as an **indicator for (minerals policy related) decision makers to act immediately** and start to develop a national minerals strategy: if these indicated minerals are compared with the **amount of imports and exports of the observed countries** (see tables 72 and 73, which are illustrating the production, import and export of mineral related commodities) and compared with the more or less **small domestic mine production** in these countries – against the background of the world market development of different commodities (prices).

As mentioned, it is the mining and minerals related industry (see also table 71 which summarizes some important European industries) which is responsible for production/supply of these commodities. However, as pointed out in Chapter 4, the industry needs support from the minerals policy framework (including its tools) to be able to secure the supply of these commodities.

In terms of the three grouped categories which have been defined according to their economic importance and supply risk, it is clear that the *required policies and actions to secure the supply of these raw materials (in the context of such a strategy)* **must be based on different minerals policy issues/aspects** (see also Chapter 4, minerals policy discussed as *cross cutting area*), that is trade policy, development policy, foreign policy, policies aiming to increase the exploration of domestic deposits, research policies aiming to increase resource efficiency, recycling etc.

Figure 154 compares the indicated minerals of the EU report with the **production, import and export** of commodities regarding the observed countries (including Norway and Switzerland, see tables 72 and 73).

Figure 154: EU critical minerals (EC, 2010) – Comparison in terms of production, import and export related to the observed countries (19 EU-countries including Non-EU countries Norway and Switzerland).

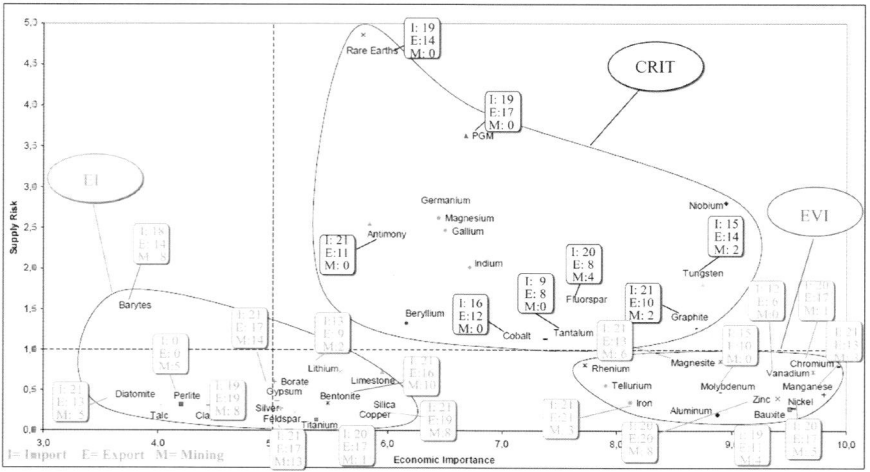

According to this figure, required mineral resources in the observed countries are listed below (according to the available BGS data information). Besides that, information is provided regarding which countries produce such mineral resources: this information indicates (again) the low contribution of metallic minerals production and the moderate contribution of industrial production. Often only a few countries are producers of certain generally needed raw materials.

CRIT – critical minerals are: antimony, cobalt, PGM, rare earth metals, tantalum, tungsten, fluorspar, and graphite.

European production in observed countries: Tungsten mining in Austria and Portugal (3.94 % global); production of fluorspar in Spain, Germany, France, UK and Italy (4.77 % global), graphite in Norway and Czech Republic (0.63 % global).

5.4 Concluding remarks

EVI: economically very important minerals are: iron, chromium, manganese, molybdenum, nickel, vanadium, aluminium, bauxite, zinc, and magnesite.

European production in observed countries (see table 67 and 69): Sweden, Austria and Norway are the main producers of iron. Finland, Albania and Greece are the main producers of chromium, Hungary, Romania and Bulgaria are the main producers of manganese. Greece, Spain, Finland and Poland are the main producers of nickel, Greece, Hungary and France the main producers of bauxite. Ireland, Sweden, Poland and Finland are the main producers of zinc. Slovakia, Austria, Greece, Spain and Poland are the main producers of magnesite.

EI: economically important minerals are: titanium, copper, lithium, silver, barite, bentonite, diatomite, feldspar, gypsum, magnesite, perlite, and talc.

European production in observed countries (see table 67 and 69): Norway is the only European producer of titanium. The main producers of copper are Poland, Bulgaria, Portugal and Sweden. Production of lithium takes place in Portugal (and Serbia). The main producers of silver are Poland, Sweden, Finland and Portugal. The main producers of baryte are Germany, Spain, Slovakia, Italy and Portugal. The main producers of bentonite are Greece, Italy, Germany, Czech Republic and Spain while the main producers of diatomite are France, Spain, Czech Republic and Romania. The main producers of feldspar are Italy, Germany, France, Spain and Czech Republic while the main producers of gypsum are Spain, Italy, France, Germany and UK. The main producers of perlite are Greece, Hungary and Slovakia while the main producers of talc are Finland, France, Austria, Italy and Spain.

The trend of needed minerals/commodities related to metallic minerals

Global perspective

As mentioned in chapter 3, commodity markets have displayed increased volatility and unprecedented movements of prices in recent years. At the heart of current developments lies a series of changes in global supply and demand patterns as well as short term shocks in key commodity and raw material markets. The years 2002 to 2008 were marked by a major surge in demand for raw materials, driven by strong global economic growth, particularly in emerging countries such as China. This increase in demand will be reinforced by the further rapid industrialisation and urbanisation in countries such as China, India and Brazil. China is already the largest consumer of metals in the world – its share of copper consumption, for example, has risen from 12 % to about 40 % over the last 10 years (World Metals Statistics Bureau – 2009 Yearbook.) Price movements have been exacerbated by various structural problems in the supply and distribution chains of different commodities, including the availability of transport infrastructure and services. These developments occur at a time when the competitiveness of European industry requires efficient and secure access to raw materials.[615]

As an example, the global top producers which considerably influence the world market of a certain commodity and thus also the European minerals market are given below (information is based on the respective diagrams in chapter 3 [appendix 2], which are illustrating the production, import and export of certain strategic metallic minerals between 1985 and 2002/2008). As the *above* listed tables indicate, these raw materials are required strongly by the European Economy (table 72), and a great amount must be imported from countries outside Europe. However, such countries either *tend* to use the produced materials mainly for the domestic market or more for the export or equally for domestic purpose and export.

China for instance is the largest producer of rare earths and is also the top producing leader of iron but uses it mainly for the domestic market. India is one of the top producing leaders of iron, but is reducing increasingly the export due to increased domestic consumption; the same is true for zinc and chromium production. Bolivia is one of the largest producers of tin but needs it increasingly for domestic consumption (moderate export). The same is true for Mexico regarding lead production. Chile, (one of) the largest copper producer, offers a moderate export of copper ore and concentrate (additionally, export of refined products). In turn, Brazil (iron ore), Peru (gold) and Russia (nickel) still are large producers and exporters (compared to domestic consumption).

At least, the comparison shows that developing and emerging economies need an increasing amount of their minerals production for domestic use, going along with their growing economic development and domestic consumption. Russia and some Latin American countries are still exporters at a high level, but it can be expected that this will change in the near future. The more raw materials remain in the country of their production, the less are available on the world market – one more reason to build strategies for a secure minerals supply for Europe.

European perspective
(BGS sources)

Production of chromium ores and concentrates in Albania has decreased to 100.000 t in 1998 from 1.100.000 in 1985. In turn, in Finland the production of

---

[615] Information from: http://ec.europa.eu/enterprise/policies/raw-materials/public-consultation-ip/index_en.htm#h2-1.

chromium ores and concentrates remained between 1985–2008 at a constant level of 500.000 t – 600.000 t. Tungsten production in Austria decreased to about 1.100 t in 2008 from 1.500 t in 1985 (considerable up- and downturns). Bauxite production in Greece remained at a constant level of 2.200.000–2.400.000 t between 1985 and 2008. Zinc production in Ireland increased to 400.000 t in 2008 from about 380.000 t in 1985. Titanium production in Norway increased to 900.000 t in 2008 from about 720.000 t in 1985 (considerable up- and downturns). Copper production in Poland decreased to about 370.000 t from about 440.000 t in 1985, then increased to about 570.000 t in 2004 and decreased again to 430.000 t in 2008. Iron ore production in Sweden remained at a constant level until 2001, and then increased constant to 25 Mt in 2008.

Recycling issue

There is a remarkable scrap trade in the European countries in relation to imports and exports.

Some European countries are producing considerable amounts of recycled metallic products, for instance Belgium, Germany, UK, Austria (copper and lead).

Trend of needed minerals/commodities related to industrial minerals

Table 66 and 68 illustrates the importance of mineral producers (between 1998 and 2008) in Europe and globally: Germany is the top producing leader of barite, kaolin, potash and salt; production level is constant between 1998 and 2008. Greece is the top producing leader of bentonite (constant production level) and perlite (moderate increasing).

Finland is the top producing leader of phosphate (strongly decreased) and talc (constant production level). France is the top producing leader of diatomite (constant production level). Czech Republic and Norway are the top producing leader of graphite, however production strongly decreased. Poland is the top producing leader of sulphur; production remained at a constant production level. Spain is the top producing leader of gypsum; production strongly increased. Italy is a top global producing leader of feldspar; production strongly increased to 5 Mt in 2008 from some 1,2 Mt in 1985. Spain is the European producing leader of fluorspar; production remained between 1991 and 2008 at a constant level of 120.000 t–150.000 t. Slovakia is the European top producing leader of magnesite; production increased constant to nearly 1,4 Mt in 2008 from some 80.000 t in 1993.

In turn, there is less production of industrial minerals in several European countries (strong import dependency). Considerable *Diamond trade* in terms of import and export exists in many of the European countries, particularly in Belgium, The Netherlands, United Kingdom, Spain and Switzerland.

In many European countries the increasing need of industrial minerals for domestic consumption induce that the same minerals that are produced and exported in a country, also have to be imported in considerable amounts.

This is the case e.g. with feldspar in Italy (which is the largest producer of feldspar worldwide) and fluorspar in Spain. The United Kingdom exports and imports at the same time practically all kinds of industrial minerals.

As discussions usually focus mainly on metallic minerals, the high dependency on industrial minerals has also to be notified. Although some European countries register considerable production and others produce small amounts of industrial minerals, each of the countries listed in table 73 has to import industrial minerals for their own needs.

Trend of needed minerals/commodities related to construction minerals

Aggregates are an essential ingredient of the key building components that make up the residential, social and commercial infrastructure of modern European society.

According to UEPG, Europe currently needs more then 3 billion tonnes of aggregates a year, equivalent to over 6 tonnes per capita. Overall about 90% of these aggregates come from naturally-occurring deposits while the remaining 10% come from recycled materials, marine and manufactured aggregates.

Based on the above listed diagrams (particularly such which are illustrating aggregates data development per capita and GDP of 10 years) and figures it can be expected that the demand for aggregates *continues* to grow with economic development at national and European levels. Empirical evidence shows that advanced economies can demand up to 12 tonnes/capita (e.g. Austria). Therefore, it is reasonable to anticipate that European demand for aggregates will reach 4 billion tonnes in the medium term, driven mainly by economic growth in Central and South-Eastern Europe (where there is an indication of high need of infrastructure, see also www.sarmaproject.eu). Therefore this growing demand for aggregates needs to be addressed by national minerals policies and planning systems (see remarks below).

In 2008, the top producers (between 200 Mt and 550 Mt) were Poland, UK, Italy, Spain, France and Germany. The middle producers (between 60 Mt and

5.4 Concluding remarks

120 Mt) were Denmark, Norway, Belgium, Czech Republic, Finland, Portugal, Sweden, Austria and Netherlands, The smaller producers (between 25 Mt and 50 Mt) were Romania, Slovakia, Bulgaria, Greece, Switzerland and Ireland.

Crushed rock replaces sand and gravel increasingly, if land use regulations do not allow profitable subsurface extraction. Production of crushed rock will be growing in the future because of utilization conflicts (also due to lack of aggregates planning based on land use planning). Probably, also the marine aggregates production will be growing in the future. Currently, the Netherlands are the largest marine aggregates producer (54 Mt in 2008), the second largest producer is UK (12 Mt); followed by France (7 Mt), Denmark (5 Mt) and Belgium (4 Mt).

Moreover, the production of recycled aggregates is becoming more and more important in Europe. Currently, United Kingdom (UK >25%), Netherlands (>20 %) and Germany (>10 %) are the leading recycling producers in Europe. France produces more than 5 % recycled aggregates. Most of the other observed countries are below 5 % (Sweden, Switzerland, Austria, Czech Republic, Romania and Slovakia). It is worth to mention that the production of primary aggregates in the UK has continually decreased in recent years while the amount of recycled aggregates has strongly increased (partially due to aggregates taxes).

The production of recycled and marine aggregates will continue to grow, however, in the longer-term about 85% of the demand will still need to come from natural aggregates. As aggregates are heavy and bulky, it is imperative for economic and environmental reasons that these are sourced locally to the main markets, particularly where transport by rail or ship is not possible, as is usually the case. Therefore access to local aggregate resources is a key, fundamental and critical issue both for the aggregates industry and for European society.

Specific remarks on minerals policy issues

In summary, the European economy relies to a *considerable* extent on mineral raw materials; this **concerns demand and supply as well.** The high **relevance of minerals supply and minerals policy** therefore is obvious, however, certain indications make deficiencies evident. As indicators and criteria for minerals policy are used here: national raw materials strategy, national minerals planning policy, legal regulatory framework and investment policy (provided by table 74).

Issue: national raw materials strategy

(see table 74)

Most of the observed countries do **not have a national raw materials strategy** which can affect to a certain degree the effectiveness of a minerals policy of a country. As mentioned in section 5.1, a published, *clearly defined national policy/ strategy is a very useful regulatory tool that serves two important functions*. Firstly, it provides the minerals industry with a clear statement of the government's expectations and intents towards the mining activities. Secondly, it provides legislative and regulatory bodies with broad guidance. In most European countries, however, minerals policies rather exist in selective forms, based on different sources, which means investors and undertakers have to use and interpret diverse sources of information about essential aspects of a minerals policy.

Moreover, as discussed in Chapter 4, a national minerals policy/strategy should include **objectives and action proposals /plans** which in turn must be **based on a detailed analysis** of the minerals raw material economy of a country. **Such detailed analysis is required to define, to develop and to justify** the needed (e.g. financial) resources and also the (e.g. technical) support from different stakeholders in terms of **the respectives minerals policy actions.**

The structure of raw materials economies can be regarded and differentiated considering production, imports and exports of mineral resources respectively minerals related commodities (as illustrated in table 72 and 73). In terms of the described required **raw materials mix:** Usually, a country is only self-sufficient in the production of construction minerals (aggregates) from its own deposits potential. Mineral deposits occur dependent on its geological potential, so the country is able to produce some metallic minerals and a certain amount of industrial minerals. In many cases, European countries must import many of the needed metallic and industrial minerals from external sources, inside and outside Europe (see table providing the ranking issue).

Production, import and export has to be jointly analysed using different methodologies, for instance material flow analysis, recycling potential analysis, analysis of raw material markets regarding price development of the concerned raw materials. The **geological potential must be analysed** in terms of the deposits potential of a country: the relation of discovery potential to the required mineral resources mix can be good, moderate or low. The **analysis of the international raw material markets and prices is essential**: the higher the prices for certain commodities, the *higher* the possibilities to use the deposit potential of a country , in order to justify the exploration, extraction and processing costs.

5.4 Concluding remarks

Demand forecast including different scenarios of the needed raw materials is crucial to determine the strategic objectives and actions of the raw material strategy of a country. It is recommended to develop differentiated scenarios for three periods: long-term (30–50 years), mid-term (20–30 years) and short term (10–15 years).

Each strategy has to be checked in certain time intervals to match the objectives and actions with the present development and if necessary, to **update/upgrade** the content of the strategy itself and the *respective scenarios*. As for every *iterative process,* a **monitoring tool based on appropriate indicators is required for the strategy.**

A national raw materials strategy can – as discussed in chapter 4 – determine *different* **priorities**, and state *different* **actions,** which for instance aim to *increase* the exploration and production possibilities in a country and reduce the import needs. Possible **priorities** in relation to *exports* can foster the domestic minerals potential to use the mineral wealth for the development of a strong mining industry and create export possibilities (as this is for instance the case in Finland, as a matter of active investment policy, see below). In turn, the rights for foreign investors to use national strategic mineral reserves can also be limited (as it is the case in Russia).

Finally, a national raw materials strategy should **strengthen the R&D possibilities** to increase **resource efficiency,** which in turn increases the domestic supply and reduces the import dependency for certain metallic minerals (this issue is for instance included in the raw material strategy of Germany, see below).

For the different non-energy mineral categories (metallic, industrial and construction minerals) **different policies/actions** are needed because of their **different nature** (compare also the definitions in Chapter 1). The availability of domestic mineral deposits, which concerns *mostly the aggregates industry* (discussed in chapter 4), needs to be secured by coordinating the extraction process with other utilization claims – particularly in the densely populated areas of Europe – and should be considered in the national minerals planning of every country as a part of the raw material strategy (see below).

Another issue, particularly concerning basic and specific metallic minerals (e.g. aluminium, rare earths) and specific industrial minerals (e.g. graphite, diamond), is whether the required mineral resources can be found *inside or outside the European Union*. The latter is the case for *many* metallic minerals and industrial minerals.

Trade policy, foreign policy and development policy of EU and its Member States have to secure the access to minerals/commodities outside the EU, while production concentration in the source countries of specific raw materials is increasing (see figure 155, below). Trade policy is a full policy competence of the EU, foreign and development policy are part competences. However, all these policies require also **the acting of the EU**. Additionally, the acting of EU in terms of increasing the resource efficiency by boosting R&D is also indispensable (as discussed in chapters 6 and 7).

Figure 155: production concentration of the 'critical' raw materials by source country (EC, DG Enterprise, 2010)

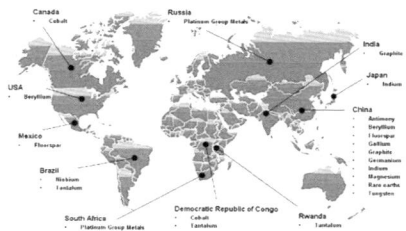

EU countries which provide a national raw materials strategy are France, Finland and Germany (also mentioned in http://ec.europa.eu/enterprise/policies/raw-materials/documents/index_en.htm). Finland and Germany published their strategies in 2010. Some other countries are preparing a national raw materials strategy (e.g. Bulgaria) or are updating their strategies (e.g. Czech Republic). Serbia considers the exploration of new deposits and approaches to sustainable mineral resource use in its National Development Strategy. Moreover, also Albania provides a raw materials strategy.

For the countries which have established a national minerals strategy it is important whether such a strategy is included in or published as an official document, which implies that the strategy has been developed in cooperation with the relevant stakeholders and ratified by the responsible minister. It is also significant whether such a strategy covers **all relevant** non-energy mineral categories **regardless if minerals are extracted in a country or are imported/exported**. In some countries only part of the minerals are considered, for instance energy minerals. According to the available information, a comprehensive national raw materials strategy seems hardly to exist in any of the observed countries.

The following table provides a comparison between the national raw materials strategies of Finland, Germany and France.

5.4 Concluding remarks

Table 75: Comparision of the national minerals strategies of Finland, Germany and France

	National	Mineral	Strategy
	Finland	Germany	France
Publication	Minister of Emplyment and Economy	Minister of Economy and technology	Minister of State, Minister for Ecology, Energy, Sustainable Development and Sea
Strategy development	Different stakeholders	Different stakeholders	Different stakeholders
Minerals	Metallic, industrial and construction minerals	Metallic, industrial and construction minerals	Metallic minerals (additionally substrategies for the other minerals)
Objectives	Three strategic objectives: Promoting domestic growth & prosperity, solutions for global mineral chain challenges and mitigating environmental impact.	Nine objectives: Reduction of trade barriers and distortions Support of German economy to diversify their raw materials sources Support of economy in the development of synergies through sustainable management and increased material efficiency Further development of technologies and tools to improve the recycling framework, Development of bilateral commodity partnerships with selected countries Opening of new substitution and materials research options Focus on resource-related research programmes Transparency and good governance regarding minerals extraction Cooperation with EC in terms of European minerals policy.	Access to strategic metalic minerals in good conditions is needed to ensure French industry conditions

407

Chapter 5 View of the minerals policies in selected states of Europe

	National Mineral Strategy		
	Finland	Germany	France
Actions	Four themes of action proposals: Strengthening minerals policy, securing the supply of raw materials, reducing the environmental impact of the minerals sector and increasing its productivity, and strengthening R&D capacities and expertise. Moreover: 12 defined action proposals	Actions not explicitely listed. Actions are contained in the context of the objectives	Action plan focuses following issues: Improving knowledge of strategic metals Extension of geological knowledge by exploration campaigns Development of new exploration tools Recycling policy of strategic metals will be coordinated.
Specific organisations		A raw materials agency was installed which is responsible for specific issues, e.g.: Establishment of a raw materials information system: to provide the German economy a better basis for securing minerals supply. Technical support for Federal Government concerning exploration, extraction, material efficiency. Research and development projects (R&D) Cooperation with resource-rich countries	A senior official will be responsible to strengthen governmental action and to intensify the dialogue between government and industry to secure the supply of metallic minerals.
Additional remarks	• *Analysis of the minerals deposits potential*/discovery potential (also considering the EU report on critical minerals) • Demand/production forecasting • Analysis of minerals industry • Improve regulatory framework …		

5.4 Concluding remarks

Issue: minerals planning policy

(see table 74)

As described in sections 4.4.2 and 5.1, a **minerals planning policy** should be part of the national minerals policy framework. A minerals planning policy, based on land use planning, is important to protect and make available economically attractive deposits of metallic minerals, industrial minerals and particularly aggregates: ensuring the adequate supply of aggregates to expanding urban centres is essential and should be a priority area in regional and local government land use planning process. Most of the observed countries do not have a national minerals planning policy, or in other words, there is **no (strategic) national mineral resources planning system to protect economically attractive deposits**. Exceptions are Austria (Austrian minerals resources plan), UK (national minerals planning guidelines) and Kosovo (mineral resources management plan at national level which is related to the land use planning).

The national minerals planning policy in many countries therefore needs to be **improved**. According to the definition used here (protection of mineral deposits by land use planning) it is either not existent or only recognisable in general outlines. A categorisation of land use planning together with the protection of deposits hardly takes place.

Considering the import problem, the increased use of the domestic mineral potential ought to be a central task of a national minerals planning policy. Though the Geological Surveys accomplish ambitious activities, it is not possible to sustainably protect the mineral deposits without comprehensive implementation of the data into land use planning structures. Mineral protection based on a national mineral resources plan is rather exceptional. Moreover, an interesting approach can be found in Germany, though the development of strategic minerals planning at national level was interrupted.

Issue: legal regulatory framework

(see table 74)

The missing of a national raw materials strategy and national minerals planning policy in most of the observed countries affects to a certain degree the legal regulatory framework, especially the licensing and permitting procedures (information collected from two main sources: one from the study on Mineral Planning Policies of 2004, the other from the consultation process of LUC of 2010). Generally, the **complex licensing procedures often seem not to be coordinated well between the different authorities**. The licensing procedures are not only time-consuming and expensive, but often produce uncertainty of a positive procedure result. Legal costs and the delayed beginning of the raw material extraction are not a cost-effective contribution to the GDP of a state, but signal a **passive minerals policy**. (Improvements regarding this issue are discussed in chapter 7).

As mentioned in chapter 5.1, a national minerals policy shall be based on consolidated cooperation between the legal and administrative authorities and institutions involved. The Finnish Raw Material Strategy refers (in action 6) to this issue: "permitting processing times shall be reduced and permitting procedures shall be refined. This shall be achieved in part by improved cooperation between different authorities."

The **regulatory framework must establish a solid legal basis**. If the entrepreneur invests, it must be clear, under which conditions the exploration and exploitation rights can be acquired. All countries have legislation for the regulation of the mining industry. For the entrepreneur the fact is essential that the exploration and exploitation licenses are stable rights, and if necessary, can be sued for at an independent court. Above all, the license duration set in relation to the amortization of the cost is an important factor; the investments are capital-intensive (see table 74, Summarisation of different issues of the Minerals Policy Framework in selected European countries). For instance, a positive example of an active minerals policy is Sweden: the Swedish Mining Law provides an extension to the extraction permit after a 10 year period without considerable administration expense. This signals that the state is interested in the long-term contribution to added value by the minerals industry.

Issue: investment policy

(see table 74)

The investment policy of a state regarding the **promotion of exploration, production and processing** by both natives and foreigners is rated as a *substantial* element of an active minerals policy. Here some European countries provide constructive bases. It is striking that in particular the countries of South-Eastern Europe, here particularly the non-European Union countries, such as Albania, Serbia and Kosovo, set considerable impulses. Moreover, Greece, the Scandinavian countries Finland and Sweden, Germany and Spain play an active role. In many other countries the investment policy is obviously of minor importance.

5.4 Concluding remarks

EU Member States

The operating environment in Finland is generally favourable for exploration and mining. The country's political and economic stability, the mining and environmental legislation, and the pro-mining attitude are important factors drawing exploration investment. Mining and taxation laws are favourable to the industry. The potential of the Fennoscandian Shield to host undiscovered mineral deposits has been attracting international mining and exploration companies to Finland. Worth mentioning in that regard is the Finnish Raw Materials Strategy, action 3 and 4:   Action 3 aims to **improve the minerals sector's financing opportunities. Institutional investors and the government have a key role** in this area through continuing **public support** for infrastructure investments and through lending and loan guarantees for mine investments. (TEM, Finnvera plc, Finnish Industry Investment Ltd, investors, financing institutions). Furthermore, according to action 4, the **potential of using tax incentives to promote** exploration for natural resources and for efficient use of resources shall be investigated.

Also Sweden and Greece pursue an active investment policy. In Sweden foreign companies explore actively for base metals, diamond, and gold. The quantity of profitable ores in existing mines is likely to be increased by effective and successful exploration in the vicinity of the mines. The global role of Sweden as an iron producer may increase considerably.

Northern Greece contains significant amounts of exploitable mineral resources and receives much attention for exploration activities. Greece is a major supplier of industrial minerals in the international market. The Government could be involved in planning investment programmes to improve the existing installations and lower operating costs. Germany supports investment actions by useful bank credits; applicable for domestic and foreign investors.

Bulgaria continues to develop its whole minerals sector, including quarries. Exploration for precious metals is expected to increase. The Law on Transformation and Privatisation of State and Municipal-Owned Enterprises, and the Mining Law were adopted to promote private enterprise and foreign investment. The principle of equal treatment of foreign and domestic investors was promulgated in law in the Constitution and in the Law on Encouragement of Investments.

Based on the Strategia Industriei Miniere pentru perioada 2004–2010, the Romanian mining industry is intended to be made competitive. Romania is expected to continue to develop its industrial minerals sector, including quarries and processing facilities for the production of construction materials. The effort of the copper mining industry enterprises Compania nationala REMIN S. A. as well as Compania Nationala Minvest which are property of the government is considerable. Both enterprises force the integration of investors based on regulation 590/2006, namely creation of joint ventures and the establishment of subsidiaries.

EU Non Member States

In the Investor Guide 2008 the mining industry and energy sector of Kosovo was emphasized in the list of sectors interesting for investment. Kosovo has a liberal trading system, which focuses on a good investment climate with the neighbour states and on export options. Regarding the imported goods, mineral raw material products are of paramount importance, while base metals and their refined products account for more than half of the exports. In the export sector the other mineral products take second place, which makes the mining industry the driving force of Kosovo. The investment climate in Kosovo is promoted particularly by incentives of the government (in the tax and tariff range).

The development of mineral exploration and development projects in Albania will depend on the country's ability to attract foreign investment and further its privatisation programme. Improvements to the business environment could increase foreign investment and increase domestic economic activity, which could result in increases in mineral production.

In 2007 the Ministry for Mining and Energy in Serbia started with the execution of the "Study on a Master Plan for the Promotion of the Mining Industry in the Republic of Serbia". The Japan International Cooperation Agency was involved in this study. The master plan shall cover the resource potential which has not yet been developed. Similarly a strategy shall be established for restructuring the mining industry and the development of efficient technologies. This master plan includes a reconstruction and a renewal strategy for the Serbian mining industry sector as well as an investment plan for domestic and foreign companies.

Closing remarks

In **summary** there are (not necessarily negligible) deficiencies in the national minerals policies. The elimination of these deficiencies would presuppose appropriate measures by the countries and particularly by the European Union. For this purpose, the European Union itself would need comprehensive structures of a minerals policy. In this respect there is still much work to be done, as is pointed out in the discussion in chapter 6 and chapter 7.

Chapter 6 EU minerals policy status quo – critical reflections

Chapter 5 refers to some EU and non-EU countries at national level. The following reflects directly on EU level.

While the present European Union had one of its roots in the European Coal and Steel Community of the 1950s,[616] securing the supply of raw materials for the European economy has not been a primary aim of common policy in the past decades. This has been mainly left to the respective national policies. Some countries have pursued protectionist policies to favour their own national enterprises and change them reluctantly.

Political priorities at the European Union level are generally discussed in the context of European Union policy and national policies. For further understanding, a short overview of the (complex) European Union structure will be given. The European Union (EU) is a joint federation of countries, an economic and political union of 27 member states primarily located in Europe. As an international organisation sui generis, the EU operates through a hybrid system of supranationalism and intergovernmentalism. In certain areas it depends on mutual agreement between the member states; in others, supranational bodies are able to make decisions without unanimity. Through a standardised system of laws, which applies to all member states and ensures the freedom of movement for people, goods, services and capital, the EU has formed a single market; it maintains common policies on trade and regional development. The EU furthermore developed a restricted role in foreign policy, in the WTO, G8 summits, and in the UN. Important institutions and bodies of the EU include the European Commission, the Council of the European Union, the European Council, the European Court of Justice and the European Central Bank.

---

616 The European Coal and Steel Community (ECSC), often called the Montan Union was a European trade association and a predecessor of the EC. It rendered all member countries access to coal and steel without paying customs. The founding members of the treaty were Belgium, Federal Republic of Germany, France, Italy, Luxembourg and the Netherlands. The ECSC Treaty, which was concluded for a period of 50 years, expired on 23 July 2002. It was not renewed, and its regime was henceforth attributed to the EC Treaty (http://de.wikipedia.org). Interestingly, there were already in the 1940s, various approaches to European minerals policy. See in this respect: Labour Science Institute of the German Labour Front (1942): Thoughts on a European agricultural and raw material policy: first attempt at a foundation. – Berlin, 1942].

Chapter 6 EU minerals policy status quo – critical reflections

6.1 General

In the last two decades little importance was attached to the European mineral policy[617] or, in other words, it was not in its whole complexity noted by the policy/decision-makers in the last decades.[618]

Hence, the publishing of the Communication "The Raw Materials Initiative – Meeting our critical needs for growth and jobs in Europe" in November 2008 which is based on the preceding consultation process (from January to March 2008), was a welcome improvement. However, it must be stated that, against the background of recent international developments, this happened considerably late.[619]

6.2 Public raw material awareness

At the European Conference on Mineral Planning in 2002 a **general deficit in public awareness regarding the importance of raw materials** for every individual, the society and national economy was recognized.[620] This

---

617 Faller, Peter (1978): Braucht die EG eine Rohstoffpolitik?: Das Rohstoffproblem, Versuch seiner ökonomisch-politischen Lösung unter besonderer Berücksichtigung der Nichteisen-Metalle. [Does the EC need a minerals policy?: The mineral problem, attempt of its economic-political solution under special consideration of the non-ferrous metals] – Aachen, Techn. Hochsch., Philos. Fak., Diss., 1978. Humphreys, David (1990): Towards an EEC minerals policy? – Resources policy, 16.1990,1, pp 35–46.
618 Cf. Faber, Gerrit (1981): Commodity trade and security of supply with minerals: EC policy towards Latin America. – In: International commodity trade. – Amsterdam: CEDLA, 1981, S. 131–170. Also European Commission (1984): Selected papers arising from the EEC primary raw materials programme (1978–81). – Luxembourg: EC, 1984 (Environment and quality of life series; 1984) (EUR/European Commission; 8617). In this sense also: List, F. K. (1986): Evaluation of the Community's primary mineral raw materials programme. – Luxembourg: EC, 1986. – (EUR/European Commission; 10191) (Research evaluation – report; 16). European Union/European Commission/Directorate-General XII, Science, Research and Development (1996): Proceedings of Workshop on industrial minerals, Athens (Greece), 25–27 September 1995: industrial and materials technologies programme (Brite-EuRam III) – (1994–1998)/ed by M. Grossou, A. Adjemian. – Luxembourg: EUR-OP, 1996 (EUR/European Commission; 16936).
619 European Commission, DG Enterprise (2008): The Raw Materials Initiative – Meeting our critical needs for growth and jobs in Europe. Communication from the Commission to the European Parliament and the Council, COM(2008)699. This is discussed briefly in the preface.
620 Cf. Hilden, H.D., Chairman of the European Conference on Mineral Planning (2002): After conferences in Zwolle, Netherlands, in 1997 and in Harrogate, England, in 1999 a third European conference on "Planning Mineral Resources" was held from 8th to 10th October, 2002 in Krefeld, Germany. The conference in which 270 representatives of the administration, the industry and the environmental associations of 17 European nations took part stood under the motto "Mineral planning in Europe. Changed basic

6.2 Public raw material awareness

was also clearly expressed in the notification of the results of the Raw Materials Initiative consultation process on April 16, 2008 in the European Parliament.[621]

Indicators are

- the decrease of exploration activity
- a heterogeneous base data
- the decrease of competitiveness of the mining industry in Europe
- the termination of the mining sponsorship and reduction of raw material research policy[622]
- a deficient mineral planning policy

Figure 156: Summary and outlook – EU's mining, Trends in global metal mining since 1850 (Data by Ekdahl 2008)

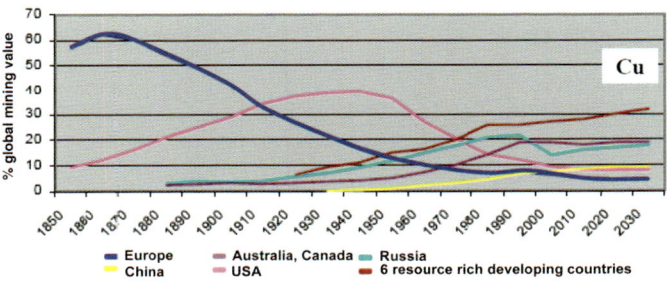

The **increased globalization of world commodity** markets has reduced the perception of policy makers that it is a necessity to achieve national self-sufficiency in minerals.[623] In addition, economic diversification and increased

---

conditions! – New perspectives?" Central subject was "sustainability" at the production of and the supply with mineral raw materials and before the background of the fact that mineral raw materials are not indefinitely available and are renewed only in geological periods.
621 See Appendix.
622 Cf. Fettweis, G., Wagner, H., Kann es die "Europäische Union" verantworten, sich aus der Bergbauforschung zurückzuziehen? Berg- und Hüttenmännische Monatshefte, [Can the "European Union" take responsibility for withdrawal from mining research? (Journal of Mining, Metallurgical, Material, Geotechnical and Planned Engineering)] S.155. Rossmann, H. (1996), Anrainer- und Umweltschutz im Bergrecht, Wien [Local residents' and environmental protection in mining law, Vienna]. Tiess, G. (2003): Gibt es eine europäische Rohstoffpolitik?, Berg- und Hüttenmännische Monatshefte [Is there a European minerals policy? (Journal of Mining, Metallurgical, Material, Geotechnical and Planned Engineering)], pp. 307–315.
623 Otto (1999), l.c., p. 5.

Chapter 6 EU minerals policy status quo – critical reflections

job opportunities in other sectors have made it easier for workers in the minerals sector to find alternative employment. The substantial growth of environmental awareness has made mining less popular to both the public and politicians, and has resulted in the development of some of the world's most extensive (and costly) environmental laws. For these and other reasons, policies that provided various subsidies, protection and economic incentives to the mineral sector have been gradually eliminated or substantially scaled back.[624]

In recent decades, the competitiveness of the European extractive industry remarkably lessened due to the decrease of deposit availability and the complex approval procedures (see below).[625]

Figure 157: Effects of globalization

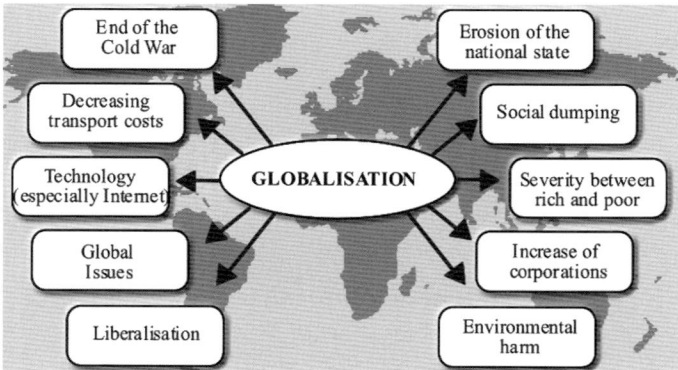

The term "mineral raw materials" cannot be found in the primary legislation of the EU;[626] neither does an EU comprehensive body of legislation for raw materials exist.[627] In EU policy there is no sector for raw materials policy[628] and no basis relevant for raw materials planning policy can be found in the natural resources documents.

It is inevitable to notice that various elements needed for the definition and realization of a policy of security of raw materials supply are **missing** or are

---

624 Otto (1999), l.c., p. 5f.
625 Cf. Wagner, H., Tiess, G., Solar, S., Nielsen, K. (2005): Minerals planning policies in Europe, International Mining Symposia AIMS, Aachen , pp. 523–538.
626 ECSC expired in 2002. See also Christman (2008): Lecture at the European Parliament. (http://www.europeanmineralsfoundation.org).
627 Cp. on the contrary for instance the Water Framework Directive.
628 Cp. on the contrary for instance the agricultural policy and water policy at EU level.

6.2 Public raw material awareness

dealt with atomistically rather than holistically.[629] In particular the following are needed:

- an adequate instrument for the **analysis of the present situation** (referring to the status quo of the EU's materials policy) as well as prospects and for defining arrangements which make it successful;
- an **institution adequate to Europe for coordinating and homogenizing data relevant to raw materials**;
- **adequate focusing of coordinated research work at EU level**;
- an **instrument for funding** which would be able to induce the required investment flows.
- a comprehensive trade agreement for product and a systematic search for multilateral solutions.

The supply with raw materials of the Union is a **matter of foreign affairs** in three respects:

- the European Union is dependent on imports from countries outside the EU for almost all metallic raw materials as well as for a considerable part of the industrial minerals;
- the future of a European raw materials policy is codetermined by the development of foreign affairs, especially of those with developing countries;
- mineral policy is a continuous debate in **international committees setting standards for raw materials markets** and therefore a part of foreign affairs.

For all the issues listed a concrete contribution of the EU is required. The question arises whether the Union is able to initialize the **instruments needed** for the preparation, coordination and conception of a long-term policy, which are not available yet.[630]

---

629 Or have been missing so far: With the publication of the communication on November 2008 "Proposal of a Mineral Raw Materials Initiative" the situation was improved. The intentions and measures addressed there concern the principle of consistency. The matter of implementing will show whether an actual contribution to the raw material supply policy is possible.
630 Again the reference: With the publication of the new communication on 4 November 2008 "Proposal of a mineral raw materials initiative" the situation has improved. The goals and measures indicated there apply on the principle of the coherence. In this connection the question of implementing will show whether an actual contribution to the long-term raw material supply policy is possible. Basically this means that for a start the discussions have begun.

Figure 158: European mining industry on a global scale (Data by Ekdahl 2008)

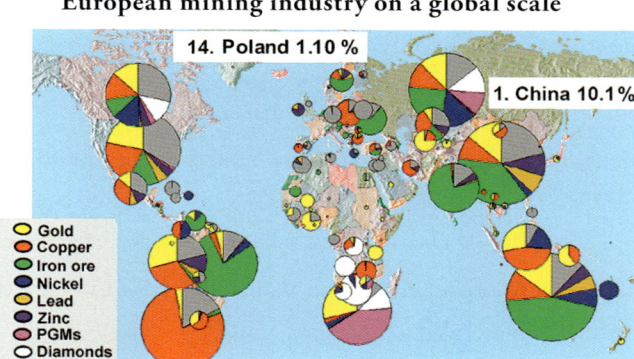

The significance of raw materials security for the European economy **has not been perceived as a whole and in all of its complexity in policy so far** or at least 2008. The economic costs of poor raw materials supply, the present and future disturbances of raw materials security and the politic and economic risks require an intense engagement of various policies with the topic of raw materials supply, **not only emphasizing on national economic and environmental policy**.

Missing raw materials strategy

Political priorities on the EU level are generally treated within different EU and national policies. The secured access to raw materials and stable raw materials prices are a main factor for the competitiveness of the European raw materials industry and the European economy. In this respect, the EU *at least until 2008* has not taken considerable action. Some of the *consequences* are:

- the lack of an active raw materials strategy.[631]

- an inconsistent legal basis concerning the raw materials industry[632]

- a downward trend concerning exploration and raw materials production development[633]

---

[631] Also addressed as part of the consultation process of the EU Raw materials initiative 2008. The proposal of a Raw materials strategy has been discussed since November 2008 by the Commission.

[632] Cf. auch Hamor (2004), Sustainable Mining in the European Union: The Legislative Aspect, Environmental Management, p.252.

[633] Cf. Ekdahl (2008), l.c. Also: Consultation process of the raw materials initiative 2008.

6.2 Public raw material awareness

The implementation of any policy requires strategies and concepts on adequate legal bases. Secured access to raw materials as well as stable raw materials prices are essential issues for the competitiveness of European industry. Prior to the implementation of the EU Raw Materials Initiative in November 2008, the European minerals policy was characterized mainly by independent and uncoordinated policies for subsections or on a national level. This might be understood considering the historic development and steady growth of the European Union, which reached its present number of 27 member states only in 2006. Several treaties between states and, later, communities, formed the present European Union, which has been undergoing a process of increasing integration and homogenization but even now has not introduced a common currency (the Euro) in all of its member states.

The European Union currently is beginning to develop a common minerals policy. In contrast to that, most EU partners in commodity trading have already established their raw materials strategies, which means a clear advantage in global competition. The USA, Australia, the so-called BRIC states Brazil, Russian Federation, India and China, Japan and Korea as well as Indonesia, several Latin American and African states and many other developing countries have clearly defined mineral policies. In some cases there have been measures that have limited access to foreign raw materials resources for the EU industry.

Proposals for a European raw materials policy require that the advantages for the Union will be noticed on one hand, and the will to solve the problems collectively on the other hand. Preliminary concerted action concerning European raw materials policy might not be easy due to a clash of interests among the different EU countries. The following reasons can be mentioned: the different interest in the maintenance and stabilization of good political relations with the raw materials countries; the varying degree of dependency on raw materials imports along with the varying risk an interruption of supply brings about; the affiliation to different preferential areas.

Figure 159: Conflicts of interest in an EU raw materials policy

Potential **obstacles** are divergent attitudes regarding foreign policy and therefore divergent mineral economic interests of the member states.[634] Moreover, an increasing inability to maintain stability of price, a high level of employment and a balanced account of payments should be mentioned at the same time.

As already mentioned the trade partners and competitors of the European Union have already been pursuing the development of a purposeful raw materials strategy for years. Altogether, a **strong increase of political measures by countries** outside the European Union means a strongly reduced access of EU countries to these raw materials.[635]

Figure 160: China's share in global consumption, 2005 (Data by Ekdahl 2008)

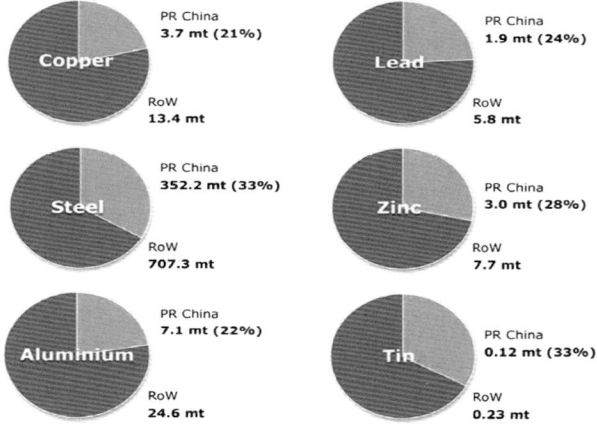

State economies like China buy in on international mining through vast share ownership to secure long-term raw materials supply for their growing industrial production.

---

634 Cf. for example, Doha Work Process, 2008. PRESSE, 31 July 2008: WTO: The consequences of failure. The end of the Doha Work Process leads to more trade disputes, promotes protectionism and prevents necessary reforms. In this sense: "There is no political support, neither from India nor from Argentina or South Africa", according to Carin Smaller, Institute for Agriculture and Trade Policy (PRESSE, 21 July.2008).
635 EC, DG Enterprise and Industry, (2008): Public Consultation on Commission Raw Materials Initiative (2008).

6.3 Knowledge basis of raw materials

EU-wide the raw materials data situation is heterogeneous. Data concerning the geological potential, particularly strategic raw materials, market analyses, and raw materials balances on the European Union level are presently not available. The latest update of the European Minerals Yearbook dates back to 1997 and is based on data of 1995. Also, a comprehensive European analysis of demand predictions is not available so far.[636] The narrowness of available raw materials data is explicitly mentioned in the Commission Staff Working Document "on the European Union non-energy extractive industry".[637] For instance, official statistics for industrial minerals and building raw materials are rather incomplete. This problem can be traced back, among other reasons, to the structure of these sectors: in most of the member states the building raw materials sector includes a large number of medium, small and very small enterprises. These are not covered by national statistics and thus neither by EUROSTAT, so the importance of these sectors is not reflected adequately. Compiling data coming from different sources to useful EU minerals statistics would be very important.

The rapidly changing dynamics of non-energy minerals supply and demand, together with the EU Sustainable Development and Consumption and Production Policies require an up-to-date knowledge base, enabling timely and adequate analysis and a coherent EU policy coordination, and the dissemination of information to other public and private organisations and the public.

Most important sources of information on European mineral resources and production are national statistics published in almost all EU member states.

For example, the British Geological Survey as well as the Austrian Federal Ministry for Economic Affairs and Labour ("World mining data") publish comprehensive annual statistics on production of certain minerals.

World wide statistics on most metallic minerals are available from the World Bureau of Metals Statistics.

The United States Geological Survey (USGS) gathers statistical information related to country production and specific commodities.

---

636 EC, DG Enterprise and Industry, (2008): Public Consultation on Commission Raw Materials Initiative (2008).
637 EC, DG Enterprise and Industry (2007), lc.

Trade federations and individual mining and quarrying companies in turn provide data. Detailed commodity related statistics are available from private consultants and Institutes against considerable payment.

Relevant statistics on specific topic-related issues may also be obtained from commercial companies (examples are including Minecost. com and Raw Materials Data) DG-Trade is in the process of developing a database on trade restrictions for a number of commodities.

Qualitative and quantitative data on the origin and use of mineral raw and base materials can be obtained from the (national) official industry and foreign trade statistics.[638] Economic data are published by the Statistical Office of the European Union (Eurostat) which compiles and reports data provided by the National Statistical Offices (NSO) and other national statistical authorities in each Member State.

Another problem for instance is the European iron and steel statistics. The collection of data rests with the national authorities. The statistic was initiated by the ECSC treaty and was considerably diminished with the expiration of this contract. The primary focus of these statistics was the steel industry as to production, capacity and imports of raw materials. Parts also dealt with iron ore mining in Europe (production, employment etc.). The aim of the statistics was to establish the basis for a common iron and steel policy, as well as to provide the steel industry with information relevant to competition. After the expiration of the ECSC contract, parts of the statistics were incorporated into other statistics or simply not collected anymore. Moreover, since then the legal basis for the collection of data is missing, i. e. companies are not obliged to release relevant information. This can be seen as an example for how to reverse useful approaches for a comprehensive raw materials statistics.

An aggregation of data coming from different sources for useful EU raw materials statistics is highly relevant, but might be rather problematic due to different nomenclature and definitions in the base statistics.[639] Moreover, data

---

638 Relevant but partly incomplete information is contained in other official statistics. Public institutions like Economic Chambers as well as individual mining companies (e.g., Minecost.com) provide alternate data. Detailed mineral raw material statistics are also available from non-public institutions against considerable payment. International compilations of production data of mineral resources are available from the British Geological Survey (BGS), Austrian Federal Ministry for Economic Affairs and Labour (World Mining Data), the U.S. Geological Survey (USGS). All 3 institutions publish yearly and differentiations are recognizable. Global statistics on ore production are available from the "World Bureau of Metals Statistics".

639 Cf. Heimburg, J., (2008): Rahmenbedingungen für das ANTAG-Projekt in Österreich, Bachelor Arbeit, Fachbereich: Bergbaukunde. [Basic conditions of the ANTAG-Project in Austria, bachelor thesis, department: Mining Engineering.]

are often gathered collectively, making a breakdown of single products impossible. This often applies to sensitive materials which, although in small quantities, are necessary for the processing industry. Another general problem is the fact that an exact description of the supply with mineral raw and base materials is complicated due to insufficient statistical coverage.

The necessity of improving the **instruments for raw and base materials statistics**, as well as **market analyses and raw materials balances on EU-level** in order to become an efficient base of decision shall be underlined.[640] Such efforts were made in the 1970s and early 80s:[641] Eurostat developed a raw materials balancing system for nonferrous metals for the indication of the supply situation, which accounted for the primary degree of self-sufficiency, the degree of self-sufficiency including domestic recycling, the technical raw materials import dependence, the economic raw materials dependence and the degree of recycling (see appendix).

6.4 Access to minerals outside Europe

6.4.1 Trade policy

The EEC treaty commits the Union to a **common trade policy** towards third countries (Art. 3 Para. 1 lit b EEC). The main instrument is the common customs tariff. Traditionally, there is a distinction between autonomous and contractual trade policy. Autonomous trade policy comprises all unilateral actions towards third countries without making agreements, whereas contractual trade policy includes multilateral and bilateral agreements with third countries. The most important forum for multilateral trade policy is the GATT.

Global trade of raw materials presently is obstructed by **trade restrictions**, rebates and absorptions, allowances and credits as well as singular interventions and interventions within market regulations. Thus, positive WTO negotiations concerning the GATT would have been of great importance for the terms under which the Union would be able to supply itself with raw materials on the global market.[642]

---

640 As discussed in Chapter 1, Item 1.5.3 (aspects of material criticality), it is necessary at first to determine the critical raw materials for a state. The same would apply for the EU: For the compilation of mineral resource trade accounts, the EU-critical raw materials have firstly to be identified in joint cooperation of the Commission and Member States. This is expressed again under point 6.2. A similar proposal is mentioned in the Commission COM (2008) 699 "Raw Material Initiative", l.c, mentioned on page 6.
641 Gocht (1983), l.c., p. 227. Michaelis (1976), l.c., pp. 30, 35f.
642 The discussions of the Doha Work Process, July, 2008 have failed, as is well-known.

Chapter 6 EU minerals policy status quo – critical reflections

EU-external risks and problems of the European raw materials supply exist due to instabilities in producing countries and an increasing conglomeration of mining areas and businesses on the international commodity markets. Politically induced, i. e. intended, restrictions of raw materials access/availability in form of trade distorting measures belong to the central problems in raw materials supply security.[643] At the international commodity markets there are tendencies to favour trade distorting national measures both on the supply and on the demand side of domestic commodities, making raw materials export more difficult. Distortion happens for example by recovering value added tax on imports, discriminating license systems or prohibitive high export duties, which in fact equal an export prohibition. This can be seen as an attempt to provide strategic industrial-political advantages at the expense of trade partners. Unsatisfactory political competition control mechanisms are another problem.[644]

China, for instance, subsidizes the import of raw materials, conferring advantages to its industry. At the same time, China limits exports of certain domestically abundant natural resources, such as rare earths (a group of 15 metallic elements of which cerium, lanthanum and neodymium are the most commonly used), which can distort availability and price of raw materials in world markets (double pricing). Lately, China introduced an export duty of 120 % on yellow phosphorus. India recently introduced an export tax on iron ore.[645] This also affects scrap trade.

Figure 161: Metal scrap

The most important source of metallic raw material to the EU, metal recycling, is directly affected. Metal scrap accounts to 40–60 % of the EU metals production.

---

643 Cf. Schorsch, Louis L. (1989): Minerals trade and commercial policy: the case of steel. – Resources policy, pp 169–187. Sowie (Also): BDI (2007), l.c., p 10.
644 BDI (2007), l.c., p. 10.
645 European Commission, DG Enterprise and Industry, (2008), Public Consultation on Commission Raw Materials Initiative (2008).

6.4 Access to minerals outside Europe

Russia applies prohibitive taxes of up to 50% on the export of scrap. Illegal scrap trade is another problem.[646]

Figure 162: WTO structure (WTO)

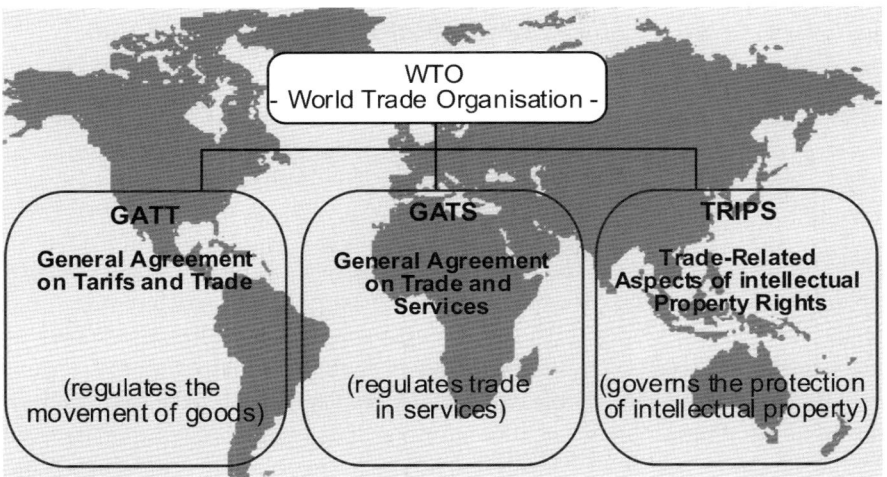

The WTO legal system currently does not comprise a ban on export duties. This legal loophole should be filled as well as possible.[647] The basic requirement for the change of the WTO system of rules and regulations is the revival of the WTO round table. Improved WTO rules and regulations concerning "double pricing", where raw materials inland are cheaper than when exported, would also make sense.

Moreover, important companies and countries, especially China, pursue a tight strategy of state-enforced backward integration. Increased shares in foreign mining companies secure China's world wide access to deposits. For example, China was acquiring foreign bauxite shares amounting to 15% of the world extraction, for ferrochromium they have already obtained 10%, for iron ore 7%. Furthermore China does not only explore its own territory, but also overseas. In case of success, parts of their own supply with raw materials are secured through investments and shareholding.[648]

---

646 Cf. COM (2008) 699.
647 Cf. WTO, Negotiation Group on Market Access, Communication from the European Communities, 27. April 2006, TN/MA/W11/Add. 6.
648 BDI (2007), l.c., p. 10.

Regarding trade policy there is an **urgent need for action**, particularly from the WTO.[649] Trade-distorting measures in the raw materials sector might have vast effects on the competitiveness of a country and therefore should rank high on WTO-level as well as in bilateral relations. The validity of trade-restricting measures depends on the kind of measure and law applied, which again depends on the parties involved.[650]

The European Commission (DG Trade) hinted on bringing a claim to the WTO against China's export restrictions on raw materials.[651] Raw materials trade is a pivotal question in further economic dialogue with China. Other states which apply export duties for the promotion of their domestic industry and as a result bring dumping products to the EU market also will have to face WTO examination.

There are over 450 export restrictions on more than 400 different raw materials (e. g. metals, wood, chemicals, hides and skins) including secondary raw materials (e. g. metal scrap). One example of such trade distorting restrictions is Russia's export duties on wood materials, which have already disturbed established production lines and affected thousands of jobs in wood related EU industries. Moreover, the Commission points to the Russian, Argentinean and Indian export barriers for raw materials products. The EU imports about 70% to 80% of its primary raw materials. On average, these import costs add up to one sixth or even one third in plastics, paper and chemicals industry, of the price for the finished product. An increase of export duty can bankrupt a European business.[652]

The same happened in 2009:[653] For several years China has applied export restrictions – quotas and export duties – to key raw materials of which China is the leading extractor and exporter. These export restrictions distort competition and increase global prices, as some of these resources cannot be found elsewhere. The latter also means that EU companies that are dependent on such resources as inputs for further processing risk having to close down business,

---

649 BDI (2007), l.c., p 22. With the publication of the new communication on 4.11.2008 "Suggestion of a Raw Material Initiative" also proposals to change the situation in this sense are pointed out.
650 Cf. BDI (2007), l.c., p 22: Thus, the regulations of the GATT apply only between the WTO members. In the relationship to non-member states of WTO as well as concerning the disciplines of the WTO-exceeding regulations, bilateral or regional arrangements play therefore a determining role. According to arrangement of the contracts export restrictions can be excluded partially or totally.
651 Dow Jones & Company, Inc. (2008): EU threatens China with WTO action, 29.09.2008, Brussels. – According to press reports the U.S. plan a similar move.
652 Press Release, European Commission, Nov 2008.
653 EU request for a WTO panel on Chinese export restrictions on raw materials. Factsheet – Brussels, 4 November 2009.

6.4 Access to minerals outside Europe

with implications for downstream industries. Export quotas are made more restrictive every year, increasing the supply issue industry is facing. Moreover, for many of the raw materials under export quotas, China also imposes export duties. Downstream industries in China therefore have access to cheaper materials than their competitors outside China. That is not a level playing field, and the EU, Mexico and U.S. have now asked for the establishment of a dispute settlement panel at the World Trade Organisation.

Moreover, China applies export duties on 373 tariff lines at 8-digit level. The imports of these products to the EU in 2009 were worth an estimated EUR 4,5 billion in 2008. While China applies export restrictions to a broad range of products, the EU is at the moment challenging the following policies:

- China imposes quantitative restrictions on the export of bauxite, coke, fluorspar, silicon carbide and zinc.

- China imposes export duties of 10 to 15 percent on bauxite (depending on product), 40 percent on coke, 15 percent on fluorspar, 10 percent on magnesium, 15 to 20 percent on manganese (depending on product), 15 percent on silicon metal and 5 to 30 percent on zinc (depending on product).

- Restricting the right to export based on, for example, prior export experience.

- Establishing criteria that foreign-invested enterprises must satisfy in order to export that are different from those that domestic entities must satisfy.

China administers this system and these requirements through its ministries as well as chambers of commerce. The products concerned are used by the steel, aluminium and chemical industries. The industries in the EU that are potentially affected represents about 4% of EU industrial activity and around 500.000 jobs. They serve a multitude of sectors such as automotive, construction and fire retardants.

Export quotas without justification are prohibited under Article XI of the General Agreement on Tariffs and Trade (GATT) from 1994. In its accession to the WTO, China agreed to restrict the number of products subject to export tariffs, as set out in Annex 6 of its Accession Protocol. Annex 6 lists a total of 84 products at 8-digit level with a maximum duty rate that can be applied. Bauxite, coke, fluorspar, magnesium, manganese, silicon metal and some zinc products are products that do not appear on the Annex 6 list, but on which China nonetheless maintains export duties. Section 11.3, Part I of the Accession Protocol prohibits the application of export duties on products not appearing on the Annex 6 list. EU industries are dependent on imports of raw materials, as shown in Chapter 5 and are therefore vulnerable to distortions in world commodities markets. Export taxes accentuate vulnerability and directly reduce European industry's competitiveness and in some cases can even cut off European industry from essential inputs. Currently there is no level playing field for the European industry with their Chinese competitors.[654]

Trade methods of EU-partners, especially emerging markets, **increasingly distort competition on international commodity markets**.[655] The main problem for competition regulators within the EU is effectively working against these restraints of competition. According to the effects doctrine in competition law, a competition authority can only take action against those restraints of competition which show effects within its area of application. A practical restraint comes up in cross-border prosecution of cartel and misuse cases. European law prohibits the misuse of a market-dominating position, as is the case for market shares of more than 40%. Market shares between 50% and 100% can be deemed market-dominating without further ado, whereas shares below 40% can only be considered such if there are further indications.[656]

The Global Europe Strategy discusses this kind of restriction on the access to raw materials; these problems were also discussed at the G8 Summit at Heiligendamm in 2007. To ensure equal competitive conditions for all economic sectors affected, appropriate measures have to be taken.[657]

Moreover, **mergers of businesses can considerably decrease competition on the market**.[658] This can lead to higher prices or narrowed choice for the consumer and prevents innovation.[659] A good example is the former intended takeover of Rio Tinto by BHP Billiton.[660] Difficulties emerge when for the

---

654 The Council of the European Union endorsed on 28 May 2009 the Commission's Communication on a Raw Materials Initiative which puts access to raw materials at the top of the EU agenda. The trade strategy developed to implement this initiative urges for action to enforce international rules, if necessary through dispute settlement (se below).
655 BDI (2007), l.c., p. 10: This artificially produced pulling effect of trade flows of primary and secondary raw materials into third countries has already led to shortages, which entailed capacity reduction and workplace losses. Restraints on the world raw material markets which are initiated by enterprises as well as countries are naturally of transnational nature.
656 BDI (2007), l.c., p. 12.
657 EC, DG Enterprise and Industry, (2008), Consultation on Commission Raw Materials Initiative (2008). Concrete proposals to solve this problem are listed in COM (2008) 699 "Raw Material Initiative".
658 BDI (2007), l.c., p. 12.
659 Cf. also chapter 1.
660 E.g. Takeover of BHP by Rio Tinto, 1 Oct. 2008. Already referred to in chapter 3.

merging of mining companies different competition authorities work within their jurisdiction, which may lead to divergent decisions.⁶⁶¹

Figure 163: European Commission, Brussels

Mergers are inspected by the Commission, which acts as European competition authority. A merging can be refused for certain reasons: if the businesses involved in the merging realize sales revenues within the EU which exceed certain limits, as well as if the merging would lead to a market-dominating position (irrelevant of where the companies involved are based in).

6.4.2 Development policy

Development policy is very relevant to the topic discussed here. Many important mineral occurrences are located in developing countries. Political interest in minerals and the problems involved arose in the colonial era and have persisted to the present day.⁶⁶²

---

661 Cf. BHP-Rio Tinto: Australian antitrust authorities approve the proposed takeover. The decisions of the European, Canadian and South African antitrust authorities are still pending.
662 Schwarz, Johannes (1975): Rohstoffprobleme und Rohstoffpolitik der Entwicklungsländer. [Mineral problems and mineral policy of the developing countries] – Wien, Hochschule für Welthandel, Dipl.-Arb. (Diploma thesis), 1975. Thoma, Gottfried; Renner, Heinrich (1979): Die Bedeutung der Rohstoffpolitik in den Beziehungen zwischen Industrie- und Entwicklungsländern. [The meaning of minerals policy in the relations between developed and developing countries] – Nürtingen, Fachhochschule, Dipl.-Arb (Diploma thesis., WS 1978/79. Kohler, Wolf-Dieter (1979): Multinationale Konzerne und die Chance von Entwicklungsländern auf eine eigenständige Rohstoffpolitik: das Beispiel der CIPEC. [Multinational corporations and the chance of developing countries for an independent minerals policy: the example of the CIPEC] – Konstanz, Diss., 1979. Bomsel, Olivier (1987): Do the mining countries of the Third World have a future? – Natural resources forum, vol 11, no 1, Feb. 1987, p 59–65. Cf. weiters (Furthermore) Harms, Uwe [et al.] (1975): Berggesetzgebung und Rohstoffpolitik in Entwicklungsländern unter Berücksichtigung regionaler Schwerpunkte – Brasilien, Peru, Sambia, Zaire, Indonesien: Gutachten im Auftr. des Bundesministeriums für wirtschaftliche Zusammenarbeit. [Mining legislation and minerals policy in develop-

In the 1970s cooperation structures between the industry and developing countries were established. In particular, the Lomé Convention[663] helped develop this cooperation and can be seen as an exemplary model for future global or regional agreements for securing sales and stabilizing export prices for raw materials under maintenance of market-economic elements and limiting the financial commitment of industrialised countries. A significant number of developing countries admit to this kind of cooperation.[664]

Figure 164: Mining in development countries, example Africa
http://images.google.at/imgres?imgurl=http://www.frameweb.org/koimage.php

The First Lomé Convention (Lomé I), which came into force in 1976, provided a new framework of cooperation between European Union and developing Africa-Caribbean-Pacific (ACP) countries and specifically addressed mining. In the course of Lomé II, III and IV the Sysmin scheme was created, which provided compensatory finance to ACP states for adverse fluctuations of global mineral export prices.

In 1999 the ACP-report stated that Sysmin had in general suffered from the procedural difficulties handicapping both the evaluation of projects and the implementation of projects and programs. The new ACP-EU agreement,

---

ing countries under consideration of regional priorities – Brazil, Peru, Zambia, Zaire, Indonesia: Reports in behalf of the Federal Ministry for Economic Cooperation] – Hamburg: ite, Institut zur Erforschung technologischer Entwicklungslinien (Institute for the Study of technological development lines), 1975. Hentschel, Thomas; Hruschka, Felix ; Priester, Michael (2003): Artisanal and small-scale mining: challenges and opportunities. – London: IIED, International Institute for Environment and Development, 2003. Petterson, M.G., Marker, B. R., MCEVOY, F., Stephenson, M. & Falvey, D. A. (2005), Sustainable Minerals Operations in the Developing World, London. – See also: Maxwell, P. (2005): Thoughts on sharing Australian mining education with Latin America. AusIMM Bulletin, Issue 5 (2005), pp. 41–44 – Auty, R. M. (1998): Mining as a generator of wealth : potential conflicts and solutions. Minerals and energy – raw materials report, Vol. 13 (1998), 2, pp. 4–12 – Auty, R. M., ed. (2001): Resource abundance and economic development. Oxford Univ. Press, Oxford and New York, 2001, ISBN 978-0199246885 – Davis, G. A., Tilton, J. E. (2005): The resource curse. Natural resources forum, 29 (2005), 3, pp. 233–242.

663 Cf. Michaelis (1977), l.c., p. 78f.
664 Michaelis (1977), l.c., pp. 83, 84.

6.4 Access to minerals outside Europe

signed on June 23, 2000 in Cotonou was enacted for a twenty-year period from March 2000 to February 2020. Recognizing that instability of export earnings, particularly in the mining sector, may adversely affect the development of the ACP States, the new ACP-EU Partnership Agreement (2000) established a system of additionally granted support within the European Development Fund. This support substituted the traditional Stabex and Sysmin, previously operated to compensate the instability of export earnings in agriculture and mining sectors.

However, since the termination in 2000 of the 8th EDF SYSMIN special financing agreement, no further projects for geological exploration or examination of the mineral resources potential were funded.

Consequences are reduced investment into the mineral resources sector of ACP countries, lack of level playing field for investment on the base of sustainable development, transparency and accountability.[665]

It should also be stated that ACP countries are not among the leaders in most mineral raw materials.[666] As measured by the global production or even by the low production within the EU, the production of the ACP countries is marginal and **cannot** be seen as a fundamental contribution to mineral supply security.[667]

---

[665] Christmann (2008). – In the COM (2008) 699 "Raw Material Initiatives" is mentioned that the European Investment Bank (EIB) allots an annual support of about 140 million euros for mining projects in developing countries, especially in those developing countries where the projects are determined on the base of agreed government action plans. This should create transparency for the mineral industry] (p.8, 9). – See also: Mapouga, O., Maxwell, P. (2001): The fall and rise of African mining. Minerals and energy – raw materials report, 16 (2001), 3, pp. 9–26.

[666] Weber, L. (2008): Rohstoffsicherungsaktivitäten der EU und ihre Auswirkungen auf nationaler Ebene, Berg- und Hüttenmännische Monatshefte, (Journal of Mining, Metallurgical, Material, Geotechnical and Planned Engineering). p. 295. – See also Michaelis (1977) p. 78f.

[667] Ibidem.

Figure 165: Africa – Caribbean – Pacific (ACP) raw materials exporting countries 1990–1999 in % of total exports (Data by World Bank, Christman, 2008, Presentation, EU-Parlament)

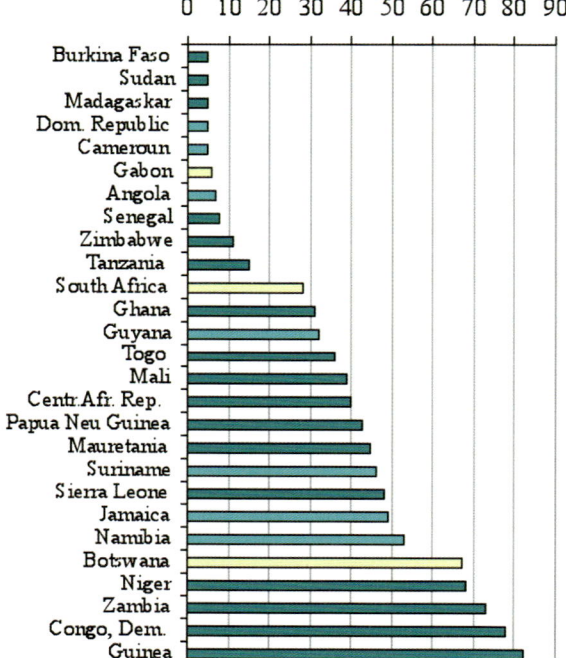

The colors represent the gross national income categories (GNI as of 2006)
- ■ low income (GNI < 2,5 $ per day per person
- ■ low to medium income (GNI < 10 $ per day per person)
- ■ medium to high income (GNI < 30,5 $ per day per person)

Figure 118 shows the 27 most important export countries of the 79 ACP countries. 17 of these 27 countries had a GNI of less than 2,5 $ per day per person, which is insufficient for developing institutions and capacities needed in the mineral raw materials sector.

An EU development policy would have to be compatible with the objectives set at the Earth Summit 2002 in Johannesburg.[668] The participating nations of this meeting highlighted the growing significance of mining for many developing countries concerning the demand for raw materials of the

---

[668] World Summit on Sustainable Development, Johannesburg, South Africa 2002 (WSSD).

industrial countries. The contribution of the mining sector to the sustainable improvement of developing countries lies in the political stabilization and the development of infrastructure. The so called Plan of Implementation, Chapter IV (§§ 24–46) is meant to take this into account. It aims to "protect and manage the natural resources base for economic and social development". § 46 (c) Plan of Implementation especially discusses the significance of mining within the development policy: Priority is given to the promotion of sustainable practices in the extraction and processing of raw materials by financial, technical and capacity-promoting means for developing and emerging countries. Furthermore, the input of scientific and technological information is meant to be increased. The reclamation and reuse of exhausted sites is important as well.

For the promotion of the realization of the Plan of Implementation, the Intergovernmental Forum on Mining, Minerals, Metals and Sustainable Development was established in 2005.[669] It is meant to strengthen the structures of a global dialogue.[670] Moreover, an evaluation of the Plan of Implementation (with an emphasis on raw materials) shall be executed in 2011.[671] The EU ought to join this global process within its development policy based on a comprehensive dialogue and appropriate actions.[672]

---

[669] Basically, an annual meeting of the Member States is planned. The last but one meeting of the Forum was held in Geneva at the Palais des Nations from 17 to 19 September 2007. The participants in the conference are international and come from Argentina, Bolivia, Brazil, Burkina Faso, Burundi, Canada, the Dominican Republic, Ethiopia, Gabon, Ghana, Jamaica, Kazakhstan, Kenya, Kirghizistan, Madagascar, Malawi, Mali, Mauretania, Mexico, Mongolia, Morocco, Niger, Nigeria, the Philippines, Rep. of Guinea, Romania, Russian Federation, Senegal, South Africa, Surinam, Swaziland, Tanzania, Uganda, Uruguay, Great Britain and Zambia. The European Commission, Germany, as well as multilateral commissions (UNCTAD, UNDESA, UNEP, ILO, World Bank and the metal study groups have participated as observers. The Annual General Meeting 2008 of the Forum was held in Geneva, at the Palais des Nations, jointly with UNCTAD, 24 to 26 November 2008.
[670] www.globaldialogue.info
[671] Cf alsoMojarov, A., Commodities Branch, United Nations Conference on Trade and Development (UNCTAD), Sustainable Development – an opportunity to bring it all together, World Mines Ministries Forum 2008 (Toronto, Canada, February 29 - March 2): The Johannesburg Declaration on Sustainable Development as well as the Johannesburg Plan of Implementation (JPOI) have been recognized by the UN General Meeting.
[672] Cf also European Commission (2008a), "Raw Materials Iinitiative", section 2.1, p. 6: A reinforced co-operation of the EU in the Johannesburg implementing plan is intended.

Chapter 6 EU minerals policy status quo – critical reflections

Figure 166: Mineral wealth of Africa including trade partners (Data by Die PRESSE, 2008)

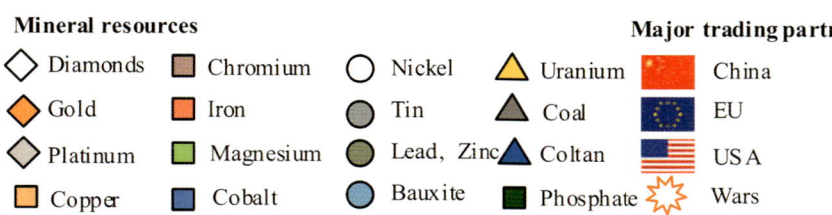

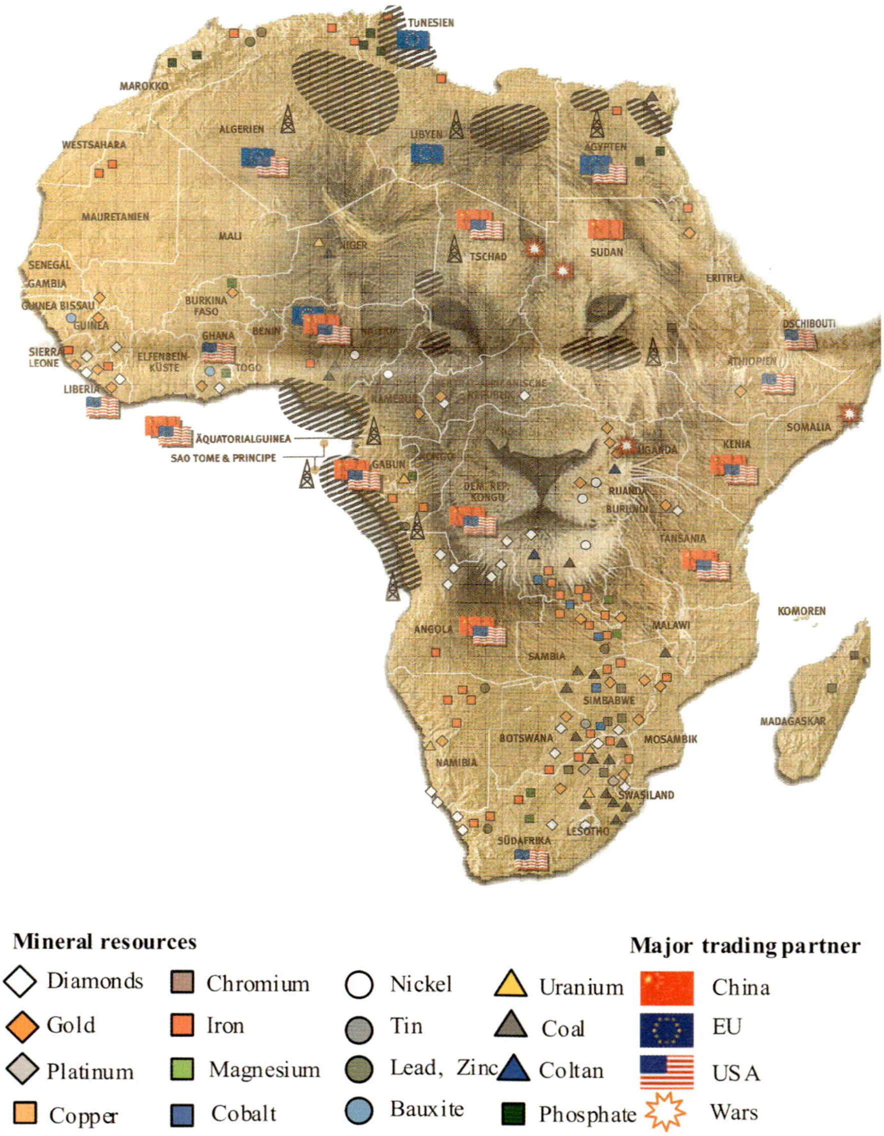

Mineral resources

◇ Diamonds   ■ Chromium   ○ Nickel   ▲ Uranium   🇨🇳 China
◆ Gold   ■ Iron   ● Tin   ▲ Coal   🇪🇺 EU
◇ Platinum   ■ Magnesium ● Lead, Zinc ▲ Coltan 🇺🇸 USA
■ Copper   ■ Cobalt   ○ Bauxite ■ Phosphate ✸ Wars

Major trading partner

6.5 Access to minerals inside Europe

In the last few years exploration and mining particularly of ores have been decreasing, as has generally the availability of mineral raw materials.[673] Moreover, licensing procedures have become very complex.[674]

6.5.1 Decreasing of exploration

Extraction of mineral resources inevitably leads to their eventual exhaustion at the site concerned. The industry therefore needs to find, and gain access to, new deposits to replace those that are depleted. However, it is a fact that exploration for deposits in Europe **has decreased heavily** in the past two decades. One major reason was the relatively cheap and reliable raw materials import situation.

Figure 167: Global exploration 1995–2008 (Data by Ekdahl 2008)

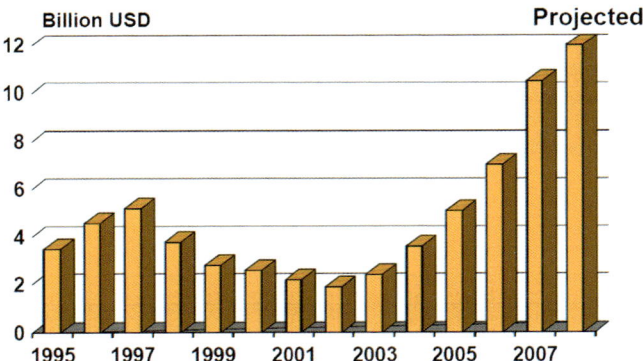

Since 1997 expenditures for explorations have been reduced continuously worldwide; in 2002 they reached a low at around 2 billion dollars. Exploration expenditures have been on the rise for a fairly long time now. In 2005 and 2006 they were at about 5 and 7,5 billion dollars respectively. An increase of production in global

---

673 Cf. u.a. Wörmann, C. (2006): Probleme der Verfügbarkeit von Rohstoffen. [Problems of mineral availability.] International markets for raw materials: the challenges ahead, Kali und Steinsalz, pp. 6–11.
674 Cf. Tiess, G. (2008): Need of coherent minerals policy in Europe – present discussion and approaches, Geología y Minería, number 3, Vol. 24, Moa, Cuba.

mining is mainly due to Asia with a plus of 1,2 billion tons; whereas Europe's contribution has been rather low.

Moreover, investments in exploration in Europe are amongst other reasons hampered by the lack of a clear national minerals policy and tedious permitting procedures and practices. As a consequence, the development of the European resources potential has not been able to keep up with international developments.[675]

The European raw materials industry finds itself under increasing pressure to find **additional reserves** at existing sites in order to prolong the lifespan of the infrastructure and to justify investments and expansions.

The ability of Europe's non-energy extractive industry of supplying existing markets and of contributing to global growth will depend on **additional resources becoming available** and is also important to reduce the EU's import dependence.

Furthermore interesting is the comparison between the Geological Surveys of the USA and the Eurogeological Survey.[676] The USGS is well in funds with 51 million dollars budget for the assessment of the minerals resources potential including 16 million dollars for the propagation of information to the US government and economy. EUROGEOSURVEYS however does not have such capacity and no funds available for such purpose. The national geological surveys are only loosely connected. The most recent European Minerals Yearbook was published in 1997, with data from 1995 and older (as mentioned above).

Table 76: Comparison – USGS and EuroGeoSurveys (Data by Christmann, 2008)

USGS	EuroGeoSurveys
Federal geological survey (USGS) with 51 M$ 2007 budget for the assessment of mineral resources potential including 16 M$ (160 full-time positions) for the provision of minerals information to US government and economy	No EU capacity, no budget. Only weakening, loosely coordinated capacities in Member States. Last edition of European Minerals Yearbook was in 1997, using data up to 1995

---

675 EC, DG Enterprise and Industry (2008), Public Consultation on Commission Raw Materials Initiative (2008).
676 Christman (2008), l.c., Geological Survey, comparison between USA and EU.

6.5 Access to minerals inside Europe

USGS	EuroGeoSurveys
Decades of federal attention to mineral resources issues	No competence given to EU Council conclusions calling for the development of a coherent political approach with regard to raw materials supplies for industry, including all relevant areas of policy

Recent trends in global exploration and mine production of most metallic and industrial minerals are upwards, as shown earlier, and this is expected to continue as the world's population is growing and countries such as China and India develop and demand more materials per capita (see chapter 3).

Figure 168: Global exploration in 2006 (Data by Ekdahl 2008; RMG, MEG)

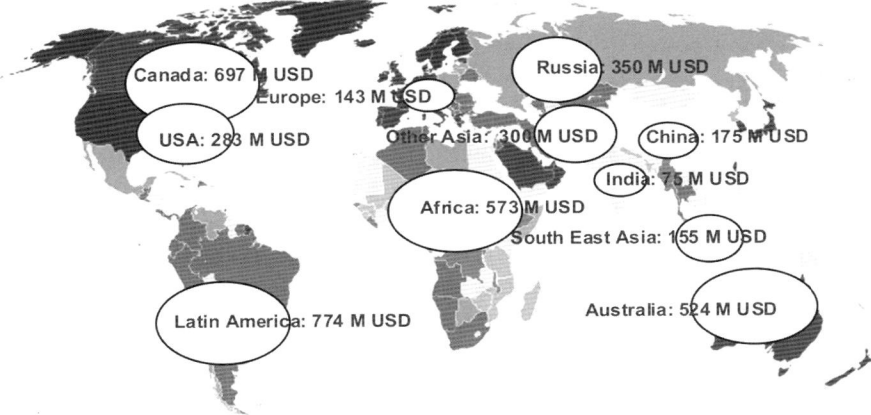

Presently, deep interest lies in the exploration of rare earth minerals. As mentioned above, these are special metals for certain applications such as hybrid cars, powerful magnets and fibre optic cables. Demand for several of these rare earth minerals has increased strongly because their use in technology has multiplied in the past decades.[677]

Toyota intended to produce at least one million hybrid cars in 2010; worldwide production might reach three million in 2012. Based on the current growth

---

[677] Braune, G. (2008): Begehrte Spezialmetalle – Seltene Erden Sind wegen der wachsenden Nachfrage nach Hybrid-Autos gesucht [Special metals in demand – Rare earths are requested because of the growing demand for hybrid cars], Handelsblatt, 12.8.08, Nr. 155. It is drawn upon the estimation of the US technical authority of the Geological Survey (USGS).

rate, the demand for rare earth minerals will surpass global production by 2011. The main problem is that China has almost a monopoly on the extraction of these minerals and limits exports by means of an export duty. Of China's total production of around 100.000 tonnes, only 40.000 are meant for exporting, which is about the amount the Japanese automotive industry would need. The growth potential for hybrid vehicles is very high due to increasing fuel prices and the climate change discussion, but there is no security of supply.[678]

Prices are rising considerably according to USGS and metal-pages.com. One example is dysprosium for magnets and electronic devices: its price rose from about 70 dollars per kg in January 2006 up to 110–115 dollars in 2008. Western industrialized countries now realize that they have not tapped new resources. Even the US Geological Survey states that the USA were able to meet their own demand ten years ago, but now are dependent on Chinese imports. Beyond that it should be mentioned that the potential for world-class deposits for some strategic minerals in Europe does exist. Most relevant is that the EU raw materials policy actively supports the development of this potential. Present developments at the Fennoscandian Shield are appreciable.[679] However, it remains to be seen whether total exploration in EU countries will increase again once the problems of the past years have been solved.

6.5.2 Decreasing of the availability of deposits

In the past years availability of deposits has continuously decreased. This phenomenon particularly occurs in densely populated countries in Western Europe (e.g. Belgium, France, Germany, Netherlands, and Austria).[680] A low awareness of mineral raw materials, especially poor raw materials planning policy,[681] and **an increasing number of prohibitions and requirements** of the EU environmental protection law as well as complex licensing procedures are responsible for this development.

---

678 www.kaiserbot-tomfish.com
679 See also chapter 5.
680 Cf. Kündig et al, (1997), Die mineralischen Rohstoffe der Schweiz [The mineral resources of Switzerland], Zürich. Dingethal et al, (1998), Kiesgrube und Landschaft [Gravel pit and landscape], München. Tiess, G. (2000): Rohstoffgewinnung im Spannungsfeld, Sand & Kies, [Mineral extraction in the area of conflict, Sand and gravel] H 48. Letouzé, G., Rossmann, H., Tiess, G., (2000), Mineralrohstoff-Gewinnung ohne Planung? [Mineral raw material exploitation without a plan?], Raum, H 39.
681 Raw materials supply security by land use planning: This means no comprehensive, according to spatial planning, coordination of mining areas with interests of human settlement, nature conservation, water economy as well as agriculture and forestry (see Chapter 4).

6.5 Access to minerals inside Europe

The availability of deposits is an essential indicator for minerals planning policy.

In general a continuous increase in land use density and land use rules and regulations in the EU countries can be observed.[682] Many of these regulations include prohibitions, restrictive rules for the access to raw materials. This reduces the availability of deposits; particularly if no comprehensive raw materials planning policy is involved.[683] Information about deposits and their categorization usually exists on part of the member states; however, often it is not taken into account by land use regulation. Consequences are excessive planning and building as well as an increase of imports (giant quarries in Scotland and Norway).

Access to land issues

Figure 169: Reduction of gravel reserves **due to other land use utilisations** (Data by Kündig et. al. 1997)

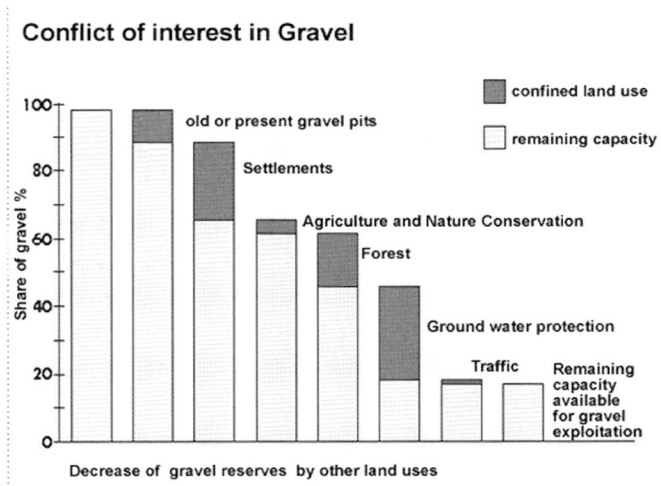

From a global geological perspective, there is no indication of imminent physical shortage of the majority of aggregates in Europe. However, geological availability does not necessarily mean access to these raw materials for the mining companies. Access to land (i. e. deposits) is increasingly influenced by policy issues, for instance, lack of mineral planning policies or obstacles like taxes and charges. Reduced availability of deposits is a phenomenon existing

---
682 E.g. Water Framework Directive as well as other spatial law-relevant EU directives.
683 Cf. Department of Mining and Tunnelling, University of Leoben (2004), Minerals Policies and Supply Practices in Europe, Final Report.

in all of Europe (and is impacting the competitiveness of the aggregates industry). However, the availability of aggregates from regional and local sources is essential for economic development, in view of logistical constraints and transport costs. Kündig et. al. (1997) stated that in Switzerland from an initial 100% of sand and gravel reserves about 10% is deducted in favour of outdated or present gravel pits, 18% in favour of residential areas, 5% for protection of nature and environment, 16% for forests, 30% for groundwater, 1% for traffic and only about 20% remain available for exploitation (see figure 169).

Figure 170: Aggregates situation in Germany (Data by BGR, 2009)

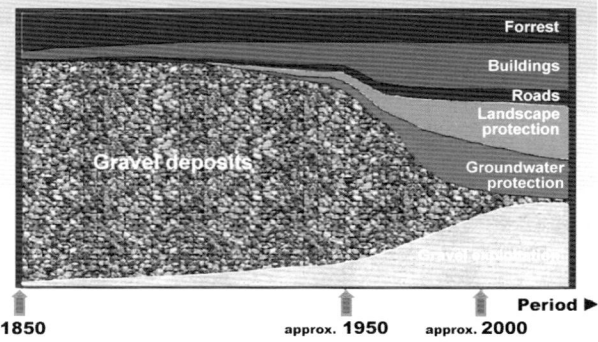

The importance of deposits protection (and the reconcilement of mining with other forms of use) by mineral planning policy is discussed in chapter 4. At the EU level so far no clear definition of raw materials planning policy has been found.[684] In terms of the Communication COM 2000/265 concerning the promotion of sustainable development in the extractive industry, the Commission considers "the need for land access to be an essential prerequisite for the further development of the industry and its relationship with regional and spatial planning that impact on this need".[685] Furthermore an EU-coordinated prioritization of the access to strategic raw materials with influences on land use planning is not available.[686]

---

684 At least until 2010: A suggested definition f national minerals planning policy was provided in the "Abridged report of the ad-hoc Working Group on "Exchanging Best Practice on Land Use Planning, Permitting and Geological Knowledge Sharing", presented at the European Minerals Conference in Madrid (June, 2010).
685 Cf. EC (2000): "Promoting sustainable development in EU non-energy extractive industry", COM/2000/265 Brussels.
686 EC, DG Enterprise and Industry (2008): Public Consultation on Commission Raw Materials Initiative (2008).

6.5.3 Complex permitting procedures

On an international scale Europe has developed very strict environmental standards.[687]

The **growing significance of the EU environmental legislation** since the 1990s has considerably affected the European raw materials industry. Time and expenses spent on various licensing and permitting procedures have increased substantially. In particular the uncertain outcome of the procedures has to be emphasized; this represents a serious problem for the investment security for the mining operator.[688]

This development must be questioned: on one hand there is a great demand for mineral raw materials, whereas on the other hand the industry's options for access to raw materials have been reduced significantly. In this sense, criticism has been expressed within the consultation process of the Raw Materials Initiative 2008.[689]

Table 77: Relevant EU-environmental directives concerning the raw material industry (extract)

1	Directive 85/337 – Effects of certain public and private projects on the environment
2	Directive 92/43/EC – Conservation of Natural Habitats and wild Flora and Fauna (FFH-Directive)
3	Directive 96/61/EC – IPPC: Integrated Pollution Prevention and Control
4	Directive 96/82/EC: On the control of major-accident hazards involving dangerous substances amended by Directive 2003/105/EC (Seveso II)
5	Directive 2000/60/EC – Water Framework Directive
6	Decision 2001/118/EC On the list of wastes (European Waste Catalogue)
7	Decision 2455/2001/EC On the list of priority substances in the water policy
8	Directive 2004/35/EC – Environmental liability with regard to the prevention and remedying of environmental damage
9	Directive 2006/11/EC On pollution caused by dangerous substances in aquatic environment

---

687 As noted above: Cf. Otto (1999), l.c., p.5.
688 Cf. Department of Mining and Tunnelling, University of Leoben (2004): Minerals Policies and Supply Practices in Europe, Final Report. Also: Grob (2005), l.c., p. 45.
689 The criticism focuses on the complex EU raw materials legislation and inefficient licensing procedures.

Chapter 6 EU minerals policy status quo – critical reflections

10	Directive 2006/21/EC on the management of waste from the extractive industries
11	Regulation 166/2006/EC on the European Pollutant Release and Transfer Register

The problems concerning complex and tedious procedures originate in the restrictive and partially inconsistent EU environmental protection law (related to the extractive industry). Also problematic are the permission procedures which often seem contradictory and are tainted with unsure results. Considering and assessing conflicts between mining and environmental protection, the latter is often favoured.

As a consequence, out-of-scale conditions might be imposed upon the entrepreneur.[690]

Figure 171: Impact of the EU environmental protection law on the raw materials industry

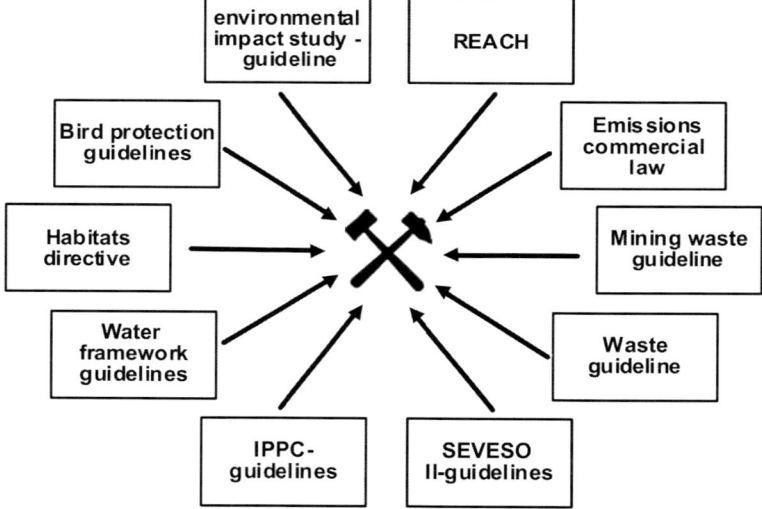

This is a problem because the raw materials industry is very capital intensive, which was also underlined during the consultation process for the Raw Materials Initiative 2008: It mentions the issue of lengthy procedures, unsure

---

690 EC, DG Enterprise and Industry (2008): Public Consultation on Commission Raw Materials Initiative (2008).

6.5 Access to minerals inside Europe

results and enormous costs against the background of an increased demand for raw materials.[691]

In the following, several problems are discussed more closely:

Directive on the conservation of natural habitats and of wild fauna and flora

Directive 92/43/EEC aims at the creation of a coherent European ecological network for the restoration or maintenance of a favourable conservation status of natural habitats and species. For that purpose, special protection areas have been nominated by the member states for designation. As reason three of the FFH-Directive, the Council of the European Union defines the support of the "preservation of bio-diversity" while simultaneously taking into account "economic, social, cultural and regional needs", as the main objective of the Directive.

Figure 172: Illustration of the proportion of land of the FFH-areas in the EU-27 (http://ec.europa.eu)

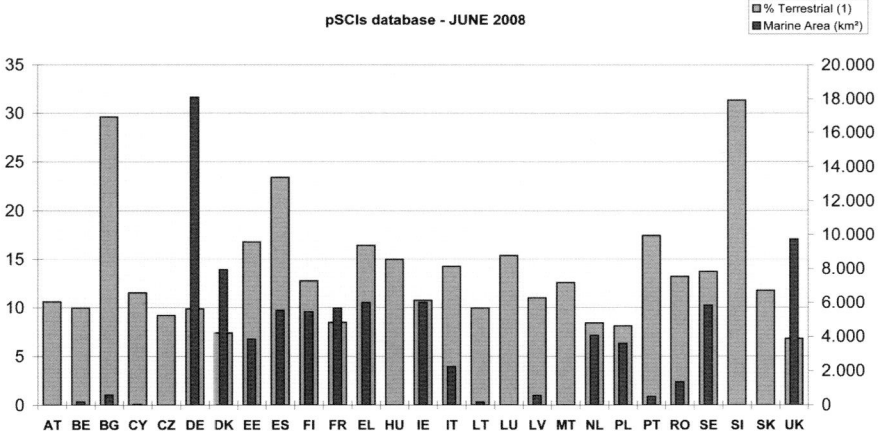

Dir. 92/43/EEC commits the countries to declare 10–20% of their territory FFH-areas.

---

691 EC, DG Enterprise and Industry (2008): Public Consultation on Commission Raw Materials Initiative (2008).): For example, the revision of the IPPC Directive should not lead to a double coverage of the mining waste directive (Directive 96/61/EC of the Council of 24 September 1996 about the Integrated Prevention and Decrease of the Environmental Pollution [IPPC]).

Large parts of FFH-areas contain potential mineral deposits.[692] According to Article 6 of Directive 92/43/EEC, any plan or project likely to have a significant effect thereon shall be subject to *appropriate assessment* of its implications for the site in view of the site's conservation objectives. In the light of the conclusions of the assessment of the implications for the site, the competent national authorities shall agree to the plan or project only after having ascertained that it will not adversely affect the integrity of the site concerned and, if appropriate, after having obtained the opinion of the general public.

Whilst the EU Directive does not create an absolute exclusion to activities such as mineral extraction, the implementation of the Directive by local authorities has lead in numerous cases to the sterilization of viable ore bodies and mines extensions. The Commission has produced a number of guidance documents on the application of Article 6 and also a few Member States (e. g. Finland) are producing guidance for their minerals industries on how to approach the problem. However, these do not seem to be applied by the local authorities.[693]

In view of the member states' obligation to designate protection areas on a national level as a part of the "coherent European ecological network of special protection areas with the title Natura 2000" according to Article 3 (1) Directive 92/43 the Directive at least has an indirect impact on the *future availability of raw materials*.[694]

Natura 2000 protection areas strongly compete with the raw material industry in the field of land utilization, because deposits that can be used for mining are often found in combination with undeveloped, mostly natural areas. Extractive activities depend on geology and the particular location of mineral deposits. As a result, access to suitable deposits is of crucial importance for

---

692 European Commission (2005): Evaluation of the 'Communication on Promoting sustainable development in the EU non-energy extractive industry', Interim Report , p. 34f
693 To improve the problematic situation, the following suggestion was made: The Commission and member states have undertaken the commitment to put up guidelines for industry and authorities. These shall indicate how mining can be balanced in or near to Natura 2000 areas regarding environmental protection. These were completed in 2010.
694 Cf. Christner, T. und Pieper, T. (1997): Bedeutung und Stellenwert 'nachhaltiger Entwicklung' bei der Gewinnung oberflächennaher Rohstoffe: ein Beitrag zur Wirkungsweise des umweltpolitischen Leitbildes eines 'sustainable development' auf planerische Abwägungsvorgänge und Genehmigungsentscheidungen im Rahmen der Rohstoffgewinnung [Meaning and significance of ‚sustainable development' at the exploitation of near-surface materials: a contribution to the effect of the environmental concept of a 'sustainable development' on planning assessment and approval decisions in the context of mineral extraction], Berlin, pp. 26, 27.

the future and competitiveness of the EU extractive industry. The designation of areas of land as Natura 2000 sites will usually prevent the extractive industry from exploiting any mineral resources on that land.[695] The comparison of protection areas with country size (see Fig. 103) indicates that between about 10% and 20% of the total national area in each Member State will ultimately be designated as a Natura 2000 site.

In terms of the access to land, the Habitats Directive is a space-oriented regulation that serves as a way to protect species and natural habitats. Thus it automatically impacts on *national spatial planning*[696], i. e. on *national minerals planning policies*.

Additional costs can result as a consequence of an impact assessment, which is obligatory for mining projects in (or in immediate vicinity of) Natura 2000 areas. Such costs could be a problem especially for smaller businesses. Thus a crucial aspect is to carry out an appropriate assessment (according to Article 6 Directive 92/43) in an *efficient way*. The approach to appropriate assessment adopted in the Baden-Württemberg Nature Conservation law is considered to be a good example. This has proven true several times already.

Table 78: Summary of responses from the non-energetic extractive industry and national geological surveys to the request for information on the current impact of Natura 2000 on permits for minerals extraction (Data by EC, DG Enterprise and Industry, Commission Staff Working Document, 2007)

Austria – A few aggregates companies are affected, although it is not clear if these are existing sites or proposals. Guidance has been produced explaining the requirements for activities affecting such sites.

Czech Republic – Current mining operations have not been affected by Natura 2000. However, a significant number of gravel and rock resources are within or adjacent to Natura 2000 sites and any future applications for a permit are expected to be rejected.

Finland – There are no exact figures on how the situation has already influenced exploration or mining, but companies will not risk money on exploration if the chances of starting mining are small.

France – A few proposals for extensions to existing sites and one renewal of a permit have been delayed, while in one case the area sought was reduced. The key concerns appear to be the time and costs emerging through this situation.

---

695 The Habitats Directive does, however, allow for reasons of national importance development to impact on the integrity of notified sites.
696 Christner/Pieper, 1997, p 27.

Germany – In one region 80 cases are said to be affected, although the nature of the difficulties has not been stated, and the outcome has yet to be decided. Another region indicated that eight potential sites had been excluded from the Regional Plan because of Natura 2000 designations. Problems were also recorded with a gypsum quarry. At another site, initial difficulties in extending a permit were resolved although this had required much time and money.

Greece – While no existing permits were found to have been affected, there is concern that large areas of the Greek mountain chains (e. g. Olympus-Pindos in central/western Greece and Rhodope in north-eastern Greece) contain important mineral deposits but have been designated as Natura 2000 sites. A number of permit applications have been rejected.

Ireland – No active industrial extraction site has reported any negative impact as a consequence of Natura 2000.

Lithuania – New permits are not being granted on land notified as a Natura 2000 site. However, extraction can continue where a permit already exists. No permits have been revoked or suspended as a result of an area being notified as a Natura 2000 site.

Portugal – There are no legal constraints on exploration or extraction activities in protected Natura 2000 areas if the activity takes place in previously defined areas. However, the existence of Natura 2000 areas makes mining activities on such land impossible, because the national institution that manages the network considers that the environmental values are incompatible with mining. Mining activity in these areas is therefore limited because of the investment risks.

Slovakia – The national list of Natura 2000 sites is still being produced. No conflicts with extraction have been noted.

Spain – The industry expresses the opinion that site selection was hurried and lacked a consistent approach across the autonomous regions. It has been suggested that the boundaries of some notified sites were set to stop future industrial development, including mineral extraction, but this has not been substantiated.

UK – Constraints have been placed on the extraction of clay resources which are overlain by peatland

One international company which had carried out a quick survey of its members concluded that none of its quarries had been directly affected. There were examples of boundaries of designated areas being very close to the limit of the mineral concession (in France and Sweden).

Proposal for an EU Soil Protection Framework Directive

One example for the continuative development of the EU environmental protection law is up for discussion: the proposal for a soil protection directive. This proposal includes measures for avoiding the worsening of soil quality and for restoring and rehabilitating damaged soil. Because of the different targets of soil protection and raw materials extraction this directive is of great importance. The extraction of raw materials is inevitably connected to claiming Earth surface. Impairments, however, are only temporary because the surface can be reused after the extraction of resources. Still this proposal might bring about further conditions for mining operators. An EU Soil Protection Framework Directive with respect to subsidiarity seems unjustified.[697] Furthermore, a European soil protection law would have to be oriented towards use.[698]

Environmental Impact Assessment Directive

Directive 85/337/EEC: On the assessment of the effects of certain public and private projects on the environment amended by Directive 97/11/EC (EIA – Directive)

The main objective of this Directive is to ensure that projects or proposals that are likely to cause significant effects on the environment are carefully considered in a publicly transparent manner before the competent authority issues a permit.

An environmental impact assessment (EIA) is compulsory for extensive projects in the extractive industry (Annex I).[699] All other projects, which fall under Annex II, are not obliged to undergo such assessment. Art. 4 Dir. 97/11/EC stipulates verification of individual cases and threshold values and criteria defined by the member states which decide on the necessity of an EIA.

The time it takes to receive permission for an EIA can vary considerably. Main reasons for delay are the involvement of several public authorities and the involvement of the public.

Experience shows that *especially* the preparation of EIA is a complex issue and tends to take up much time and also much of management's attention. It can lead to long approval periods and costs for mineral projects. There are two cost aspects. The first concerns the cost of the preparation of the project proposal and the cost of the environmental impact assessment and associated investments.[700] The second aspect is the capital cost. A feature of minerals projects is that considerable costs are incurred prior to the production phase.

---

697 Cf. VRB (2008), l.c., pp. 45, 46. It likewise represents this view (including the German industry).
698 After the European Parliament had at first reading called for far-reaching, intensified demands the consultations were continued in the council of ministers of environment. In particular, the negative attitude of Germany, France and the United Kingdom led to the fact, that the council did not achieve the desired agreement by end of 2007.
699 See Directive RL 97/11/EG, appendix I, item 19.
700 In the case of the diabas quarrry referred to on the text this cost amounts to 10 % of the project cost.

These include the cost of prospecting and exploration as well as the cost of acquiring property rights and site preparation and establishment. Any delays in production result in a higher cost of interest on the capital already spent. For small operators in particular, this can be a serious issue. In some instances long observation periods may be necessary to determine seasonal aspects of the environment.[701] The time taken to complete EIAs *can* result in regional shortages of mineral reserves, thus necessitating the transport of minerals over greater distances with the associated higher costs to the user and impacts on the environment. The diabase mining operation in western Austria mentioned earlier is such a case.

A key element of the authorization process is whether or not a project application requires an *Environmental Impact Assessment*, *Directive 97/11/EC*, which lays down criteria for environmental impact assessments, is ambiguous as far as the extractive industry is concerned. Annex I, (19) specifies for quarries and surface mines a threshold value of 25 hectares and for peat production a value of 150 hectares. In Annex II, (2), which covers other quarries and surface mines, specific threshold values are not given, leaving it to member states to set threshold values and criteria in order to decide on the need for an environmental impact assessment.

Directive 97/11/EC can lead to a distortion of competition. A survey of member states has shown that there exists no common pattern as far as environmental assessments are concerned. The span of threshold values ranges from 5 hectares in Ireland and Portugal up to 500 hectares in the case of state owned minerals in the Netherlands. With regard to marine aggregates, Ireland and the Netherlands make an EIA compulsory for all project applications. Irrespective of defined threshold values it has become practice in some member states to subject all applications for extraction licenses to an EIA. Examples are Greece, Norway, Portugal and quarrying operations in France.[702]

The member states' attitude is therefore crucial and determines whether there is a competitive environment for mineral producers from different member states. Although Directive 85/337 is similar in all Member States, significant divergences in the field of threshold values and criteria can be observed. Different threshold values, set by the member states, can cause unfavourable competitive market conditions between and within member states in case of

---

701 Stempkowski R.: Beitrag zur Weiterentwicklung der UVP (Contribution to the development of EIA) [online]. ÖIAV-Arbeitskreis. 2002, Available from: http://www.oiav.at/pdf/uvp-2002.pdf [Accessed 27.05.2004].
702 Department of Mining and Tunnelling (2004): Minerals Policies and Supply Practices in Europe, Final Report, University of Leoben.

such minerals which are traded on international markets (e. g. in terms of time and money needed to carry out an Environmental Impact Assessment).[703]

Thus, in reference to the transposition of the Environmental Impact Assessment Directive into national legislation, there can be a negative attitude of potential investors, who fear long duration, high cost EIA procedures of uncertain outcome. This can distract from investment in minerals projects

REACH – Regulation and Waste framework directive

REACH is a new legal basis for the registration, assessment, permission and regulation of chemicals. This regulation came into effect on July 1, 2007. Its main goal is the improvement of the protection of human health and environment from possible chemicals-related risks. According to REACH, the industry is responsible for assessing and minimizing such risks and supplying the consumer with appropriate safety information.

This regulation might apply to many mining products including imported mineral raw materials as well as mining and processing waste. REACH might classify many industrial minerals and metals (including metals contained in scrap) as dangerous chemicals which so far have been considered harmless by the producers. This might lead to certain problems, such as for example the possibility of classifying scrap as hazardous material.[704]

Figure 173: Important raw material – metal scrap

---

703 Department of Environment: Mineral Planning Policy and Supply Practices in Europe – Main Report. HMSO, London, 1994, pp 61–64: Member States that link the requirement to carry out an environmental assessment to specific threshold values and criteria do again differ as far as transposition is concerned. – Compare also: De Lespinay, Y., Toward a European policy for the access to mineral access, in: Mineral Planning in a European Context, 1997, pp.13–14.
704 Together with other raw materials industry representatives, German steel producers tried to make clear that for example scrap is not defined as a hazardous substance under REACH: Alliance of German steel recycling companies and the German Steel Federation, 2006, in: USGS (Germany) (2008), l.c., p. 3.

The Waste Framework Directive brings up even more problems concerning waste. So far metal scrap has been classified as "waste" and belongs to the regulatory area of the European Waste Framework Directive. The classification as waste brings about great administrative efforts, severe restrictions and therefore high costs for the businesses.[705]

Against the background of the enormous importance of scrap for metal recycling – 40 to 60% of the total production – this is particularly absurd. The European Waste Legislation should aim at the environmentally friendly processing of waste as a contribution to the sustainable use of resources. Moreover, nationally different ways of implementing European standards distort competition. The implementation and organization of European regulations for product waste, like scrap cars, often are the member states' responsibility.[706] The current revision of the Waste Framework Directive should result in a change of this structure. Environmental criteria require a uniform legal basis.

The revised version of the EU Waste Framework Directive now seeks to eliminate existing legal uncertainties in the classification of materials as waste and in the definition of recovery and disposal operations.[707] Raw materials extraction companies are particularly affected. Whether mining wastes (e. g. tailings) accumulating in the extraction process are regulated by the Waste Framework Directive, depends on the definition of "waste". The perpetuation of the definition of waste is of great importance because it ensures that mine residue used for rehabilitation is not classified as waste. This specification is also relevant for the definition of mining wastes. Within the revision of the Waste Framework Directive a five-grade waste hierarchy including prevention, reuse, recycling, other recovery and disposal shall be introduced.

EU emission trading legislation

Emission trading is a major aspect of climate policy and poses fundamental problems for the European minerals industry.[708] The European Emission

---

705 BDI (2007), l.c., p. 20.
706 BDI (2007), l.c., p.20: Thus German metal recyclers must pay value added tax for the purchase of scrap, unlike their counterparts in Spain or Italy. This represents a significant distortion of competition. In 2006 19,6 Mt of steel scrap was needed by the German steel recycling companies, the price depending on the specification on average is around 200 €/t.
707 VRB (2008), l.c., p. 43. After the approval of a mutual point of view by the Council, the European Parliament held consultations in close coordination with the Council and approved of the draft with amendments in a second reading.
708 Hartung, M. (2008): Rohstoffe und Energie im Spannungsfeld von Klimaschutz und Versorgungssicherheit. [Raw materials and energy in the interaction between climate protection and supply security.] Bergbau: Zeitschrift für Rohstoffgewinnung, Energie, Umwelt (Journal of raw materials, energy, environment), p. 338–342.

Trading System (ETS) is a political instrument of the European Union to reach its climate protection targets, the reduction of greenhouse gas emissions, as stated in the Kyoto Protocol. It covers electricity production and several industrial sectors like cement or steel production in 31 European countries, which together are responsible for about half of the European CO_2 emissions. The first multinational trading system for emissions came into force on January 1, 2005 and precedes a possible global system. At the moment (2008/09) negotiations for Phase III are conducted.

Since 2005 businesses have had to show the so-called CO_2-certificates, which are assigned free of charge until 2013, then they will have to be purchased.[709]

The directly affected industrial sectors relevant to raw materials in Austria are the magnesite,[710] cement, steel and paper industries because of their considerable consumption of energy and materials. For the analysis of Austrian emission trading from 2013 to 2020 Kearney generated two scenarios.[711] The first one assumes a "business as usual"-situation, i. e. the acquisition of certificates for about 122 million tonnes of CO_2 (71 % of the certificates needed). The second scenario includes the EU target which leads to the purchase of certificates for 78,6 million tonnes of carbon dioxide. However, these figures do not include energy industry. Kearney predicts additional costs of 3 to 4,7 billion euro for the Austrian industry from 2013 to 2020, 1,7 to 2,7 billion of which fall upon the iron and steel industry, 0,4 to 0,6 billion upon cement and 0,3 to 0,5 billion upon paper industry. Analysis of these figures predicts a drop of business success and profitability by 45 % in the iron and steel industry, 40 % in the cement industry and severe losses in the paper industry. Therefore the industry may have to put up with losses in competitiveness of 5 to 9 % unless regulations will be adapted to these branches by 2020. According to Kearney, the EU should find a balance between competitiveness and climate protection in order to prevent the migration of businesses to foreign countries.

---

709 For further details, see Reisinger, M., Tiess, G. (2007): Das Protokoll von Kyoto und sein potentiellen Auswirkungen auf den internationalen Bergbau [The Kyoto Protocol and its potential impact on international mining], Berg- und Hüttenmännische Monatshefte, (Journal of Mining. Metallurgical, Material, Geotechnical and Planned Engineering), pp. 108–114.
710 Dr. Thomas Drnek, RHI AG (2008a), Magnesit und CO2 [Magnesite and CO_2], Österreichische Akademie der Wissenschaften [Austrian Academy of Sciences], Wien 15. April 2008.
711 Cf. Seminararbeit von Heise T. (2008) (Term paper by Heise T.): Emissionsrechtehandel, Vorlesung Umweltaspekte des Bergbaus, Montanuniversität Leoben. He refers to A.T. Kearney (2008): "CO_2 certificates cost Austrian industry up to 4,7 billion euros", Press release.

Figure 174: RHI AG – magnesite extraction (Data by Drnek, 2008)

More than 1,2 million tonnes of magnesite and dolomite are produced in Austria, Turkey and Italy every year. About 50% of the production is extracted by RHI in its own sintering and smelting furnaces.[712]

RHI produces about 50% of the raw materials in its own sintering and smelting furnaces. Recently RHI started a new smelter in Isithebe, South Africa. Shortly the first stage of development of RHI's new Chinese raw material factory starts the production of magnesia-raw materials in Dashiqiao (80%:20% joint-venture).

The EU's post-Kyoto target is a reduction of carbon dioxide emissions by 20% compared to 1990. Currently there is only little potential for further reduction of greenhouse gas emissions by RHI in Austria. Presently a reduction of emissions seems technologically almost impossible or connected to disproportionately high expenses.[713] The target value of CO_2 reduction will only be reached through a reduction of production unless legislation creates fair terms and conditions concerning competitiveness for energy-intensive business. Therefore, production costs would no longer be covered and Europe might lose its position as a production site of refractory materials. Production of these materials will be shifted to countries without emission trading.[714]

---

712 Drnek (2008a), l.c. – In addition, as mentioned in Chapter 5, but in this context again worth mentioning: Austria takes world rank 6 with its magnesite mining.
713 Ibidem.
714 Ibidem

Figure 175: Cement plant Retznei (Ehrenhausen), Lafarge Perlmooser Gmbh (Data by http://www.lafarge.at)

With the EU regulations in effect as planned, in 2020 domestic cement businesses would not even be able to purchase the certificates needed for supplying domestic demand according to the Verein für Österreichische Zementindustrie (VÖZ).[715] The missing materials would have to be balanced with imports. Therefore the cement industry demands for the acceptance into the EU protected industries to prevent imports from third countries. Major cement companies like Lafarge question their investments in Europe.

The EU plans for CO_2 reduction could also pose major problems for Voestalpine.[716] As of 2013 an additional cost of more than 200 million euro annually would have to be spent. Voestalpine emits eight Mt of CO_2 per year, which is one-tenth of Austria's total emission. In steel industry reduction possibilities are low for physical reasons. The current CO_2 discussion has lead to a reconsideration of VOEST to relocate a steel plant planned to be built in Aus-

---

[715] 21. Mai 2008 – Die Presse "2020 können wir Zementbedarf nicht selbst decken" – "Zwischen 2013 und 2020 soll die Zementindustrie schrittweise keine Gratiszertifikate mehr bekommen". ["By 2020 we cannot meet demand for cement by ourselves" – "Between 2013 and 2020 the cement industry should gradually get no more free certificates".]

[716] 20. März 2008 – Wirtschaftsblatt – CO_2-Zertifikate würden Voest mehr als 200 Millionen € kosten. [CO_2 certificates would cost the Voest more than 200 millions €.]

tria. Ukraine would be a possible candidate when considering environmental restrictious having important iron ore deposits. Companies need planning reliability. Thus, according to VOEST, the EU would have to come up with clear ideas for solutions.

Final comments regarding the legal approval process

The complex and partially inconsistent EU environment law leads to complex approval procedures. These procedures often tend to be inefficiently coordinated, which also affects their duration, which can be seen as an indicator for the *competitiveness* of the raw materials industry.[717]

The time required for extraction permission varies *considerably*. It ranges from a few months to several years and usually exceeds the time specified. Reports from Member States indicate that the time required for obtaining extraction permission is significantly shorter if the application concerns a mineral deposit that is situated in a designated mineral extraction area. The main reasons for time delays are the involvement of many different authorities in the licensing procedure and the involvement of the public in certain elements of the approval process. Experience shows that the preparation of Environmental Impact Assessments (EIAs) is a complex issue and tends to take up much of the time. In summary the main reasons for time delays are the following

- the involvement of many different authorities in the licensing procedure
- inefficient procedures
- minerals not being considered in land use plans
- the involvement of the public in certain elements of the approval process.
- appeal procedures
- Environmental Impact Assessments EIAs

Some examples are given below.[718] The time required for obtaining a mining permit in Austria may take 1–10 years. The time required to obtain a mining permit in Germany may take 1–8 years depending on the federal state and the sensitivity of the area. The procedure for a mining permit in France can take several years. In Greece, expanding an existing mining operation is preferred to a new mining venture, especially for construction materials. The time required for obtaining the mining permit depends on many factors. Currently

---

717 High costs, expenditure and, in addition, an doubtful procedure outcome are a fact meanwhile: Consultation process to the Raw Material Initiative in 2008.
718 All information from Department of Mining and Tunnelling, University of Leoben (2004): Minerals Policies and Supply Practices in Europe, Final Report (see also chapter 5).

average times for permitting existing operation frameworks, i. e. opening a new mine next to an old one may take 3 to 5 years. Expanding an existing mining operation is preferred to a new mining venture, especially for construction materials. The time required to obtain the mining approval/permit is process dependable. It is important that the process (including public engagement [or consultation]) is complete and that consensus is built.

In Italy a number of laws regarding raw materials exist. This leads to a complex approval procedure and many contradictions. In practice, since the holder of a mining license (state-owned raw materials) has to request further permissions from national, regional or local authorities the actual start of business is much later than stated in legal regulations.

The time required to obtain permission to extract minerals in Spain can be lengthy, and frequently exceeds the specified times. Part of the difficulty is the involvement of many different agencies in the authorization process.

In the Netherlands before the revision of the Excavation Act in 1996, it was necessary to obtain a mineral permit and then to adjust the local land use plan (municipal level). The regional spatial plan (provincial level) also had to be amended. As a consequence, the process was very time consuming. Since 1996, the procedures mentioned above operate in parallel. The local government is asked to adjust their local land use plan if "extraction sites" are designated in the regional spatial plan. If the local government refuses to cooperate they can be forced to. The estimated time for this procedure is about 45 months. Even if the local government cooperates, the estimated time for this procedure is also 45 months. If no "extraction sites", but "extraction zones" are designated in the regional spatial plan, the estimated time for this procedure might take up to 5,5 years. Much more time was needed before the revision of the Excavation Act. The use of procedures in parallel has shown to be very beneficial.

6.5.4 Consultation process of the Raw Materials Initiative 2008

The above remarks were also discussed during the consultation process of the Raw Materials Initiative in 2008, accomplished by the European Commission (DG Enterprise).[719]

The consultation process is based on the Commission (DG enterprise) paper "Analysis of the competitiveness of the non-energy extractive industry"

---

[719] http://ec.europa.eu/enterprise/newsroom/cf/itemlongdetail.cfm?item_id=1249. Has been and is going to be referred to several times within this paper.

(2007).[720] After the publishing of this paper, the Competitiveness Council demanded the Commission to develop a coherent concept to guarantee secure minerals supply for the EU industry including all relevant policies (foreign affairs, trade, environmental development and research policy).[721] Furthermore, suitable measures for a reliable access to and production of natural resources, secondary raw materials and usable waste materials, particularly concerning the markets of third countries, are to be specified. The international dimension of the "raw materials question" finally was discussed at the G8 summit in Heiligendamm from June 6 to 8, 2007. As a result, a declaration on the "Responsibility for Raw Materials: Transparency and Sustainable Growth" was published.[722] After that, an online consultation process (from January to March 2008) started, in which five challenges were defined:

The G8 Summit declaration of Heiligendamm stated that *"free, transparent and open markets are fundamental to global growth, stability and sustainable development"* and called for a strong commitment to the principles of free trade and a further strengthening of the multilateral trading system. Furthermore, it announced the intention to promote global applicability of and compliance with WTO rules, also with regard to trade in primary and secondary mineral raw materials and calls on their trading partners to refrain from trade restraints and competition distortions in contravention of WTO rules.

Figure 176: EU Commission, Brussels (Data by Wikipedia)

---

720 Available on http://ec.europa.eu/enterprise/steel/index_en.htm.
721 Vice-president Verheugen (2007), Securing raw material supply for EU industries; Press release, June 5, 2007, Brussels.
722 http://www.g-8.de/Webs/G8/EN/Homepage/home.html

In 2008, the consultation process of the Raw Materials Initiative took place before the background of the international mineral markets. In consideration of the last 50 years this is a remarkable fact.

Various organizations (e.g. member organizations of BUSINESSEUROPE, the Confederation of European Business) suggested a Communication of the European Commission dealing with the issue of European raw materials supply security. At the same time the options for action in trade, foreign, development, environment and research policy should be presented. As a result, an online consultation process was started, which stated **five challenges**:[723]

1) improving supply with raw materials from domestic deposits on a sustainable basis
2) securing sustainable and transparent supply from third countries
3) encouraging capacity building in third countries
4) improving efficiency in the use of resources
5) establishing an adequate EU knowledge base on raw materials

The evaluation of the consultation process confirms feedbacks from 240 stakeholders, i.e. 172 organizations and 68 individual persons from the public and private sector.[724] 24 of the 27 European Union countries participated. Challenge 1 and challenge 5 were given first priority; and then challenge 4, challenge 2 and challenge 3 were given normal priority.

Regarding challenge 1, almost one-third of the participants call the lack of raw materials planning policies very problematic (i.e. access to land, environmental issues) because it reduces raw materials production from domestic deposits. Furthermore, the access to raw materials (i.e. use of deposits) is deemed important by 32% of the participants. In comparison with raw materials disproportionately much importance is attached to environmental protection, making access to raw materials hard or even impossible. Moreover, time-intense and costly requirements and conditions within the approval processes and their unsure outcome are criticized.

---

723 Cf. auch Weber (2008): Rohstoffsicherungsaktivitäten der EU und ihre Auswirkungen auf nationaler Ebene. Minerals safeguarding activities of the European Union and their impacts on national level. – BHM Berg- und Hüttenmännische Monatshefte, (Journal of Mining. Metallurgical, Material, Geotechnical and Planned Engineering). 153. 2008, H. 8, pp. 289–295.
724 http://ec.europa.eu/enterprise/newsroom/cf/document.cfm?action=display&doc_id=778&userservice_id=1&request.id=0

The accumulation of global production (48.8%) is considered very problematic. Also alarming is the distortion of competition (30.4%), insufficient effectiveness of international means (20.8%) and the lack of market transparency (50%). Political instability and risks and insufficient infrastructure in developing countries are relevant. The lack of long-term policy (62.9%), the consumers' unawareness of the advantage of high-quality products (52.1%) and high production costs for the implementation of new technologies (44.2%) were mentioned. Regarding challenge 5 the heterogeneous (43.3%) and incomplete data records (41.3%) are considered problematic.

Chapter 7 Towards a European minerals policy

7.1 General view

The numerous present problems should make us realize that we must not neglect the problems to come in near future. They should make us understand how much an adequate preparation can make easier the solution of difficult problems if it is not possible to avoid them. The point is to recognise that the size of the problem exceeds the national framework of the individual member states and that its importance requires creating an "anchorage" for all the wide-spread activities and actions concerning mineral raw materials which are being carried out all over the European Union. Necessarily a coherent minerals policy should be positioned on the EU level, providing frameworks for the establishment of national minerals policies. The member states are responsible for the organisation of their individual minerals policies

In comparison to current status quo the European Union should aim for an active minerals policy. This is to be understood in an economic sense, i. e. (with reference to the definition presented in Chapter 4) to take measures which produce a cost-optimal contribution to the EU Gross Domestic Product. To acknowledge this fact, the EU should decide to develop **structures of an active minerals policy**. This means to carry out a paradigm change, away from a passive role to the active "minerals policy-protagonist". For the purposes of a **coherent approach** (as well as according to the definition presented in chapter 4 minerals policy) has to be meshed sufficiently with the component policies concerned.[725]

Issues of minerals supply security have to be treated at three levels. The supply of raw materials primarily is a task of the enterprises. The responsibility of the EU Member States (national level) is to establish an optimal minerals

---

[725] With the publication of the communication COM (2008) 699 on 4.11.2008 (EC, DG Enterprise and Industry) effective accents have been set on part of the Committee. The suggestion of a raw material initiative is stimulated, including aims and measures. – The following conceptions have also directed their focus upon sustainable protection of the raw materials supply of the Community. However, they have been developed from **a personal point of view of a distinct perspective** and should be understood as **complementary suggestions to the** Communication.

policy framework, while it is the duty of the European Commission (EU level) to establish a coherent approach for raw materials protection in Europe, which exceeds the field of responsibility of the Member States.

Regarding non-energy mineral raw materials – metallic minerals, industrial minerals, construction minerals, **adapted minerals policies must be considered** and developed. As a consequence of their respective geological origins, the occurrences of the main types of non-energy mineral raw materials are not regionally distributed in a homogenous way, neither is their availability always granted. There is also a discrepancy between the location of reserves and the places of consumption, either for primary needs of the population or for the demands of manufacturing industries (particularly related to metallic minerals and partly to industrial minerals). Whereas Europe is to a high degree self-sufficient in construction minerals, it is a major importer of metallic minerals concentrates, as its domestic production is limited to about 3 % of global production. Concerning industrial minerals, in 2006, for instance, the European Union was the world's largest producer of feldspar (60 %), perlite (54 %), kaolin (31 %), gypsum (23 %) and salt (22 %). Nevertheless, the EU is a net importer of industrial minerals as a whole.[726]

An adequate minerals policy must respond to the manifold conditions for the supply situation of each commodity:

The exploitation of **construction minerals**, which are usually easily available and widely spread in Europe, often leads to conflicts in land-use and landscape protection – particularly if not sufficient considered by land-use planning management, as such mining sites require (temporarily) relatively large areas and may come along with various emissions near human settlements. Consequences are lengthy licensing procedures, which influence the feasibility of deposits and may even lead to sterilisation. On the other hand, the availability of construction minerals, avoiding too long ways of transport, is of great importance for the economic development of all countries. Therefore it is important to define priority aims in land-use planning, incorporating the environmental factor, and thus to provide a positive climate for investment in the extraction of construction minerals.

Industrial minerals are needed in large volumes for various manufacturing and production industries. In some cases, as already mentioned above, the EU is a globally important producer of such minerals, whereas the supply with others (e. g. graphite) from European deposits is by far not sufficient. This serious dependence on imports from non-European countries leads to uncertainty

---

726 EC, DG Enterprise and Industry, Commission Staff Working Document, 2008.

of supply, rises in price, high taxes and disadvantages against other competitors. Developing economies show an increasing demand of their domestic raw materials and are inclined to make access more difficult for European companies. Securing a constant supply at reasonable and foreseeable prices is particularly important for European production industries. This cannot be achieved by business companies and governments of the countries alone, but is a greater task which requires support by the WTO and other international organisations. At the same time it is necessary to strengthen exploration activities to recover new deposits and foster development of domestic mining extraction, which consequently may lead to conflicts in land-use and access to mineral reserves. In this field, an active minerals policy on national as well as on the EU level should cover many issues.

Metallic minerals are of particular importance for Europe, as total consumption of 30 % of all metallic minerals is contrasted with just 3 % of world production. Therefore, particularly large quantities have to be imported for the demand of European industries and consumers. Developing and emerging countries have recently been intensifying their economic interrelations by the so-called South-South links, China, India and many other Asian and certain African states (the "African Lions") appear as serious competitors to Europe. The novel technologies of the 21st century are dependent on high-tech materials such as rare earths, which are needed only in relatively small quantities but cannot be easily substituted at all. As mentioned before, for many other raw materials, China is the world's largest producer of rare earths, while there is a tendency to use them for Chinese domestic purposes.

The different types of mineral raw materials with their specific occurring problems have a central and crucial issue in common: the only way to cope with the present situation of mineral supply and demand is to cooperate beyond regional and national boundaries, to bundle all efforts and produce a strategy for a satisfying coherent minerals policy on the European level.

7.1.1 Targets

The present raw materials problems should be the impetus to reconsider economic policy, enhancing the **rank of the objectives of securing European minerals supply**.

A process for transformation and advancement should be possible, a process which considers different approaches to supply policy.

A precondition for proposals of an EU minerals policy is that the advantages expected for the European Community have to be recognisable and at the same time there is a will to **solve the problems at Community level if necessary**.

Primarily, it is substantial to create basic conditions for public raw materials awareness. It is essential to indicate the importance of raw materials at the EU Level: Issues of raw materials should be sufficiently considered in the context of the EU policies. At the EU level, the importance of the reliable long-term supply of mineral resources in the national economic context needs to be clearly recognized (providing a minerals statement), and all Member States should be encouraged to implement measures to ensure this longer term supply from internal and external sources as far as is practicable. **In that way to implement a national minerals policy/strategy in Member States is crucial.** Each national minerals policy must

- Create an awareness of society's dependence on minerals, i.e. metallic minerals, industrial minerals and construction minerals.

- Point out the importance of the secure supply of minerals for society, and promote a balanced approach in the assessment of conflicting interests between minerals development and other policy issues.

- Reduce consumption of raw materials apart from providing a secure mineral supply (i.e. protection of primary mineral reserves) – increasing of raw materials efficiency is also a relevant target by careful use of raw materials and by reducing the intensity of raw materials also means applying the principle of sustainability

If possible a **coherent minerals policy at the EU Level should be established.** Such policy shall provide *objectives, strategies and action plans coordinated by the competent institutions at EU Level*.

In addition a substantial objective should be the intensified networking between the European Union and the non-member states in Europe. These have not only remarkable deposit potentials but also interesting minerals policy approaches.

An important input to the EU economy of the wide range of minerals and their derived products is essential. Global competition for access to and control of vital mineral resources is developing, and likely to become even fiercer as the world population is expected to grow towards 9 billion by 2050. Huge economic imbalances persist between developed countries and populous emerging countries, such as China, India, Brazil or (different states of) Africa. Therefore it is necessary to foster the sustainability of the EU non-energy extractive industry, to provide a European source of (network) information on global minerals industry including an organisation and an action plan to implement the network (European Minerals Intelligence Network).

Goals can be pursued by actions. These actions as means of the achievement of objectives can be formulated again as goals, which can be pursued by other actions (means). If goals are interconnected by such means-purpose relations, a **system of objectives** or a hierarchy of objectives develops. A condition for the formation of such a hierarchy of objectives is that the superior goal and the subordinated goal aim at the same direction (are complementary). In this sense the primary objective of an EU minerals policy must be the (long-term) securing of the raw materials supply of the Community. Thereby the EU should to consider external aspects (outside EU, EG trade issues) and internal aspects (inside EU, EC access to minerals). Subordinated goals may include the supply goal, the cost minimizing goal, the resource protection goal, the re-use goal and the ecology protection goal.[727]

The sequence of the subordinated goals (hierarchy of objectives) is to be decided politically. It should be noted that the minerals policy related objectives have changed their value: All these changes reflect the change of politico-economic orientation. The safeguarding of the supply and the preservation of the domestic raw material economy has received basic precedence before price worthiness. **Internal economic objectives of a supply policy will take a substantial ranking in the future**. The objectives regarding economic use and ecology as well as substitution and recycling have recently gained greatly in importance. In addition the special meaning of the external economical objectives is evident.

List of objectives:

1) **Supply goal**

    Sufficient and secure supply of mineral raw materials (i. e. metallic minerals, industrial minerals and construction minerals) for the European economy in a long-term perspective is crucial.

2) **Cost-minimization goal**
    EU-external aspect: securing/supporting mineral imports from countries outside EU, based on acceptable prices. EU-internal aspect: Increasing the competitiveness of the EU raw materials industry, i. e. secured access to land, efficient permitting procedures and steady prices for the industry is very important.

---

[727] On the basis of the objectives for raw material importers mentioned in chapter 4.

3) **Resources protection goal**
   Including the increase of resource efficiency based on new technologies, considering substitution is also important.

4) **Re-use goal**
   Increasing of recycling based on new technologies and appropriate market conditions is a challenging goal.

5) **Ecology protection goal**
   Guaranteeing environmental protection in a sustainable manner is a must in modern operations.

The objectives mentioned above can only be realised in the context of a comprehensive raw materials strategy (mid- and long-term).

7.1.2 Strategy

As noticed in chapter 4, a strategy is a **long-term plan striving for an aim**. It aims at the effective employment of definite means in time and space, generally concerning a superior aim. "Strategy" is the **"superior plan above all"** or the "basic pattern of actions". In this sense the following can be stated: The aim of sustainable security of mineral raw materials supply for the European Union requires for its realisation a comprehensive strategy, which has to manifest itself in economic policy, foreign policy, development policy, industrial, regional and social policy, research policy and environmental and consumer policy.

This can only be realised based on common guidelines, which can serve likewise for motivation as well as coordination instruments. In which way should the determination of an EU strategy be accomplished? In order to meet the principle of sustainability, an EU raw materials strategy should be integrated into the "European Union strategy for sustainable development." and "Europe 2020 Strategy".[728] Furthermore, an EU raw materials strategy must be compatible with the "EU Thematic Strategy on the Sustainable Use of Natural Resources", which outlines a long term strategy aimed at a decoupling of resources use and economic growth.[729]

---

[728] EC, DG Enterprise and Industry (2001): Communication of the Commission (No.264) "A sustainable Europe for a better world: a European Union strategy for sustainable development" concentrating on four main aspects: Reduction of the climate change and increase use of clean energy; handling of threats to public health; responsible use of natural resources; improvement of traffic system and land use. – Europe 2020 Strategy; published by EC 2010 (see section 7.2).

[729] Published by the European Commission on December 21, 2005 in Brussels (COM 670).

7.1 General view

The European Union is to be classified a **mineral raw materials importer** (particularly metallic minerals, but also a considerable part of industrial minerals). In order to reduce the import dependence from non-EU countries, the internal economic conditions must be improved. In particular the extension of the knowledge of European geology and raw materials potential, as well as the development of existing instruments, such as recycling and substitution, can **substantially** increase the security of EU minerals supply. Additionally, an improvement of external economical conditions is necessary, i. e. securing the supply of raw materials coming from countries outside the EU, in particular based on an active EU foreign policy, EU trade policy and development policy.

As it is possible to divide the overall goal in sub-goals, also an overall strategy can be divided in *sub-strategies* which are *interdependent* and influencing each other. Sub-strategies will be **different** for metallic minerals, industrial minerals and construction minerals regarding:

1) **Strategy for realising the supply goal**

Strategy 1) includes internal and external economic aspects and is influenced by strategies 2), 3), 4).

2) **Strategy for realising the cost-minimising goal**

3) **Strategy for realising the resource preservation goal**

4) **Strategy for realising the re-use goal**

5) **Strategy for realising the ecology preservation goal**

In the following the different strategies will be discussed. Discussion is done on the one side in a general way and is linked on the other side to the different categories of the non-energy mineral sector, i. e. metallic minerals, industrial minerals and construction minerals.

1) Strategy for realising the supply goal

a) Increasing of the internal economic component

The establishment of an EU raw materials strategy requires first of all an *analysis of the existing demand and supply situation* of the Member States including metallic minerals, industrial minerals and construction minerals. Based on the analyses (*amongst others*) "critical" raw materials have to be identified. Critical minerals are commodities necessary to the economic well-being, but subject to possible supply interruption or restriction, so it is of great importance to *identify new and feasible deposits* in Europe.

Also an **EU raw materials balance system** shall be developed, for instance by Eurostat in close cooperation with the member states.[730] Such an EU raw materials balance system may be based on an analysis of the present and future demand and supply situation, *material flow analyses* demand forecast and focus the *balance between demand and supply* of minerals needed by the EU economy. In that context, to observe the development of international raw material markets is a prerequisite. With regard to the supply offer, it must take into account the primary degree of self-sufficiency, considering domestic recycling, the technical raw materials import dependency, the economical raw materials import dependency and the recycling rate. Based on such raw materials balance system, consequently the EU may act in terms of **influencing the (sustainable) secured supply of mineral resources** (metallic minerals, industrial minerals and construction minerals) required from the EU economy and initiate **different policy actions** for instance in terms of foreign policy, trade policy.

However, first of all the statistics on non-energy minerals including production, imports from and exports to countries outside the EU have to be improved. Present data are fragmentary, often inconsistent, partly not covering the whole EU region, particularly regarding (industrial minerals and) aggregates. Data refer to different definitions and terminology of material groups like construction minerals, building materials or simply minerals, all of them including aggregates to a large extent (problem of inconsistent definitions and data). The responsibility to improve this issue rests with the European Commission (EUROSTAT) and the Member States.

Detailed information on mineral raw materials is limited, in particular information is lacking on the geographical distribution and geological conditions of their deposits potential. It will be necessary to develop an interoperability of geographic information about mineral deposits in line and beyond the *INSPIRE Directive requirements*;[731] also collecting, organising and making

---

[730] As mentioned in Chapter 6, in the 1970er Eurostat has developed a raw material balancing system for nonferrous metals for the indication of the supply situation.

[731] The Directive 2007/2/EC of the European Parliament and of the Council of 14 March 2007 establishing an Infrastructure for Spatial Information in the European Community (INSPIRE) was published in the official Journal on the 25th April 2007. The INSPIRE Directive entered into force on the 15th May 2007. To ensure that the spatial data infrastructures of the Member States are compatible and usable in a Community and transboundary context, the Directive requires that common Implementing Rules (IR) are adopted in a number of specific areas (Metadata, Data Specifications, Network Services, Data and Service Sharing and Monitoring and Reporting). These IRs are adopted as Commission Regulations/Decisions. The Commission is assisted in the process of adopting such rules by a regulatory committee composed by representatives

EU knowledge on (not only) critical mineral resources available to the public, to policy-makers and industry; planning of further research on the basis of identified knowledge gaps (ETP SMR, EuroGeoSurvey-research issue including also other research institutions).

The development of structures of a comprehensive EU mineral knowledge base is crucial. This requires increased exploration of the domestic deposit potential taking into account (not only) the EU mining industry regions. Implementation of raw materials information in land use planning structures is necessary to protect deposits in a long-term perspective. Also **funding of a financial- and guarantee instrument** is needed which can provide the required *investment flows*, i. e. promoting of exploration through EU public funding (e. g. appropriate EU-research programmes like FP7/8). Development of mining clusters in EU mining regions should be the further consequence.

As mineral resources are well known at the surface in EU countries the first step will be for example, to identify resources through remote sensing exploration technologies including satellite and airborne geo-remote sensing exploration targeting deposits below 1000 m for underground and surface geo-exploration by means of new instruments and new methods (based on ETP SMR, EuroGeoSurvey-research including also other research institutions). This could limit European minerals import dependency and increase the supply from domestic resources (i. e. extension of existing resources, operations with continued economic growth, developing new resources) and provide worldwide leadership by

- In-situ on-line resource diagnosis for continuous extraction/recovery,
- New drilling technology, fully integrated intelligent mining processes,
- Sensor development for grade control, rock mechanics,
- Integrated and automated data processing and analysis for 3D mine planning.
- Elaboration of actualized metallogenetic models

Metallic minerals

Demand from manufacturing industry and the construction sector in the EU has been strong and increasing in recent decades. Many of the required metallic minerals by the EU economy must be imported, thus increased exploration of

---

of the Member States and chaired by a representative of the Commission (this is known as the Comitology procedure).

the EU deposit potential is very important to reduce the import dependence. Significant deposits shall be identified (e. g.) based on the GMES (Global Monitoring of Environment and Security)[732] with respect to national or European consumption. As metallic minerals potential is well known at the surface in EU countries the challenge will be to explore and extract deposits in 1000 m depths and more. Deep mineral resource *extraction* (+ 1500 m) addresses an issue of European, but also global dimension: the safe and very deep resource extraction. This is a pre-requisite for the extractive business of the future addressing increasingly rising issues of accessibility of the resources and the pressure on reduced environmental footprint by extractive operations through: 1. appropriate, new technologies; 2. rock mechanics and ground support for static and dynamic loading at +1500 m; extreme rock stress and temperature conditions; 3. comprehensive automation processes to reduce human underground exposure. However, with mining activities below 1000 m depth, the cost per ton of ore, especially when the ore grades are marginal, increase considerably. The exploration of mineral deposits in deeper geological units is mostly cost intensive despite of innovative exploration methods. The world market price is crucial for the profitability of the exploitation.

There is still world class mining potential in Europe. This occurs for example in the Fennoscandian shield (e. g. Finland) with its geological conditions similar to the shield areas of Australia and Canada. Also the Iberian Pyrite Belt in Western Europe (about 85 % belongs to Spain, the other part to Portugal), including volcanic-hosted massive sulphide deposits (e. g. copper, zinc, gold and silver), is noteworthy. For instance (as mentioned in Chapter 5) regarding Portugal, in 2006, the Neves Corvo Mine was one of the highest-grade copper mines in the world. The mine had proven copper reserves of 6.835 Mt at an average grade of 5,73 %, probable copper reserves of 9.975 Mt at an average grade of 5,29 %, and probable zinc reserves of 10.626 Mt at an average grade of 7,96 %.

Industrial minerals

Industrial minerals have to possess certain chemical and/or physical properties for special applications. This leads to limitations in the range of products and choice of prospection and exploration areas in this category of raw materials. For businesses, it is common to carry out exploration near known deposits, as access to land presently is restricted in the Member States. Important industrial minerals (e. g. magnesite) required by the EU economy must be imported,

---

732 GMES (Global Monitoring for Environment and Security) is the European Initiative for the establishment of a European capacity for Earth Observation (see http://www.gmes.info/).

also some the extracted industrial minerals in the EU are only mined in one (e. g. phosphor) or two Member States. Demand from manufacturing industry in the EU has been strong and increasing in recent decades. To reduce import dependency, increased exploration not only limited to known deposits is very important. As was shown in chapter 5, the Member States (for instance Germany, Italy, France, Spain, and Greece) have significant deposit potential used for the EU economy but also for exports to countries outside the EU. The economic importance of industrial minerals has to be taken into account that industrial minerals are needed not only for the local and regional level but also for Europe as a whole. The economic importance of domestic uses and of exportation of industrial minerals should equally be considered (development of international markets).

What will be important in the future is the existence of geological information from various sources of industrial minerals deposits of the EU countries available to various actors. Significant deposits shall be identified (e. g.) based on GMES[733] with respect to national or European consumption, and in full respect of the data ownership rights.

Construction minerals

Demand from the construction sector in the EU has been strong and increasing in recent decades particularly concerning the new EU Member States.[734] Supply is secured in terms of deposit potential which mostly exists (although for certain minerals in one or another) in different parts of a country. However, generally speaking, aggregate resources are not mapped in detail in land use plans (which affect the access to land) unless the local aggregates association has specifically made inputs to the national or regional development plans. Sometimes even when this has been done, access requirements can be ignored by the planning authorities. This situation needs to be addressed and rectified both at the EU level and at the national (strategic minerals planning) level.

b) External economic component

In the context of a coherent foreign policy, trade policy as well as a development policy transparent supply structures with countries rich in resources shall be created. Responsibility to act rests primarily with the European Commission

---

[733] GMES (Global Monitoring for Environment and Security) is the European Initiative for the establishment of a European capacity for Earth Observation (see http://www.gmes.info/).
[734] Countries like Poland, Hungary, Czech Republic, Slovakia, Slovenia, Estonia, Lithuania, Latvia, Malta, and Cyprus joined the EU in 2004, Bulgaria and Rumania joined the EU in 2007.

(in close cooperation with the Member States); i. e. trade policy is a fully competence, foreign policy and development policy is a partial competence of the European Union. Some issues are:

- Comprehensive consideration of raw materials in international product agreements (between EU and other states).
- Increased dialogue between EU and international raw materials forums
  - United Nations Conference on Trade and Development (UNCTAD), World Bank, Group of Eight (G8), World Mines Ministries Forum, World Trade Organisation (WTO), Organisation for Economic Cooperation and Development (OECD)
- Participation of EU in international organisations respectively initiatives
  - e. g. Extractive Industries Transparency Initiative (EITI)[735], – Participation in the realisation of the Johannesburg implementation plan 2011
- EU support concerning capacity building in developing countries
  - Coordination may be done by EuroGeoSurveys. The new ACP-EU Partnership Agreement (2000) established a system of additional support within the European Development Fund.
- Bilateral discussions on foreign minister and state secretary level, bilateral mixed commissions as well as agreements on partnership and cooperation
- Supporting securing minerals supply by mining investments, cooperation agreements and diversification of supply sources

Metallic minerals

In 2006, for instance, the percentage of the EU's self-sufficiency concerning the supply with metallic minerals was rather low, for instance iron 1,99 %, chromium 5,36 %, nickel 2,03 %, manganese 0,11 %; cobalt, molybdenum, niobium-tantalum, vanadium 0 % respectively.[736] Securing the supply of the EU economy with metallic minerals requires transparent supply structures with countries rich in such resources.

---

[735] Initiatives like EITI can equally serve a positive development of resource-rich developing countries as well as a transparent international raw material market. A precondition is that raw material policy succeeds in encouraging more governments to join and participate actively in consistent application of the rules of transparency (Cf. BDI (2007), p. 27.
[736] Weber (2008), l.c.

Critical metallic raw materials at EU level recently have been identified as follows: antinomy, beryllium, cobalt, gallium, germanium, indium, niobium, platinum group metals (PGMs), rare earths, tantalum and tungsten (EC, DG Enterprise and Industry, 2010).

The question arises, which countries could be ideally EU partners? In so far not only the criteria whether a country provides such resources is relevant but also for instance the **political stability** of a country or the transport issue (distance and logistic) will be a decision criteria. For example, as mentioned in Chapter 3, Africa is an important supplier with metallic minerals, particularly high tech minerals (e.g. Cobalt, most part of cobalt is produced in Congo, the global production in 2008 was 31.000 t [nearly 50 %])[737]. In that regard, the discussion process has started: In 2010, within the context of the EU-African Union partnership, the European Commission and the African Union Commission recently agreed to develop a bilateral co-operation in the field of raw materials and to work together, taking fully into account the Africa Mining Vision of February 2009 and the EU Raw Materials Initiative of December 2008, in particular on issues such as governance, infrastructure and investment and geological knowledge and skills.[738]

Industrial minerals

Industrial minerals markets are internationally structured, however not so much as is the case for metallic minerals (a matter of scarcity). Critical industrial (and economical important) minerals at EU level are fluorspar, graphite and magnesite (EC, DG Enterprise and Industry, 2010), although fluorspar is produced in some of the Member States (e.g. Italy), the same applies for magnesite (e.g. Austria). Transparent supply structures with countries rich in such resources shall be developed.

The three worldwide most important countries producing magnesite are China (global production in 2008 was 10 million t), besides that Russia ([Europe] global production in 2008 was 2,3 million t), and Turkey (global production in 2008 was 2,14 million t). The three worldwide most important countries producing fluorspar are China (global production in 2008 was 3,2 million t), Mexico (global production in 2008 was 1 million t) and South Africa (global production in 2008 was 0,3 million t). The three worldwide most important countries producing graphite are China (global production in 2008 was 0,8 million t, besides that India (global production in 2008 was 0,13 million t)

---

737 Weber et al. (2010), l.c.
738 http://ec.europa.eu/enterprise/policies/raw-materials/files/docs/questionnaire-raw-mat-pc-b_en.pdf

Chapter 7 Towards a European minerals policy

and Brazil (global production in 2008 was 0,08 million t).[739] In *each case*, China by far is the country producing most magnesite, fluorspar and graphite. To find a trade arrangement with China would be very important, however, problems with issues like increasing prices, market distortion are affecting the trade possibilities between EU and China (see also strategy (2), external component).

Construction minerals

Construction minerals markets are local or at least regionally based. An exception to a certain grade is for instance the trade with natural stone between EU and USA. For construction minerals (in particular aggregates) Europe is self sufficient.

(2) Strategy for realising the cost-minimisation goal

Strategy (2) includes internal and external economic aspects.

a) External economical component

As mentioned, due to change of paradigm more of half of the world (increasing) is using mineral resources for development of their economy. Not only the BRIC-states but also the leading African markets are pushing the demand of mineral resources. The consequences are limited availability of certain mineral resources, the increased prices particularly related to metallic minerals and industrial minerals influenced by other issues, e. g. trade distorting measures. As mentioned in Chapter 3, from the beginning of 2003 to the end of 2006, prices for iron ore and scrap steel increased by 100%; prices for nonferrous metals increased by 128%, while prices for some particular metals even rose by 500%.

One of the central problems of mineral supply security is **politically induced restrictions** of access to and availability of raw materials by means of national trade distorting measures. On the international raw materials markets there are tendencies to favour such measures both on the supply and on the demand side of domestic raw materials, making raw materials export more difficult. Distortion happens, for example, by recovering value added tax on imports, discriminating licence systems, or prohibitive high export duties, which equivalent to an export prohibition. The industries in the EU that are potentially affected represents, as mentioned in Chapter 6, about 4% of the EU industrial activity and around 500.000 jobs.

---

[739] All data from Weber, W., Zsak, G., Reichl, C., Schatz, M. (2010): World Mining Data 2010.

Supply of raw materials from countries rich in resources, however should be based on economical prices and thus, based on (an accelerated) co-operation between European Union and these states (see also a supply goal/strategy 1 b). This requires consolidation strategies at the EU level, particularly regarding trade policy; responsibility rests with DG Trade (European Commission [EC], as mentioned, trade policy is a fully competence at the EU level) in close cooperation with DG Enterprise and Industry (EC).

Metallic minerals

It is crucial to provide a consolidated trade policy at the EU level for many of the strategic metallic minerals needed by the EU economy and which are imported from countries outside the EU introducing high taxes, etc. A number of states subsidize the import of raw materials and thereby grant advantages to their own industries. Several states limit the export of certain domestic raw materials. The European Commission pointed out examples of export restrictions by China (concerning aluminium, copper, ferroalloys of chromium, nickel, molybdenum and tungsten, high tech metals like rare earths, manganese, molybdenum, nickel and India (iron ore).[740]

For instance, regarding wolfram (identified as critical metallic raw materials at the EU level), in 2006 EU's self sufficiency concerning the supply with tungsten was about 4,27 % (Weber, 2008). The global wolfram production in China in 2006 was 45.000 t (*more than 85 %*), in 2008 it was 43.500 t; besides that Russia (Asia) with a global production in 2006 of 2.465 t (6 %) and 2.720 t in 2008 or Canada (2.608 t), Austria (1.122 t), Bolivia (1.150 t), Portugal (900 t) in 2008.[741] A comparison between the different countries shows that China by far is producing the largest part of tungsten.

Market distortion problems leading to increasing prices of strategic mineral resources may be solved primarily by (foreign policy and) trade policy in the context at least of the *World Trade Organisation*.[742]

Industrial minerals

To provide a consolidated trade policy at EU level is crucial for some of the critical industrial (and economical important) minerals needed by the EU economy

---

740 European Commission, 2008c, Working Staff Document SEC(2008)2741, as mentioned in Chapter 6.
741 All data from Weber et al. (2010), l.c.
742 As it happened in 2008: The European Union, the USA and Mexico has been asking for a dispute settlement panel on Chinese export restrictions at the World Trade Organisation. However optimal would be to find a political agreement.

and imported from outside the EU (i.e. fluorspar, graphite and magnesite). Presently there are many problems related to countries introducing high taxes, etc., for instance regarding fluorspar and magnesium imported from China. China imposes export duties of 15 % on fluorspar and 10 % on magnesite.

As mentioned, in 2008 China produced 10 million t magnesite (about *50 % of the global production*). Though the EU is an important producer of magnesite (Slovakia as fourth world wide producer 1,44 million t [4], Austria 837.476 t [6], Greece 529.546 t [7], Spain 443.000 [8], Poland 60.000 t [15]), the EU must import a considerable part of these raw materials from China. Due to problems with taxes, etc. prices are very high. Market distortion problems leading to increasing prices of industrial mineral resources may be solved primarily by (foreign policy and) trade policy in the context at least of the *World Trade Organisation*.[743]

b) Internal economic component

Access to land is crucial for the competitiveness of the EU mining industry but since many years strongly limited, amongst other reasons not sufficiently considered by land use planning (lack of mineral planning policies) and affected by (restrictive) environmental law, as mentioned in Chapter 6. In terms of the import dependency, but also in terms of the importance of the EU mining industry for the EU economy (added value), this issue must be solved both at EU level (appropriate *recommendations* by the European Commission) and primarily at national level; defining a national minerals policy as mentioned in chapter 5 (improving mineral [planning] policies, updating the repulatory framework i.e. related laws and permitting procedures). In other words: Responsibility to act primarily rests with the Member States, as mineral planning policy and permitting procedures are competences at national level (subsidiarity principle). However, recommendations from the European Commission maybe drawn to the Member States in terms of exchanging best practices and monitoring actions.

Consistent EU law and efficient permitting procedures related to mineral resources are needed. This is important for the non-energy extracting minerals industry. For instance, development of guidelines for improving the situation of mining in Natura 2000 areas ([FFH-] Directive 92/43/EC) is necessary. A Guideline document on "Non-Energy Extractive Industry and Natura 2000" aims to provide clarification. The Guideline document was published by end of July 2010 and is available via the web site of Environment Directorate General.

---

743 l.c. – All used data here: Weber et al., 2010.

The following issues are related to (national) minerals policy and regulatory framework considering metallic minerals, industrial minerals, and construction minerals.

Establishment of a national minerals policy

Each Member State should define a national minerals policy strategy including a mineral statement, appropriate mineral planning policies and a coherent regulatory framework. As mentioned in Chapter 4 and 5, minerals planning policy (based on land use planning) is seen as part of the national minerals policy of a country. As mentioned in Chapter 5, only a few Member States have a clear, structured national minerals policy strategy. This issue is affecting both minerals planning policies *and* thus, also the permitting processes (inefficient, ineffective and time-consuming). Mineral resources are not considered in land use planning in most countries, and even where they are, there is an unbalanced pre-disposition against mining activities. Given the geologically-determined locations of mineral resources (i. e. metallic minerals, industrial minerals and construction minerals), these deserve the same status in land use planning as other issues, such as water or other environmental resources, to ensure long-term access to mineral resources. To improve the situation the following requirements have to be met, i. e. *appropriate recommendations by the European Commission to the Member States shall be made*:[744]

There is need of definition of the spatial and temporal priorities for the mining areas with regard to economic and policy criteria. Development of land planning shall be based on the established priorities to preserve the future mining areas. It is a fundamental requirement when balancing the need for the mineral extraction and environmental considerations to have a clear policy statement on the importance of mineral resources, i. e. metallic minerals, industrial minerals and construction minerals. A reliable long term minerals planning is an essential condition for investments as well as for a sustainable supply of national economies with raw material.

A minerals planning policy should address **strategic (minerals) planning** if possible at a national or at least a regional level (including *all relevant* mineral resources of a country, i. e. metallic minerals, industrial minerals and construction minerals) **and operative (minerals) planning** based on (de-

---

[744] In the following also issues from the report "Planning Policies and Permitting Procedures to Ensure the Sustainable Supply of Aggregates in Europe", Department of Mineral Resources and Petroleum Engineering, University of Leoben, l.c. and also comments from IMA (Industrial Minerals Association – Europe) and EUROMINES, done in the context of the EU Consultation Process on Raw Materials (2008), l.c., are considered.

mand forecasting and) land use plans. At strategic level it should be decided which planning strategy will be the best for a country. At regional and/or local level land use plans shall include metallic minerals, industrial minerals and construction minerals by taking into account the specific issues of the related extractive industries. The planning horizon shall be both mid-term and long-term to ensure that access to local resources is really secured: Minerals planning should allocate proven and consented reserves for at least 30 years (mid-term) and 50 years (long-term). In correlation to this long term planning preserving existing deposits, an obligation to restore the mined areas in the shortest possible delays and to reduce to a minimum the mined areas depending on the short term productions is required.

National, regional and local coordinated *aggregates planning policies* need to take account of local geology, whether or not hard rock or sand and gravel are present geologically; whether or not the deposits are in potentially sensitive areas due to being protected areas (Natura 2000); the distance from urban, highly populated or industrial areas where there would be large aggregates demand, the road, rail or waterway infrastructure for transporting the aggregates from the point of exploitation to the point of usage.

Regulatory framework

As mentioned in Chapter 5 (national level) and Chapter 6 (EU level), few Member States have efficient permitting systems. In many Member States, multi-body permitting regimes exist for historical reasons, often with differing perspectives and areas of responsibility. In the absence of coordination, conflicts may happen between several administrations in charge of protecting different interests. The authorisation process is complex and therefore very slow in most countries, taking typically 5–10 years to obtain authorisation for a new production site, and furthermore permissions are often granted for only similar timescales, too short to justify capital investment. Indeed, the fragmentation of government departments is a real problem, with risk of conflicting messages coming out of each department. To improve the situation the following requirements have to be met, i. e. these issues should be addressed in the EU Raw Materials Initiative respectively *appropriate recommendations by the European Commission to the Member States shall be made*:[745]

---

[745] In the following issues from the report "Planning Policies and Permitting Procedures to Ensure the Sustainable Supply of Aggregates in Europe", Department of Mineral Resources and Petroleum Engineering, l.c. and also comments from IMA (Industrial Minerals Association – Europe) and EUROMINES, done in terms of the EU Consultation Process on Raw Material (2008), l.c., are considered.

First of all, incorporation of minerals deposit information in land use planning data banks is (absolutely) necessary to **facilitate** efficient permitting procedures. Furthermore, all permitting considerations have to be linked to the geological presence of mineral resources, and the physical ability to get access. When granting permissions, the duration of these should always be in line with the lifetime of the deposit: sustainability requires the extraction of the total deposit.

In principle, each Member State should have a permitting system that allows efficient and timely granting permissions for projects, which entails clear and appropriate legislative structure, with clear designation of authorities and competences. Creating a unified "mining code" where all the laws are assembled definitely helps simplifying understanding and procedures.[746] An ideal mining law should provide, that mining projects should have at least the same importance as other (spatial) interests (like e.g. nature conservation, house building etc.). In no case should mineral extraction be excluded a priori (e.g. in protected areas). A case by case decision is required as to verifying if extraction and the protected aims of the area are compatible. The mining law should also expressively highlight that raw materials supply is not only a private but also a public interest (supply of national economy with raw materials). Not only local economic interests should be considered, but also the added value of the minerals extraction for the general economy and the EU competitiveness. Often import is promoted, even though this may be not sustainable, while export is considered to have a negative effect, while it contributes to the EU economy (see Chapter 5). Finally, such mining law should also provide good investment conditions, fair fees and taxes, royalties for both domestic and foreign operators.

The application process has to be rationalised through one authority (as an "one-stop-shop")[747], or at *least well co-ordinated procedures* between all authorities. Time-limited procedures for clarification by all stakeholders of applications, such that the overall process has to be completed within a 3 year times-

---

[746] In that regard, also reference is made in Tiess (2010): Legal basics of Mineral Policy in Europe, l.c., Chapter 1.
[747] The "One stop shop" principle means a procedural concentration and attracts increasingly interest in Europe. For a simplification of procedure, regulations for the consolidation of administrative proceedings concerning the same object are installed: If several approvals are necessary for a project and are applied at the same time, the authority has to combine these to joint procedures and decisions and to co-ordinate them with the procedures executed by other authorities. Such concentration of proceedings shall make sure that the necessary (several) permits for one project are issued by a single authority.] (Cf. Stolzlechner, H. (2004): Einführung in das öffentliche Recht [Introduction to public law], Wien).

cale.[748] Organising the management of natural resources including minerals through one central government department responsible for it and providing clear and concise policy and separate guidance is a must. The one stop shop implies a one consent covering all operations taking place at a mineral site, with one regulatory body responsible also for monitoring the operations. As mineral extraction crosses domains that depend on several governmental authorities, the one stop shop does not mean that just one administration is in charge of the whole process, but that **one administration should be in charge of coordinating the work of every involved department**, and act as one regulatory body vis-à-vis the applicant.

Regarding applications related to large metallic minerals or industrial minerals operations, or at least also large construction minerals operations, there should be one (effectivily applied) environmental impact assessment established covering all aspects of a project, including water protection, mining waste, Natura 2000, renaturation measures. Different advising governmental bodies may deliver conflicting advice. Long lasting and costly procedures to cope with conflicts between mining projects and environmental preoccupations have to be avoided. There is a need of such procedures which settle conflicts based on objectives criteria, particularly clarity in the procedures of settlement of conflicts with surrounding communities shall be provided. A reasonable balanced approach conserving the environment, biodiversity, etc, but equally recognising the need for mineral resources, and the regional benefits needs to be created. Extraction projects should have at least the same importance as other spatial interests, and in no case should extraction be prohibited *a priori* even in protected areas.

Regarding metallic minerals and industrial minerals (as matter of scarcity): The decision about mining project permissions should be taken at the *highest possible level* (national or if appropriate regional). Authorities established at high level have the overview about the global aspects of a mining project and its importance for certain industries. The permitting decision should be primarily in the competence of national/regional authorities and not in the competence of local authorities who very often are influenced by various political (local and economical) interests. Due to these political influences (and the political dependency of local decision makers), the global aspects of raw material supply are often out of focus.

---

748 As mentioned, there are many situations now which take 5–10 years, which few companies can afford (see Chapter 6).

On the contrary, aggregates supply is done at local level (issue transport) and requires particularly a local focus. In that case the permitting decision is competence of local authorities.

It seems also to be important that local authorities are included in this process as interested parties under EIA procedures. Moreover, when granting permissions, for hard rock quarries a 50-year timescale should typically be considered. No permissions should be less than 15 years otherwise the major capital investment cannot be justified. For sand and gravel pits, the permission timescale should be 15–50 years depending on the scale of the deposit, with further renewals anticipated also proportionate to the scale of the deposit.

(3) Strategy for realising the resources protection goal (resource efficiency)

This strategy is strongly interlinked with strategies (4) (aimed at realising the re-use goal [recycling]) and strategy (5) aimed at realising the ecology preservation goal.

Strategy (3) requires the establishment of comprehensive structures for consistent EU mineral research; also close co-operation between research institutions, particularly ETPSMR in cooperation with different platforms at EU level including national research organisations. *Technological improvement* and *innovative concepts* to increase resource efficiency are needed.

Responsibility to act primarily rests with the European Commission (i. e. DG Research, providing research programmes [funding issue] also considering issues related to mineral resource efficiency), DG Enterprise, DG Environment and the Member States.

An important challenge will be the improvement of efficiency **throughout the life cycle of mineral resources**, beginning at extraction and processing: On the one hand it is necessary to continue developing of new technologies for better extraction and utilisation of minerals resources. On the other hand it is important to continue developing new recovery technologies for secondary and waste materials, thus *reducing the loss of these resources* for the economy (in that context, see also (5) Strategy for realising the ecology preservation goal).

Main research targets are: improving stewardship and operational acceptability through using minerals to save precious and rare resources (substitution); using alternative raw materials including secondary raw materials and waste materials e. g. heaps, decontation basins, where appropriate; promoting the Life Cycle Contribution of industries. An issue related to resource efficiency also is the creation of new product functionalities.

Metallic minerals

Novel technologies for metals extracting and processing are required. Some examples are heat management, replacement of energy intensive pyrometallurgical technologies with less intensive methods, new technologies in hydrometallurgy, optimal chemical, physical and high temperature processes for production of precious metals, nanotechnologies.

Industrial minerals

Novel technologies for industrial minerals extracting and processing are required. Some examples are mineral particle engineering, i. e. optimal chemical, physical and high temperature processes for industrial minerals.

Construction minerals

To ensure on one hand the demand for construction raw materials and on the other hand to improve resource efficiency *and* solve the problems of land use, innovative approaches on the part of entrepreneurs are in greater demand than ever. Thereby one approach is a combination of the **working of primary and secondary raw materials in the same operating establishment**. This can take place as follows: First, the exploitation of the primary raw materials takes place (e. g. sand and gravel) for example on community-owned ground; afterwards the resulting hollow spaces can be used as dumping grounds for the communities, e. g. as dump site for construction and demolition waste. Here, at the same time a reduction of landfill volume is reached by extraction of secondary raw materials from construction and demolition waste (using the *same processing plant* as for processing of primary raw materials): entrepreneurs and the community will benefit jointly. At the same time, the concept of sustainability is observed as natural resources and raw materials are preserved for future generations. A *multiple use of soil and area takes place with the use of synergy effects*. Ultimately, one can speak of a sustainable extraction of mineral construction materials. The benefits should be highlighted again: reduction of waste bulk, which should have been disposed of as well as a reduction of primary raw materials. At the same time this induces a lesser land use, which is especially important for densely populated countries such as Belgium, Germany, Netherlands, Switzerland, Austria and Liechtenstein.

(4) Strategy for realising the re-use goal (recycling)

Increased use of secondary (i.e. recycled) raw materials *contributes to security of supply* (and energy) efficiency. However, today many end-of-life products do not enter into sound recycling channels, resulting in an irremediable loss of valuable secondary raw materials. This mainly concerns exports of end-of-life vehicles and electronic equipment, which leave Europe as reusable products but end up being dismantled abroad. To counter these trends, the need to reinforce the Waste Shipment Regulation and related legislation is evident.[749]

An enhanced European Union research effort is required, also a framework setting for economy and effective (recycling) market. Of the three sub-sectors i. e. metallic minerals, industrial minerals and construction minerals, metals provide the highest potential for using recycled materials (e. g. scrap). Direct recycling issue for industrial minerals is usually not feasibly since the mineral forms an intrinsic part of end-use application (e. g. glass). However, there may be exceptions (but also new possibilities when economically beneficial, as may be the case for recycling of magnesite materials. Issues related to (improvement of) recycling of metallic minerals and construction minerals are: Increasing of recycling rates by optimised research (technological issue [ETP SMR and other platforms]) *and* improvement of conditions for recycling markets (political/legal issue [European Commission]).

Responsibility to act primarily rests with the European Commission (i. e. DG Research, providing research programmes related to recycling [funding issue]), DG Enterprise, DG Environment and the Member States.

Metallic minerals

Many metals, including iron and steel, copper, tin, lead and aluminium, are relatively simple to recycle as they can be melted and recast without losing their important characteristics (additionally, in general less energy is needed compared with the production of primary metallic minerals). As mentioned in Chapter 3, lead scrap (e. g. from car batteries), for example, accounts for around 64 % of lead consumption in the EU. Recycled aluminium, steel and copper also make significant contributions to total supply within the EU. While recycled metal can make an important contribution to meeting demand, in a growing economy there is a **limit to the extent** to which it can contribute to materials supply. It will be affected by the available amount of material originally used (limiting for instance recycling of rare earth's) and by its lifetime in use.

---

749 http://ec.europa.eu/enterprise/policies/raw-materials/files/docs/questionnaire-raw-mat-pc-b_en.pdf

A challenge (and therefore to be considered in research) would be the *complete* utilization of resources from secondary materials and scrap treatment from non-ferrous metals industry by development of combined highly-efficient technologies for metals recovery for example leeching processes from scraps (and multi-metallic and multi-material waste). Metals contained within articles with a short life and high recovery rates maybe will satisfy more of the demand for a particular material than those present in longer lived articles.

Furthermore, a challenge will be the **management of (access to) scrap**. Prices of some recovered materials have reached record levels due to the high demand from third countries. A considerable part of it is traded illegal or limited due to export restrictions. Actions are necessary to be taken to measure the extent of illegal trade in products containing these secondary materials but also the extent of export restrictions. In November 2008, as mentioned in Chapter 6, the European Commission stated considerable export restrictions concerning metal scrap, including China (non-ferrous scrap), Ukraine (ferrous scrap), Russia (ferrous and non-ferrous scrap), India (non-ferrous scrap), and Pakistan (non-ferrous scrap). In each case, there is the need for an improvement in European statistics on secondary raw materials.

A challenge also will be the *recycling of high tech minerals like rare earths*. As mentioned, rare earths are needed by the high tech industry but access to these minerals is increasingly limited due to different (e. g. political) reasons. Estimation indicates that up from 2012 the supply offer for certain minerals of rare earths could be very problematic.[750] Recycling of rare earths depends not only on technologies but also on the required recycling material.

Construction minerals

Recycling can make an essential input. In 2008, recycled aggregates in Europe reached 216 million tonnes, which although very significant was only 6,1% of total aggregates demand. This already represents very high levels of recycling (about 20%) in some countries (for example UK, Belgium and Netherlands), corresponding to almost full (90%) recycling of all demolition materials available. Other countries still have apparently low levels of recycling (such as France). In general, in less densely populated countries, the economics of recycling are less attractive compared with densely-populated regions. In the medium term, the average rate of recycling across Europe is therefore unlikely to exceed 8–10%. Likewise marine and manufactured aggregates together in 2008 comprised only 4,3% of European total output: this could grow to 5–6%

---

750 See http://www.clingendael.nl/resourcescarcity/presentations/(Precious minerals for the production of goods and food).

in the medium term. Therefore some 85 % of all aggregates production in the medium term will still need to come from natural aggregate resources.[751]

(5) Strategy for realising the ecology preservation goal

The establishment of comprehensive structures for a sustainable mineral economy is required. Besides that efficient cooperative structures regarding research related to environmental protection and mining issues at EU level are required, i. e. cooperation between ETP SMR, the different environmental platforms (e. g. water platform) at the EU level.

Responsibility to act primarily rests with the European Commission (i. e. DG Research, providing research programmes [funding issue] addressing appropriate topics), DG Environment and the Member States.

Research issues include: enhancing energy efficiency, providing new technologies to reduce environmental impacts, promoting Life Circle Analysis approaches, promoting material flow analyses. Environmental criteria will be different for metallic minerals, industrial minerals and construction minerals. For instance, metallic minerals production implies the waste issue (e. g. tailings). Industrial minerals production (e. g. magnesite production) implies high energy consumption. Construction minerals production and transportation includes noise, dust in surrounding municipalities.

Eco-efficient in-situ extraction for instance, addresses the need to extract resources in-situ by biochemical/chemical solution technologies in order to provide novel methods for extraction with minimal environmental foot-print; and an increased reserve base by providing viable methods for deposits so far not possible to extract (ETP SMR-research issue). Novel technology for selective extraction concerns small-scale mechanical excavation and backfill systems by developing mobile mineral processing plants with closed process systems. This can minimise the environmental footprint and increase the resource base. It also can increase small and middle enterprises (SME's) in the sector which could provide a substantial contribution to the development of eco-efficient mining in the developing world (ETP SMR-research issue). Exploitation of large deposits of high production rate needs a modified approach.

Regarding *metallic minerals*, a challenge will be the development of new technological processes for treatment of polymetallic materials by implementing of new methods for recovery of currently not recovered metals, as well

---

751 All information/data from UEPG, in: Department of mineral resources and petroleum engineering (2010), l.c.

as new technologies to improve efficiency of already recovered by-product metals. Innovative methods shall be developed for valorizing waste by creating of sustainable recycling systems of metallurgical wastes and effluents to increase production of basic metals by launching (e.g. leaching) technologies for processing of low-quality raw materials, tailings, waste.

Regarding the resource and energy efficiency issue both from the economic and ecological view: As mentioned in Chapter 1, the estimated energy costs in the EU as proportion of overall site operating costs amounts for construction minerals 3%, for metallic minerals 15%–17% and for industrial minerals 11%–19%. Optimising extraction and processing of industrial and metallic mineral resources is necessary. The aim is to reduce energy consumption by intensification of processes and implementation of new equipment. Reuse of heat and more synergistic processing could also help reduce energy consumption. A reduction of energy consumption of 5–10% could be targeted issue, including other research institutions (ETP SMR-research).

Development of new technologies will also allow to eliminate the main sources of pollution and determine the most suitable measures to meet the highest environmental and economic effectiveness, limit bioactive and carcinogenic emissions and improve utilisation of hazardous waste, provide a multi-criteria analysis covering the development of tools to evaluate technological impact on the environment, including reduction of maintenance and repair costs (ETP SMR-research issue).

7.1.3 Conception

The implementation of an EU raw materials strategy requires a **coherent conception** to coordinate different EU policies. As mentioned in Chapter 4, a conception is a *comprehensive compilation of objectives and subsequent strategies and measures* for the realization of a higher goal. It includes all the information needed as well as time, measures and resources plans (time, costs, material, personal). Conceptions are usually put into writing and should be checked for relevance and topicality regularly. It is necessary to determine the measure with the **highest degree of success** in the **realization of the goal** by comparing between the conceivable alternatives. Usually the optimal measure is a "programme" which inserts several means in different dosage and in temporal gradation. An *exemplary compilation* of goals and strategies and measures related to European minerals policy) derived thereof is given in table 27.

7.1 General view

Table 79: Concept of a European minerals policy

Targets	Overall Strategy	Action	Action plan (Assignment of priorities)
Overall Target Sustainable supply with Mineral resources for the European Economy	**Domestic economical component (inside EU):** Reduce import dependence of raw materials substantially by development of domestic economy possibilities/interference	Action which provides an optimal cost input for the (EU) GDP Creating a framework (and instruments) for a raw material sector policy at EU level Raw material sector policy combined with other sector policies. Referring to various interactions between raw material policy functionaries	When creating a coherent concept interactions have to be coordinated, both at EU level and at national level. Homogenisation of relevant data for raw materials at EU level • Establishment of an adequate EU-raw material knowledge base • Realisation of an EU width raw material analysis EU-raw materials balance system
	Foreign economical component (outside EU): ensuring supply of raw materials for the EU-Economy by development of foreign economical options / interventions.		

Sub-target	Sub strategy	Action	Action plan
Supply goal Sufficient and secured supply of mineral resources	**Domestic economical component** Establishment of an EU raw material knowledge base Increased exploration and implementation of deposit information in land use planning	Inventory of raw materials in the EU. Criteria: Import dependence Substitution Recycling Political risk of supply, security of supply Urgency of actions	1. Presentation of comprehensive raw material statistics
2. Characterisation of the EU supply situation
3. Recognizing the critical raw materials
4. Establishment of a raw material accounting system
5. Consideration by land use planning (involvement in ESDP, „Land service of Kopernikus)
	Foreign economical component: coherent EU foreign and trade policy, development policy for creation of reliable and transparent supply systems with countries outside the EU	Adequate consideration of the raw materials for product agreements. Increased dialogue with international fora (OECD, World Bank, UNCTAD …) Capacity-building in developing countries with Eurogeosurvey based on research	1. Structural development at EU level, linked up dialog
2. Clear direction of trade and foreign policy

Chapter 7 Towards a European minerals policy

Sub-target	Sub strategy	Action	Action plan
Cost-minimization goal Stable raw material prices, predictable prices for the industry	**Foreign economical component** Raw material supply from countries outside EU at reasonable prices on the basis of a strengthened cooperation of the EU and non-EU member states		
	Domestic economical component Increase of the competitiveness of the EU extractive industry by establishing efficient structures for permitting procedures	• EU Environmental Law – Analysis • Expand the One Stop shop – approach	1. Increase awareness of raw materials 2. Adaptation of EU environmental law consistency 3. Exchange of best practices between EU Member States
Resources protection goal Increase in raw material and material efficiency	Creation of comprehensive structures for consistent raw material research, co-ordination between the EU research institutions in cooperation with the ETPSMR	• Increase in raw material and material efficiency through research.	1. Structural development for consistent raw material research at EU level 2. Improvement of the basic conditions of raw material research (e.g. FP7/8-program)
Re-use goal Increase of recycling rates	Increased EU research (ETPSMR);	• Increasing recycling rates through research Options and improvement of market conditions	1. Adaptation of EU environmental law 2. Structural development for improving the recycling market
Ecology protection goal Ensure a sustainable environmental protection	Establishment of a sustainable raw material economy; efficient cooperation between environmental related platforms at EU level and ETPSMR	Increasing energy efficiency, promotion of LCA research in industry	1. Suitable structures for dialogue at EU level 2. Developing effective cooperation structures between environmental technologies and raw material industry

The development of a coherent concept of an EU minerals policy requires the comprehensive attention of relevant representatives both on EU and national level. Table 80 shows relevant policies for the EU mineral supply.[752] In addition, policies from *outside the EU community*, particularly the mineral policies of developing countries, must be taken into account.

Table 80: Policies relevant to the EU mineral supply (Data by Michaelis, 1977 with additionally remarks from Tiess)

Global and horizontal policies	Foreign trade policies	Sector policies
1) Economic and financial policy	1) General (autonomous or contractual) trade policy Development policy	1) **Policy of securing mineral raw materials supply**
2) Industrial policy	2) Foreign policy Policy towards the different groups of countries	2) Environmental policy
3) Research and technology policy	3) Currency policy	3) Energy policy
4) Regional and social policy	4) Investment policy	

The **framework for a coherent EU minerals policy including a mineral raw materials sector policy on the EU level** has to be established first. A raw material sector policy needs *defined and consolidated* structures and interrelations to a variety of other policies. Here arises the question which EU institution would be responsible for the coordination of the different relations. A clear picture is needed.

In view of the importance of these interactions the following model demonstrates the interrelations of the component policies and organisations including EU level. Reference is made to the model in Chapter 4, **"Mechanisms of minerals policy"**.

The structure of the European Union is complex: Interactions therefore must be effective and efficient, and transparency both on the EU and the national level must be given.

---

752 Michaelis (1977), l.c., p. 99.

Chapter 7 Towards a European minerals policy

Figure 177: Interrelations of EU minerals policy

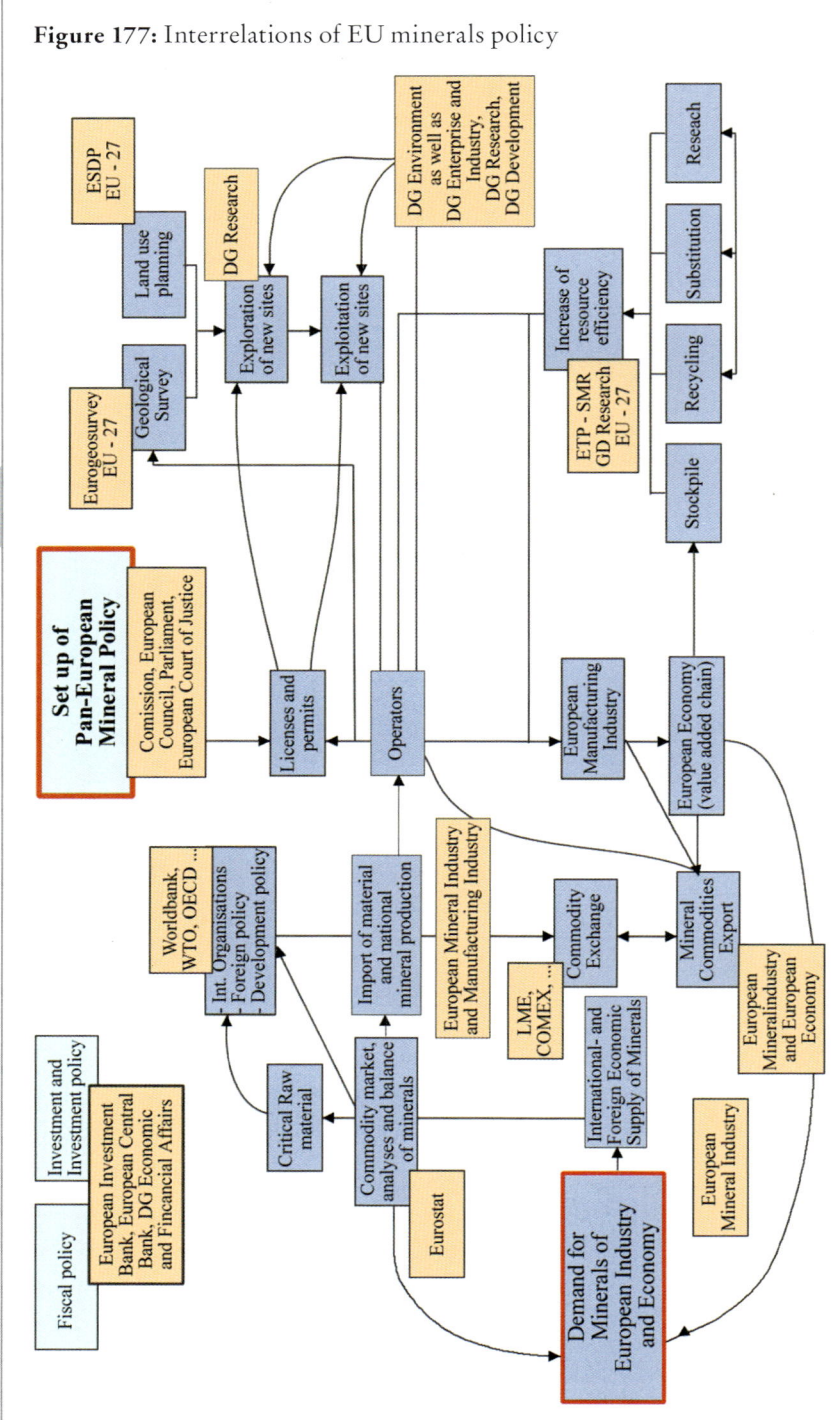

These interactions are only realisable on the basis of a comprehensive cooperation and **consultation process** on the EU level as well as between the EU level and national levels. Therefore, appropriate consultation structures of Council, Commission, and Parliament (as well as other EU institutions significant for the topic of mineral raw materials) are to be established. Furthermore, consultation structures are needed between the various General Directorates of the Commission (DG Enterprise, DG Trade, DG Development, DG Environment, DG Research etc). Finally, such structures are to be established for the cooperation of the Commission on one hand and the specialist representatives of Member States (e. g. for geology, economics, technology) on the other hand.

Consultation structures between diverse European mining organisations should be homogenised. Through the above described way of cooperation and consultation processes it is possible to compile information and knowledge necessary for the establishment of an efficient European minerals policy. For example:

EU raw material base knowledge
An adequate EU raw materials knowledge base should be established; this would require a comprehensive coordination between Geological Surveys and EuroGeoSurveys.[742]

EU raw material analysis
Accomplishment of an EU wide raw material analysis for the identification of the EU supply situation would need a comprehensive co-operation between the Member States and EUROSTAT

State of the art technology
Intensified co-operation between specialized universities and EU research centres can develop the state of the art regarding mineral and material efficiency, substitution and recycling effectively.

The security of supply with raw materials for the production industry has always been of utmost significance for the prosperity of every economy and, against the background of increasing globalisation, will gain even more importance for Europe in the years to come. The European Union has to cope with the fact that there is not a centralised government to manage economic problems by forcing legislation on the member countries. The EU consists of independent states persuing their own policies, being responsible to their respective citizens, but united by common agreements, conventions and treaties and the will to cooperate. Consequently, the way from having a good idea to

---

753 The appendix contains an summary of the Geological Surveys of the European countries.

actually starting an initiative sustained by all member states, is often cumbersome as a mere problem of information and communication.

Concerning the secure supply with raw materials, plenty of EU departments and other organisations are involved, which, following the organigram, can be illustrated by three pillars:

Responsible for the sector policies are, first and foremost, DG Enterprise and Industry (for all minerals policy matters), evidently followed by DG Environment, DG Energy, DG Maritime Affairs and Fisheries (e.g. responsible for maritime spatial planning and coastal zone management, Arctic hydrocarbon resources – should be extended to Arctic mineral resources!), and DG Mobility and Transport.

Global and horizontal policies are covered by DG Economic and Financial Affairs, DG Budget, DG Research and Innovation, DG Regional Policy.

Foreign policies are assigned to DG Foreign Policy Instruments Service, DG Trade, DG Development and Cooperation, DG Taxation and Customs Union.

All these EU institutions have to cooperate and communicate not only with each other, but with a plethora of scientific and commercial institutions (in the organigram, pointed out in yellow colour) and, beyond that, with national and regional administrative bodies and organisations (in green colour, at the bottom of the organigram).

To enable an effective and swift way of communication within this complex mesh of institutions involved, it is suggested to constitute a "Preparatory commission for a European minerals policy", a kind of steering committee, who has to gather comprehensive information concerning the field of raw materials in order to prepare reports and recommendations for the European Commission and might be located in the DG Enterprise and Industry. For this purpose, a staff of about twenty international experts in the field of mineral resources and policies should tie the manifold information sources and communicate with all parties envolved. As an outcome, the European Commission should be provided with concise papers to facilitate their decisions and enable them to create a common minerals policy framework. (An analogue institution is the Bureau of European Policy Advisers, BEPA, who has the task to provide the President of the Commission with background analysis and strategic advice. Particularly the Chief Scientific Adviser, CSA, is responsible for high level, independent and timely scientific advice directly to the President.) It is to be expected that both the Commission and the national administrations and decision makers would highly profit from frequent contacts with such an expert group, who would perform a kind of "diplomatic mission" networking between scientific, technolog-

7.1 General view

ical, economic, environmental and financial experts, competent contact persons and decision makers from all institutions of the organigram. Public awareness of the crucial importance of the raw material matter would be increased by an open discussion on various levels, incorporating small industries as well as multinational businesses, regional space-planning authorities as well as national administrations, research institutions and universities, and last, but not least the information media. Only bundling all efforts to establish an effective European minerals policy can secure economic and social prosperity and make Europe fit to tackle the challenges of our globalised world.

A possible organigram chart of a coherent EU mineral policy is shown here:

Figure 178: Organigram chart of a coherent EU minerals policy

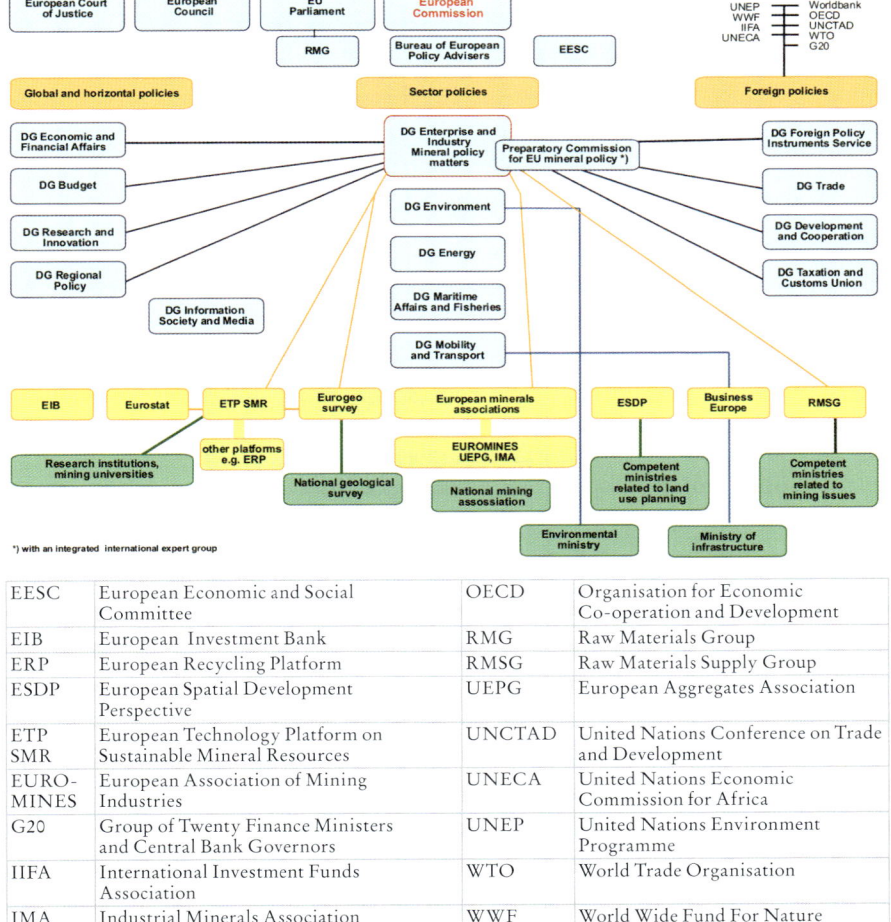

EESC	European Economic and Social Committee	OECD	Organisation for Economic Co-operation and Development
EIB	European Investment Bank	RMG	Raw Materials Group
ERP	European Recycling Platform	RMSG	Raw Materials Supply Group
ESDP	European Spatial Development Perspective	UEPG	European Aggregates Association
ETP SMR	European Technology Platform on Sustainable Mineral Resources	UNCTAD	United Nations Conference on Trade and Development
EURO-MINES	European Association of Mining Industries	UNECA	United Nations Economic Commission for Africa
G20	Group of Twenty Finance Ministers and Central Bank Governors	UNEP	United Nations Environment Programme
IIFA	International Investment Funds Association	WTO	World Trade Organisation
IMA	Industrial Minerals Association	WWF	World Wide Fund For Nature

7.2 Present developments at the EU level
The Raw Materials Initiative

The global political and economic situation implies that the multiple and complex problems of sustainable supply of non-energy raw materials for the European Union need a decisive European response. The European Commission states in its Communication on "The raw materials initiative – meeting our critical needs for growth and jobs in Europe" that the issue requires *"an integrated EU strategy that ties together various EU policies and promotes further cooperation between the Member States where appropriate"*.[754] Accordingly, the proposed Raw Materials Initiative was launched in November 2008. The Commission will report to the Council after two years on its implementation. The Communication itself is a political declaration stating the necessity of a coherent EU minerals policy, focussing on non-energy minerals. A major topic is to create basic conditions for public awareness of raw materials issues and to make sure that they are sufficiently considered in the context of other relevant EU policies. The relevance of an analysis of supply and demand of non-energy raw materials and, consequently, of an integrated raw materials strategy are stated.

The EU Raw Materials Initiative includes a **strategy** based on the three following pillars:

(1) Ensuring access to raw materials on international markets under the same conditions as other industrial competitors;

(2) Setting appropriate framework conditions within the EU in order to foster sustainable supply of raw materials from European sources;

(3) Boosting resource efficiency and promote recycling to reduce the EU's consumption of primary raw materials and decrease the relative import dependence.

The first pillar focuses the external economic component; pillar 2 and 3 concern the internal economic component of the European Union.

Conception

Only in the context of an efficient cooperation between the European Union and its Member States can a coherent concept be developed and implemented.

---

[754] European Commission (2008b), l.c. The Communication "The raw materials initiative – meeting our critical needs for growth and jobs in Europe" already is mentioned in the preface and on several parts in Chapter 6.

7.2 Present developments at the EU level|The Raw Materials Initiative

An interservice group on raw materials, chaired by DG Enterprise,[755] aims at ensuring coherence and consistency. The EU Raw Material Initiative introduces the **following concept**: ten actions based on the three strategic pillars (table 2). Pillar I covers actions 2–5 (EU external economic component); pillar II covers actions 6 and 7. pillar III includes actions 8–10 (EU internal economic component). Action 1 is not linked to a specific pillar; it is relevant for all pillars. Thus, external and internal supply risks are considered and both primary and secondary raw materials are covered. The Commission is to report to the Council on the implementation of the Raw Materials Initiative in two years.

The Raw Materials Initiative regulates the respective responsibilities of European Union, Member States and industry as follows: The supply of raw materials is a task primarily of the enterprises. The responsibility of the EU Member States (national level) is to establish an optimal framework for minerals policy, whereas it is the duty of the European Commission (EU level) to establish a coherent approach for raw materials protection in Europe, which exceeds the Member States' field of responsibility.

---

755 Regarding the EU Raw Material Initiative, DG Enterprise and Industry has the overall coordination; whereas key DGs such as Trade, Development and Environment account for implementing actions that resort specifically under their competence. For instance, DG Trade focuses on trade restrictions which do not correspond with WTO/bilateral rules.

Table 81: EU Raw Materials Initiative – 10 actions (Data by European Commission, 2008b) WG: working group; EC: European Commission

		Level of response			"Work in Progress"		
					WG 1	WG 2	–
		EC	Member States	Industry	Criticality	Best practices	Recycling
1	Define critical raw materials	x	x	x	x		x
2	Launch of EU strategic raw materials diplomacy with major industrialized and resource rich countries	x	x		x		x
3	Include provisions on access to and sustainable management of raw materials in all bilateral and multilateral trade agreements and regulatory dialogues as appropriate	x	x		x		x
4	Identify and challenge trade distortion measures taken by third countries using all available mechanisms and instruments, including WTO negotiations, dispute settlement and the Market Access Partnerships, prioritising those which most undermine open international markets to the disadvantage of the EU. Monitor progress by issuing yearly progress reports on the implementation of the trade aspects, drawing, as appropriate, on inputs from stakeholders	x		x	x		x
5	Promote the sustainable access to raw materials in the field of development policy through the use of budget support, cooperation strategies and other instruments	x	x		x	x	x

7.2 Present developments at the EU level The Raw Materials Initiative

		Level of response			"Work in Progress"		
					WG 1	WG 2	–
		EC	Member States	Industry	Criticality	Best practices	Recycling
6	Improve the regulatory framework related to access to land by: promoting the exchange of best practices in the area of land use planning and administrative conditions for exploration and extraction and developing guidelines that provide clarity on how to reconcile extraction activities in or near Natura 2000 areas with environmental protection	x	x		x	x	
7	Encourage better networking between national geological surveys with the aim of increasing the EU's knowledge base		x		x		
8	Promote skills and focussed research on innovative exploration and extraction technologies, recycling, materials substitution and resource efficiency	x	x	x	x	x	
9	Increase resource efficiency and foster substitution of raw materials	x	x	x			x
10	Promote recycling and facilitate the use of secondary raw materials in the EU	x	x	x	x		x

In the following, a **short description of the** particular **actions** related to the three pillars of the EU Raw Materials Initiative is given:

Referring to Pillar I: Securing the supply of raw materials in the international markets under equal conditions.

Action 2 concerns foreign policy (responsibility: European Commission and Member States). The EU should actively pursue **raw materials diplomacy** with regard to securing access to minerals. This includes coordination on EU level in the management of EU strategic partnerships (EU strategic partners include Brazil, Canada, China, India, Japan, Russia and USA) and political dialogues with third partner countries. In particular with:[756]

- Africa, by reinforcing dialogue and actions in the field of access to raw materials and natural resources management as well as transport infrastructure, within the implementation of the Joint Strategy and Action Plan 2008–2010 (compare Action 5, development policy, below);

- Emerging economies rich in resources, such as China and Russia, by reinforcing dialogue;

- Resource-dependent countries such as the US and Japan, by identifying common interests and devising joint actions and common positions in international forums, e. g. joint projects with the US Geological Survey in areas open to international cooperation.

Moreover, the European Commission also intends to support the raising of awareness in forums such as the G8, OECD, UNCTAD, UNEP and explore possibilities of cooperation with international organisations such as the World Bank and the International Seabed Authority.

Action 3 and 4 are related to trade policy. Action 3 includes provisions on access to and sustainable management of raw materials in all bilateral and multilateral trade agreements and regulatory dialogues. Action 4 shall identify and question trade distorting measures taken by third countries (i. e. countries outside the EU) using all available mechanisms and instruments, including WTO negotiations, dispute settlement and Market Access Partnerships, prioritising those which mostly undermine open international markets to the disadvantage of the EU. In this regard, progress shall be monitored by issuing annual progress reports on the implementation of the trade aspects, drawing, where appropriate, on inputs from stakeholders

---

756 European Commission (2008b), l.c., p 3.

7.2 Present developments at the EU level The Raw Materials Initiative

Action 5 is related to development policy. Action 5 shall promote sustainable access to raw materials in the field of development policy through the use of budget support, cooperation strategies and other instruments. A European Union development policy has to be compatible within the context of the World Summit on Sustainable Development in Johannesburg 2002. The nations participating in this meeting underlined the increasing significance of the mining industry for many developing countries regarding the industrialised countries' need of raw materials. The contribution of the mining industry sector to a stable economic development in developing countries also depends on political stability as well as infrastructural improvement. The Implementation Plan, chapter IV (article 24–46) must consider this issue. In order to do so, the Implementation Plan was established in 2005 in the Intergovernmental Forum on Mining, Mineral, Metals and Sustainable Development. The Forum was meant to strengthen the structures of a global dialogue.[757] Beyond that, an evaluation of the Implementation Plan (with emphasis on raw materials) has to be accomplished until 2011 (Mojarov, 2008). This, in principle, is part of action 2 of the EU Raw Materials Initiative (see above).

Referring to Pillar II: Setting the appropriate framework conditions within the EU in order to foster sustainable supply of raw materials from European sources.

Actions 6 and 7 pertain to pillar II. Establishing an EU raw materials strategy requires an analysis of the current supply situation of the European Union, i. e. defining critical raw materials. For this purpose, an ad hoc working group was established in May 2009, consisting of members of the European Commission, representatives of the Member States, experts from universities, geological surveys, the industry and NGO's. The Commission is responsible for criticality at the EU level, taking into account the following three criteria: significance for the EU economy, supply risks (external and internal) and lack of substitutes. Results shall be discussed in the planned meetings at the EU level.

Furthermore, access to land should be carefully regulated through land use planning, so as to avoid sterilisation of deposits, and more efficient licensing and permitting procedures, i. e. shortening of approval procedures.

This is covered by action 6: According to the principle of subsidiarity, the Commission will provide a platform for Member States to exchange best practices in the area of land use planning and administration and other important framework conditions for the mining industry. For this reason, a second ad

---

[757] www.globaldialogue.info. Unless the situation has changed in the past few months, the only developed nations that are formally participating in the Forum are Canada and the UK.

hoc working group was established in May 2009, consisting of members of the European Commission and Member States. This group will also be occupied with the recommendation by the European Commission to promote better networking between the national geological surveys to facilitate the exchange of information and improve the interoperability of data and their dissemination (action 7).

Referring to Pillar III: Boosting resource efficiency and promote recycling to reduce the EU's consumption of primary raw materials and decrease the relative import dependence.

Actions 8, 9 and 10 focus on the internal economic component of the EU. Action 8 is relevant for pillar II promotion of skills and research on exploration/extraction) and for pillar III (research resource efficiency, recycling, substitutes). Action 9 focuses on increasing resource efficiency and foster substitution of raw materials. Action 10 promotes recycling and facilitates the use of secondary raw materials in the EU.

Further development in 2009 and 2010

On 3 February 2009, the European Economic and Social Committee published the "Opinion of the European Economic and Social Committee on the 'Non-energy mining industry in Europe'", underlining in its conclusions and recommendations that the main pillars for the future security of raw materials supply in Europe are *domestic supply, international supply, capacity building and resource efficiency*.[758]

Regarding the external angle of the RMI, a first milestone was achieved with the publication of **DG Trade's activity report 2009 on raw materials**.[759] The report summarizes the progresses accomplished along the *three axes of the trade raw materials strategy* as follows:

Include, as appropriate, the relevant trade disciplines on sustainable supply of raw materials in bilateral and multilateral trade agreements.

Identify illegitimate trade distortive measures taken by third countries and tackle them using all available instruments, including through bilateral consultations, the Market Access Partnership process or, if necessary, the WTO dispute settlement; while delimitating more clearly permissible exceptions for e. g. development purposes.

---

758 European Economic and Social Committee (2009/C 27/19).
759 http://trade.ec.europa.eu/doclib/docs/2010/june/tradoc_146207.pdf

7.2 Present developments at the EU level The Raw Materials Initiative

Reach out to third countries to show that the question of sustainable raw materials supply is an issue relevant to all countries, developing or developed, resource-rich and resource-poor alike as the uncontrolled, unregulated multiplication of trade restrictions can lead to a generalized beggar-thy-neighbour policy detrimental to most countries; while recognising the importance of respecting internationally agreed rules on the subject.

In May 2009, the Competitiveness Council endorsed the major objectives set out by the EU Raw Material Initiative (RMI) and invited the Commission, Member States and stakeholders to act swiftly in the implementation of various lines of action outlined by the RMI. It also welcomed the Commission's intention to report back on the implementation of the RMI by the end of 2010. The launch of the RMI coincided with the full onset of the financial and economic crisis. The evolution of the international raw material markets has confirmed the structural nature of the issues at stake and thus reinforced the need to further pursue the objectives of the EU RMI.

Luleå Declaration 2009

The importance of sustainable mineral resources supply was also highlighted in the Luleå Declaration. From October 12 to 14 2009, a conference on sustainable mineral resources within the European Union was held in Luleå, Sweden, under the title "European higher education and research on metallic and mineral raw materials". The organisations present at the conference strongly supported the EU Raw Materials Initiative, it being the right step at the right time. As a result, the Luleå Declaration was adopted, which was meant as a response from the European extractive industry, governmental institutions and academia across Europe. It emphasises the necessity of a common agenda concerning research and higher education in the field of sustainable supply of metallic and non-metallic raw materials. The organisations present at the Conference, listed below, agreed on the following declaration:

- *The past and future society without minerals and metals is unthinkable.*

- *The global growth of the population and the growth of the world economy put strong emphasis on securing future mineral supply.*

- *There is great potential for a sustainable supply of raw materials from EU resources, but we need access to land, an improved knowledge base and R&D to improve methods for exploration, extraction and recycling.*

- *European industry is highly competitive and a high-tech technology provider for the world. It is important that the mineral sector is recognized in the EU land access planning and in the EU R&D programmes for maintaining this leadership.*

Chapter 7 Towards a European minerals policy

- *The future mineral supply is a great challenge for the society at large and the extractive industry is committed to achieve a sustainable mineral supply to meet future challenges by excellence in research through EU funding.*

- *A sustainable supply of minerals and metals also involves balancing the impact on the environment and climate. This could be reached by improvement of resource and energy efficiency and by increased use of secondary raw materials. It is important to stimulate innovation and R&D in all these areas.*

Table 82: Organisations present at the Luleå Conference

Universities	Organisations and Institutes	Industry
INPL, Nancy	Centek	Atlas Copco
Luleå University of Technology	Cewik	Boliden AB
RWTH, Aachen	Euromines	Destia Ltd
University of Leoben	Georange	KGHM
University of Oulu	Inst. Mech. Constr. & Rock	LKAB
University of Oslo	MinFo	Lappland Goldminers AB
	Mintek	Nordic Rock Tech Centre AB
Geological Surveys	MIRO	Outotec Oyj
BRGM, France	MITU	Raw Materials Group
EuroGeoSurveys	North Sweden European Office	RHI
Geological Survey of Finland	ProMine	
Geological Survey of Norway	SveMin	
Geological Survey of Sweden	Sveriges Bergmaterialindustri	
IGME, Spain		

The annex of the Luleå Declaration describes the way forward: The involved stakeholders in the ETP-SMR are committed to contribute through a major research effort to secure critical and essential resources for Europe and achieve higher resources efficiency. In order to achieve this, the ETP-SMR is seeking support from the European Commission and the Member States

7.2 Present developments at the EU level The Raw Materials Initiative

through complimentary funding of research as well as active involvement through a related ERANET[760] which the industry will apply for.

For this purpose it was recommended that the Framework programmes 7 and 8 clearly identify topics related to the ETP-SMR Strategic Research Agenda. For the FP8, the necessary actions relating to the EU Raw Minerals Initiative should be clearly identified in the programme. The structure of the programme should be suitable for the process industry where IT, energy, environment and materials issues are strongly linked. The organisations present at the Luleå Conference supported increased transparency and simplified processes of planning and implementation of the framework programmes.

The following items are seen as essential in order to implement the Raw Minerals Initiative research strategy:[761]

Figure 179: Raw Materials Initiative research strategy – relevant items (Data by http://ec.europa.eu)

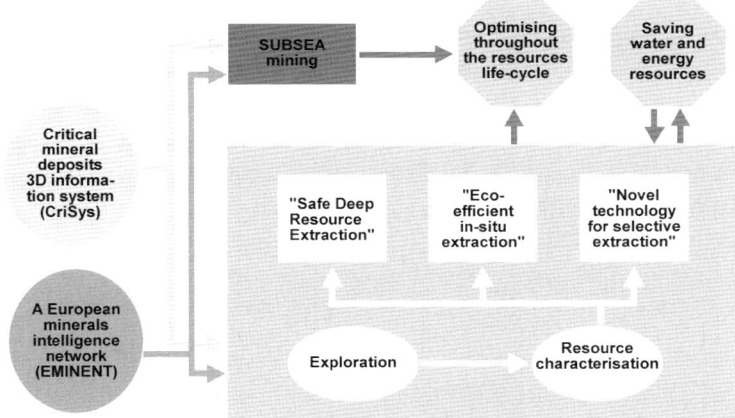

Finally, the annex of the Luleå Declaration is considering the strong links between high quality research and development and high quality education. It is important to have a clear definition of European minerals programmes, a framework for funding Pan-European education, funding of mobility of stu-

---

760 The objective of the ERA-NET scheme is to step up the cooperation and coordination of research activities carried out at national or regional level in the Member States and Associated States through the networking of research activities conducted at national or regional level, and the mutual opening of national and regional research programmes (http://cordis.lu/coordination/era-net.htm).
761 Most of these research topics are also mentioned in section 3.

dents and teachers, international networking and integration of European education with other leading schools internationally.

http://ec. europa. eu/enterprise/policies/raw-materials/files/best-practices/lulea_decl_2009_en. pdf (Luleå – Declaration)

European Minerals Conference in Madrid 2010

June 2010 a European Minerals Conference was held in Madrid (Spain). At the conference, the "Raw Materials Initiative Working Groups Documents" from ad hoc working group on "Best practices in the area of land use planning and permitting and geological knowledge sharing"[762] and ad-hoc working group on "defining critical raw materials" were presented. Finally the Madrid Raw Materials Declaration 2010 was published.

The "Report of the Ad-hoc Working Group on defining critical raw materials" is assessing the term criticality, includes results and list of critical minerals (see table 27) and provides recommendations.[763] The report presents the outcome of this cooperation achieved through an expert working group ("the Group") which was active between April 2009 and June 2010 under the umbrella of the Raw Materials Supply Group. This report sets out an approach to determining criticality. In particular, it takes into account the substitutability between materials, i. e. the potential for substitution of a restricted raw material by another that does not face similar restrictions. It deals with primary and secondary raw materials, the latter being considered as similar to an indigenous European resource. Several recommendations are made. In the following an excerpt is provided:

"The Group recommends that the list of EU critical raw materials should be updated every 5 years and that the scope of the criticality assessment should be increased."

"The Group recommends policy actions to improve access to primary resources aiming at supporting the findings and recommendations resulting from the work carried out by the ad hoc working group on "Best practices in the area of land use planning and permitting" with a view to securing better access to land, fair treatment of extraction with other competing land uses and to developing a more streamlined permitting processes."

---

[762] EC, DG Enterprise (2010): Best practices in the area of land use planning and permitting and geological knowledge sharing (http://ec.europa.eu/enterprise/policies/raw-materials/sustainable-supply/index_en.htm).

[763] EC, DG Enterprise (2010): Report of the Ad-hoc Working Group on defining critical raw materials, p. 5–9. (http://ec.europa.eu/enterprise/policies/raw-materials/critical/index_en.htm)

"The Group recommends that policy actions, with regard to trade and investment as defined in the trade raw materials strategy, be pursued."

"The Group recommends that policy actions are undertaken to make recycling of raw materials or raw material-containing products more efficient."

"The Group recommends that substitution should be encouraged, notably by promoting research on substitutes for critical raw materials in different applications and to increase opportunities under EU RTD Framework Programmes."

"The Group recommends that the overall material efficiency of critical raw materials should be achieved by . . . fundamental measures."

Table 83: List of critical raw materials at the EU level (EC, DG Enterprise, 2010)

Antimony	Indium
Beryllium	Magnesium
Cobalt	Niobium
Fluorspar	PGMs (Platinum Group Metals)
Gallium	Rare earths
Germanium	Tantalum
Graphite	Tungsten

Madrid Raw Materials Declaration 2010

The ‚Madrid Raw Materials Declaration 2010' is the final document of the European Minerals Conference Madrid 2010. The Industry's collective viewpoints are described in detail in its Madrid Declaration, the key points of which may be summarised are:[764]

"At European level, a Raw Materials Policy needs to be promoted, defining the strategy to ensure that Europe in future will have sufficient supplies of imported raw materials and sufficient access to indigenous raw materials.

There is a need to develop corresponding Raw Materials Policies at national, regional and local levels to ensure good present and future access to the raw materials geologically present.

There is a need to develop associated Land Use Planning Policies, to ensure that land use development for minerals extraction around these geologically-present resources is preferentially treaded.

---

[764] Madrid Raw Materials Declaration 2010, press release, June 2010.

There is a need to adopt best practices in Permitting Procedures following good examples from other Member States, to ensure permits are granted in a timely and efficient manner and for duration that justify the significant capital investments involved."

The Minerals Industry recommends the Raw Materials Initiative to adopt these proposals and to incorporate them in the **final communication on the Initiative expected by end-2010**. Because of the ongoing importance of the issue of access to raw material resources, the Industry also suggests that the Initiative be followed up by regular reviews over the next 5 years and also be **part of the Commission's 2020 Agenda and Strategy**.[765]

New developments

All information below is taken from European Commission documents.

Europe 2020 Strategy

Europe 2020 is the EU's growth strategy for the coming decade and was published in 2010.[766] In a changing world, the EU aims at developing into a smart, sustainable and inclusive economy. These three mutually reinforcing principles shall support the EU and the Member States to achieve high levels of employment, productivity and social cohesion. Concretely, the Union has set five ambitious objectives - enhanced employment, R & D/innovation and education, measures responding to climate change/energy, reduction of poverty and social exclusion – to be reached by 2020. Each Member State shall transform these goals into its own national targets. Concrete actions at both EU and national levels will underpin the strategy. Seven so-called flagship initiatives were presented as new engines to boost growth and jobs, where both the EU and national authorities have to coordinate their efforts to make them mutually reinforcing.[767] Related to the topic of this book, two initiatives of the Europe 2020 flagship have to be pointed out: "An Industrial Policy for the Globalisation Era" and "A resource-efficient Europe".

---

[765] Note: Meanwhile the EU Raw Material Initiative has gathered extra momentum with adoption of the Europe 2020 Strategy that includes as one flagship "An industrial policy for the globalisation era" and that foresees the setting up of a framework for a modern industrial policy that will "address all elements of the increasingly international value chain from access to raw materials to after-sales service" (see: http://ec.europa.eu/enterprise/policies/raw-materials/files/docs/questionnaire-raw-mat-pc_en.pdf).
[766] http://ec.europa.eu/europe2020/index_en.htm
[767] http://ec.europa.eu/europe2020/tools/flagship-initiatives/index_en.htm

Europe 2020 flagship: An Industrial Policy for the Globalisation Era[768]

End of October 2010, the Commission published the Communication on "An Integrated Industrial Policy for the Globalisation Era Putting Competitiveness and Sustainability at Centre Stage".[769] It sets out a strategy that aims to boost growth and jobs by maintaining and supporting a strong, diversified and competitive industrial base in Europe offering well-paid jobs while lessening the carbon footprint. The core message of the Communication is, industry must be placed centre stage if Europe is to remain a global economic leader. Coordinated European policy responses are needed, looking at the whole added value chain, from infrastructure and raw materials to after-sales service. Promoting the creation and growth of small and medium-sized enterprises has to be at the core of EU industrial policy. Moreover, the transition to a sustainable economy has to be seized as an opportunity to strengthen competitiveness. Only a European Industrial Policy targeting competitiveness and sustainability can muster the critical mass of change and coordination needed for success.

Ten key actions for European industrial competitiveness are listened; action 6 is referring to minerals: *A new strategy on raw materials will be presented to create the right framework conditions for sustainable supply and management of domestic primary raw materials.*

Section 6.2 (page 18/19) is referring to "Ensuring access to raw materials and critical products". Secure, affordable, reliable and undistorted access to raw materials is essential for industrial competitiveness, innovation, and jobs. Security of supply highly depends on the degree of diversification of suppliers and their reliability, whether they are located inside the country or abroad. **Well-functioning global markets** for raw materials and commodities are essential for an efficient allocation of global resources and to enable technological progress. However, short-term movements in prices necessitate hedging of substantive risks. At the same time, global competition for resources has been sharpened by the rise of the emerging countries in the world economy. To address these issues, the Commission launched the **Raw Materials Initiative** in November 2008, followed by a detailed analysis of the demand and potential scarcity of defined key raw materials in June 2010. These initiatives have pio-

---

768 http://ec.europa.eu/enterprise/policies/industrial-competitiveness/industrial-policy/index_en.htm

769 Communication from the Commission to the European Parliament, the Council, the European Economic and Social Committee and the Committee of the Regions
An Integrated Industrial Policy for the Globalisation Era Putting Competitiveness and Sustainability at Centre Stage, COM(2010) 614, {SEC(2010) 1272}, {SEC(2010) 1276}
http://ec.europa.eu/enterprise/policies/industrial-competitiveness/industrial-policy/files/communication_on_industrial_policy_en.pdf

neered the development of an EU strategy on raw materials emphasizing the concept of "added value chain", which continues to pursue the 3-pillar strategy to: (i) ensure a level playing field in access to resources in third countries; (ii) foster sustainable supply of raw materials from European sources, and (iii) reduce consumption of primary raw materials by increasing resource efficiency and promoting recycling.

New multilateral rules and agreements on **sustainable international management and access to raw materials** have to be made; while policy actions are needed to address export restrictions and constraints on exploration and extraction by third countries, in particular strategic partner countries and Africa. In cases of anti-competitive agreements or market concentration threatening to endanger access to raw materials, it is also essential to apply vigorously the EU's existing competition rules. Promoting **mining and processing technologies** to strengthen resource efficiency, recycling, substitution and use of renewable raw materials, are efficient tools to reduce the critical dependence of the EU on primary raw materials. This is also an important contribution to improve the environmental balance, e.g. through increased use of secondary raw materials (scrap), end of life electronic equipment and vehicles exported to third countries; enforcement of the Waste Shipment Regulation; reuse or recycling of products and materials based on agreed minimum standards.

A **sustainable supply and management of raw materials within the EU** requires adequate framework conditions: boosting the efficient use of EU's own resources, recycling and increased substitution of raw materials wherever this is possible. So far unknown European mineral reserves, located at terrestrial as well as marine deposits, may be discovered if sufficient investment is provided for new and innovative exploration and extraction techniques under favourable administrative conditions also for land use and maritime spatial planning, thus ensuring sustainability. The Commission intends to: *present a Strategy on Raw Materials including proposals on fostering better framework conditions for sustainable supplies of domestic primary raw materials, increased recycling, and finding substitutes for other raw materials.*

A resource-efficient Europe – Flagship initiative of the Europe 2020 Strategy

End of January 2011 the Communication on "A resource-efficient Europe" was published.[770] The flagship initiative for a resource-efficient Europe under the

---

[770] Communication from the Commission to the European Parliament, the Council, the European Economic and Social Committee and the Committee of the Regions: A resource-efficient Europe – Flagship initiative under the Europe 2020 Strategy, COM(2011) 21. http://ec.europa.eu/resource-efficient-europe/pdf/resource-efficient_europe_en.pdf; http://ec.europa.eu/resource-efficient-europe/index_en.htm

7.2 Present developments at the EU level The Raw Materials Initiative

Europe 2020 strategy shall support the shift towards a resource-efficient, low-carbon economy to achieve sustainable growth. Our economy and our quality of life are based on the availability of natural resources. Their efficient and sustainable use is necessary to secure economic growth and jobs for Europe. Targets are improved productivity, reduced costs and increased competitiveness. The flagship initiative for a resource-efficient Europe shall provide a long-term framework for actions in many policy areas, besides *raw materials* also covering the areas of climate change, energy, transport, industry, agriculture, fisheries, biodiversity and regional development. This is expected to increase certainty for investment and innovation and to ensure resource efficiency of all relevant policies in a balanced manner.

Global economy and our high-level standard and quality of life are inconceivable without access to and availability of natural resources. These resources include raw materials such as fuels, minerals and metals but also food, soil, water, air, biomass and ecosystems. By 2050, the global population is expected to have grown by 30% to around 9 billion – all of them competing for naturally limited resources. Inhabitants of developing and emerging economies will legitimately strive for similar welfare and consumption levels as are standard in developed countries. To enjoy the benefits of a resource-efficient and low-carbon economy, three conditions have to be fulfilled (page 2,3 of COM (2011)21):

- *"First, we need to take coordinated action in a wide range of policy areas and this action needs political visibility and support.*
- *Second, we have to act urgently due to long investment lead-times. While some actions will have a positive impact on growth and jobs in the short-term, others require an upfront investment and have long pay-back times, but will bring real economic benefits for the EU economy for decades to come.*
- *Third, we have to empower consumers to move to resource-efficient consumption, to drive continuous innovation and ensure that efficiency gains are not lost."*

This flagship initiative aims to create a framework for policies to support the shift towards a resource-efficient and low-carbon economy which shall help us to:

- *"boost economic performance while reducing resource use;*
- *identify and create new opportunities for economic growth and greater innovation and boost the EU's competitiveness;*
- *ensure security of supply of essential resources;*
- *fight against climate change and limit the environmental impacts of resource use."*

Resource efficiency for Europe needs technological improvements and significant changes in the fields of energy, industrial production, transport systems and agriculture. This means changes in our behaviour as producers and consumers, too. Businesses need certainty to invest now with a view to benefit for future generations. To provide long-term stability, the establishment of a regulatory framework must be started immediately. Improving resource efficiency implicates keeping costs under control by reducing material and energy consumption and goes along with boosting future competitiveness.

This flagship initiative is intended to help build a strategic and integrated approach to achieve the longterm goals. Conrete actions already decided for the period until 2020 shall pave the way to the next stage, the appropriate development until 2050 and beyond. The synergies inherent in such a broad based strategy need to be optimised, and any problems and trade-offs as part of well-informed policy making have to be identified and tackled. A coherent analysis of the reasons why some resources are not used efficiently must be the starting point for making resource efficiency an indispensable principle for a wide range of policies. The development of a set of adequate tools to allow policy makers to drive forward and monitor progress shall encourage the clear support and involvement of national, regional and local authorities, stakeholders and citizens.

In the framework of the Europe 2020 flagship initiative on resource efficiency, the Commission presents a roadmap for a resource efficient Europe.[771]

Communication tackling the challenges in commodity markets and on raw materials

At beginning of February 2011 the „Communication tackling the challenges in commodity markets and on raw materials" was published.[772] The Communication considers the issues of the Raw Materials Initiative 2008 (see below) and is structured as follows:

1. Introduction
2. Developments on global commodities markets
2.1 Developments on the physical markets
2.1.1 Energy (oil, electricity, gas)
2.1.2 Agriculture and food security

---

[771] Annex 1: Initiatives foreseen in 2011 to deliver on the resource-efficient Europe flagship10.
[772] Communication from the Commission to the European Parliament, the Council, the European Economic and Social Committee and the Committee of the Regions tackling the challenges in commodity markets and on raw materials, com (2011) 25, http://ec.europa.eu/enterprise/policies/raw-materials/files/docs/communication_en.pdf

2.1.3	Raw materials
2.2	The financialisation of commodities markets
3.	EU Policy response to developments on commodities markets
3.1	Physical markets
3.1.1	Energy (oil, electricity gas)
3.1.2	Agriculture and food security
3.2	Regulation of financial markets
3.3	The interaction between physical and financial commodities markets
4.	The European Raw Materials Initiative
4.1	Identifying critical raw materials
4.2	Implementing the EU trade strategy for raw materials
4.3	Development instruments
4.4	New research, innovation and skills opportunities
4.5	Guidelines on the implementation of Natura 2000 legislation
4.6	Increased resource efficiency and improved conditions for recycling
5.	Future orientations on the Raw Materials Initiative
5.1	Monitoring critical raw materials
5.2.	Fair and sustainable supply of raw materials from global markets (pillar 1)
5.2.1	A trade strategy
5.2.2	Sustainable supply of raw materials and development needs
5.3	Fostering sustainable supply within the EU (pillar 2)
5.4	Boosting resource efficiency and promoting recycling (pillar 3)
6.	Way forward

Markets are experiencing the growing impact of finance, with a significant increase in financial investment flows into commodity derivative markets in recent years. Between 2003 and 2008, for example, institutional investors increased their investments in commodities markets from 13 billion euro in 2003 to between 170 and 205 billion euro in 2008. While the financial crisis interrupted the upward trend, financial positions approached or even exceeded their 2008 peaks on many markets in 2010 and investment by index traders in particular has increased strongly. The challenges of commodity prices and raw materials are closely intertwined and touch on policies in the areas of financial markets, development, trade, industry and external relations. This Communication presents an overview of what has been achieved in each of these areas and of the steps which are planned to take the work forward. As a part of the Europe 2020 strategy it intends to ensure smart, sustainable and inclusive growth and is closely linked to the flagship initiative for a resource efficient Europe. It shall feed into the work of the G20 which agreed at the Pittsburgh summit "to improve the regulation, functioning, and transparency of financial and commodity markets to address excessive commodity price volatility". This commitment was reinforced in November 2010 by the G20 summit in

Seoul, which pledged to address food market volatility and excessive fossil fuel price volatility.

Future orientations of the Raw Materials Initiative (RMI)
Section 5 of the new Communication deals with future orientations of the Raw Materials Initiative. While significant progress has already been made in implementing the RMI, further improvements are necessary. An integrated approach based on the three pillars is essential, as each contributes to the objective of ensuring a fair and sustainable supply of raw materials to the EU.

Monitoring critical raw materials (section 5.1)

Securing supplies of raw materials is defined as essentially the task of companies, while public authorities have to establish the right framework conditions for companies to carry out this task. The Commission intends to explore with the extractive, recycling and user industries the potential for targeted actions, notably with regard to recycling. It is also ready to examine with Member States and industry, the added value and feasibility of a possible stockpiling programme of raw materials. Moreover, the Commission will:

- *Monitor the issues of critical raw materials to identify priority actions, and will examine this with Members States and stakeholders.*
- *Regularly update the list of critical raw materials at least every 3 years.*

Fair and sustainable supply of raw materials from global markets – pillar 1 (section 5.2)

The EU will actively pursue a "raw materials diplomacy" with a view to securing access to raw materials, in particular the critical ones, through strategic partnerships and policy dialogues.[773]

Development policy and sustainable supply of raw materials
Sustainable mining is a tool to support sustainable development. However, in many developing countries – especially in Africa – governance issues like regulatory frameworks or taxation symstems have turned out to be obstacles for the realisation of sustainable and inclusive growth, though the richness of existing mineral reserves would allow for that . Enhancing governance and transparency, as well as encouraging trade and investment in the raw materials sector, is essential for achieving inclusive growth and sustainable development in resource rich countries. European development policies and partnership with developing countries can create benefit from the sustainable supply of raw materials for both developed and developing countries. Financial revenue

---

773 Already mentioned in the Raw Material Initiative 2008.

from the mining sector can be used for the improvement of social conditions, supporting the objectives of inclusive growth and poverty reduction.

The Commission will consider further these issues in the context of the Green Paper consultation process on the future of EU development policy and budget support as well as in its public consultation on country-by-country reporting. The EU will encourage partner governments to develop comprehensive reform programmes that clearly identify objectives such as improving mining taxation regimes or enhancing revenue and contract transparency, or enhancing the capacity for using revenues to support development objectives. Greater transparency will help society at large and national supervisory bodies to hold governments and companies to account for revenue payments and receipts, and thus decrease fraud and corruption and ensure a more predictable trade and investment climate.

In Addis Ababa in June 2010 the Commission agreed with the African Union Commission (AUC) to establish bilateral co-operation on raw materials and development issues based on the RMI and the AUC's policy on mining and minerals, i.e. the 2009 'African Mining Vision'.[774] This co-operation will focus on three areas: governance, investment and geological knowledge/skills. Under the Africa-EU Joint Strategy 2011–2013, agreed at the Africa-EU Summit held in November 2010, actions on raw materials are foreseen under the Trade, Regional Economic Integration and Infrastructure Partnership. The EU and its Member States will work jointly on these issues. The Commission proposes to:

– *enhance European financial and political support for the Extractive Industries Transparency Initiative (EITI), and help developing countries to implement it;*

– *share best practice with international organisations such as the World Bank, IMF, and the African Development Bank;*

– *examine ways to improve transparency throughout the supply chain and tackle in coordination with key trade partners situations where revenues from extractive industries are used to fund wars or internal conflicts;*

– *promote more disclosure of financial information for the extractive industry, including the possible adoption of a country-by-country reporting requirement. The Commission will take into account progress made by the International Accounting Standards Boards on an International Financing Reporting Standard for extractive industries, as well as the current status of legislation of third countries active in the region;*

---

774 Also mentioned above.

- *promote the application of EU standards by EU companies operating in the developing countries and the application of the Best Available Technique Reference document and by developing a code of conduct of EU companies operating in third countries; and*

- *support the work by the OECD on due diligence in the mining sector;*

- *continue to assess – with African countries – the feasibility of assisting further co-operation between both continents' geological surveys and to promote co-operation in this area in multilateral fora such as UNESCO's Geosciences Programme.*

Many developing countries possess considerable mineral wealth but cannot use it for the benefit of their populations, because the poor state of transport, energy and environmental infrastructure limits their activities. The European Commission, the European Investment Bank (EIB), and other European development financing institutions, in co-operation with African national and regional authorities, shall continue to assess how to promote the most appropriate infrastructure, and related governance issues, that can contribute to the sustainable use of the resources of these countries and facilitate raw materials supply, using respective sector dialogues to steer this process. In particular, the European Commission will assess (a) the feasibility of increasing lending (which may include grant-loan elements) to industry, including mining and refining projects and in particular post-extractive industries and (b) investigate the possibility of promoting financial instruments that reduce risk for operators on the basis of guarantees supported by EU, including by the European Development Fund. The existing EU-Africa Infrastructure Trust Fund (the purpose of the Trust is to benefit cross-border and regional infrastructure projects in sub-Saharan Africa.) could also assist African countries in this task.

Development policy shall also target the creation of linkages from the extractive industry towards local industry, by improving the value chain and maximising diversification. Therefore, an enabling business capacity building should be fostered and trade agreements provide the necessary flexibility to achieve this aim. The EU can also help developing countries increase their geological knowledge (for example, the AEGOS project brings the EU's and Africa's geo-surveys together to improve the level and quality of resource data available for Africa) to allow them to better estimate national mineral reserves, better plan budgets based on expected revenues from these reserves and give increased bargaining power vis-à-vis mining firms.

7.2 Present developments at the EU level|The Raw Materials Initiative

Reinforcing the raw materials trade strategy

The Commission intends to reinforce the Raw Materials Trade Strategy (see above) as set out in section 4,2 in line with development and good governance objectives. The Commission considers that the EU should:

– continue to develop bilateral thematic raw materials dialogues with all relevant partners, and strengthen ongoing debates in pluri – and multilateral fora (including e.g. G20, UNCTAD, WTO, OECD); carry out further studies to provide a better understanding of the impact of export restrictions on raw materials markets, and foster a dialogue about their use as a policy tool.

– further embed raw materials issues, such as export restrictions and investment aspects, in ongoing and future EU trade negotiations in bilateral, plurilateral and multilateral frameworks.

– pursue the establishment of a monitoring mechanism for export restrictions that hamper the sustainable supply of raw materials, and will continue to tackle barriers distorting the raw materials or downstream markets with dialogue as the preferred approach, but using dispute settlement where justified.

– encourage in OECD activities the inclusion of relevant non-OECD members in the work on raw materials, and explore further multilateral and plurilateral disciplines including consideration of best practices.

– use competition policy instruments to ensure that supply of raw materials is not distorted by anti-competitive agreements, mergers or unilateral actions by the companies involved.

– take forward the above mentioned actions, and further analyse priorities for raw materials in relation to third countries through autonomous measures, bilateral and multilateral frameworks and dialogue; and continue to pursue a consistent EU trade policy on these priorities.

Fostering sustainable supply within the EU – pillar 2 (section 5.3)

The Europe 2020 Strategy underlines the need to promote technologies that increase investment in the EU's natural assets. In this context, extractive industries often have to face land use conflicts and are hindered in their development by a heavy regulatory framework yielding competition with other land uses. Many regulatory issues in this area are the competence of Member States. The Commission therefore acts mainly as a facilitator for the exchange of best practices.

At the same time, extraction in the EU must occur in safe conditions. This is important both for the image of the sector and as a precondition for the public acceptance. The Commission considers that the following practices ("Improving framework conditions for extracting minerals for the EU". Report of the RMSG Ad-hoc working group on exchanging best practices on land use planning, permitting and geological knowledge sharing. June 2010.) are particularly important in promoting investment in extractive industries:

- *defining a National Minerals Policy, to ensure that mineral resources are exploited in an economically viable way, harmonised with other national policies, based on sustainable development principles and including a commitment to provide an appropriate legal and information framework;*

- *setting up a land use planning policy for minerals that comprises a digital geological knowledge base, a transparent methodology for identifying mineral resources, long term estimates for regional and local demand and identifying and safeguarding mineral resources (taking into account other land uses) including their protection from the effects of natural disasters;*

- *putting in place a process to authorise minerals exploration and extraction which is clear, understandable, provides certainty and helps to streamline the administrative process (e.g. the introduction of lead times, permit applications in parallel, and one-stop-shop).*

The Commission proposes to assess with the Member States, in full respect of the subsidiarity principle, the feasibility of establishing a mechanism to monitor actions by Member States in the above area, including the development of indicators. A sufficient supply with relevant data accessible in a knowledge base is necessary to build a raw materials strategy. In the short term the Commission proposes to assess with the Member States the scope for increased synergies between national geological surveys, that would allow for economies of scale, reduced costs and increased potential to engage in joint projects (e.g. harmonised minerals database, European Raw Materials Yearbook). In the medium term, any synergies should contribute to an improved European raw materials knowledge base in a co-ordinated way, in particular taking into account future opportunities within the GMES programme.

Boosting resource efficiency and promoting recycling – pillar 3 (section 5.4)

As worldwide demand for raw materials increases, while raw materials reserves are limited, greater efforts will have to be made on recycling. Higher recycling rates will reduce the pressure on demand for primary raw materials, help to reuse valuable materials which would otherwise be wasted, and reduce energy consumption and greenhouse gas emissions from extraction and processing.

'Urban mining' is the process of extracting useful materials from urban waste and is one of the main sources of metals and minerals for European industry. The use of secondary raw materials contributes to resource efficiency and helps reduce greenhouse gas emissions and preserve the environment. However, the full potential of many of these resources is not being exploited and although recycling of municipal waste in the EU has doubled in 10 years, there are large differences in the situation in the Member States. In spite of pressures to reduce carbon emissions, protect human health and reduce external dependence, there are still barriers to tackle which prevent recycling. The Commission considers that these barriers fall into three broad categories: 'leakage' of waste to sub-standard treatment inside or outside the EU; obstacles to the development of the recycling industry; and inadequate innovation in recycling. Better implementation and enforcement of existing EU waste legislation is essential for promoting a more resource-efficient Europe. The Commission proposes therefore to:

– *review the Thematic Strategy on waste prevention and recycling in 2012 to develop best practices in collection and treatment of key waste streams, in particular those which contain raw materials with a negative impact on the environment. When necessary, the availability of recycling statistics will be improved;*

– *support research and pilot actions on resource efficiency and economic incentives for recycling or refund systems;*

– *carry out an an ex-post evaluation of the EU waste acquis, including an assessment of areas where legislation in the various waste streams could be aligned to improve coherence. This would include the effectiveness of deterrents and penalties for breaches of EU waste rules;*

– *review the action plan on sustainable consumption and production in 2012 to identify what additional initiatives are necessary in this area;*

– *analyse the feasibility of developing ecodesign instruments (i) to foster more efficient use of raw materials, (ii) ensure the recyclability and durability of products and (iii) promote the use of secondary raw materials in products, notably in the context of the Ecodesign Directive; and*

– *develop new initiatives to improve the competitiveness of EU recycling industries notably by introducing new market based instruments favouring secondary raw materials.*

"Innovation"– a cross-cutting issue (section 5.5)

Raw materials are essential inputs for the competitiveness of industry and for the development of many environmentally-friendly, clean-technology applications. Innovation is key to the European Union Network for the Implementation and Enforcement of Environmental Law EU's potential in this area and can play a role in addressing the challenges of the three pillars of the RMI. There is a need for innovation along the entire value chain, including extraction, sustainable processing, eco-design, recycling, new materials, substitution, resource efficiency and land use planning.

European Innovation Partnership (EIP) on raw materials
The Council Conclusions of 10th of March 2011 "INVITES the Commission to further promote innovation and research and development efforts in the raw materials value chain, to assess the case for launching a European Innovation Partnership (EIP) on raw materials and to come forward with proposals for this as appropriate". The proposed partnership shall contribute to the mid and long term security of supply with and access to raw materials essential for the competitiveness of the EU industries, to increase resource efficiency in the EU and to the development of new European based recycling activities (largely SME-based).

Meanwhile, a *public consultation process has started.*[775] *This public consultation concerns a proposal for an Innovation Partnership, for which a decision will need to be taken by the European Commission in the framework of the Innovation Union.*

---

775 From 15/04/2011 to 27/05/2011. http://ec.europa.eu/enterprise/policies/raw-materials/public-consultation-ip/index_en.htm#h2-1

7.2 Present developments at the EU level — The Raw Materials Initiative

Table 84: Minerals policy - Global and EU relevant activities between 1950 and 2011 (some selections)

	Global	European Union
50ies		Treaty of Paris, 1951: Foundation of the European Coal and Steel Community (ECSC). Founding members were Belgium, France, Germany, Italy, Luxembourg, The Netherlands
		Treaty of Rome, 1957: Foundation of the European Economic Community (EEC) and the European Atomic Energy Community (EURATOM). Members: Belgium, France, Germany, Italy, Luxembourg, The Netherlands
60ies	The decades from the 1960 saw an economic decline in the output of the more developed nations of Europe, particularly western Europe. Several Asian nations specialized in producing certain goods, utilizing comparably cheaper labour forces (Japan and the four "Asian tigers", South Korea, Taiwan, Hong Kong, Singapore). –	Treaty of Brussels, 1965: ECSC, EEC and EURATOM merged to European Communities
	Latin American Free Trade Association (LAFTA) was created in the 1960 Treaty of Montevideo by Argentina, Brazil, Chile, Mexico, Paraguay, Peru, and Uruguay. It should foster mutual regional trade among the member states, as well as with the US and the European Union.	
	1963 establishment of Organisation of African Unity (OAU)	
	1967 Bangkok Declaration: foundation of Association of Southeast Asian Nations (ASEAN), including Indonesia, Malaysia, Philippines, Singapore, Thailand (10 member states by now).	
70ies	LAFTA expanded to 11 members.	1975: Communication (COM) of the European Commission (EC) regarding supply of non-ferrous metals
	1st Oil crisis 1973	
	Georgetown, 1975: Foundation of African, Caribbean and Pacific Group of States (ACP). Lomé I Convention about common trade policy with EU states	
	2nd Oil crisis 1979	
80ies	In 1980, LAFTA reorganized into the Latin American Integration Association (ALADI) by the Montevideo Treaty.	Foundation of Raw Materials Supply Group (RMSG). –
	By the latter half 1980ies the shift of industrial production began in the newly industrialising countries: firstly in cheaper, lower technology products (e.g. textiles), then in higher technology goods (refrigerators, automobiles). The shift of international industrial production out of Europe is a key outcome of globalisation.	1984: Selected papers arising from the primary raw materials programme (1978–81) of EC
90ies	End of Soviet Union and its east European satellite states	
	Rio Declaration on Environment and Development, 1991	Treaty of Maastricht, 1993: European Union
	From 1993 till present: increasing engagement of China in Africa (amongst other issues) to secure China's raw materials supply	
	World Trade Organisation (WTO) commenced 1995 under the Marrakech Agreement, replacing the General Agreement on Tariffs and Trade (GATT), which commenced in 1948	By 1995, EU enlargement (to 15 countries). – RMSG publishes European Minerals Yearbook
	Asian Financial Crisis 1997 – 1998 (Southeast Asia, Japan). – 1997: joint declaration on EU-Australia relations (concerning e.g. agriculture, trade, climate, environment)	1997: European Conference on Mineral Planning in Zwolle, Netherlands. –
		RMSG publishes revised edition of European Minerals Yearbook.
		EU-Russia: Partnership and Cooperation Agreement, 1997 (to create four Common Spaces)
		1999: ACP-report stated that Sysmin had in general suffered from the procedural difficulties. –
		1999: European Conference on Mineral Planning, Harrowgate, England
		1999 – 2002: introduction of the EURO as a common currency in EU states (today: 17)

Chapter 7 Towards a European minerals policy

	Global	European Union
00ies		2000: ACP (African, Caribbean and Pacific Group of States) and EU agreement: Lomé IV Convention expired in 2000. – New ACP-EU agreement in Cotonou (2000 – 2020)
		2000: EC, COM 265: Promoting sustainable development in EU non-energy extractive industries (COM/2000/265); Safe operations of mining activities (COM/2000/664)
	The Doha Development Round (WTO) started in 2001 and continues 2011	2001: EC, COM 264: EU Strategy for sustainable development
	2002: Earth Summit, Johannesburg. –	ECSC Treaty expired in 2002. –
	2002: African Union, 53 member states (successor of Organisation of African Unity, OAU)	2002: European Conference on minerals planning, Krefeld, Germany.
		St Petersburg Summit, 2003: reinforce cooperation Russia – EU (Four Common Spaces, e.g. common economic space)
		2004: EU-enlargement by 10 countries (to 25 countries). – 2004: EU-Study on **Mineral Planning Policies** and Supply Practices, commissioned by EC.
	Moscow Summit, May 2005, adopted a package of road maps for the creation of the Four Common Spaces; partly implementation of the road maps at London Summit, Oct. 2005	2005: European Commission (EC), EU thematic strategy on the sustainable use of natural resources, COM 670. – 2005: Establishment of European Technology Platform on Sustainable Mineral Resources (ETPSMR)
		2006: publishing of Mining Waste Directive (2006/21/EC). – 2006: Statement of European Economic and Social Committee on risks and problems associated with the supply of raw materials to European industry (2006/C309/16)
	Beginning of Global Financial Crisis in USA (2007, housing bubble). –	2007: EC, DG Enterprise and Industry, SEC (2007)771: Commission Staff working paper. Analysis of the competitiveness of the non-energy extractive industry in EU. –
	Global economic recession 2008. –	EU enlargement 2007: 27 Member States
		2008: ETPSMR officially recognised.
		2008: Consultation Process of the Raw Material Initiative (RMI); publication of its results in the EU-parliament.
		End of 2008: publication of the Raw Material Initiative, COM (2008)699: The Raw Material Initiative – Meeting our critical Needs for Growth and Jobs in Europe.
	2009 a Free Trade Agreement with the ASEAN regional block of 10 countries and New Zealand and its close partner Australia was signed, it is estimated that this FTA would boost aggregate GDP across the 12 countries by more than US $ 48 billion over the period 2000–2020.	Treaty of Lisbon, 2009
		2009: DG Trade's activity report 2009 on raw materials
		Oct 2009: Luleå Raw Material Declaration 2009

7.2 Present developments at the EU level The Raw Materials Initiative

	Global	European Union
10ies	End of 2010: China becomes the world's second largest economy (surpassing Japan)	2010 Euro Crisis (European sovereign debt crisis). –
	Sept 2009: G 20 summit in Pittsburgh	June 2010: Madrid Raw Materials Declaration 2010.-
	Nov 2010: G 20 summit in Seoul	2010: Europe 2020 Strategy
		June 2010: EC, DG Enterprise and Industry, Report on exchanging best practices on land use planning, permitting and geological sharing. –
		June 2010: EC, DG Enterprise and Industry, Report **of the Ad-hoc Working Group** on defining critical raw materials.
		2010: An integrated Industrial Policy for the Globalisation Era Putting Competitiveness and Sustainability at Front Stage (COM (2010)614). –
		2010: EU-Africa agreement: EU-Africa Joint Strategy for 2011–13 including mining as instrument for development and poverty reduction.
	2011: G 20 summit in France	Jan 2011: A resource-efficient Europe, Flagship initiative under the Europe 2020 strategy, COM(2011)21 final. –
		Feb 2011: Tackling the challenges in commodity markets and on raw materials, COM(2011)25. –
		Feb 2011: European Parliament ratifies EU-South Korea Free Trade Agreement to come into force July 2011

Chapter 8 Conclusions

Raw materials mark the beginning of a complex value added chain. In times of increasing globalization, they are **prerequisite for the functioning and the possibilities for development, prosperity and growth of any national economy.** Adequate minerals policy is an absolute necessity for safeguarding a secure raw materials supply.

The decision makers in the European Union have acknowledged the importance of this issue: **measures aiming at the development of a common European Union minerals policy are being discussed**. The publication of the Communication (COM(2008)699) and the proposal to launch a **European Union Raw Materials Initiative** at the beginning of November 2008 underline the significance of the matter. Nevertheless, it has to be emphasized that these activities of the EU decision makers have started rather late, given the relevance of the topic.

All appropriate contributions that lead to the creation of a European Union minerals policy and accelerate this process are of particular importance. However, the consequences of the delay have to be realized. All EU partners in commodity trading have already established their raw materials strategies, which mean a clear advantage in global competition. The USA, Australia, the so-called BRIC states Brazil, Russian Federation, India and China as well as Indonesia, several Latin American and African states and many developing countries have clearly defined mineral policies. Should the European debate on minerals policy again encounter any obstacles and cause further interruptions, consequences might be augmented and intensified.

Evidently, the raw materials industry is of essential significance for the European Union economy and it's GDP. As a consequence, securing the minerals supply by means of an effective minerals supply policy is of utmost relevance. A minerals policy on EU level on one hand and national mineral policies on the other hand have to be distinguished. The implementation of appropriate structures is essential for both levels. The particular importance of a coherent common European minerals policy is to be emphasized with respect to the legal structure of the European Union.

Public awareness of the significance of raw materials has been insufficient as yet, adequate structures have been lacking. In particular, there is no comprehensive sectoral policy for the securing of minerals supply on EU level (in contrast, for example, to agricultural or water policy).

First of all, the frame conditions and instruments for a sectoral minerals policy on EU level are to be created, a policy for securing the supply with mineral raw materials. This means the establishment of a **consistent and coherent policy** for the supply of the European Union with raw materials and commodities *within the frame of the general EU policy*, considering the background of changing global supply conditions.

Establishing such a policy seems the more necessary, as the European Union becomes increasingly dependent on mineral-rich countries at low stages of economic development and is confronted with offensive economic policies of developing countries. The Communication of November 2008 could be the beginning of a discussion of such a sectoral policy. This is also important considering the fact that minerals relevant partial policies on the EU level are usually aligned in sectoral and not integral manner.

Actually, the per capita raw materials demand of EU citizens will remain at a high level. To cope with the global financial crisis, economic stimulus packages are implemented which aim at the improvement of market conditions by boosting the **growth of industrial goods production.** Global raw materials demand as a whole will continue to rise. At the same time, the **problems of access to mineral resources within and outside of the EU countries** are increasing. It is important to mention that there does not exist any natural limitation of the minerals potential. The **problem lies in the availability of the minerals potential,** particularly with respect to ecological and legal conditions. This issue will be faced in all European Union countries.

To create structures of a secure and economically profitable raw materials supply, the European Union must pursue an **active raw materials policy**. In opposition to the past years and in the sense as discussed in Chapter 4, the EU has to implement structures of such a policy which render a cost-optimal contribution to the GDP of the Union.

What would be the **consequences** if no consistent EU raw materials policy will be established? **Price fluctuations** and **distortions of competition** on the international commodity markets could cause further serious consequences for the EU industries. **Rises in price** of strategic raw materials such as iron ore could actually persist, particularly considering the increasing tendency of mergers of market controlling mining companies. Here actions have to be taken immediately.

Unless an agreeable balance between ecological and mining interests can be established, the frame conditions concerning **cost-intensive and extended permitting processes** will deteriorate even more (no cost-optimal contribution to the EU GDP, passive minerals policy). This would cause a **further reduction of competitiveness of the raw materials industry**; produce precarious legal conditions and boost shifting and closing of industrial plants. Shifting abroad industrial sites, e. g. cement or steel plants, brings about increased imports and ecological impact.

The establishment of a coherent comprehensive European Union raw materials policy is the need of the moment. In view of the complexity of the raw materials matter, observing the principle of coherence is of particular importance.

A coherent European Union raw materials policy should provide a **framework** for the Member States within to create their own national mineral policies. The **implementation of an EU minerals strategy** needs a comprehensive concept. It must be highly efficient to yield results which can be realized within a determined period of time, provided the political power is granted for this purpose. In this context, the question of a consistent and coherent **EU minerals sector policy** arises again. Without it, the development and implementation of a raw materials strategy based on a coherent concept can hardly be accomplished.

The complex matter of European Union raw materials issues covers a wide range of technical, scientific and legal aspects and needs consolidated political structures. Accordingly, effective consultation processes between officials at EU level as well as between EU and national levels should take place.

Lacking an existing EU regulation system for raw materials, this will not be realizable in a simple way – particularly before the background of the international competition for raw materials. Most of the appearing competitors are in possession of not only a coherent raw materials policy, but also a comprehensive regulation system, which contributes enormously to a functioning minerals policy.

Chapter 9 References

A

Acciarito, G. (1985): La politica mineraria nel contesto delle relazioni economiche. II. (Italia). – Milano. – L'Industria mineraria, anno 6, no 5, settembre-ottobre 1985, p 1–28

Ad hoc Working Group of German Geological Departments – Mineral resources protection in the Federal Republic of Germany-Current Status", Geological Department Rheinland-Pfalz, Mainz, September 2001.

Adams, P. (2010): Africa: a political minefield. Mining journal special publication – Indaba

Aksjuk, N.N. (1984): politika razvivajuščichsja stran Afriki [Rohstoffpolitik in den entwickelten Ländern Afrikas]. – Moskva: Izdat. Nauka, Glavnaja Red. Vostočnoj Literatury, 1984

Anderson, E.W.; Anderson, Liam D. (1998): Strategic minerals: resource geopolitics and global geo-economics. – Chichester, England: Wiley, 1998

Anciaux, P. (2005): Die Wettbewerbsfähigkeit des Europäischen Rohstoffsektors, Berg- und Hüttenmännische Monatshefte, BHM (The Competitiveness of the European raw material sector, Journal of Mining. Metallurgical, Material, Geotechnical and Planned Engineering).

[Anonymous] (2007): Rohstoffpolitik ist Daueraufgabe und Zukunftssicherung/ BDI Kongress. Metall – Internationale Fachzeitschrift für Metallurgie, [Mineral policy is a permanent task and guarantee for the future/International Journal of Metallurgy] 61. 2007,H. 4,pp. 230–231

[Anonymous] (2011): Mining contributes high wages; New report shows nearly 1.8 million owe their jobs to mining. Industry Newswatch, Mining Engineering, Vol 63 (2011), 1, p. 10

[Anonymous] (2011): Rare earths are not so rare; USGS determines that there are 14 rare earth deposits in the US. Industry Newswatch, Mining Engineering, Vol. 63 (2011), 1, p. 26

Chapter 9 References

Antrekowitsch, H.: (2006), Sustainable Technologies in Metal Production and Processing, Berg- und hüttenmännische Monatshefte, BHM (Journal of Mining, Metallurgical, Material, Geotechnical and Planned Engineering), pp. 266–269

Antrekowitsch, H.; Biedermann, H.; Buchmayr, B.; Ebner, F.; Eichlseder, W.; Harmuth, H.; Kepplinger, W.; Kessler, F.; Krieger, W.; Lorber, K.; Ludwig, A. (2006): Universitärer Forschungscluster „Sustainable Technologies in Metal Production and Processing" (STMP), in: Berg- und hüttenmännische Monatshefte, BHM (Journal of Mining, Metallurgical, Material, Geotechnical and Planned Engineering) 151, pp. 263–265

Apel, H. [et al.] (1975): Zukunftsorientierte Energie- und Rohstoffpolitik. Internationaler Fachkongress zum Thema Probleme einer Zukunftsorientierten Energie- und Rohstoffpolitik [Forward-looking energy and raw materials policy. International conference about problems of a future-oriented energy and raw materials policy], 13./14. Oktober 1975, Bonn, veranst. von der Friedrich-Ebert-Stiftung (organized by the Friedrich-Ebert-Stiftung.). – Bonn – Bad Godesberg: Verlag Neue Gesellschaft, 1976

Arbeitswissenschaftliches Institut der deutschen Arbeitsfront (1942): Gedanken über eine europäische Agrar- und Rohstoffpolitik: erster Versuch einer Grundlegung [Considerations about an European Agrarian and Raw Materials Policy: First Attempt for a Basis]. – Berlin, 1942

Aron, J. (1992): Economic policy in a mineral-dependent economy: the case of Zambia. – University of Oxford, 1992

Atzenhofer Ch., Pressler N. (2007): Bergbau Kirunavaara – Raumordnung, Präsentation, Vorlesung Raumordnung [Mining Kirunavaara – land use planning, presentation; lecture land use planning], Montanuniversität

Auty, R. M. (1998): Mining as a generator of wealth: potential conflicts and solutions. Minerals and energy – raw materials report, Vol. 3 (1998), 2, pp. 4–12

Auty, R. M., ed. (2001): Resource abundance and economic development. Oxford Univ. Press, Oxford and New York, 2001, ISBN 978-0199246885

B

Bachmann, H. [et al.] (1980): Rohstoffpolitik der achtziger Jahre zwischen Strategie und Alibi. [Raw material policy of the eighties between strategy and alibi] – Zürich: Schulthess, 1980. – Aussenwirtschaft, Jg. 35,4 = Sondernummer, pp. 287–405

Bansal, R. (2010): Iron ore future – the next decade. Journal of mines, metals and fuels, 58 (2010), 3–4, pp. 70–73

References

Baron, S.; Glismann, H. H., Stecher, B. (1977): Internationale Rohstoffpolitik: Ziele, Mittel, Kosten. [International minerals policy: Objectives, means, costs] – Tübingen: Mohr, 1977 (Kieler Studien; 150)

Bassani, A. (1993): Steps to a Market Economy. – Paris. – OECD Observer, no 180, February– March 1993. p. 15–18

BDI (Bundesverband der Deutschen Industrie) (2007): Rohstoffsicherheit – Anforderungen an Industrie und Politik, Ergebnisbericht der BDI – Präsidialgruppe "Internationale Rohstofffragen" (Mineral security – requirements for industry and policy, Summary report "International raw material issues") pp. 3–4

Bender, F. [ed.] (1977): The importance of the geosciences for the supply of mineral raw materials: proceedings of an International symposium, held in Hannover at the Federal Institute for Geosciences and Mineral Resources, Oct. 25–26, 1976. – Stuttgart: Schweizerbart, 1977

Bermejo G. R. (1990): L'Antarctique et ses ressources minérales: le nouveau cadre juridique. – Paris: PUF, 1990. – (Publications de l'Institut Universitaire de Hautes Études Internationales de Genève.)

BGR (Bundesanstalt für Geowissenschaften und Rohstoffe [Federal Institute for Geosciences and Natural Resources]) (2007): Rohstoffwirtschaftliche Steckbriefe für Metall- und Nichtmetallrohstoffe [Economical profiles for metal and nonmetall raw materials], Hannover

BGR (Bundesanstalt für Geowissenschaften und Rohstoffe; Federal Institute for Geosciences and Natural Resources), 2009, Hannover, Germany, http://www.Bgr.bund.de.

Bilardo, U.; Mureddu, G. (1983): Situazione e prospettive del riciclo dei metalli. – Roma. – Energia e materie prime, no 33–34, settembre – ottobre 1983, p 43– 53

Bleischwitz, R., Bahn-Walkowiak, B. (2006): Sustainable Development in the European aggregates industry: a case for sectoral strategies. p. 5–7. www.coleurop.be/.../Bleischwitz_Aggregates_Bruges2006.pdf

Boettcher, R. (2003): Thesen zur nachhaltigen Rohstoffsicherung. – Bergbau, [Theses for the sustainable raw material supply] 54. 2003, H. 5, pp. 199–201

Bolz, R. [Hrsg.] (1975): Kooperation oder Konfrontation?: Materialien zur Rohstoffpolitik. [Cooperation or confrontation?: Materials for the Raw material policy] – Bonn: Progress Dritte Welt, 1975

Bomsel, O. (1987): Do the mining countries of the Third World have a future? – Natural resources forum, vol 11, no 1, Feb. 1987, pp 59–65

Chapter 9 References

Bonazza, P. (1976): Die Rohstoffpolitik: Studie. – Bruxelles: Agence Européenne d'Informations, 1976. Teil 1 u. Bothe, M. (2005): Environment, development, resources. – Recueil des cours/ Académie de Droit International de La Haye, Vol. 318. 2005(2007), pp 333–516

Bradley Jr., R. L. (2007): Resourceship: an Austrian theory of mineral resources. Review of Austrian economics, Vol. 20 (2007), 1, pp. 63–90

Brandstätter, W. (1989): Der Einfluß von Steuern auf die Planung von Bergbaubetrieben, Diss., Leoben [The influence of taxes on the planning of mining operations, Diss, Leoben] p. 22f.

Bourrelier, P.; Callot, F. [et al.] (1975): Matières premières minérales et relations internationales. – Lausanne: Centre de recherches européennes, 1975

British Geological Survey (2004): World mineral statistics 1998–2002: production, exports, imports/compiled by L. E. Taylor [et al.]; British Geological Survey. – Keyworth, Nottingham: British Geological Survey, 2004

British Geological Survey (2008): European mineral statistics 2002–06: a product of the World mineral statistics database/by L. E. Hetherington [et a.] – Keyworth, Nottingham: British Geological Survey, 2008

British Geological Survey (2008): World mineral production 2002–06/L. E. Hetherington [et al.] – Keyworth, Nottingham: British Geological Survey, 2008

Braune, G. (2008): Begehrte Spezialmetalle – Seltene Erden sind wiegen der wachsenden Nachfrage nach Hybrid-Autos gesucht [Special metals in demand – Rare earths are requested because of the growing demand for hybrid cars], Handelsblatt (www.handelsblatt.com), 12.8.08, No. 155.

Bülow, A. v. [et al.] (1982): Energie- und Rohstoffpolitik: Probleme, Trends u. Perspektiven in e. sich wandelnden Weltwirtschaft. Internationaler Fachkongress [Energy and raw materials policy: problems, trends and perspectives in a changing global economy. International Conference], 7./8. Dezember 1981, Bad Godesberg, veranst. von der Friedrich-Ebert-Stiftung. Mit Beiträgen von Andreas von Bülow (organized by the Friedrich-Ebert-Stiftung. With contributions by Andreas von Bülow) – Bonn: Verlag Neue Gesellschaft, 1982

Bundesministerium für Handel, Gewerbe und Industrie (1981), Konzept für die Versorgung Österreichs mit mineralischen Roh- und Grundstoffen (Draft for supply of Austria with mineral raw and base materials) p. 11.

Bundesministerium für Wirtschaft, Referat Mineralische Rohstoffe, Geowissenschaften, Bergwirtschaft <Bundesrepublik Deutschland> (1976): Bericht zur Rohstoffpolitik [Report on raw materials policy]. – Bonn, Bundesministerium, 1976 (BMWI-Dokumentation; Nr 227)

Bundesministerium für Wirtschaft und Arbeit (2005), Thesen zur Rohstoffpolitik, Berlin, 9. Februar 2005 [Theses for a raw material policy, Berlin]

Bundesministerium für Wirtschaft und Arbeit (BMWA) (2008) [Austrian Ministry of Commerce, Trade and Industry (2008)]: Österreichisches Montanhandbuch, 82. Jahrgang [Austrian Mining Handbook, vol. 82], Wien

Bundesverband der Deutschen Industrie/Arbeitskreis Rohstoffpolitik (1981): Rohstoffversorgungspolitik – Ziele und Instrumente: Fachtagung des Arbeitskreises Rohstoffpolitik des Bundesverbandes der Deutschen Industrie, 20. Mai 1981 in Köln. – Köln: BDI, 1981 [Federation of German Industry/Working Group raw material policy (1981): raw materials supply policy – objectives and instruments: Symposium of Working group raw material policy of the Federation of German Industry, Cologne, Germany.]

C

Chapman, G. R. et. al., British Geological Survey (2003), European Mineral Statistics 1997–2001

China's policy on mineral resources (2003). Chinese Government's Official Web Portal gov. cn, Official publications 2003

Christian, J. M., (2008): The BRIC's Impact on Commodity Markets, World Mines Ministries Forum 2008 (Toronto, Canada, February 29-March 2)

Christmann, P. (2008), Lecture at the European Parliament, 16. April 2008 (http://www.europeanmineralsfoundation.org)

Christner, T. und Pieper, T. "Bedeutung und Stellenwert 'nachhaltiger Entwicklung' bei der Gewinnung oberflächennaher Rohstoffe: ein Beitrag zur Wirkungsweise des umweltpolitischen Leitbildes eines 'sustainable development' auf planerische Abwägungsvorgänge und Genehmigungsentscheidungen im Rahmen der Rohstoffgewinnung." Meaning and significance of 'sustainable development' at the exploitation of near-surface materials: a contribution to the effect of the environmental concept of a 'sustainable development' on planning assessment and approval decisions in the context of mineral extraction] Erich Schmidt Verlag, Berlin, 1998.

Clain, Y. (2001): Matières premières minérales. – Les notes bleues de Bercy, 2001, n. 212, 1er-31 août, [8 p.]

Clement, H. (1981): Minerals requirements in the Soviet Union, 1981–1990: Paper prep. for the conference on "The Soviet Strategic Mineral Posture: Self-Sufficiency or Global Search?" The Johns Hopkins Foreign Policy Inst., May 21–22, 1981. – München, 1981 (Arbeiten aus dem Osteuropa-Institut München; 80)

Crowson, P. (1977): Non fuel minerals and foreign policy. – London: Royal Institute of International Affairs, 1977 (In: British foreign policy to 1985)

Crowson, P. (1978): British foreign policy to 1985: 3. Dependence on non-fuel minerals. – International affairs, 54 1978,1, pp 48–59

Crowson, P. C. F. (1984): Non-renewable mineral resources and the European Community. – Brussels: Centre for European Policy Studies, 1984 (CEPS Working documents; 1984/10 (Economic))

Crowson, P. (2002): Pandora's box: economic policy issues for the mining industry. Minerals and energy – raw materials report, 17 (2002), 1, pp. 3–9

D

Dams, T.; Grohs, G. [Hrsg.] (1977): Kontroversen in der internationalen Rohstoffpolitik: ein Beitrag zur Rohstoffpolitik der Bundesrepublik Deutschland nach UNCTAD IV. [Controversies in the international raw material policy: a contribution to the raw material policy of the Federal Republic of Germany according to UNCTAD IV] – München: Kaiser; Mainz: Matthias-Grünewald-Verlag, 1977 (Reihe Entwicklung und Frieden [Series development and peace]: Materialien; 7)

Daniel, P. (1990): Economic policy in mineral exporting countries: what have we learned? – University of Sussex, Institute of Development Studies, 1990

Daul, Johannes (2008): Auswirkungen der aktuellen Klimapolitik auf die österreichische Mineralrohstoffindustrie [*Impact of current EU climate change policy on the Austrian mineral industry*]. BHM – Berg- und Hüttenmännische Monatshefte, Vol. 153 (2008), 8, pp. 296–301

Davis, G. A., Tilton, J. E. (2005): The resource curse. Natural resources forum, 29 (2005), 3, pp. 233–242

De Wit, Maarten J. (1985): Minerals and mining in Antarctica. Science and technology, economics and politics. – Oxford: Clarendon Press, 1985 (Oxford science publications)

Department of the Environment (1995): Minerals planning policy and supply practices in Europe: main report, London

Department of Mining and Tunneling, University of Leoben (2004): Minerals Policies and Supply Practices in Europe, Final Report, S. 17, Commissioned by the European Commission Enterprise Directorate General under Contract n° ETD/FIF 2 003 0781, Leoben – Brüssel, November 2004

Department of Minerals and Energy, Republic of South Africa (1998): A minerals and mining policy for South Africa. October 1998, Pretoria. http://www.dmegov.za/minerals/min_whitepaper.stm

Department of Environment and Natural Resources (2004): NATIONAL POLICY AGENDA ON REVITALIZING MINING IN THE PHILIPPINES, Issuance of E. O. 270

Department of Mineral Resources and Petroleum Engineering, University of Leoben (2010): Planning Policies and Permitting Procedures to Ensure the Sustainable Supply of Aggregates in Europe, commissioned by UEPG.

Deutsche Bundesregierung (2007): Elemente einer Rohstoffstrategie der Bundesregierung, Berlin [German Federal Government: Elements of a mineral strategy of the Federal Government.]

Deutscher Industrie- und Handelstag (1976): Rohstoffpolitik ohne Illusion: zur Reform der Weltwirtschaftsordnung [Raw materials policy without illusion: to the reform of the international economic order]. – Bonn: Dt. Industrie- u. Handelstag, [ca.] 1976

Deutsches Institut für Wirtschaftsforschung (1987): China industrialisiert auf breiter inländischer Rohstoffbasis. (German Institute for Economic Research (1987): China industrialises on a broad domestic raw material base) – Berlin. – Wochenbericht/Deutsches Institut für Wirtschaftsforschung, Jg 54, Nr 21, pp. 289–294

Devaney, J.; Eden-Green, M.; Hargreaves, D. (1994): World index of resources and population. – Aldershot: Dartmouth, 1994.

Dingethal et al, (1998), Kiesgrube und Landschaft [Gravel pit and landscape], München

Dingethal, F. J. (2002): Neuere Entwicklungen in der Rohstoffsicherung. [Recent Developments in Raw Material Security]– Erzmetall, 55. 2002, H. 4, pp. 247–253

Dixon, K. (2008): Is it possible to predict how long our mineral resources will last and is there anything we can do to slow their inevitable decline?, Mining Environmental Management, July 2008

Donges, J. B. (1976): Kritik der Pläne für eine neue internationale Rohstoffpolitik. – Kiel: Inst. für Weltwirtschaft an der Universität Kiel, 1976 [Criticism of the plans for a new international raw material policy, Inst. for World Economy at the University of Kiel, in 1976] (Kieler Sonderdrucke; 37)

Dorian, J. P.; Borisovich, V. T. (1992): Energy and minerals in the former Soviet republics: distribution, development potential and policy issues. – Resources policy, 18. 1992,3, pp 205–229

Dragadze, T. (1992): Transcaucasia in transition. – London. – Central European, no. 14, July – August 1992, pp. 37–42

Drnek, T. (1995), Die wirtschaftliche Bedeutung der Steine und Erdengewinnung in Österreich, Berg- und Hüttenmännische Monatshefte, [The economic significance of industrial minerals extraction, in Journal of Mining. Metallurgical, Material, Geotechnical and Planned Engineering 142. Jg. (1997)] pp. 447–453

Drnek, T., (2008a), Magnesit und CO2 [Magnesite and CO2], Vortrag, Österreichische Akademie der Wissenschaften, Wien, 15. April 2008

Drnek, T. (2008b): Mineralwirtschaft, Spezielle Mineralwirtschaft, (minerals economy/specific minerals economy), Leoben 2008 Vorlesungsunterlagen

(Dutch) Ministry of Transport, Public Works and Water Management (2003): Construction Raw Materials Policy and Supply Practices in Northwestern Europe, Delft

E

Ebensperger, A., Maxwell, P., Moscoso, C. (2005): The lithium industry: its recent evolution and future prospects. Resources policy, Vol. 30 (2005), 3, pp. 218–231

Ebner, F., Weber, L., Rohstoffpotential Österreich, Die Hoffnungsgebiete – und die Forschungsdefizite [Raw materials potential of Austria, areas of hope – and the shortcomings in research], in: Sand & Kies, H. 35, Wien 1998

Eggert, R. G. (2010): Critical minerals and emerging technologies. Issues in science and technology, Vol. 26 (2010), 4, p. 14. ISSN 07485492

Ekdahl, E. (2008): Mineral resources in Europe, Presentation, International Symposium on the Planet Earth, Trondheim, 7–8 February 2008

Engel, J. R. (2007): Entwicklung des Bergbaues in der Slowakischen Republik. The mineral industry of Slovakia. – BHM Berg- und Hüttenmännische Monatshefte, 152. 2007, H. 12, pp. 410–412

Enzer, H. (1981), Decision making from the administrative viewpoint, Materials and Society, Bureau of Mines, Washington, DC

Ericsson, M.; Tegen, A. (1990): The evolving structure of the European mining industry. – Natural resources forum, vol 14,1. 1990, pp 14–21

Ericsson, M. (2010): African countries prepare for the next mining boom. Mining journal special publication – Indaba

Great Britain/Scottish Office /Environment Dept. (1994): Land for mineral working. – Great Britain, Scottish Office, Environment Dept., 1994 (National planning policy guideline; NPPG 4)

Grob, Jacques (2005): Die schweizerische Sand- und Kiesindustrie. (The Swiss sand and gravel industry) – BHM Berg und Hüttenmännische Monatshefte [Journal of Mining, Metallurgical, Material, Geotechnical and Planned Engineering], 150. 2005, H. 2, pp. 42–45

Gronwald, Leo (2008): Rohstoffversorgung der Stahlindustrie am Beispiel ThyssenKrupp Steel AG. [Supply of raw materials for steel industry, at the example of Thyssen Krupp Steel AG] – Bergbau (Mining): Zeitschrift für Rohstoffgewinnung, Energie, Umwelt, Jg 59,7. 2008, pp. 318–321

Gulley, David A. (1987): The new future of world metals: a conference report. – Natural resources forum, vol 11,1. 1987, pp 67–76

Guzmán, J. I., Nishiyama, T., Tilton, J. E. (2005): Trends in the intensity of copper use in Japan since 1960. Resources policy, Vol. 30 (2005), 1, pp. 21–27

Gylfason, T. (2001): Natural resources, education, and economic development. European economic review, Vol. 45 (2001), 4–6, pp. 847–859

H

Hamor (2004), Sustainable Mining in the European Union: The Legislative Aspect, Environmental Management, Vol. 33. No. 2, pp. 252–261.

Harms, Uwe [et al.] (1975): Berggesetzgebung und Rohstoffpolitik in Entwicklungsländern unter Berücksichtigung regionaler Schwerpunkte – Brasilien, Peru, Sambia, Zaire, Indonesien: Gutachten im Auftr. des Bundesministeriums für wirtschaftliche Zusammenarbeit [Mining legislation and minerals policy in developing countries under consideration of regional priorities – Brazil, Peru, Zambia, Zaire, Indonesia: Reports in behalf of the Federal Ministry for Economic Cooperation]. – Hamburg: Institut zur Erforschung technologischer Entwicklungslinien (Institute for the Study of technological development lines), 1975

Hartung, Matthias (2007): Rohstoffe und Bergbau – Positionen und Perspektiven. (Raw materials and mining – positions and perspectives) – Bergbau: Zeitschrift für Rohstoffgewinnung, Energie, Umwelt (Journal of Raw Material Extraction, Energy, Environment), 58. 2007, H. 11,pp. 486–490

Hartung, Matthias (2008): Rohstoffe und Energie im Spannungsfeld von Klimaschutz und Versorgungssicherheit [Raw materials and energy in areas of conflict with climate protection and supply security]. – Bergbau: Zeitschrift für Rohstoffgewinnung, Energie, Umwelt (Journal of Raw Material Extraction, Energy, Environment), 59,7. 2008, pp. 338–342

Hartung, Matthias; Schächter, Norbert (2007): Die Vereinigung Rohstoffe und Bergbau – VRB. – Glückauf, 143. 2007, H. 7/8, pp. 347–351

Heinrich M. (2007): Lockergesteine in Österreich, Geo-Atlas Österreich: Die Vielfalt des geologischen Untergrunds [Unconsolidated rock in Austria, Geo-Atlas of Austria: The variety of geological subsurface]/Hrsg.: Hofmann, T. & Schönlaub, H. P., Wien

Heimburg, J. (2007): Heimburg, J., (2008): Rahmenbedingungen für das ANTAG – Projekt in Österreich, Bachelor Arbeit, Fachbereich: Bergbaukunde. [Basic conditions of the ANTAG – Project in Austria, bachelor thesis, department: Mining Engineering.]

Hennecke, H. P. (1993): Die Steine- und Erden-Industrie in Europa unter besonderer Berücksichtigung der Kalkindustrie. – Zement, Kalk, Gips [Industrial minerals industry in Europe with particular emphasis on lime. – Cement, lime, gypsum], 46. 1993, H. 1, pp. 1–8

Hentschel, T.; Hruschka, F.; Priester, M. (2003): Artisanal and small-scale mining: challenges and opportunities. – London: IIED, International Institute for Environment and Development, 2003

Ho, P. (2006): Trajectories for greening in China: theory and practice. Development and change, Vol. 37 (2006), 1, pp. 3–28

Hoffmann, H.-G. (2006): Entwicklungen an den Rohstoffmärkten belasten die Wettbewerbsfähigkeit der deutschen Metallindustrie [Developments in the raw material markets strain the competitiveness of the German metal industry]. – World of Metallurgy – Erzmetall, 59. 2006, H. 4, pp. 216–219

Hruschka, F. (2008): Aufbau durch Abbau – Gewinnung mineralischer Rohstoffe und Nachhaltige Entwicklung. Construction through extraction – Mineral raw materials extraction and sustainable development. BHM Berg- und Hüttenmännische Monatshefte, Vol. 153,2. 2008

Humphreys, D. (1990): http://gso. gbv.de/DB=2.1/SET=1/TTL=13/MAT=/NOMAT=T/REL?PPN=062 271 377Towards an EEC minerals policy? – Resources policy, 16 1990,1, pp 35–46

I

Ibrochim, Azim, Dzanobilov, M. D., Mamadvafoev, M. M., Fachrutdinov, R. S., Gafarov, A. R. (2009): Mineral and raw material base of Zeravshansky mining region [in Russian]. Gornyj zurnal, Moscow, Vol. 185 (2009), 8, pp. 12–16

Ike, P. (2007): Minerals planning in the Netherlands compared to the U. K. – BHM Berg- und Hüttenmännische Monatshefte, 152. 2007, H. 12, S. 403–409

India/Ministry of Mines): National minerals policy, 1993, for non-fuel and non-atomic minerals. (Last updated 29/11/2004). http://mines. nic. in/nmp. html

Institut für Weltwirtschaft und Internationales Management <Bremen> (1987): Projekt Internationale Rohstoffpolitik und Entwicklungsländer: Wintersemester 1985/1986 bis Wintersemester 1986/1987 [Project International raw materials policy and developing countries: winter term 1985/1986 till 1986/1987 winter term]/Univ. Bremen, Fachbereich Wirtschaftswissenschaften. – Bremen: Inst. für Weltwirtschaft u. Internationales Management, 1987 (3 Bände – 3 Volumes). Bd 1. Grundfragen internationaler Rohstoffmärkte [Volume 1: Fundamental questions of international raw material markets], Bd 2. Märkte und Akteure [Volume 2:Markets and players], Bd 3. Länder und Strategien [Volume 3: Countries and strategies]

Institution of Mining and Metallurgy <Great Britain>; Jones, Michael J. [ed.] (1985): Role of government in mineral resources development. Proceedings of the conference organized by the Institution of Mining and Metallurgy and the Department of Mineral Resources, Government of Thailand and held in Bangkok, Thailand from 5 to 8 December 1983. – London: Institution of Mining and Metallurgy, 1985

Ippolito, F. (1986): Report on the exploitation of the Community's mineral resources/EC, European Parliament, Committee on Energy, Research and Technology. Rapporteur: Felice Ippolito. – Luxembourg: European Parliament, 1986. – (EP Documents; 1986/0032 A2)

J

Jamaica/Ministry of Agriculture and Lands: The National minerals policy: ensuring a sustainable minerals industry. 2nd draft (for discussion purposes), August 2006. www.moa.gov.jm/land/minpolicy/national_min_policy_draft_2_sep06.pdf

Jin, D., Seo, H., Choi, S. (2010): Arctic governance and international organization: a focus on the Arctic council [In Korean]. Ocean and polar research, Vol. 32 (2010), 1, pp. 85–95. ISSN 1598-141x

Johansen, H. E.; Matthews, O. P.; Rudzitis, G. [eds] (1987): Mineral resource development: geopolitics, economics and policy. – Boulder, London: Westview, 1987 (Westview special studies in natural resources and energy management)

Johnson, C. J.; Clark, A. L. (1985): Potential of Pacific Ocean nodule, crust, and sulfide mineral deposits. – Natural resources forum, Vol. 9,3. 1985, pp 179–186

K

Keyzer, M. (2010): Towards a closed phosphorus cycle. Economist, Vol. 158 (2010), 4, pp. 411–425

Klaffke, G. (1993): L'or: un redressement durable? – Bruxelles. – Bulletin financier/ Banque Bruxelles Lambert, année 66, no 2274, septembre 1993, pp. 1–10

Kleingeld, W. J., Nicholas, G. D. (2007): Diamond resources and reserves – technical uncertainties affecting their estimation, classification and valuation. In: Conference proceedings, Orebody modelling and strategic mine planning, Perth, WA, 2007. Australasian Institute of Mining and Metallurgy publication series, 2007, pp. 227–233, ISBN 978-192080677-4

Klump (2006), Wirtschaftspolitik, Instrumente, Ziele und Institutionen [Economic policy, instruments, objectives and institutions], München

Kohler, W.-D. (1979): Multinationale Konzerne und die Chance von Entwicklungslaendern auf eine eigenstaendige Rohstoffpolitik: das Beispiel der CIPEC [Multinational corporations and the chance of developing countries for an independent minerals policy: the example of the CIPEC]. – Konstanz, Diss., 1979

Köhler-Schnura, A. (1983): Konturen bundesdeutscher Rohstoffpolitik [Outline of West German raw material policy]. Köln: Deutscher Instituts-Verlag, 1983 (Beiträge zur Wirtschafts- und Sozialpolitik [Contributions to economic and social policy]; 113 = 1983,2)

Kommission der Europäischen Gemeinschaften (1975), Die Rohstoffversorgung der Gemeinschaft, Bulletin der Europäischen Gemeinschaften, Beilage 1/75

Krüger, H. J. (2002): Die Rohstoffversorgung der deutschen Stahlindustrie unter dem besonderen Aspekt der Versorgungssicherheit. – Stahl und Eisen [The raw material supply of the German steel industry under the particular aspect of supply security. – Steel and Iron], 122. 2002, H. 6, pp. 15–20

Kündig et al, (1997), Die mineralischen Rohstoffe der Schweiz [The mineral resources of Switzerland], Zürich

L

Lagos, G. (1997): Developing national mining policies in Chile: 1947–96. Resources Policy, Vol. 23 (1997), 1–2, pp. 51–69

Land Use Consultants (LCU) (2010): consultation process regarding Ad-hoc Working Group on Exchanging Best Practices on Land Use Planning and Geological Knowledge Sharing (for European Commission, DG Enterprise and Industrie).

Laruelle, M. (2007): Asie centrale: le "retour" de la Russie. – Politique internationale, 2007, n. 115, pp 377–391

Land Use Consultants (LCU) (2010): consultation process regarding Ad-hoc Working Group on Exchanging Best Practices on Land Use Planning and Geological Knowledge Sharing (for European Commission, DG Enterprise and Industrie).

Letouzé, G., (1996): Projekt Harmonisierungsmodell, Schritte zu einer bundesweiten Harmonisierung der Materie Mineralrohstoff-Vorsorge, Gutachten zum Fachbereich Rohstoffgeologie, Anhang 2, Wien 1996

Letouzé, G., Rossmann, H., Tiess, G., (2000), Mineralrohstoff-Gewinnung ohne Planung? [Mineral raw material exploitation without planning?], Raum, H 39.

Letouzé-Zezula, G., Tiess,G. (2002): Im Spannungsfeld zwischen Umwelt und Raumordnung: Die Mineralrohstoffgewinnung [In the area of conflict between environment and spatial planning: The mineral exploitation], in: Raum und Ordnung H1, S. 15–17

Lévy, J.-P. (1979): The evolution of a resource policy for the exploitation of deep sea-bed minerals. – Ocean-management, 5.1979,1, pp 49–78

Liebernickel, W. (1989): Qualitativ veränderte Kapitalverwertungsbedingungen in der internationalen kapitalistischen Nichteisenmetallwirtschaft und imperialistische Rohstoffpolitik [Qualitative changes of conditions of capital utilization in the international capitalist non-metall economy and imperialistic raw material policy]. – Berlin, Akad. für Gesellschaftswiss. beim ZK d. SED, Diss. B, 1989

Linden, E. von der (1997): Marktrelevante Überlegungen zur Rohstoffversorgung und zu Beteiligungen im internationalen Bergbau [Market-relevant considerations for raw material supply and participation in international mining, Erzmetall]. – Erzmetall, 50. 1997, H. 12, pp. 761–768

Linden, E. von der (2004): Der Bergbau weltweit im konjunkturellen Aufwind – Verkäufermarktbedingungen für Rohstoffe [Mining industry worldwide in economic ascent – Sellers' market conditions for raw materials]. – Bergbau, 55. 2004, H. 10, pp. 441–442, 444

Linden, E. von der (2004): Natural resources: market change to sellers' conditions. – World of Mining – Surface & underground, 56. 2004, H. 3, pp. 217–221

List, F. K. (1986): Evaluation of the Community's primary mineral raw materials programme. – Luxembourg: EC, 1986. – (EUR/European Commission; 10 191) (Research evaluation – report; 16)

Long, Keith R., Van Gosen, Bradley S., Foley, Nora K., Cordier, Daniel (2010): The principal rare earth elements deposits of the United States – a summary of domestic deposits and a global perspective. U.S. Department of the Interior, U. S. Geological Survey, Scientific Investigations Report 2010–5220

Ludwig, M. (1957): Internationale Rohstoffpolitik [Internationl raw material policy]. – Zürich: Polygraphischer Verlag, 1957

K

Koziol et. Al. (2008), production of aggregates in EU (http://www.min-pan.krakow.pl/Wydawnictwa/GSM2443/koziol-kawalec-kabzinski.pdf).

M

Macqueen, M.; Nötstaller-R (1997): Langfristige Entwicklung der Metallnachfrage – Tendenzen und Paradigmen. Long-term evolution of metal demant – trends and paradigms. Berg- und Hüttenmännische Monatshefte, 142 1997, 8, pp. 352–357

Maier, A. (2006): Österreichischer Bergbau – Aktuelle Entwicklungen 2006. Mining in Austria – Current trend 2006. – BHM Berg- und Hüttenmännische Monatshefte, 151. 2006, H. 8, pp. 314–319

Maier, A.; Weber, L. (2008): Der Österreichische Rohstoffplan. – In: Sorger, Veit [Hrsg.] [et al.]: Herausforderung Verwaltungsreform: Best Practice Beispiele für eine effiziente Verwaltung (The Austrian Raw Materials Plan, in: Sorger, Veit [Ed] [et al.]: Administrative challenge: best practice examples of efficient management. Vienna: Austrian Industry, 2008). Wien: Industriellenvereinigung, 2008

Magno, C.: „The extractive industry – from the supply control model to a regulatory model geared to the challenges of sustainable development" Mining Bulletin (Boletim de Minas), Vol. 38, No. 4.

Mamaev, J. A., Van-Van-E., A. P. (2009): Mining-industrial potential of the Far East and perspectives of its development [in Russian]. Izvestija Vyssich Ucebnych Zavedenij, Gornyj zurnal, Vol. 52 (2009), 2, pp. 32–38

Maponga, O., Maxwell, P. (2001): The fall and rise of African mining. Minerals and energy – raw materials report, 16 (2001), 3, pp. 9–26

Markandya, A.; Mason, P.; Friedrich, R.; Hacker, M.; Gressmann, A.; Wagner, H.; Nötstaller, R. (2000): SAUNER – Sustainability And the Use of Non-Renewable Resources: Summary final report. University of Bath, UK; IER, Universität Stuttgart, Germany; Montanuniversität Leoben, Austria. Research funded by the European Commission, DGXII Environment and Climat Programme. Nov. 2000

Marois, M. [ed.] (1977): Towards a plan of action for mankind: needs and resources; methods of forecasting. Proceedings of the World conference held in Paris 9–13 Sept. 1974/ed. by M. Marois, Institut de la Vie, Paris. – Oxford: Pergamon, 1977

Maskovcev, G. A., Korotkov, V. V. (2009): Mineral and raw material potential of metallurgical facilities in the Urals, Siberia and Far East [in Russian]. Gornyj zurnal, Moscow, Vol. 185 (2009), 3, pp. 29–33

Matheu, M.; Imauven, C.; Moreau, G.; Legrand, B. [et al.] (1991): Minerais et métaux en France et dans le monde. I. Panorama de l'année 1989 et statistiques Réalités industrielles, no 2, Février 1991, p 5–46

Maull, H. W. (1984): Western Europe's non-fuel mineral vulnerability: how serious, how vulnerable? Atlantic quarterly, no. 4. 1984, pp 337–358

Maull, H. W. (1988): Versorgungsrisiken bei ‚strategischen' Rohstoffen. [Supply risks in ‚strategic' raw materials] – Glückauf <Essen>, Jg. 124, nr 10, 19. Mai 1988, pp. 572–577 http://dispatch.opac.d-nb.de/DB=4.1/SET=1/TTL=4/REL?PPN=131 785 907

Maxwell, P. (2004): Chile's recent copper-driven prosperity: does it provide lessons for other mineral rich developing nations? Minerals and energy – raw materials report, Vol. 19 (2004), 1, pp. 16–31

Maxwell, P. (2005): Thoughts on sharing Australian mining education with Latin America. AusIMM Bulletin, Issue 5 (2005), pp. 41–44

Maxwell, P., Al Rawashdeh, R. (2005): Are minerals a blessing or a curse? Some reflections on the recent debate. Resources, energy and development, Vol. 2 (2005), 2, pp. 107–123

Maxwell, P., Mather, D. (2006): Sustainable or transitory? Assessing the economic future of the Australian mineral industry. Australasian Institute of Mining and Metallurgy Publication series, 2006, pp. 179–186

Mayer, Peter (2006): Macht, Gerechtigkeit und internationale Kooperation: eine regimeanalytische Untersuchung zur internationalen Rohstoffpolitik [Power, justice and international cooperation: a regime-analytical study on international raw material policy.]. – 1. Aufl. Baden-Baden: Nomos, 2006. (Weltpolitik im 21. Jahrhundert; Bd 13)

Chapter 9 References

Menezes, J. J. (1988): Perspectivas de desenvolvimento de actividade mineira em Portugal. – Lisboa. – Boletim de Minas, vol 25,4. 1988, pp. 331–341

Mezger, D. (1977): Konflikt und Allianz in der internationalen Rohstoffwirtschaft, das Beispiel Kupfer [Conflict and alliance in the global raw material economy, example copper]. – Bonn: Progress-Dritte-Welt-Verlag; Bremen: Bremer Afrika-Archiv, 1977 (Reihe Rohstoffpolitik und neue Weltwirtschaftsordnung [Series: raw material policy and new global economic order])

Michaelis, H. (1975): Memorandum über die Europäische Rohstoffpolitik: Gutachten, Juli 1975; vorläufige Fassung. Erstellt im Auftrag der Kommission der Europäischen Gemeinschaften. [Memorandum on European raw materials policy: report, July 1975, preliminary version. By order of the Commission of the European Communities.] – Brüssel, 1975

Michaelis, H. (1976): Europäische Rohstoffpolitik [European mineral policies]. – Essen: Verlag Glückauf, 1976 (Bergbau, Rohstoffe, Energie; Bd 13)

Ministère de l'Industrie <France> (1979): Les chiffres clés des matières premières minérales. – Paris: Ministère de l'Iindustrie, 1979

Ministry of Energy and Minerals (1997): The Mineral Policy of Tanzania (http://www.tanzania.go.tz/pdf/themineralpolicyofTanzania.pdf)

Ministry of Mineral Resources (2003): Core Mineral Policy, Sierra Leone (http://docs.google.com/viewer?a=v&q=cache:iwastyxj1GAJ:www.daco-sl.org/encyclopedia/4_strat/4_2/mmr_mineralpolicy.pdf+Mineral+Policy&hl=de&gl=at&sig=AHIEtbTKrAHeOJovTjMWhsCPPadWubnHPA).

Mojarov, A. (2008), Commodities Branch, United Nations Conferenz on Trade and Development (UNCTAD), Sustainable Development – an opportunity to bring it all together, World Mines Ministries Forum 2008 (Toronto, Canada, February 29-March 2)

Moore, D. J., Tilton, J. E., Shields, D. J. (1996): Economic growth and the demand for construction materials. Resources policy, Vol. 22 (1996), 3, pp. 197–205

Morgan, C. L. (2009): The status of marine mining worldwide. In: Proceedings of the International Conference on Offshore Mechanics and Arctic Engineering OMAE, Vol 4, issue PART B, 2009, pp. 1489–1494

Morozov, A. F., Lipilin, A. V., Petrov, O. V., Kiselev, E. A., Feoktistov, V. P. (2007): State of predicting resources of mineral raw materials in Russian Federation [in Russian]. Gornyj zurnal, Moscow, Vol. 183 (2007), 10, pp. 47–51

Muchabbatov, C. M. (2009): Mineral resources potential of mountain regions of Tajikistan [in Russian]. Gornyj zurnal, Moscow, Vol. 185 (2009), 8, pp. 5–8

Müller-Ohlsen, L. (1981): Die Weltwirtschaft im industriellen Entwicklungsprozess. Kieler Studien [World economy in the industrial development process. Kiel Studies], Tübingen

Müller, W. and Schulz, P.-M. „Handbuch Recht der Bodenschätzegewinnung [Legislation manual for exploitation of resources]." Baden-Baden, Germany, 2000

Müller, H. (2007): CRU 13th World Steel Conference – from BRICs to Africa. – Steel Times International, 31. 2007, H. 3, pp. 53–54, 56

Müller, W.; Werthebach, E.; Schäfer, V. (2000): Rohstoffversorgung und technologischer Fortschritt: Grundelemente einer nachhaltigen Entwicklung in unserer Gesellschaft (Raw material supply and technological progress: basic elements of sustainable development in our society). – Bergbau, 51. 2000, H. 12, pp. 546–548, 550–554

N

National Economic and Social Council <Ireland> (1981): National Minerals policy. – Dublin: Stationery Office, 1981 (National Economic and Social Council publications; 60)

National Minerals Policy Review Group <Ireland> (1995): Report of the National Minerals Policy Review Group to the Minister for Transport, Energy and Communications: A new minerals policy. – Dublin: Stationery Office, 1995

Nilsson, M. (2002): Merger policy and the EU: how can the airtours decision affect minerals and energy markets? – Basingstoke: Taylor & Francis, 2002. Minerals & energy, 17. 2002, 4, pp. 18–24

Ninni, A. (1985): L'industrializzazione resource-oriented nei paesi in via di sviluppo e le nuove forme di coinvolgimento estero delle imprese. – Milano. – Economia e politica industriale, no 47, Settembre 1985, pp. 161–174

Nötstaller, R. (2000): Zur Entwicklung der Nachfrage nach mineralischen Rohstoffen – Zusammenhänge und Folgerungen. On the evolution of mineral demand – relationships and conclusions. BHM – Berg- und Hüttenmännische Monatshefte (Journal of Mining, Metallurgical, Material, Geotechnical and Planned Engineering), 145 2000, H. 8, pp. 314–318

Nötstaller, R. (2002): Patterns of mineral demand and supply global and regional perspectives. Entwicklungen von Nachfrage und Angebot mineralischer Rohstoffe – globale und regionale Perspektiven. BHM – Berg- und Hüttenmännische Monatshefte, 147 2002, H. 12, pp. 402–407

Nötstaller, R. (2003a), Österreichischer Rohstoffplan, Arbeitskreis 2, Rohstoffwirtschaft und Bergwesen [Austrian Mineral Resources Plan, Working Group 2, raw material economy and Mining], Institut für Bergbaukunde, Bergtechnik und Bergwirtschaft Montanuniversität (Institute of Mining Engineering of the University of Leoben) Leoben

Nötstaller, R.; Wagner, H. (2003b): Zur langfristigen Entwicklung der Nachfrage nach Baurohstoffen in Österreich – Rückblick und Vorschau. Longterm trends in the demand for building materials in Austria – review and forecast. BHM – Berg und Hüttenmännische Monatshefte (Journal of Mining, Metallurgical, Material, Geotechnical and Planned Engineering), 148 2003, H. 8, pp. 316–320

Nötstaller, R.; Wagner, H. (2007): Überlegungen zum Rohstoffbedarf und zur Rohstoffpolitik. Reflections on resource consumption and resource policy. BHM – Berg- und Hüttenmännische Monatshefte (Journal of Mining, Metallurgical, Material, Geotechnical and Planned Engineering), 152 2007, H. 12, pp. 383–390

O

Oberhänsli, H. (1982): Internationale Rohstoffpolitik im Zeichen von Instabilität und Inflation: Stabilisierungsmöglichkeiten unter besonderer Berücksichtigung von Indexklauseln [International raw material policy in criterion of instability and inflation: possibilities of stabilisation under special consideration of index clauses]. – Diessenhofen: Rüegger, 1982. (SIASM-Schriftenreihe; Bd. 4. Sankt Galler Dissertation)

OECD, Development Centre (2010): Perspectives on global development 2010.

Onillon, J. (2007): Why shareholders are requesting steel industry to consolidate. (Folienpräsentation). – Stahlmarkt, Handelsblatt-Jahrestagung für die Stahlindustrie, 11. 2007, p. 1 [plus 27 Folien]

Ortiz R., A. (1994): Dossier: El reto industrial del medio ambiente. VIII. La industria minera y el medio ambiente: ideas para una interpretación ecológica amplia. – Madrid. – Economia industrial, no. 297, Mayo – junio 1994, pp. 105–116

Otto, J. (1992): A Global Survey of Mineral Company Investment Preferences, Mineral Investment Conditions in Selected Countries of the Asia-Pacific Region, United Nations ST/ESCAP/1197, pp. 330–342.

Otto, J. M. (1997): A national minerals policy as a regulatory tool. Resources policy, 23 (1997), 1–2, pp. 1–7

Otto, J. M. (1999): Mining, Environment and Development, United Nations Conference on Trade and Development, USA

P

Parry, A. (1981): Towards a minerals policy for Britain. – London: Foreign Affairs Research Institute, 1981

Pereira, A. (1989): Aspectos gerais do desenvolvimento do sector extractivo. – Lisboa. – Boletim de minas, vol 26,2. 1989, pp. 175–179

Petterson, M. G., Marker, B. R., MCEVOY, F., Stephenson, M. & Falvey, D. A. (2005): The need and context for sustainable mineral development, in: Sustainable Minerals Operations in the Developing World, Edited by Marker, B. R. et al, London

PricewaterhouseCoopers (2010): The economic contributions of U.S. mining in 2008, a report prepared by PricewaterhouseCoopers for the National Mining Association. Washington DC, Oct. 2010

Prillhofer, R., Prillhofer, B., Antrekowitsch, H.: Verwertung von Reststoffen beim Aluminium-Recycling, Berg- und hüttenmännische Monatshefte, BHM (2008) 3, pp. 103–108

Pticyn, A. M., Ljudin, J. K., Polonskij, G. V. (2004): State and measures for strengthening of mineral and raw material base of Russian metallurgy. Gornyj zurnal, Moscow, Vol. 180 (2004), 3, pp. 45–53

Puckov, L. A., Petrov, V. L. (2005): Development of mining and high-altitude mine foundations in the Urals, in Siberia and in the Far East of Russia [in Russian]. Izvestija Vyssich Ucebnych Zavedenij, Gornyj zurnal, Vol. 48 (2005), 4, pp. 125–147

Pulvermacher, K. (2010): The balance of power. Mining journal special publication – Indaba

R

Radetzki, M. (1986): Structural change in world copper supply. – Natural resources forum, vol 10,3. 1986, pp 281–291

Radetzki, M., Eggert, R. G., Lagos, G., Lima, M., Tilton, J. E. (2008): The boom in mineral markets: how long might it last? Resources policy, Vol. 33 (2008), 3, pp. 125–128

Ray, G. F. (1992): Le riserve mondiali di minerali e metalli. – Milano. – Innovazione e materie prime, no 3. 1992, pp. 80–94

Rheinisch-Westfälisches Institut für Wirtschaftsforschung (Rhine-Westphalian Institute for Economic Research) (RWI Essen) (2006), Trends der Angebots- und

Nachfragesituation bei mineralischen Rohstoffen (Trends in supply and demand situation for mineral resources), Essen

Reisinger, M.; Tiess, G. (2007): Das Protokoll von Kyoto und sein potentiellen Auswirkungen auf den internationalen Bergbau [The Kyoto Protocol and its potential impact on international mining]. – BHM Berg- und Hüttenmännische Monatshefte (Journal of Mining, Metallurgical, Material, Geotechnical and Planned Engineering), 152. 2007, H. 4, pp. 108–114

Rensburg, W. van (1978): Mineral supplies from South Africa: their place in world resources. – London: Economist Intelligence Unit, 1978 (EIU Special report; 059)

Republic of Uganda/Ministry of Energy and Mineral Development: The minerals policy of Uganda. September 2000. Executive summary. www.energyandminerals.go.ug/minpol00.pdf

Richards, J. P. (2009): Mining, Society and a Sustainable World, Heidelberg 2009

Roberts, P. W.; Shaw, T. (1982): Mineral resources in regional and strategic planning. – Aldershot: Gower Technical, 1982

Roithner, T. (2008): Von kalten Energiestrategien zu heißen Rohstoffkriegen?: Schachspiel der Weltmächte zwischen Präventivkrieg und zukunftsfähiger Rohstoffpolitik im Zeitalter des globalen Treibhauses [From cold energy strategies to hot wars for raw materials?: Game of chess of world powers between preventive war and sustainable raw materials policy in the era of a global greenhouse]/Österreichisches Studienzentrum für Frieden und Konfliktlösung [Austrian Study Center for Peace and Conflict Solution] (Hg.). Projektleitung (project management): Thomas Roithner. – Wien (u. a.): Lit, 2008

Rolshoven, H. (1968): Das Industriedreieck Saar-Lothringen-Luxemburg: Rohstoffwirtschaft und Rohstoffpolitik [The industrial triangle Saar-Lorraine-Luxembourg: Raw material economy and policy]. – Karlsruhe: Braun, 1968 (Welt am Oberrhein; 5/1968, Das Saarland)

Rossmann, H. (1996), Anrainer- und Umweltschutz im Bergrecht [Residents and environmental protection in mining law], Wien.

S

Safonov, Yu. G. (2010): Mineral potential of Russian Arctic: state and efficient development. Russian geology and geophysics, Vol. 51 (2010), 1, pp. 112–120

Schächter, Norbert, Johannes, Dieter, Einenkel, Betty (2006): Bergbau und nachhaltige Entwicklung – Ressourcen für heute und morgen [Mining and sustainable development – resources for today and tomorrow]. Bergbau, Vol. 57 (2006), 11, pp. 503–508

Schächter, N. (2008): Internationale Entwicklungen auf dem Rohstoffmarkt [*International developments on the raw material market*]. – Bergbau, 59. 2008, H. 6, pp259–262

Schorsch, L. L. (1989): Minerals trade and commercial policy: the case of steel. – Resources policy, 15. 1989,2, pp 169–187

Schuver, B. (2001): Winstheffing op het Nederlandse continentaal plat. – Weekblad voor fiscaal recht 2001, v. 130, n. 6452, p. 1406–1415

Schulze, Gerit (2010): Russlands Metallurgie-Sektor kommt glimpflich durch die Krise. Sinkende Schuldenlast und große Zukunftspläne. Germany Trade & Invest, Berlin, Bonn, Datenbank Länder und Märkte, 18.01.2010, https://www.gtai.de/DE/Navigation/Datenbank-Recherche/Laender-und-Maerkte/ (accessed 12.03.2011)

Schwarz, J. (1975): Rohstoffprobleme und Rohstoffpolitik der Entwicklungsländer [Raw material problems and raw material policy of developing countries]. – Wien, Hochschule für Welthandel, Dipl.-Arb., 1975

Schwarz, C., Helmar, D. (2008), Geschichte einer Ausbeutung (History of an exploitation), Presse, November 2008

Seebacher, H., Sunk, W., Antrekowitsch, H.: Recycling von Aluminium in der Automobilindustrie [Recycling of aluminum in the automobile industry], in Aluminium (Fachzeitschrift der Aluminium-Industrie; international journal for industry, research and application) 1/2 (2006) 82, pp. 24–31

Seitz, K. (1979): Die internationale Rohstoffpolitik: Rückblick und Ausblick [The international raw material policy: Review and Outlook]. – Bonn: Bundeszentrale. Aus Politik und Zeitgeschichte: APuZ, 1979, (280 479)B17, pp. 3–15

Shields, D., Solar, S., and W. Martin (2004): The role of values and objectives in communicating indicators of sustainability. Ecological Indicators.

Shields, D. J., Solar, S. V. (2005): Sustainable development and minerals: measuring mining's contribution to society. Geological Society Special publication, 250 (2005), pp. 195–212

Siebert, H. (1981): Strategische Ansatzpunkte der Rohstoffpolitik der Industrienationen nach der Theorie des intertemporalen Ressourcenangebots. – Mannheim: Institut für Volkswirtschaftslehre und Statistik der Univ., 1981 (Discussion papers/ Institut für Volkswirtschaftslehre und Statistik, Department of Economics, Universität Mannheim; 194) (Beiträge zur angewandten Wirtschaftsforschung)

Chapter 9 References

Siebert, H. (1983): Ökonomische Theorie natürlicher Ressourcen. [Economic theory of natural resources] – Tübingen: Mohr, 1983

Siebert, H. [Hrsg.] (1980): Erschöpfbare Ressourcen: Verhandlungen auf der Arbeitstagung des Vereins für Sozialpolitik, Gesellschaft für Wirtschafts- und Sozialwissenschaften [Exhaustible resources: negotiations at the workshop of the Association for Social Policy, Society for Economic and Social Sciences], 1979. – Berlin: Duncker & Humblot, 1980 (Schriften des Vereins für Sozialpolitik, Gesellschaft für Wirtschafts- und Sozialwissenschaften [Writings of the Society for Social Policy, Society for Economic and Social Sciences]; N. F. 108)

Siebert, H. [Hrsg.] (1986): Angebotsentwicklung und Preisbildung natürlicher Ressourcen. (Supply development and pricing of natural resources. Series of the Institute of Energy Economics) – München: Oldenbourg, 1986 (Schriftenreihe des Energiewirtschaftlichen Instituts; 30)

Skinner, B. J. Keynote presentation to the 31st International Geological Congress. Rio de Janeiro, 2000.

Slotta, R. (2003): Bergbau – Es geht nicht nur um die Kohle! [Mining – It's not just about the money!]– Bergbau, 54. 2003, H. 12, pp. 551–555

Spagni, D.; Ford, G.; Simnett, J.; Cameron, H.; Georghiou, L. (1984): Les ressources minérales des mers. (Dossier). – Paris. – La rechereche, vol. 15, no 159, octobre 1984, pp 1308–1318

Steine- und Erden-Industrie (2001): Baustoffe 2001, Stein-Verlag Baden-Baden GmbH [Construction Minerals 2001, Stein-Verlag, Baden GmbH]

Steger, W. (1980): Ziele und Instrumente internationaler Rohstoffpolitik [Objectives and instruments of international minerals policy. – Vienna University of Economics, diploma thesis], – Wien, Wirtschaftsuniversität, Dipl.-Arb., 1980

Sterk, G., Rohstoffpolitik und Rohstofforschung in Österreich [Raw materials policy and research in Austria], Raumordnung aktuell, H. 3, Wien 1982

Stolzlechner, H. (2004): Einführung in das öffentliche Recht [Introduction to Public Law], Wien

Streit, M. E. (1975): Einige alte Überlegungen zu neuerlichen Schwierigkeiten in der internationalen Rohstoffpolitik. – Mannheim: Institut für Volkswirtschaftslehre und Statistik der Universität Mannheim, 1975 (Discussion papers/ Institut für Volkswirtschaftslehre und Statistik; Department of Economics, Universität Mannheim; 69) (Beiträge zur angewandten Wirtschaftsforschung (Contributions to applied economic research))

Stribrny, B.; Vasters, J.; Brinkmann, K. (2006): Die Entwicklung auf den internationalen Märkten metallischer Rohstoffe. – World of Metallurgy – Erzmetall, 59. 2006, H. 4, pp. 191–201

Svedberg, P., Tilton, J. E. (2006): The real, real price of nonrenewable resources: copper 1870–2000. World development, Vol. 34 (2006), 3, pp. 501–519

Swedish Committee on Minerals Policy (1980): Industrial minerals: summary of a report. – Stockholm, 1980

T

Tanzania National Website [Mineral policy of Tanzania] http://www.tanzania.go.tz/mining.html

Tatarkin, A. I., Kozakov, E. M., Selomencev, A. G., Strovskij, V. E. (2006): About development perspectives of the resource basis for the black iron metallurgy sector in the Urals [in Russian]. Izvestija Vyssich Ucebnych Zavedenij, Gornyj zurnal, Vol. 49 (2006), 3, pp. 120–127

Tienhaara, K. (2006): Mineral investment and the regulation of the environment in developing countries. – International environmental agreements, Vol. 6, 4.2006, pp. 371–394 DOI: 10 1007/s10 784–006–9010–6

Thoma, G.; Renner, H. (1979): Die Bedeutung der Rohstoffpolitik in den Beziehungen zwischen Industrie- und Entwicklungsländern. [The importance of the raw materials policy in relation between industrial and developing countries.] – Nürtingen, Fachhochschule, Dipl.-Arb., WS 1978/79

Tiess, G. (2000a): Rohstoffgewinnung im Spannungsfeld, [Raw materials in areas of conflict, Sand & Gravel, H 48] Teil 1 – in: Sand und Kies H 48, pp. 3–5

Tiess, G. (2000b): Rohstoffgewinnung im Spannungsfeld, [Raw materials in areas of conflict, Sand & Gravel,] Teil 2 – in: Sand und Kies H 49, pp. 3–6

Tiess, G., Rossmann, H., Pilgram, R. (2002a): Die Bedeutung des Vorsorgeprinzips bei der Gewinnung mineralischer Baurohstoffe, Teil 1" [The relevance of the precautionary principle in the extraction of mineral construction minerals, Part1] – in: Recht der Umwelt (Right of the environment) H 3, pp. 84–92

Tiess, G., Rossmann, H., Pilgram, R. (2002b): Die Bedeutung des Vorsorgeprinzips bei der Gewinnung mineralischer Baurohstoffe, Teil 2" [The relevance of the precautionary principle in the extraction of mineral construction minerals, Part2] – in: Recht der Umwelt (Right of the environment) H 4, pp. 130–136

Tiess, G. (2003): Gibt es eine europäische Rohstoffpolitik? [Is there a European raw materials policy?] – BHM Berg- und Hüttenmännische Monatshefte (Journal of Mining, Metallurgical, Material, Geotechnical and Planned Engineering), 148. 2003, H. 8, pp. 307–315

Tiess, G.; Pilgram, R. (2003): Zentrale Aufgabe der Raumplanung: die nachhaltige Sicherung mineralischer Rohstoffe (Important task of land use planning: to ensure sustainable supply with mineral raw materials). BHM Berg- und Hüttenmännische Monatshefte [Journal of Mining, Metallurgical, Material, Geotechnical and Planned Engineering], 148. 2003, H. 10, pp. 403–411

Tiess, G. (2005): Bedeutung der Sand- und Kiesindustrie in Europa. Importance of the extractive industry (construction minerals) in Europe. – BHM Berg- und Hüttenmännische Monatshefte (Journal of Mining, Metallurgical, Material, Geotechnical and Planned Engineering), 150. 2005, H. 2, pp. 33–38

Tiess, G. (2005): Sustainable supply of the European industry and society with minerals: importance of the non-energy extractive industry. BHM Berg- und Hüttenmännische Monatshefte (Journal of Mining, Metallurgical, Material, Geotechnical and Planned Engineering), 150. 2005, H. 12, pp. 415–423

Tiess, G. (2007): Environmental related Aspects of the Non-Ferrous Mining Industry in Europe, BHM Berg- und Hüttenmännische Monatshefte (Journal of Mining, Metallurgical, Material, Geotechnical and Planned Engineering), H. 10 pp. 309–316

Tiess, G. (2008): Need of coherent minerals policy in Europe – present discussion and approaches. – Geología y Minería, number 3, volumen 24, 2008. ISSN 1 993 8012, Moa, Cuba.

Tiess, G. (2009): Rechtsgrundlagen der Rohstoffpolitik [Legal basics of Mineral Policy in Europe], Fokus Europa

Tilton, J. E., Lagos, G. (2007): Assessing the long-run availability of copper. Resources policy, Vol. 32 (2007), 1–2, pp. 19–23

Tietzel, M. (1977): Internationale Rohstoffpolitik: eine Analyse der rohstoffpolitischen Aspekte des Nord-Süd-Dialogs [International raw material policy: an analysis of resource-political aspects of the North-South dialogue]. – Bonn – Bad Godesberg: Verlag Neue Gesellschaft, 1977 (Reihe Weltwirtschaft)

Tietzel, M. (1979): Primary commodities in the North-South dialogue. – Bonn: Friedrich-Erbert-Stiftung, 1979

Tilton, J. E. Economics of the Mineral Industries (1992), in: H. L. Hartmann (Ed.). SME Mining Engineering Handbook. Society of Mining, Metallurgy, and Exploration.

Treholt, A. [Red.] (1976): Norges havretts- og ressurspolitikk. [Norwegens Seerechts- und Rohstoffpolitik]. – Oslo: Tiden Norsk Forl., 1976

Trojer M. (2007), Großtagebau [Large Opencast Mining, Bachelor Thesis], Bachelor Arbeit, Leoben

U

Udd, J. (2006): Management in action – the frozen north: Arctic mining in Canada. Mining magazine, Vol. 194 (2006), 2, pp. 26–27

UK Minerals Forum (2009): Shaping UK minerals policy, http://www.mauk.org.uk/newsdocs/shaping_uk_mins_policy_final_180809.pdf (accessed: 09.03.2011)

Union Européenne des Producteurs de Granulats UEPG (2008): Aggregates production in Europe. The European Aggregates Industry – Annual Statistics 2006; (http://www.uepg.eu/uploads/documents/122–51-uepg_statistics_2005-de.xls) http://www.uepg.eu/uploads/documents/141-99-statistics2006de.xls

UNCTAD (United Nations Conference on Trade and Development) (2010): Recent developments in key commodity markets:trends and challenges. (http://www.unctad.org/en/docs/cimem2d7_en.pdf)

USGS (U.S. Geological Survey), Department of the Interior (2008): Minerals Yearbook 2006, Volume III, Albania (Brininstool, M.)

USGS (U.S. Geological Survey), Department of the Interior (2008): Minerals Yearbook 2006, Volume III, Belgium (Harold R. Newman)

USGS (U.S. Geological Survey), Department of the Interior (2009): Minerals Yearbook 2007, Volume III, Bulgaria (Mark Brininstool)

USGS (U.S. Geological Survey), Department of the Interior (2009): Minerals Yearbook 2007, Volume III, Czech Republic (Mark Brininstool)

USGS (U.S. Geological Survey), Department of the Interior (2008): Minerals Yearbook 2006, Volume III, Denmark (Harold R. Newman)

USGS (U.S. Geological Survey), Department of the Interior (2008): Minerals Yearbook 2006, Volume III, Finland (Harold R. Newman)

USGS (U.S. Geological Survey), Department of the Interior (2008): Minerals Yearbook 2006, Volume III, France (Walter G. Steblez)

USGS (U.S. Geological Survey), Department of the Interior (2008): Minerals Yearbook 2006, Volume III, Germany (Steven T. Anderson)

USGS (U.S. Geological Survey), Department of the Interior (2008): Minerals Yearbook 2006, Volume III, Greece, (Harold R. Newman)

USGS (U.S. Geological Survey), Department of the Interior (2009): Minerals Yearbook 2007, Volume III, Hungary (Mark Brininstool)

Chapter 9 References

USGS (U.S. Geological Survey), Department of the Interior (2008): Minerals Yearbook 2006, Volume III, Ireland (Harold R. Newman)

USGS (U.S. Geological Survey), Department of the Interior (2008): Minerals Yearbook 2006, Volume III, Netherlands (Harold R. Newman)

USGS (U.S. Geological Survey), Department of the Interior (2008): Minerals Yearbook 2006, Volume III, Norway (Harold R. Newman)

USGS (U.S. Geological Survey), Department of the Interior (2008): Minerals Yearbook 2006, Volume III, Poland (Mark Brininstool)

USGS (U.S. Geological Survey), Department of the Interior (2008): Minerals Yearbook 2006, Volume III, Portugal (Alfredo C. Gurmendi)

USGS (U.S. Geological Survey), Department of the Interior (2008): Minerals Yearbook 2006, Volume III, Romania (Mark Brininstool)

USGS (U.S. Geological Survey), Department of the Interior (2008): Minerals Yearbook 2006, Volume III, Serbia (Walter G. Steblez)

USGS (U.S. Geological Survey), Department of the Interior (2008): Minerals Yearbook 2006, Volume III, Slovenia (Walter G. Steblez)

USGS (U.S. Geological Survey), Department of the Interior (2008): Minerals Yearbook 2006, Volume III, Slovakia (Walter G. Steblez)

USGS (U.S. Geological Survey), Department of the Interior (2008): Minerals Yearbook 2006, Volume III, Spain (Alfredo C. Gurmendi)

USGS (U.S. Geological Survey), Department of the Interior (2008): Minerals Yearbook 2006, Volume III, Sweden (Harold R. Newman)

USGS (U.S. Geological Survey), Department of the Interior (2008): Minerals Yearbook 2006, Volume III, Switzerland (Harold R. Newman)

USGS (U.S. Geological Survey), Department of the Interior (2008): Minerals Yearbook 2006, Volume III, UK (Walter G. Steblez)

USGS (U.S. Geological Survey), Department of the Interior (2008): Minerals Yearbook 2006, Volume III,

USGS (U.S. Geological Survey), Department of the Interior (Oct. 2009):)Minerals Yearbook, China, 2008

USGS (U.S. Geological Survey), Department of the Interior (Nov. 2009): Minerals Yearbook, Africa, 2007

USGS (U.S. Geological Survey), Department of the Interior (2010): Minerals Yearbook, India 2008

USGS (U.S. Geological Survey), Department of the Interior (April 2010): Minerals Yearbook, Latin America and Canada, 2007

UEPG, 2008

V

Vajna, T. (1974): Importabhängigkeit und Rohstoffpolitik [Dependence on imports and raw material policy]. – Köln: Deutscher Instituts-Verlag, 1974 (Beiträge/Institut der Deutschen Wirtschaft; 12)

Vakalopoulou, A. (1992): Antarctica: a scientific challenge for Europe. Report drawn up on behalf of the Commission of the European Communities (DG XI) on the scientific, political and legal status of the Antarctic continent. 6 Sept. 1990/EC, Commission. – Brussels: EC, 1992 (Publication UE/CE) (Publication internationale)

van der Meulen, M. J. (2005): Sustainable mineral development: possibilities and pitfalls illustrated by the rise and fall of Dutch mineral planning guidance. Geological Society Special publication, 250 (2005), pp. 225–232

Varet, J. (2005): Les matières premières minérales: flambée spéculative ou pénurie durable? – Futuribles 2005, n. 308, mai, pp. 5–23

Verband schweizerischer Hartsteinbrüche (VHS), Auftraggeber der Studie: "Kurze Transportdistanzen begrenzen schädliche Umweltauswirkungen" [Association of Swiss hard rock quarries (VHS), which commissioned the study: "Short transport distances limit environmental effects",], Sand und Kies (Sand and Gravel), Nr. 56 (2002), pp. 4–7.

Vereinigung Rohstoffe und Bergbau (VRB) (2008): 2008: Positionen und Pespektiven.

Vondran, R. (2001): Die Konzentrationswelle erfaßt die Rohstoffmärkte. – Stahl und Eisen, 121. 2001, H. 4, pp. 111–116

W

Wagner, M, Huy, D., (2005), Schafft der Strukturwandel in der Nachfrage eine neue Dimension für die Weltrohstoffmärkte, Bundesanstalt für Wissenschaften und Rohstoffe [Does the structural change in demand accomplish a new dimension to the world raw material markets? Federal Institute for Geoscience and Raw Materials], Hannover 2005

Chapter 9 References

Wagner, H.; Fettweis, G. B. L. (2003): Main areas for future mining research and development. Hauptgebiete für die zukünftige Bergbauforschung und Entwicklung. Glückauf, 139 2003, H. 9, pp. 490–493

Wagner, H. (1997): Untersuchung der Versorgung Österreichs mit mineralischen Rohstoffen aus heimischen Vorkommen (Investigation of the supply of mineral raw materials from domestic reserves in Austria), Bd. I–V, Wien, Leoben.

Wagner, H.; Nötstaller, R. (1997): Zur Frage der Bedeutung der Versorgung Österreichs mit mineralischen Rohstoffen aus heimischen Vorkommen. To the importance of the supply of the Austrian industry with minerals from local deposits. Berg- und Hüttenmännische Monatshefte (Journal of Mining, Metallurgical, Material, Geotechnical and Planned Engineering), 142 1997, H. 8, pp. 339–349

Wagner, H.; Tiess, G. (2005): Zu Fragen der Rohstoffpolitik der erweiterten EU. – BHM Berg- und Hüttenmännische Monatshefte (Journal of Mining, Metallurgical, Material, Geotechnical and Planned Engineering), 150. 2005, H. 4, pp. 122–126

Wagner, H., Tiess, G., Solar, S., Nielsen, K. (2005): Minerals planning policies in Europe. – Aachen International Mining Symposia AIMS, vol. 4. 2005, pp. 523–538

Wagner, H., Tiess, G. (2005): Rohstoffpolitik – in vielen Ländern ein Fremdwort [Raw material policy – an unknown term in many countries], in: Stein & Kies (2005) 79, pp. 4–6

Wagner, H., Tiess, G., Solar, S. (2006): Different approaches to mineral planning policies and practices in EU countries, in: 5th Pan-European Conference on Planning for Minerals and Transport Infrastructure (PEMT'06), pp. 197–207 Proceedings: 5th Pan-European conference on planning for minerals and transport infrastructure; the way forward

Wagner, H., Tiess, G., Solar, S., Nielsen, K. (2007): National minerals policy practices: Key to minerals supply in Europe., in: Proceedings of the 3rd International Conference on Sustainable Development Indicators in the Minerals Industry (SDIMI 2007), pp. 49–53 Proceedings of the 3rd International Conference on Sustainable Development Indicators in the Minerals Industry (SDIMI 2007). Milos Conferences: Restoring the natural balance.

Wasserbacher, R., Koller, W.; Schneider, H. W., Luptáčik, M. (2007): Die volkswirtschaftliche Bedeutung mineralischer Rohstoffe in Österreich. The relevance of raw materials extraction to Austria's economy. BHM – Berg- und Hüttenmännische Monatshefte (Journal of Mining, Metallurgical, Material, Geotechnical and Planned Engineering), Jg 152. 2007, H. 12, pp. 391–396

Wasserbacher, R. (2005): Sand-Kies-Industrie im Dialog – Kommunikationsplattform Forum Rohstoffe [Sand-gravel industry in dialogue – communication platform Forum raw materials]. – BHM Berg- und Hüttenmännische Monatshefte (Journal of Mining, Metallurgical, Material, Geotechnical and Planned Engineering), Jg. 150. 2005, H. 2, pp. 52–53

Wasserbacher, R. (2006): Umweltvorschriften der EU und ihre Umsetzung in Österreich, [EU environmental legislation and their implementation in Austria] BHM Berg- und Hüttenmännische Monatshefte (Journal of Mining, Metallurgical, Material, Geotechnical and Planned Engineering), Jg 151. 2006, pp. 150–151

Weber, L. (1997): Mineralrohstoffe als Basis für die Wirtschaft, in: Österreichische Akademie der Wissenschaften [Mineral raw materials as the basis for economy, in: Austrian Academy of Sciences]., Wien, pp. 217–220

Weber, L. (1998): Die neue Metallogenetische Karte Österreichs. Das Rohstoffpotential Österreichs [The new metallo-genetical map of Austria. The raw material potential of Austria]. – In: Zeman, J. [Hrsg.]: Energievorräte und mineralische Rohstoffe – wie lange noch? [Energy reserves and mineral resources – how much longer?] – Wien: ÖAW, 1998 (Schriftenreihe der Erdwissenschaftlichen Kommission)

Weber, L. [Hrsg.] (2000): Handbuch der Lagerstätten der Erze, Industrieminerale und Energierohstoffe Österreichs [Manual of ore deposits, industrial minerals and energy resources of Austria]. (Textband u. Anlagenband). – Wien: Geologische Bundesanstalt, 1997. – (Archiv für Lagerstättenforschung)

Weber, L. (2007a): Der österreichische Rohstoffplan als Werkzeug einer langfristigen Rohstoffsicherung. – BHM Berg- und Hüttenmännische Monatshefte, [The Raw Material Plan as an instrument for long-term raw material security, Journal of Mining, Metallurgical, Material, Geotechnical and Planned Engineering)] 152. 2007, pp. 252–S. 258

Weber, L. (2007b): The Austrian mineral resources plan. Der Österreichische Rohstoffplan. – World of mining – surface & underground, Jg. 159. 2007, H. 6, pp. 442–452

Weber, L., Zsak, G. (2008): World Mining Data 2006, Bundesministerium für Wirtschaft und Arbeit, Wien

Weber, L. (2008): Rohstoffsicherungsaktivitäten der EU und ihre Auswirkungen auf nationaler Ebene (Minerals safeguarding activities of the European Union and their impacts on national level), Berg- und Hüttenmännische Monatshefte Jg. 153. 2008, H. 8, pp. 289–295

Weber, L., Zsak, G., Reichl, C., Schatz, M. (2010): World Mining Data 2008

Wegge, Njord (2011): The political order in the Arctic: power structures, regimes and influence. Polar record, Vol. 47 (2011), pp. 165–176. (Published online 11 Jun 2010, DOI:10.1017/S0032247410000331)

Weizsäcker, E. U. (2010): Factor Five. Transforming the Global Economy through 80 % Improvements in Resource Productivity, London

Wellmer, F. W. (1995): Rohstoffversorgung und Geologische Dienste – Wandel der Aufgaben in den letzten 100 Jahren. Mineral resources supply and Geological Surveys – changes in their responsibilities during the last 100 years. Erzmetall, 48 1995, H. 9, pp. 608–618

Wellmer, F. W. (1996): Resource development, land-use planning and sustainability in Germany. Entwicklung von Rohstoffvorkommen, Landnutzungsplanung und nachhaltige Entwicklung in Deutschland. Zeitschrift für angewandte Geologie, 42 1996, H. 1, pp. 62–65

Wellmer, F. W. (1998): Lebensdauer und Verfügbarkeit energetischer und mineralischer Rohstoffe. Lifetime and availability of energy and mineral resources. Erzmetall, 51 1998, H. 10, pp. 663–675

Wellmer, F. W. (2001): Rohstoffe und Energie – Auswirkungen der Globalisierung auf die Versorgungssicherheit Deutschlands [Raw materials and energy – impact of globalisation on supply security in Germany]. Bergbau, 52 2001, H. 7, pp. 315–321

Wellmer, F. W. (2003): Die Rohstoffsituation der Welt. The state of natural resources in the world. Erzmetall, 56 2003, H. 12, pp. 705–717

Wellmer, F. W.; Dalheimer, M. (1998): Trends in der Rohstoffwirtschaft – Die Rolle der ‚Junior'-Firmen und der Berggesetzgebung für die internationale Exploration. Trends in raw material management – The role of the ‚Junior'- companies and mining legislation for international exploration. Glückauf, 134 1998, H. 9, pp. 528–534

Wellmer, F. W.; Dalheimer, M. (1998): Trends in der Rohstoffwirtschaft – Verschiebungen bei Firmen im Weltbergbau und in der Exploration. Trends in the raw material industry – displacements among companies in the international mining industry and exploration. Glückauf, 134 1998, H. 7/8, pp. 415–424

Wellmer, F. W.; Dalheimer, M. (1998): Trends in der Rohstoffwirtschaft – die internationale Rohstoffversorgung. Trends in raw material management – the international Supply of raw materials. Glückauf, 134 1998, H. 6, pp. 319–324

Wellmer, F. W.; Dalheimer, M. (2000): Rohstoffe und Energie – Auswirkungen der Globalisierung auf die Versorgungssicherheit Deutschlands. Raw materials and energy – Effects of globalization on the safety of German's supply sufficiency. Erzmetall, 53 2000, H. 6, pp. 385–397

Wellmer, F. W.; Dalheimer, M.; Wagner, M. (2008): Economic evaluations in exploration. – 2. edition. – Berlin, Heidelberg: Springer, 2007

References

Wellmer, F. W.; Hennig, W. (2003): Aspects for formulating mineral resources management policies. Gesichtspunkte für die Formulierung von Unternehmenspolitik im Rohstoffsektor. Erzmetall, 56 2003, H. 1, pp. 3–10

Wellmer, F. W.; Kosinowski, Michael (2005): Sicherheit der Rohstoffversorgung unter dem Aspekt der nachhaltigen Entwicklung. Natural resources – security of supply with respect to sustainable development. BHM – Berg- und Hüttenmännische Monatshefte, 150 2005, H. 4, pp. 117–121

Wellmer, F. W.; Stein, V. (1998): Mögliche Ziele nachhaltiger Entwicklungen bei mineralischen Rohstoffen. Possible aims of effective developments on mineral raw materials. Erzmetall, 51 1998, H. 1, pp. 27–38

Wellmer, F. W.; Wagner, M. (2000): Rohstofftrends am Beginn des 3. Jahrtausends. Mineral trends at the beginning of the third millennium. Erzmetall, 53 2000, H. 10, pp. 569–582

Wellmer, F. W.; Wagner, M. (2006): Metallic raw materials – constituents of our economy. In: Gleich, Arnim von; Ayres, Robert U.; Gößling-Reisemann, Stefan: Sustainable metals management. Springer Netherlands, 2006. (Eco-Efficiency in Industry and Science; Vol. 19) pp. 41–68

Wessel, D.; Seifried, D.; Hoa, Y. T. (1979): Rohstoffpolitik. – Hamminkeln [Raw material policy – Hamminkeln]: Unctad-Kampagne, 1979 (Studienheft/Unctad-Kampagne, Informations- und Aktionskampagne zur 5. Konferenz der Vereinten Nationen für Handel und Entwicklung (Information and action campaign for the 5th United Nations Conference on Trade and Development), UNCTAD V; 2)

Wilson, E. (2007): Arctic unity, Arctic difference: mapping the reach of northern discourses. Polar record, Vol. 43 (2007), 2, pp. 125–133

Wörmann, C. (2006): Probleme der Verfügbarkeit von Rohstoffen. International markets for raw materials: the challenges ahead. – Kali und Steinsalz, 2006, H. 2, pp. 6–11

World Steel Association (2008a): International Iron and Steel Institute. Fact sheet Steel and raw materials, http://www.worldsteel.org

World Steel Association (2008b): International Iron and Steel Institute. Steel statistics January 2000 (and 2008). Production evoluation of the last months, http://www.worldsteel.org

Z

Zeller, J. R. (1981): Nationale Rohstoffpolitik. – Bern (u. a.) [National raw material policy. – Bern (among others)]: Haupt, 1981 (Beiträge zur Wirtschaftspolitik; Bd 35) [(Contributions to economic policy, vol 35)]

Zentrales Komitee des Kommunistischen Bundes Westdeutschland (1976): Rohstoffpolitik und Kriegsvorbereitung. – Mannheim: Kühl, 1976. (Kommunismus und Klassenkampf/Arbeitshefte; 2)

Chapter 10 Appendix 1 – International minerals policy approaches

10.1 Alaska

State of Alaska – Minerals Development Policies[776]

Policies on mining and mineral development are found in a number of sources that range from those outlined in Article VIII of the Constitution of the State of Alaska to a specific policy outlined in Statute (AS 44,99 110) to general policies in the Title 27 Mining Statutes and to clarifications provided in various regulations.

The Alaska Constitution – Article VIII

Alaska has a separate article in its constitution devoted exclusively to natural resources. The framers of the constitution stated "The future wealth of the State of Alaska will depend largely on how it administers the immense and the varied resources to which it will fall heir." It was important enough that the framers felt it necessary to give constitutional recognition to the policies that would guide management of the state's endowment. Statements in article 08 of the constitution related to mineral development include:

Statement of Policy – It is the policy of the State to encourage the settlement of its land and the development of its resources by making them available for maximum use consistent with the public interest.

Alaska Statute – Title 44. State Government

Chapter 99. Miscellaneous Provision and Policies. Article 2. General State Policies. Section 44 is providing a **Declaration of state minerals policy**.

---

[776] All information from: http://209.85.129.132/search?q=cache%3APw7Uc6WqAvUJ%3Adnr.alaska.gov%2Fmlw%2Fmining%2FAK_MineralPolicy.pdf+Mineral+Policy&hl=de&gl=at

The legislature, acting under article VIII, section 1 of the Constitution of the State of Alaska, in an effort to further the economic development of the state, to maintain a sound economy and stable employment, and to encourage responsible economic development within the state for the benefit of present and future generations through the proper conservation and development of the abundant mineral resources within the state, including metals, industrial minerals, and coal, declares as *the minerals policy of the state that*:

(1) mineral exploration and development be given fair and equitable consideration with other resource uses in the multiple use management of state land;

(2) mineral development is encouraged through reasonable and consistent nonduplicative regulations and administrative stipulations;

(3) mineral development and the entry into the market place of mineral products be considered in developing a statewide transportation system;

(4) mineral development be encouraged through appropriate public information and education, scientific research, technical studies, and University of Alaska program involvement;

(5) economic development with respect to the state mineral industry be encouraged with Pacific Rim nations. (§ 1 ch 138 SLA 1988)

Alaska Statute – Title 27. Mining. Chapter 05

Administration and Services. Article 1. Department of Natural Resources.

(a) The department has charge of all matters affecting exploration, development, and mining of the mineral resources of the state, the collection and dissemination of all official information relative to the mineral resources, and mines and mining projects of the state, and the administration of the laws with respect to all kinds of mining.

(b) The department is the lead agency for all matters relating to the exploration, development, and management of mining, and, in its capacity as the lead agency, shall coordinate all regulatory matters concerning mineral resource exploration, development, mining, and associated activities. Before a state agency takes action that may directly or indirectly affect the exploration, development, or management of mineral resources, the agency shall consult with and draw upon the mining expertise of the department.

10.2 Brazil

Apart from the Constitution, the Mining Code (Law No 7805 of 1989) is where the general mining policy is more clearly propounded, especially for large-scale mining. The Introduction to the Mining Code declares that the scope of Brazil's minerals policy is to:[777]

- stimulate the discovery and enlarge the knowledge on the country's mineral resources;

- make use of the mineral production as an instrument of accelerated economic and social development, by the intense exploitation of known reserves, either for internal use or for exports;

- foster the economic utilisation of mineral resources and enhance their extraction, distribution and consumption;

- guarantee the supply of the nation's market;

- encourage private investments in exploration and exploitation of mineral resources; and

- create a good legal framework, to secure mining rights and stimulate private investment.

10.3 Canada

The Minerals and Metals Policy of the Government of Canada

The Minerals and Metals Policy was developed after extensive consultations involving federal departments and agencies, provincial and territorial mines ministries, industry, environmental groups, labour, and Aboriginal communities.[778] The mining industry plays an essential role in Canada's economy. It provides jobs for more than 340 000 Canadians and is the economic foundation for some 150 communities in Canada's rural and northern regions. These contributions are an important part of the economic and social fabric of Canadian

---

[777] Carlos Augusto Vilhena Filho (1997): Brazil's minerals policy. LLM in Natural Resources Law (Dundee), Associate of Pinheiro Neto-Advogados, SCS Qd 1, BI 1 Ed Central, 6° andar, 70304-900, Brasilia/DF, Brazil, in: Resoure Policy.

[778] Cp. Minister of Natural Resources Canada, Government of Canada Policy (http://www.nrcan-rncan.gc.ca/mms-smm/poli-poli/gov-gov-eng.htm)

society. The structure of the Minerals and Metals Policy of the Government of Canada is the following:[779]

Part I. Introduction

Part II. Federal Decisions in Minerals and Metals

Part III. The Business Climate: Ensuring the Competitiveness of Canada's Minerals and Metals Industry

Part IV. Minerals, Metals and Society: Promoting Products, Markets and Stewardship

Part V. Aboriginal Communities: Promoting Involvement in Minerals and Metals Activities

Part VI. Science and Technology: Progress through Innovation

Part VII. Minerals and Metals at the International Level:

Part VIII. Measurement and Follow-Up

Part I. Introduction

Minerals and metals are of vital interest to Canada and are relevant to federal policies and programs because of their substantial contribution to Canada's social and economic well-being. Two important and inter-related developments have implications for Canada: the globalization of the industry, and the mounting need for governments around the world to collaborate in the development of solutions to environmental concerns and other challenges.

Part II. Federal Decisions in Minerals and Metals

Implementing a sustainable development approach to achieve sustainable development, environmental, economic and social considerations must be taken into account as early as possible in the decision-making process. To help the Government meet this challenge in the area of minerals and metals, the Policy enunciates a number of principles for sustainable development-based decision-making, including a responsive public policy framework; the role of the market mechanism; the role of regulation; the role of non-regulatory approaches; the importance of science; endorsement of the concept of pollution prevention.

---

[779] Minister of Public Works and Government Services Canada 1996, http://www.nrcan.gc.ca/mms/sdev/policy-e.htm

Part III. The Business Climate: Ensuring the Competitiveness of Canada's Minerals and Metals Industry

Canada must compete as never before to attract investment capital to sustain its minerals and metals industry. In this environment, all governments must work together to ensure that a positive investment climate is maintained. As a consequence, the Government makes a series of commitments in the spheres of finance and taxation, regulatory efficiency, and investment and export promotion. The Government affirms its support for the creation of a Canadian Securities Commission, in partnership with interested provinces, and establishes four principles to guide the development of all federal fiscal measures affecting the minerals and metals industry. The Government also sets out a seven-item checklist for the development of any new federal regulatory processes that affect minerals and metals. As well, the Policy states that the industry must continue to assume greater responsibility for environmental performance and for stewardship of minerals and metals throughout their life cycle.

Part IV. Minerals, Metals and Society: Promoting Products, Markets and Stewardship

The Government supports the responsible use and management of minerals and metals. Given Canada's role as a world leader in the production of these commodities, managing issues related to health and the environment is a policy priority. The Policy introduces an approach to the responsible use and management of minerals and metals called the Safe Use Principle.

The Safe Use Principle takes a life cycle-based approach to the use and management of minerals and metals, including the application of risk assessment and management strategies, in accordance with well-established stewardship practices. The Principle builds on and complements the Toxic Substances Management Policy (TSMP). In doing so, it sends the message domestically and internationally that minerals and metals and their products can be used safely and responsibly.

Recycled minerals and metals constitute an important source of secondary materials for industry, and generate environmental benefits. As a consequence, the Government will work to: enhance the efficiency and effectiveness of regulations; promote a more efficient metals recycling industry in Canada; advance recycling as a feature of product design; and, at the international and domestic levels, promote common approaches to the definition of waste (including a distinction between metal-bearing recyclables destined for recovery and wastes destined for final disposal).

The federal government has a role to play in the reclamation of mine sites within its areas of responsibility, including establishing fiscal and regulatory conditions respecting reclamation for mine development on federal lands. The Policy recognizes the need to clean up those abandoned and orphaned mine sites within federal jurisdiction that represent an unacceptable risk to the environment or human health and safety. It also acknowledges the need for site owners, where they can be identified, to pay for clean-up costs.

Land access for mineral exploration and development is necessary if the minerals and metals industry is to continue to contribute to Canada's economic and social well-being. In regard to Canada's ocean territory, that access will be determined through an integrated oceans governance strategy adopted by the Government. In addition to land access, governments must provide reasonable certainty to the industry that when it finds a mineral deposit, it can develop that deposit.

The Government affirms its commitments respecting the completion of the National Parks network and the establishment of National Marine Conservation Areas. It also remains committed to identifying and protecting terrestrial and marine critical wildlife habitat in Canada, and developing and implementing protected area strategies for federal lands and waters. In meeting these commitments, the Government will follow certain guidelines that recognize the important economic and social role played by the minerals and metals industry in Canada.

Part V. Aboriginal Communities: Promoting Involvement in Minerals and Metals Activities

Part VI. Science and Technology: Progress through Innovation

The Government will pursue the following strategic, long-term directions in S&T related to minerals and metals: providing a comprehensive geosciences information infrastructure; supporting a sustainable minerals and metals industry (through the use of S&T to promote technological innovation both in mining operations and in the safe and efficient use of minerals and metals); enhancing the health and safety of Canadians; promoting the competitiveness of the Canadian industry; and developing value-added mineral and metal products.

Part VII. Minerals and Metals at the International Level

Providing Leadership in the Implementation of Sustainable Development Canada plays a leadership role at the international level, deriving from its position as the world's largest exporter of minerals and metals and a major player in the promotion of sustainable development, including the implementation of the Rio Summit's Agenda 21. The international nature of many of the pressures on the sector and the lessons learned to date, including the potential of initiatives aimed at environmental, health and social concerns to affect the competitiveness and acceptability of minerals and metals in the marketplace, requires an effective and flexible response by the Government.

Part VIII. Measurement and Follow-Up

Part VIII focuses on the effective implementation of the Policy, noting the importance of developing sustainability criteria and indicators related to minerals and metals. The Policy recognizes the need for ongoing accountability for, and assessment of, results that flow from the Policy. To this end, the Minister of Natural Resources, in cooperation with other federal departments and agencies, will issue periodic progress reports on its implementation.

10.4 Chile

The document „Mining Policy" is the result of the work of an interdisciplinary (government) team. It includes the key results of an extensive process of discussion, including the Report guidelines issued by the Special Senate Committee responsible for the Study of Taxation Mining and priorities defined by the Ministry of Mines.[780]

The Mining Policy of Chile has its mission „to promote a competitive and sustainable mining industry which is capable of inserting successfully in global markets through strategic alliances aimed to promote new uses and products with higher added value and to ensure an adequate contribution to the development of economic and social life." Specifically, the central axes of the State Policy are:

- securing the development of the mining industry and enhancing its competitiveness

- promoting their integration and international leadership,

- Engaging the mining industry to sustainable development,

- increasing its contribution to economic and social development,

---

[780] Ministry of Mines: Política Minera del Bicentenari o, Source: http://www.mch.cl/docu mentos/pdf/Politca_Minera.pdf. – See also: Lagos, G. (1997): Developing national mining policies in Chile: 1947–96. Resources Policy, Vol. 23 (1997), 1–2, pp. 51–69.

Chapter 10 Appendix 1 – International minerals policy approaches

- performing a social economic and environmentally management that promotes responsible sustainability of the mining regions.
- building capacity in education, science, innovation and development
- increasing mining knowledge, leading the mining industry to scientific and technological international level.
- Enhancing the role, effectiveness and contribution of mining state enterprises
- Strengthening the institutional mining system to generate integral development for the mining industry.

One of the key elements for the mining sector has been to develop a regulatory framework, economic and social environment, without prejudice the dynamism of modernization. This has helped to promote foreign investment for the implementation of major mining mega-projects and consolidating state mining companies making this sector an incentive of economic development. Mining has contributed significantly to the development of the country. However, the public has demanded in recent years that the benefits of this activity grow and transcend to future generations. This includes the need to examine the future development of the sector where questions arise such as:

- What priorities should guide the development of national mining, both private and state
- How to ensure its competitiveness and global leadership,
- How to ensure a balance between sector development and the economic, social and environmental development.

The Ministry of Mines has collected these concerns in order to formulate a modern vision of sustainable mining. This new vision should optimize existing strengths and overcome weaknesses in order to submit a renewed State policy according to the new times. The "Bicentennial Independencia de Chile" is an opportunity to undertake a minerals policy to suit these new requirements that globalization places, ensuring sustainability and competitiveness of the mining sector.

The Mining Policy of Chile is structured as follows:

1. Mining Development: Competitiveness and International Linkages
2. Sustainable Development of Industry
3. Human Capital, Innovation and Technological Development

4. Public Enterprises in Mining and Hydrocarbons
5. Towards an Institutional Mining Efficiency and Dynamics

10.5 China

Great achievements have been obtained in the survey and development of China's mineral resources in the past five decades since the founding of New China.[781] A great number of mineral resources have been verified, and a fairly complete system for the supply of mineral products has been established, providing an important guarantee for the sustained, rapid and healthy development of the Chinese economy. 2003, over 92% of the country's primary energy, 80% of the industrial raw and processed materials and more than 70% of the agricultural means of production come from mineral resources.

China attaches great importance to sustainable development and the rational utilization of mineral resources, and has made sustainable development a national strategy and the protection of resources an important part of this strategy. Immediately following the UN Environmental and Development Conference in 1992, the Chinese government took the lead in formulating the "China Agenda 21 – the White Paper on China's Population, Environment and Development in the 21st Century." It approved and implemented the "National Program on Mineral Resources" in April 2001, and, in January 2003, began to implement "China's Program of Action for Sustainable Development in the Early 21st Century."

To build a well-off society in an all-round way is China's objective in the first 20 years of the new century. China will depend mainly on the exploitation of its own mineral resources to guarantee the needs of its modernization program. The Chinese government encourages the exploration and exploitation of the mineral resources in market demand, especially the dominant resources in the western regions, to increase its domestic capability of mineral resources supply. At the same time, it is an important government policy to import foreign capital and technology to exploit the country's mineral resources, make use of foreign markets and foreign mineral resources, and help Chinese mining enterprises and mineral products enter the international market. The Chinese government holds that to have foreign mining companies enter China and Chinese mining enterprises enter other countries to make different countries mutually complementary in resources is of great significance for the common

---

781 China's Policy on Mineral Resources: Information Office of the State Council of the People's Republic of China December 2003, Beijing. Updated: 2006-08-04; http://www.chinamining.org/Investment/2006-08-04/1154674314d441.html

prosperity and healthy development of world mineral resources prospecting and exploitation.

    I. The Present Situation of Mineral Resources and Their Exploration and Exploitation

    II. Targets and Principles for Mineral Resources Protection and Rational Utilization

    III. Increasing the Domestic Capability of Mineral Resources Supply

    IV. Widening the Opening of, and Cooperation in, Mineral Resources Exploration and Exploitation

    V. Achieving the Coordinated Development of Mineral Resources Exploitation and Environmental Protection

    VI. Improving the Management of Mineral Resources

10.6 Guatemala

The Ministry of Energy and Mines (MEM) is aware of the importance and potential that the mining sector may have to develop the country and the many challenges that this development implies. The Ministry is pleased to present the "Mining Policy Guidelines", which is a guiding instrument related to the main actions of both the public and civil society. Considering the private sector and community it is important to consider the international development and to ensure that mining activities include long-term goals.[782]

This document also delineate issues that the Ministry wants to boost in mining and incorporates all the consensus reached by the High-level Commission (Catholic Church Government representatives and civil organizations and environmentalists) during the months of February to August 2005. It is important to mention the active participation of all actors national and international, whose participation contributed to the formulation of this policy, support technical staff of the Ministry in drafting the document, the High Level Commission for their work in 2005 and the National Program for Competitiveness (PRONACOM) which in 2007 supported the launch of this policy.

---

[782] Ministerio de Energía y Minas, República de Guatemala (2007): Lineamientos de política minera 2008–2015 [Ministry of Energy and Mines (2007): Lineamientos de política minera, **Mining Policy Guidelines**, 2008–2015, Octubre 2007, Republic of Guatemala].

The Mining Policy Guidelines 2008–2015 are structured as follows:
- A Rectory in the Mining Sector Policy
- B International Context
- C. Current status of mining
  - C1 Scale mining projects
  - C2 Current Mineral Production
- D. Mining Policy Guidelines
  - 1. Promotion of the technical and rational development of mineral resources
  - 2. Modernizing the legal framework governing mining and strengthen the regulatory role of government in this matter
  - 3. To favour the social and economic development of the communities so that they are provided with wealth of mineral resources
  - 4. To develop activities of dialogue and agreement during the process of granting of licenses with the actors directly involved
  - 5. To strengthen the Environmental Protection

10.7 India

The National Minerals Policy of India is structured as follows:[783]

- Preamble
- Regulation of minerals
- Objectives
  Role of the state in mineral development
- Survey and exploration
- National inventory of mineral resources
- Strategy of mineral development
- Foreign trade
- Fiscal aspects

---

783 http://mines.nic.in/, 1993.

- Research and development
- Conclusions

  The National Minerals Policy was updated in 2008 and also in 2010 ("Model State Minerals Policy 2010"). The salient features of the National Minerals Policy, 2008 (NMP) generally applicable to the States are as follows:[784]

- Minerals being a valuable resource, extraction has to be optimised through scientific methods, beneficiation and economic utilization. Zero waste mining will be the goal (para 2.1 of NMP).
- The regulatory environment will be made more conducive to investment. Transparency in allocations of concession will be assured with security of tenure to a concessionaire (para 2.2 of NMP)
- 'First in time' and 'continuity or seamlessness' principles will be fully recognized (para 3.3 of NMP).
- The duration of all concessions shall be rationalised (para 5.2 of NMP).
- Data filing requirements will be rigorously monitored and concessionaires will be closely monitored in this regard (para 6.2 of NMP).
- A framework of sustainable development will be designed to ensure that mining can take place along with restoration (para 2.3 of NMP).
- Mining shall not be undertaken in ecologically fragile or biologically rich areas. Mining in forest areas will be accompanied by time-bound reclamation (para 7.10 of NMP).
- Special care will be taken to protect the interests of host and tribal populations (para 2.3 of NMP).
- Project affected persons will be protected through comprehensive R&R packages (para 2.3 of NMP).
- Old disused mining sites will be converted into forests or used in some other appropriate manner (para 7.10 of NMP).
- When mine closure becomes necessary, it should be orderly and systematic with rehabilitation of workers (para 7.12 of NMP).
- Value addition will be encouraged, as also growth of mineral sector as a stand alone industrial activity (para 2.4 of NMP). Mining as a backward

---

784 http://mines.gov.in/new/model.pdf

linkage and value addition as a forward linkage within the State will be encouraged (para 7.4 of NMP).

- Mining sectoral value addition through beneficiation, calibration, blending, sizing, concentration, pelletisation, purification and customization will be encouraged (para 7.2 of NMP).
- Mining infrastructure requires a special thrust. Infrastructure needs will be financed through innovative structures including user charges, PPP mode and viability gap funding (para 2.5 of NMP).
- An enabling environment will be created to motivate large capacity mining companies to undertake creation of transportation network (para 7.7 of NMP).
- In the public funding of infrastructure, greater thrust will be given to development of health, education, drinking water, road and other related facilities for integrated regional development (para 7.7 of NMP).
- State agencies involved in mineral sector development and regulation will be encouraged to modernize in the areas of prospecting as well as regulation (para 2.6 of NMP).
- The State Directorates will be suitably strengthened to enable them to regulate mining in the interests of conservation and scientific development of the sector (para 2.2 of NMP).
- States will be assisted to overcome the problem of illegal mining through operational and financial linkages with IBM (para 2.6 of NMP).
- Mining of small deposits will be suitably regulated to promote scientific and efficient extraction and to control illegal mining. A cluster approach shall be encouraged through formation of consortia of small scale miners (para 7.9 of NMP).
- There will be arms length distance between State agencies that mine and those that regulate (para 4 of NMP).
- Ore bodies shall be reserved for PSUs only if there are no applications from private players for the area (para 4 of NMP).
- Use of machinery and equipment which improve the efficiency, productivity and economics of mining operation and safety and health of workers and others shall be encouraged (para 7.5 of NMP).

- Emphasis will be laid on mechanization, computerization and automation of the mining units and the manpower development programmes shall be suitably reoriented (para 7.6 of NMP).

- Educational institutions will be geared to meet the needs of the sector in the medium and long-term (para 7.6 of NMP).

Rajastan

Importance of the mineral wealth for the economy of the State and the need for a well defined policy for the mineral sector is well recognized. The first Minerals Policy was announced by the State Government in 1977. Since then, significant changes have taken place in the knowledge of mineral deposits, legal regime governing mineral development, organisational structure of the Mines Department and the infrastructure availability. The process of economic reforms has led to the globalisation, privatisation and opening up of the economy and this in turn, has necessitated changes in the policy and the legal regime regulating the growth and development of the mineral sector. The move towards an open market economy based on competitiveness and international outlook has made the announcement of a New Minerals Policy imperative.[785] The National Minerals Policy, 1993 has radically altered the earlier policy. Important changes introduced are as follows:

(a) Thirteen minerals viz. Iron ore, manganese, chrome, sulphur, gold, diamond, copper, lead, zinc, molybdenum, tungsten, nickel and platinum group of minerals have been deleted from the list of minerals which had earlier been reserved for exclusive exploitation by the public sector, these minerals are now open for exploitation by the private sector.

(b) Foreign investment and technology will be encouraged. Ceiling on foreign equality in the mining industry has been raised to 50% in the equity of Indian companies engaged in mining activities.

(c) Mineral and metal processing units which wish to develop captive mines to secure assured supplies of raw material will also be allowed foreign equity participation in the manner and to the extent already permitted to such processing units. Equity participation over 50% by foreign parties in non-captive mines can also be considered on a case to case basis.

---

[785] http://www.rajasthan.gov.in/rajgovresources/actnpolicies/MINERAL.html

10.8 Jamaica

Minerals are a significant part of the patrimony of people; however, left undeveloped, they are of little benefit, particularly in the case of an emerging economy such as Jamaica's. The materials that these resources provide are needed to propel the economy, generate wealth and improve the well-being of our people. Primary industrial activities generate limited social and economic benefits. The Policy places significant emphasis on the production of value-added mineral products, which create far greater revenues for the country, increased profits and more and better paying jobs. It is also clear that poorly regulated or unregulated mineral exploitation activities are very serious potential threats as they may cause large-scale environmental, social and economic harm.[786]

Jamaica's first National Minerals Policy sets out a comprehensive and unified approach to develop the country's Minerals Industry and to ensure that the Jamaican people are the major beneficiaries. This Policy therefore constitutes an important aspect of the Government's determination to develop a broad-based industry within the construct of sustainability and signals its aim to create a modern and diversified Minerals Industry. It presents Government's vision, establishes the official framework and facilitating structures to guide the effective management and continued transformation of the industry and to ensure its harmonious co-existence with competing interests in the wider economy. It is expect that this Policy will help to fashion a positive, balanced and competitive environment which encourages investment and development with an emphasis on the manufacture of value-added products, export-led growth and effective environmental management. This will generate greater wealth, provide quality employment for larger, more skilled and more highly qualified numbers of Jamaicans and so extract optimum benefits from these resources.

Jamaica has a range of commercially exploitable minerals, including a wide variety of limestones, hard volcanic rocks, bauxite, marble, gypsum, shale, sand and gravel. These minerals are of major significance to Jamaica's economic development, particularly their contribution to the national economy, their impact on, and linkages with other sectors, and their overall contribution to Gross Domestic Product (GDP). In fact, since 1985 the Minerals Industry has annually contributed at least 4,0 % of Jamaica's GDP.

---

[786] Robertson, J. (2009): National Minerals Policy of Jamaica, Minister of Energy and Mining, May 2009.

Chapter 10 Appendix 1 – International minerals policy approaches

Rationale for a Minerals Policy

The Government of Jamaica recognizes that a properly planned, efficiently regulated, and professionally marketed Minerals Industry can make a significant contribution to national development. The National Minerals Policy seeks to establish the framework for the country's approach to managing its mineral resources and developing its Minerals Industry. It arises out of a necessity to:

- create a **single, coherent national approach** for the Minerals Industry, which will develop the enabling environment to *encourage further investment* and promote diversification and development of the industry;

- consider the rationalization of the numerous pieces of legislation governing the industry;

- revise the land-use and land management framework to allow for sequential planning and so optimize the benefits of exploiting the country's mineral resources, while minimizing negative social and environmental impacts.

The **vision** of this Policy is for a modern, diversified, efficient and attractive Minerals Industry that protects environmental integrity and socio-cultural values, adds significant value to the economy, is based largely on the manufacture and export of value-added products, has strong and properly structured institutions and co-exists with competing interests in the wider economy.

Policy Objectives and Strategies

The policy objectives which it is envisaged will result in a viable and publicly supported mineral resource sector are:

- To enhance the industry's regional and international competitiveness.

- To facilitate increased investment.

- To ensure the effective management of the Minerals Industry, mineral resources and mineral-bearing lands from the pre-mining to post-mining stage.

- To provide a framework for the increased application of science and technology within the Minerals Industry.

- To facilitate product diversification, increased levels of import substitution, improved product quality, optimized utilization of minerals resources and expansion of the industry.

- To address under-development of certain sub-sectors while removing imbalances.

- To regulate the industry to ensure effective management of the environment, and the promotion of and adherence to best practices in health and safety standards.

- To promote increased benefits from minerals operations to the host communities.

- To facilitate exports and increased market share.

- To promote and facilitate increased levels of integration with other segments of the economy.

- To increase public awareness of the country's endowment of mineral resources and their strategic role in the country's social, cultural, economic and industrial development.

- To promote the development of human resource capacity within the industry.

- To further investigate the feasibility of promoting the exploitation of offshore mineral resources.

- To reform the existing legislative framework to encourage the development of all sectors of the Minerals Industry.

In this regard, it furthers the commitments of the **National Development Plan: Vision 2030** Jamaica relating to the development of minerals as a significant pillar of the economy; and the National Land Policy as it relates to land management, and specifically the management of mineral-bearing lands, including the rehabilitation of mined lands. Importantly, it also promotes the manufacturing of value-added products, sequential land use planning and the prevention of sterilization of economically beneficial mineral resources. Particular effort has been made to ensure that sustainable development considerations, especially environmental considerations, are infused into the strategies outlined for the development of the industry and the guiding principles that will be followed.

This Minerals Policy outlines numerous new policy initiatives and is designed to effectively address the emerging issues, challenges and requirements of a growing industry within a complex financial, socio-political and environmental scenario. The **strategies and approaches** to be pursued in the effort to realize the industry's continued transformation and development are elaborated under the following headings:

- A Sustainable Approach to Mineral Development – which speaks to incorporating the guiding principles of sustainable development within the Industry.

- The Business Climate: *Ensuring Competitiveness* – which details the approaches which will be taken in presenting the country's Minerals Industry as an area in which to invest. It addresses issues such as attracting investment, ensuring the existence of an enabling regulatory environment, and a strategy of mineral development.

- Institutional Arrangement – which details proposals for the most effective arrangement to enable the policy's effective implementation? Attention is given to strengthening various state entities associated with the management of the country's mineral resources and facilitating the development of industry.

- The Role of Science and Technology – including research and development, technology and training and the use of scientific methods for exploitation and land rehabilitation.

- The Management of Mineral Resources – including the management of mineral reserves, mineral-bearing lands, land rehabilitation, the production of minerals and the manufacturing of value-added mineral products. Projected within the context of a small island developing State, with the vulnerabilities and limitations attendant upon such a status, the Minerals Policy creates necessary balances between complex competing interests pertinent to facilitate sustainable national development and to augment the goals of the National Development Plan: Vision 2030 Jamaica, the National Industrial Policy and other national policies.

10.9 Japan

In 2004, the Japanese government created the Japanese Oil, Gas and Metals National Cooperation (JOGMEC).[787] Among JOGMEC's important activities are providing financial assistance to Japanese companies for mineral exploration and deposit development, gathering and analysing information on mineral and metal markets to better understand supply risk, and managing Ja-

---

787 http://www.meti.go.jp/english/press/data/nBackIssue200803.html

pan's economic stockpile of rare metals. JOGMEC defines rare metals as those that (a) are essential to Japanese industry, sectors such as iron and steel, automobiles, information technology, and home appliances and (b) are subject to significant supply instability. JOGMEC took over and manages the Japanese rare-metal stockpiles in cooperation with private companies, with the goal of having stocks equivalent to 60 days of domestic industrial consumption. Stocks exist for seven materials: chromium, cobalt, manganese, molybdenum, nickel, tungsten, and vanadium. JOGMEC is closely observing 7 other raw materials.

Early in 2008 the Japanese government published its "**Guidelines for Securing National resources**", which includes the **statement** that the Japanese Government "*will support key resource acquisition projects by promoting active diplomacy and helping these projects to be strategically connected to economic cooperation measures, such as official development assistance (ODA), policy finance and trade insurance*". Potential projects must fulfil the criteria 1) "projects to acquire exploration or development interests" and 2) "projects related to long-term supply contracts that contribute to supplying … resources to users in Japan".

10.10 Liberia

The Ministry of Lands, Mines and Energy, Monrovia, Government of the Republic of Liberia published in August 2008 a draft of the Mineral Policy of Liberia.[788] The suggested Mineral Policy of Liberia is structured as follows:

1. Liberia Mining Vision
2. Introduction
3. Key Principles of the Mineral Policy
4. Mineral Endowment
5. Regulatory Framework
6. Equitable and Competitive Fiscal Framework
7. Competing Land Rights and Land Use Options
8. Environmental Stewardship and Social Responsibility
9. An Integrated Mining Sector

---

788 Draft 5: http://www.molme.gov.lr/doc/Liberia%20Draft%20Minerals%20Policy%20clean.pdf

10. Artisanal and Small-Scale Mining (ASM)
11. Quarrying (sand, stone, clay and laterite mining)
12. Transparent Benefits from Mining
13. Developing with Broad Participation
14. Building Capable Institutions
15. Investing for the Future (Sustainable Investment)
16. Expected Outcomes

The draft of the suggested Mineral Policy provides the following vision statement:

Towards Liberia Mining Vision for the 21st Century – Vision Statement

"Equitable and optimal exploitation of Liberia's mineral resources to underpin broad-based sustainable growth and socio-economic development". The shared vision will aim to achieve:

- A knowledge-driven mining sector that catalyses and contributes to the broad-based growth and development of, and is fully integrated into, an African market through:
  - Down-stream linkages into mineral beneficiation and manufacturing;
  - Up-stream linkages into mining capital goods, consumables & services industries;
  - Side-stream linkages into infrastructure (power, logistics, communications, water) and skills & technology development (HRD and R&D);
  - Mutually beneficial partnerships between the state, the private sector, civil society, local communities and other stakeholders; and
  - A comprehensive knowledge of its mineral endowment.
- A sustainable and well-governed mining sector that effectively garners and deploys resource rents and that is safe, healthy, gender & ethnically inclusive, environmentally friendly, socially responsible and appreciated by surrounding communities;
- A mining sector that has become a key component of a diversified, vibrant and globally competitive industrialising Liberian & African economy;

- A mining sector that has helped to establish a competitive Liberian & African infrastructure platform, through the maximisation of its propulsive local & regional economic linkages;

- A mining sector that optimises and husbands Liberia's finite mineral resource endowments and that is diversified, incorporating both high value metals and lower value industrial minerals at both commercial and small-scale levels; and

- A mining sector that is a major player in vibrant and competitive national, continental and international capital and commodity markets.

10.11 Malaysia

To support the mineral industry, the Government formulated a new National Mineral Policy (NMP) in 1994 and established the National Mineral Council (NMC) in 1998 to oversee the overall integrated development of the mineral industry and to *assure such development would meet its policy objectives*.[789] The National Mineral Council is also charged with coordinating relations between the Federal and State Governments. The NMP provides the foundation for the development of an effective, efficient and competitive regulatory environment for the mineral sector. The thrust of the policy is to expand and diversify the mineral sector through optimum exploration, extraction, and utilization of resources using modern technology as well as R&D. Emphasis is also given to environmental protection, sustainable development and the management of social impacts.

Under the National Mineral Policy, a mining lease application must include an environmental protection plan that is approved by the Department of Environment of the Ministry of Science, Technology, and Environment. The environmental aspects of mine development are regulated by the Environmental Quality (Prescribed Activities) (Environmental Impact Assessment) Order 1987, which was an amendment to the Environmental Quality Act of 1974. Under Order 1987, an environmental impact assessment is required for mining lease areas that are more than 250 hectares. The salient features of the NMP are the provisions for security of tenure, high land-use priority for mining, uniform and efficient institutional framework, regulations and guidelines, and rehabilitation and environmental control.

---

789 http://www.jmg.gov.my/en/information/legislation-and-policy/doc_details/9-national-mineral-policy-2.html

Two main legal instruments that had been formulated for the full implementation of the NMP, are the Mineral Development Act, 1994 and the State Mineral Enactment. The Mineral Development Act came into force in August 1998, while the State Mineral Enactment is currently at various stages of being adopted by the respective State Governments. The Mineral Development Act 525 of 1994 defines the powers of the Federal Government for inspection and regulation of mineral exploration and mining and other related issues.

Mineral producers are required to pay an income tax and a development tax based on the profits from their operations. Export duties on most minerals have been abolished. Most raw minerals, including ores and concentrates, are subject to low or zero level import duties. For those minerals still subject to import duties, the importer may apply to the Government for a waiver. Imported equipments for use in mineral projects are subject to the general schedule of import tariffs but an application for a waiver may be made on a case-by-case basis. Value-based royalties are assessed by some individual states on some mineral commodities. A schedule of area-based land premiums and rental fees, processing and application fees for mining lands is published in each State.

Figure 180: http://asiamining.org/countries/malaysia/policy.htm

To support the mineral industry, the Government formulated a National Mineral Policy (NMP) in 1994. Main elements are the following:

Mineral Sector Mission Statement

"To enhance Malaysia's competitiveness advantage in a globalized market for a selected mineral commodities and their value added products considered economically and socially strategic in our national development, as well as to maximise the usage of research & development, minerals and geoscientific information for socio-economic improvement of the people with due consideration for environmental protection and sustainable development."

Regulatory Objectives – Salient Features

1. To contribute to national and State development by promoting diversification and expansion of the mineral industry.

1.1 Expansion and diversification of the industry

The Geological Survey Department of Malaysia has identified occurrences of more than thirty minerals which may have commercial development potential. Given this potential and the need to supply the rapidly growing industrial sector, the Government supports and encourages the development of projects which will expand and diversify the mineral industry.

1.2 Promote regional development

Mining and mineral processing projects will promote optimal economic development.

1.3 Self-sufficiency

Self-sufficiency will be encouraged through the location and development of deposits which can produce minerals which are commercially more competitive then imported minerals

1.4 Mineral exports and imports

Malaysia encourages the development of export-oriented mineral projects. Malaysia encourages and supports the development of mineral processing facilities which utilize locally produced raw material that will contribute to import substitution and savings on foreign exchange. Where the minerals or mineral products are not available locally, then the may be imported.

2. To provide an attractive, efficient, and stable mineral sector regulatory framework

2.1 Ownership of the resource endowment

All minerals within the boundaries of a State are vested solely in the State Authority of the State. All minerals outside the mineral jurisdiction of the State but within the outer limits of the Continental Shelf or Exclusive Economic Zone of Malaysia are vested solely in the Federal government

2.2 Regulatory authority

Current Federal and Sate mineral laws are being phased out, or are being amended to conform to the respective areas of regulatory authority as defined in the Federal Constitution.

2.3 Role of the Federal Government

As provided for in the Federal Constitution, the responsibility of the Federal Government will be enhanced in areas, such as, the development of mineral resources; mines; mining; mineral; and mineral ores; and regulation of labour, safety in mines and the environment.

2.4 Role of the State Governments

With the objective of supporting and encouraging the expansion and diversification of the mineral industry, Sate Government would continue to work towards the implementation of an improved, competitive and more uniform State Mineral Enactment and regulations.

2.5 Security of tenure

The General policy of both the Sate and Federal Governments will be that the holder of a current and valid licence granting a right to prospect and explore for minerals should be given the priority or the first right to obtain a mining lease should a deposit be discovered in the licence area.

2.6 Regulatory groups of minerals and rock materials

For regulatory purposes, minerals and rock materials have been designated as belonging to one of three groups, namely, minerals, radioactive minerals, and rock materials. The exploration and exploitation of these three groups of minerals are governed by the respective laws falling within the jurisdiction of the relevant authorities i. e. the State or Federal Governments.

2.7 Mineral occurrences in the Federal Territories

The power to issue exploration and mining rights on lands in the Federal territories is within the sole jurisdiction of the Federal Government.

2.8 Mineral occurrences located within a State

The power to issue exploration and mining rights for all areas within State boundaries, including permanent reserved lands is within the sole jurisdiction of the respective State Authority.

2.9 Mineral Occurrences in privately-owned (alienated) land

Minerals underlying privately owned land located within a State's jurisdiction are vested in the State Authority.

2.10 Mineral occurrences in the Continental Shelf and Exclusive Economic Zone

Malaysia welcomes the development of mineral resources located within its Continental Shelf and Exclusive economic Zone. Authorization for the issuance of appropriate licences is decided on a case-by-case basis.

2.11 Minerals underlying international waters

Malaysia support the principals and tenets of the currently proposed United Nations Convention on Law of the Sea for the regulation of exploration, development, and exploitation associated with mineral resources underlying international waters.

3. To encourage exploration and a beneficial expansion of the mineral industry

3.1 Government equity participation

The Federal Government is proponent of private sector development of the mineral sector. However, the Federal Government reserves the option to participate on an equity basis through a corporate entity in key projects deemed to be of crucial or strategic importance to the national interest. Similarly, State Government may, at their option, require private investors to enter into an equity joint venture arrangement with the State or a State Economic Development Corporation or any of its agencies or nominees. All participation is exclusively on a paid, carried interest basis or other similar arrangement.

3.2 Foreign investor participation

Foreign entities exploring for minerals in Malaysia may do so on a 100% foreign held equity basis or can enter into arrangements with local enterprises. For projects which involve the extraction, mining or processing of mineral ores, initial majority foreign equity participation of up to 100% may be permitted.

3.3 Role of small and large scale operations

Malaysia recognises the importance of both small and large scale mining operations and supports the further growth and development of these categories of mines.

4. To provide a stable and conducive fiscal system

4.1 The fiscal system

Malaysia strives to maintain a fiscal system which is stable, fair, transparent and competitive.

4.2 Federal taxes

Mineral producers are required to pay an income tax and a development tax based on the profits from their operations. Export duties on most mineral have been abolished. Most raw minerals, including ores and concentrates, are subject to import duties. For those minerals still subject to import duties, the importer may apply to the Government for a waiver. Imported equipments for use in mineral projects are subject to the general schedule of import tariffs but an application for a waiver may be made on a case by case basis.

4.3 State assessments

Value-based royalties are assessed by some individual states on some mineral commodities. A Schedule of area-based land premiums and rental fees, processing and application fees for mining lands are published in each State.

5. To accord the mineral industry a high land-use priority in areas open for exploration

5.1 Prospecting and exploration are not a land-use

Prospecting and exploration to identify the mineral endowment on most land which have not been specifically closed to these activities.

5.2 Land-use priority for mining activities

The Policy proposes that the State and Federal Governments should accord the mining industry a high land use priority so that the minerals in the ground are not sterilised by infrastructure and other developments.

6. To enhance the development of domestic expertise for mineral resource development through research, education and training activities

6.1 Role of the public sector

The Federal and State Governments support the development of Malaysian mineral expertise through a wide variety of on-going activities including but limited to: university departments specialising in mineral related fields; in house training programmes; funded researches; statistical studies; participation in international and regional mineral organisation meetings, seminars and conferences.

6.2 Role of the private sector

The private sector in encouraged to contribute towards the development of mineral expertise in Malaysia through the support of scholarship funds, training and educational programmes, research programmes, and trough cooperation with scientific institutions. All firms engaged in mining and processing are expected to endeavour to comply with current government employment policies.

7. To provide environmental protection and management of social impact

7.1 Regulation of active project sites

Mineral projects shall employ modern and efficient method of extraction, processing and waste disposal which minimise the immediate and long term impacts on the environment. Firms must comply with the environmental standards and obligations detailed in the Environmental Quality Act 1974, and all other relevant laws.

7.2 Rehabilitation requirements

Land subject to mineral exploitation shall be rehabilitated to a state which will allow future economic use or timely reversion to a safe and, wherever realistically and economically possible, aesthetically natural-like condition.

8. To provide timely and accurate regulatory, scientific and technical information required by the industry, Federal Government and State Governments, including periodic review and publication of the national policy on the mineral sector

8.1 Publication and Public access to information

The State and Federal Governments actively support the publication of regulatory, scientific and technical information.

8.2 Mineral Resources Information Centre

To assist mineral sector investors, a Mineral resources Information Centre will be established in Kuala Lumpur to act as a repository of publicly available information pertaining to the nation's mineral endowment, as well as the laws and regulations governing the mineral industry.

10.12 Pakistan

The National Mineral Policy having consensus of all the provinces has been approved by the Federal Cabinet for announcement.[790]

The Government of Pakistan for many years has been attempting to formulate a National Mineral Policy. Despite pledges made by the governments in the

---

790 Ministry of Petroleum & Natural Resources, (September 1995) – NATIONAL MINERAL POLICY 1995, Government of Pakistan, September 1995, http://www.sindhmines.gov.pk/pdf/mineral_policy.pdf

Sixth and Seventh Five year Plans, a Mineral Policy could not be announced. Since its induction into office in October 1993, the Government of Prime Minister Benazir Bhutto has taken bold initiatives on a number of fundamental issues facing the national economy especially in the power, oil and gas and telecommunication sectors and has announced policies which have attracted large scale private investment to Pakistan. Taking note of our substantial mineral potential, the Government also assigned special attention to the Mineral Sector to facilitate the inflow of substantial local and foreign investment.

The formulation of National Mineral Policy required a clear perception of the demands of the industry and also of the aspirations of the Provincial Governments who constitutionally own this natural resource. Extensive consultations, spread over a period of nine months, were therefore held commencing with an International Conference on Mining in Pakistan in October 1994 which was inaugurated by the Prime Minister and its concluding session was chaired by the President of Pakistan. The Conference was well attended by senior executives from International mining companies and Federal and Provincial governments. The Conference was followed by setting up of a broad based Task Force on Minerals which submitted recommendations keeping in view the concerns of the industry as well as of the governments and the prevailing international practices.

The recommendations of the Task Force on Minerals were debated at a Workshop, attended by the representatives of the Federal and Provincial governmental agencies, which are concerned with its implementation. The modified recommendations were further reviewed at two interprovincial meetings attended by all the four Provincial Chief Ministers and the concerned Federal Ministers. The structure and content of the **NATIONAL MINERAL POLICY** is the following:

1. Objectives

2. Constitutional Position on Minerals

3. Consultative and Regulatory Framework

1. Objectives

1.1

The available geological information provides ample evidence that Pakistan could be blessed with large or even world class mineral deposits as are existing in similar geological environments elsewhere in the world. Mining activity in Pakistan is, however, confined to industrial and construction minerals and contributes less than one percent of Pakistan's Gross Domestic Product (GDP). This mineral potential can be well utilized for sustainable socio-economic development of local population in mineral bearing areas. The Government of Pakistan is therefore launching a major policy initiative in order to expand mineral sector activity mainly through private investment and thereby enhance the contribution made by this economic activity to GDP and also lend support to the social uplift programmes.

1.2

In seeking to attract both foreign and local private risk capital to Pakistan's mineral sector, the Government wishes to satisfy several important objectives. The most significant benefits to be derived from an expansion of mineral sector activities are:

- expansion of employment opportunities; enhancement of skills;
- sustained development of mineral bearing area; expanded
- business opportunities for local industries; increased revenue
- flow to the Provincial and Federal Governments; technology
- transfer; regional infrastructure development and an improved
- data base of Pakistan's mineral resources.

2. Constitutional Position on Minerals

2.1

Minerals are a provincial subject under the Constitution, except oil, gas and nuclear minerals and those occurring in special areas (FATA and NA). Provincial Government are responsible for development and exploitation of minerals which fall in their domain. In line with this Constitutional frame work, the Federal and Provincial Governments have jointly set out in this document Pakistan's first National Mineral Policy which provides for appropriate institutional arrangements, a modern regulatory framework, an equitable and internationally competitive fiscal regime and a programme to expand Pakistan's geological database.

2.2

The provisions of Mineral Policy distinctly convey the message that focus of all activities and decision making is at the Provincial level while the Federation would provide requisite support and advice to the Provinces to take up the challenge of achieving sustainable benefit from the development of non-renewable mineral resources.

3. Consultative and Regulatory Framework

3.1

Mining has many peculiarities of its own including quite a wide span of effects over other sectors requiring established consultative mechanisms for achieving optimal benefits of mineral resources. Long gestation period of over a decade which is generally spent in exploration of a mineral deposit and still longer period for its exploitation, require a fairly stable and equitable regulatory regime. With these characteristics in view, the Federal and Provincial Governments have agreed to put in place the requisite framework.

3.2 Mineral Investment Facilitation Authority (MIFA)

3.2.1 A Mineral Investment Facilitation Authority (MIFA) will be set up in each Province (including AJK, and Special Areas i. e. FATA & NA). The membership of MIFA will be as follows:

- Chief Minister of the Province (Federal Minister Incharge in case of AJK & NA and Governor NWFP in case of FATA) Chairperson
- Minister for Mineral Development Vice-Chairperson
- Chief Secretary/Additional Chief Secretary (Dev) Member
- Secretary Mineral Development Department Member
- Secretary Finance Department Member
- Secretary Works Department Member
- Secretary Forest Department Member
- Chairperson of the Provincial Mineral Development Corporation Member
- Representative of Environmental Protection Agency Member
- Two representatives of Business Community nominated by Chief Minister

3.2.2

The Department of Mineral Development will serve as the Secretariat of MIFA. Provincial Government may change the composition or the functions of MIFA to the extent considered necessary by it. MIFA may coopt or request presence of a representative of any government department for a particular meeting.

3.2.3

Each MIFA has to execute the following functions:

Regular monitoring and direction of mineral related activities and mineral programmes of the Government and public sector in the Province.

Periodic review of implementation of the new regulatory regime and functioning of the administrative set-up in the Province.

Review progress of approvals from the relevant agencies on grant and working of concession and other related development issues and direct for any specific support to the investors required for mineral exploration and development (such as access to land, private or public and communication).

Introduce measures to promote use of local goods and services, create opportunities for appropriate education and training of nationals in modern mining skills.

Perform as appellate forum for resolution of disputes of specified nature between Licensing Authority and the investors of specified projects or categories.

Promote establishment of secondary and tertiary processing facilities within Pakistan.

Arrange and approve mineral portfolios for attracting private investment.

Ensure adequate protection of the environment.

Any other function assigned by the Chief Minister (Federal Minister in charge in case of Special Areas).

10.13 Russian Federation

The GOVERNMENT of the RUSSIAN FEDERATION, published a "State policy on the use Mineral Resources and Subsoil Use". The Ministry of Natural Resources (MNR) jointly with interested federal executive authorities will organize the work to implement the principles of mineral state policy.[791]

Basics of the State policy on the use Mineral Resources and Subsoil Use

Russia has an important mineral potential and is one of the world's leading manufacturers of mineral materials. At the same time, the raw-material base of the country has relatively low investment attractiveness due to unfavourable geographic and economic location of many mineral deposits and relatively low quality minerals. There is a lack of competitiveness in modern economic conditions, no long-term State strategy in the field of mineral resources, reproduction mineral resource base and technological rearmament of enterprises mining and primary processing of minerals. These negative factors in the near future could lead to significant difficulties in the functioning of the mineral complex, slow economic development and the reduction of security.

Russia shall conduct an effective policy in the management of mineral resources considering the consequences of a globalized world economy, growth in consumption of mineral resources in the XXI century and the most important condition for the restructuring and modernization economy and the gradual rise of welfare. This document defines the objectives, principles of those activities:

1. Objectives and Principles of State Policy of Mineral Resources and Subsoil Use

2. Improving the legal and regulatory framework in the field of mineral and subsoil

3. Improving governance subsoil fund

4. Arrangements for the use and reproduction of mineral reserves and ensure efficient operation of the extractive industry

To address the objectives and to implement the principles of State Policy related to mineral resources and subsoil based on an action plan is necessary.

---

791 GOVERNMENT of the RUSSIAN FEDERATION (2003): "State policy on the use Mineral Resources and Subsoil Use", Government order N 494-p of 2003.

1. Objectives and Principles of State Policy of Mineral Resources and Subsoil Use

The main objectives of public policy related to mineral and subsoil are:

- ensuring effective development and reproduction of mineral resource base in Russia to ensure sustainable economic development in Russia, raising welfare of its citizens; organization management and integrated mineral resources management for the benefit of present and future generations of citizens of Russia; protection of the geopolitical interests of Russia, including world mineral market.

- Government policies on the use of mineral raw materials and subsoil is based on maintaining state ownership of the subsoil, mineral resources;

- regulation of development and utilization of mineral raw materials of the country in accordance with long-term public strategy and short-term programs of geological studies and reproduction of mineral raw materials base, developed on the basis of long-term (up to the prospect 25–50 years) prediction of consumption levels of the main minerals types;

- granting of subsoil use in a competitive (auction, tender) basis;

- formation of the federal reserve fund regarding minerals deposits, including hydrocarbons (stock mineral resources for future generations);

- improving the system of taxes and charges associated with subsoil use, in order to ensure equitable distribution between the state and public property, creating an enabling economic environment for operation of the mineral complex of Russia, maintain its competitiveness in the global market;

- establishing clear lines of authority between Russia Federation and subjects of Russia in the field of use and protection of natural resources;

- promotion of an economic transition in Russia saving technologies, rational and integrated use of mineral raw materials from its production and processing;

2. Improving the legal and regulatory framework in the field of mineral and subsoil

To implement the state policy use of mineral resources and subsoil it is necessary to ensure the improvement of mineral resources legislation, drawing attention to the legal regulation of the following aspects activities in the field of subsoil use:

- providing administrative and civil law mechanisms for the use of subsoil plots, including concession contracts and other forms of contractual relationships;
- clear division between Russia and the subjects Russia authority for management of the subsoil use;
- consolidation of licenses, license agreements and contracts to use subsoil areas considering volume and types works related to subsoil use;
- application of economic and legal sanctions to subsoil users who violate conditions of subsoil use;
- promote the use of best technologies of exploration and production mineral resources, minimize negative impacts on environment during prospecting and mining works;
- development of public monitoring of the geological environment;
- an effective system of state control over exploration and mining, the implementation of conditions of licenses and contracts for the right to use subsoil areas;
- improving public examination results regarding geological exploration and mineral inventory system;
- development of new and adjustment of existing standards, norms and regulations in the field of subsoil use and protection of the environment, considering the generally recognized principles and norms of international law;
- mandatory state examination of the project documentation on the development of mineral deposits;
- safeguarding the interests of the state and subsoil user.

3. Improving governance state fund subsoil

In order to improve the management of state subsoil fund requires a clear division of powers between federal executive authorities and bodies executive power of RF subjects by:

- consolidation of the federal bodies of executive power strategic planning functions and the major regulatory and control functions;
- allocation of executive and administrative functions between federal state authorities and bodies state power of RF subjects;
- strengthen the system of state control over geological study and use of mineral resources;
- extending the practice of holding public auctions of the provision of rights for subsoil use for geological study, exploration and mining;
- efficient use of available land for subsoil use.

4. Arrangements for the use and reproduction of mineral reserves and ensure efficient operation of the extractive industry

An important part of implementing the state policy regarding use of mineral resources and subsoil are federal geological study programs resources, including the following activities:

- policy development concerning use of mineral resources in the long term (20, 30, 50 years);
- analysis and assessment of the prospects for domestic consumption, exports, imports of mineral raw materials and development of proposals to meet deficits of the type of mineral raw materials;
- strategic public mineral reserves, including reserves of fresh groundwater water to form the necessary and sufficient storage of strategic mineral raw materials reserves;
- timely and regular funding from the federal budget geological studies carried out in terms of defence and national security, anticipation, identification and assessment of strategic mineral raw materials, meet the needs of industries in geological information on mineral resources, as well as forecasting geological processes which are hazardous to life and health;
- monitoring and evaluation of mineral resource base; promoting activities of mining companies to conduct work aimed at obtaining growth of mineral resources, including through budgetary funds.

In order to ensure the efficient operation of the extractive industry the following is important:

- promotion of mining companies with highly competiveness on the world market production and mineral processing;
- increased participation of small and medium enterprises in the field of mining activities;
- use of various forms of state support, protecting the interests of Russian mining companies on the world market, as well as Russian resource companies involved in the implementation of major investment projects related to mineral resources development, including projects outside Russia.

ANNEX- Action plan to implement the principles of State Policy in the field of mineral resources and subsoil regarding

I. Improving the legal and regulatory framework in the field of mineral resources and subsoil use

II. Improving the management of public subsoil fund

10.14 Sierra Leone

The structure and content of the Mineral Policy of Sierra Leone is as follows:[792]

1. Introduction

2. Core Mineral Policy: Vision

3. Core Mineral Policy: General Strategy

4. Core Mineral Policy: Summary of Main Strategic Objectives

5. Core Mineral Policy: Details

6. Conclusion.

1. Introduction
1.1

The importance of the minerals sector to the economy of Sierra Leone is illustrated by the fact that during the 1960's and 1970's it provided the country

---

[792] http://docs.google.com/viewer?a=v&q=cache:iwastyxj1GAJ:www.daco-sl.org/encyclopedia/4_strat/4_2/mmr_mineralpolicy.pdf+Mineral+Policy&hl=de&gl=at&sig=AHIEtbTKrAHeOJovTjMWhsCPPadWubnHP

with over 70% of foreign exchange earnings, 20% of gross domestic product and 15% of fiscal revenue.

1.2

The new Core Mineral Policy (CMP) of the Government of Sierra Leone has been designed to create an internationally competitive and investor-friendly business environment in the mining sector. The policy is expected to assist the mining industry in attracting foreign and local private sector funds and to provide benefits and protection for the people and the environment of Sierra Leone. The policy will provide an enabling legal and fiscal regime for all mining operations from large-scale mines such as those of Sierra Rutile to the small artisanal gold and diamond mines in the provinces. The policy is also intended to enable the private sector to take a lead in exploration, mine development, mineral beneficiation and marketing.

1.3

The 1995 and 1998 mining policies provided the foundation for establishing an enabling environment for attracting much needed foreign and local investments into the minerals sector. The CMP is designed to improve the goals and objectives of those previous policies by enhancing the social and economic benefits for the country and the communities affected by mining activities. The sector is expected to make important contributions towards industrial, social, economic and infrastructure development particularly in rural areas. It is also expected to provide new employment opportunities, generate foreign exchange earnings and contribute significantly to government revenue.

1.4

The minerals sector has been an essential part of the Sierra Leone economy for almost ninety years. Reform of the minerals sector and the diamond industry in particular are considered crucial because of their importance to the economic development of the country. The Government has correspondingly assigned a high priority to activities aimed at the reactivation and sustainable development of the minerals sector to assist in rebuilding the country and rectifying the devastation caused by the war. The principles and objectives outlined in the Core Mineral Policy will ensure that the development of the minerals sector is achieved in ways that are economically beneficial, socially responsible and will protect the environment.

2. Core Mineral Policy: Vision

2.1

The CMP will facilitate the creation of a fair and transparent business environment that will stimulate the development of a successful minerals sector. The Government expects private investors to re-start the former rutile and bauxite mining operations and to make possible the development of several new gold and diamond mines.

3. Core Mineral Policy: General Strategy

3.1

The Government of Sierra Leone will ensure that the mineral wealth of the nation makes a positive contribution to economic and social development throughout the country. The Government will do everything in its power to make sure that the income from minerals including diamonds will be used to the benefit of the country and ensure this will be achieved by ensuring that all licensing fees, royalties, taxes and any other forms of income from the exploitation of minerals are collected and properly recorded and that all mining, trading and exporting companies in the minerals sector operate within laws that comply with international trading protocols.

3.2

The Government of Sierra Leone is committed to a 'free market' approach and economic policies, which will ensure the development of the minerals sector in accordance with international best practice and within an enabling environment competitive with other countries with similar exploration and mining potential. The Government will ensure the sector is managed in a transparent, open and accountable manner. This approach may require changes being made to laws and procedures that are not part of the minerals sector. Examples of such laws are those used to monitor and regulate banking and foreign exchange transactions in the diamond industry and those used to monitor and regulate the import and export tariffs and procedures for mining and exploration equipment and mineral products.

3.3

The Government will stimulate and guide potential foreign and domestic investors in the minerals sector by making available at a reasonable cost, geological data and other information required by prospecting, exploration and mining companies at a specified Government institution. The Government will ensure that guidelines and information services are created to provide freely available information on licensing procedures, fees structures, royalties, taxes and other liabilities. The Government will create a publicly accessible system for publishing export, dealing, and mining licence applications which will be updated on a regular basis.

3.4

In the short term, the Government will strictly enforce current mining laws, licensing procedures and regulations. However, the government will move to make changes that are acceptable within International Law to bring Sierra Leone into a more competitive position vis a vis international mining law. The Government will review all of the laws and regulations governing the mining and marketing of minerals in Sierra Leone to ensure that these laws fall in line with best practices and are comparable with regulations in the rest of the sub-continent. The review and any subsequent amendments to these laws and regulations will ensure that the legal and fiscal frameworks are consistent, clear and unambiguous.

3.5

The Government of Sierra Leone fully supports the aims and objectives of the international trading protocols used for the import and export of rough diamonds. The trading and exporting of rough and polished diamonds from Sierra Leone will be carried out in accordance with the rules and principles of the 'Kimberley Process' and the laws of Sierra Leone. The Government will ensure compliance with the 'Kimberley Process' by improving the monitoring, licensing, mining, dealing and exportation procedures in the diamonds sub- sector.

3.6

The Government will enforce the existing laws designed to provide protection for the workforce and the environment. The Government will develop health and safety legislation, underground mining regulations and review existing environmental legislation in particular the law regarding the rehabilitation of 'abandoned' mine workings. The Government will also develop appropriate legislation to prevent the use of child labour.

3.7

The Government of Sierra Leone is committed to the continued economic development and increased efficiency of the artisanal mining sector. The Government will encourage the development of credit schemes and the formation of co-operatives designed to disseminate information on more efficient and safer mining practices and to bring artisanal miners into the sphere of influence of those trying to develop diamonds as a positive way forward within the Sierra Leone economy.

3.8

The Government will encourage the creation of new marketing opportunities, particularly for diamonds, and the development of business activities that will also add value.

3.9

The Government will do all in its power to facilitate the institutional capacity build-up and human resources development to achieve all of the foregoing.

4. Core Mineral Policy: Summary of Main Strategic Objectives

- Review and Amend Mining Laws, Regulations and Associated Laws to make them as attractive as possible for investment here rather than in neighbouring countries with similar mineral potential.
- Strengthen the Institutions that Administer, Regulate and Monitor the Mineral Industry in Sierra Leone to allow the mining industry, especially with respect to the diamond industry to be turned around to become a positive for Sierra Leone;
- Develop and Strengthen Human Resources in the Minerals Sector.
- Attract Private Investments into the Minerals Sector. Encourage private investment to use the implementation of the Kimberley Process as a positive at the forefront of selling diamonds for peace and development properly registered by the Kimberley Process;
- Ensure that Sierra Leone's Mineral Wealth Supports National Economic and Social Development
- Improve the Regulation and Efficiency of Artisanal and Small-Scale Mines
- Minimise and Mitigate the Adverse Impact of Mining Operations on Health, Communities and the Environment.
- Promote Improved Employment Practices, Encourage Participation of Women in the Mineral Sector and Prevent the Employment of Children in Mines.
- Add Value to Mineral Products and Facilitate Trading Opportunities for Mined Products.
- Improve the Welfare and Benefits of the Individuals and Communities Participating in and Affected by Mining.

5. Core Mineral Policy: Details

Objective 1: Review and Amend Mining Laws, Regulations and Associated Laws to make them as attractive as possible for investment here rather than in neighbouring countries with similar mineral potential.

Objective 2: Strengthen the Institutions that Administer, Regulate and Monitor the Mineral Industry in Sierra Leone to allow the mining industry, especially with respect to the diamond industry, to be turned around to become a positive for Sierra Leone

Objective 3: Develop and Strengthen Human Resources in the Minerals Sector.

Principle: The Government will encourage the provision of opportunities for improving the qualifications and experience of people involved in the minerals industry. Support will be given to initiatives aimed at providing new job opportunities and that will increase the knowledge and skills of the workforce.

Objective 4: Attract Private Investment into the Minerals Sector. Encourage Private Investors to implement the Kimberley Process as appositive for peace and development.

Objective 5: Ensure that Sierra Leone's Mineral Wealth Supports National Economic and Social Development.

Objective 6: Improve the Regulation and Efficiency of Artisanal and Small-Scale Mines.

Objective 7: Minimise and Mitigate the Adverse Impact of Mining Operations on Health, Communities and the Environment.

Objective 8: Promote Improved Employment Practices, Encourage Participation of Women in the Mineral Sector and Prevent the Employment of Children in Mines.

Objective 9: Add Value to Mineral Products and Facilitate Trading Opportunities for Mined Products.

Objective 10: Improve the Welfare and Benefits of the Individuals and Communities Participating in and Affected by Mining.

6. Conclusion

The Government's role is that of promoting, monitoring and facilitating investment in the minerals sector. The Government also provide formal assaying and valuation services in particular to artisanal and small-scale miners. The

Government will develop public information programmes to raise awareness nationally and internationally of the existence of business opportunities in the minerals sector in Sierra Leone and to broaden opportunities for private sector participation.

10.15 South Africa

South Africa's mining industry is supported by an extensive and diversified resource base, and has since its inception been a cornerstone of South Africa's economy. The changes which have come about in the country make it necessary to prepare the industry for the challenges of the twenty-first century.[793]

The review process has taken account of the problems and opportunities confronting the mining industry against the backdrop of changes in the country's policy and institutional environment. In particular, the passage to the Mine Health and Safety Act of 1996 will have far-reaching impacts on the industry in the areas of health and safety and human resource development. Changes in labour legislation and the introduction of employment equity legislation, as well as the reform of the environmental regulatory system, create a dynamic context for this policy review. Beyond the borders increasing competition, both in commodity markets and for investment, from mineral-rich countries that have liberalised their economic and political systems to attract investment are significant influences on the policy reform process.

The mineral policy review process took account of problems and opportunities facing the mining industry. The gold mining sector, particularly, is re-examining its production techniques in the light of a static gold price, deep levels of working and higher operating costs. Undoubtedly some of the older mines are reaching the end of their lives, leading to job losses and the other attendant negative effects of downscaling, but these problems are being tackled energetically within the sector, through restructuring of mining groups, technological advances and innovative methods of improving productivity. Apart from gold mining, there are many other minerals being produced, for some of which South Africa is the leading producer and holder of reserves. The White Paper also has a chapter on small-scale mining which is intended to encourage the small and medium sized operator, to the benefit of employment and the overall economy.

---

793 Department of Minerals and Energy, http://www.dme.gov.za/minerals/min_whitepaper.stm (3 of 60)10/3/2007 1:05:47 PM.

Government mineral policy had to take account of the international nature of the mining industry in order to ensure the continuing prosperity of our own mines. In September 1995, the Mineral Policy Process Steering Committee was formed consisting of representatives from both the executive and legislative branches of Government, as well as organised business and organised labour. The mandate given to the Steering Committee was to conduct an extensive consultative process to canvass stakeholder opinion for the preparation of a new minerals and mining policy for South Africa. In November 1995, a Discussion Document on Minerals and Mining Policy for South Africa was published and extensive written comments were received. Four hundred people attended public mineral policy workshops held in March 1996, at which a wide range of issues were debated.

Bilateral meetings were held with inter alia provincial governments, ministries, departments, investment analysts, foreign-owned mining companies and environmental interest groups. In addition, written submissions were received from several interested parties during the consultative process. The end result of this, the most comprehensive consultative process yet conducted for a review of a minerals and mining policy in South Africa, was a document containing proposals that have been drafted after careful consideration of a very broad range of views. The submission of the document to the Minister of Minerals and Energy concluded the task of the Steering Committee.

The Minister requested the Department of Minerals and Energy to consider certain adjustments to the document in line with his budget speech in the National Assembly on 21 May 1997. The views of stakeholders, such as small-scale miners, environmental groupings and communities, who felt that they were not properly consulted by the Steering Committee, as well as the outcomes of other policy processes, were also considered in the final editing of the document. The document was then ratified by the Minister of Minerals and Energy and Cabinet as a Green Paper on Minerals and Mining Policy for South Africa.

The Green Paper was published on 3 February 1998 and the public was invited to respond not later than 31 March 1998. The Department of Minerals and Energy received more than a hundred written submissions from the public, and in addition, submissions were received from interested parties during public hearings held by the Parliamentary Portfolio Committee on Minerals and Energy. The Department of Minerals and Energy established working groups to consider the various inputs and to effect appropriate amendments. The document was then ratified by the Minister of Minerals and Energy as a Draft White Paper. The Cabinet Committee for Economic Affairs requested further amendments; whereafter Cabinet on 23 September 1998 approved the document as a White Paper on Minerals and Mining Policy for South Africa.

The Draft White Paper was organised into six main themes:

- Business Climate and Mineral Development, which looks at the continuation of policy conducive to investment and includes a section on Mineral Rights and Prospecting Information which presents changes to the system of access to, and mobility of, mineral rights;

- Participation in Ownership and Management, which examines racial and other imbalances in the industry;

- People Issues, which looks at health and safety, housing needs, migrant labour, industrial relations and downscaling;

- Environmental Management;

- Regional co-operation; and Governance.

Each chapter and subchapter contains a general background to the particular issue, a statement of intent (policy objective), the views of the different stakeholders and, finally, the policy statements by Government. Policy making occurs in a dynamic setting, and minerals and mining policy, which is necessarily broad in its scope, needs to be co-ordinated with other policies which properly fall within the remit of other forums and spheres of government. Reference is therefore made in the document to matters that are being considered by other policy forums, and policies developed by other spheres of government.

10.16 Tanzania

The structure and content of the **Mineral Policy of Tanzania** is as follows:[794]

Foreword

Mission and Vision

1.0 Introduction

2.0 Tanzania's Mineral Endowment and Potential

3.0 The Mineral Policy of Tanzania

---

[794] Ministry of Energy and Minerals, The United Republic of Tanzania (1997): The Mineral Policy of Tanzania (http://www.tanzania.go.tz/pdf/themineralpoli cyofTanzania.pdf.).

3.1   Challenges
3.2   Objectives
3.3   Mineral Policy Directions and Strategies
3.3.1 The Macro-economic Environment
3.3.2 The Fiscal Regime
3.3.3 The Legal Regulatory Framework
3.3.4 The Institutional Framework
3.3.5 Financial Services
3.3.6 Rationalizing Artisanal and Small-scale Mining
3.3.7 Mining Sector Support Services and Facilities
3.3.8 Creating and Maintaining a Viable Infrastructur
3.3.9 Establishing Formal Marketing System
3.3.10 Integrating Mining into the National Economy
3.3.11 Human Resources Development
3.3.12 The Environmental and Social Sustainability of Mining Development
4.0   The Role of the Government
4.1   Government as Regulator
4.2   Government as Promoter and Facilitator
4.3   Government as a Service-provider
5.0   Communication Strategy for the Mineral Policy
5.1   Marketing Tanzania's Mineral Potential
5.2   Identifying the Target Groups
6.0   Conclusion.

Mineral Policy Directions and Strategies

In order to achieve the sector policy objectives, the following policy strategies have been set:

The Macro-economic Environment

A sound and stable macro-economic environment is an important prerequisite for private sector development and investment in mining. Key elements for such policy include: appropriate market-based exchange rates and monetary policies, trade and fiscal policies. In effect, a sustainable development of the mineral sector requires a stable, transparent and predictable macro-economic environment. The Government is committed to pursuing these policies consistently for sustainable sectoral and overall development.

The Fiscal Regime

The Government will formulate and implement a mining taxation regime which is conducive to investment in exploration and mining development. The tax package. should recognize the investor's need to recover exploration and development outlays, to achieve a rate of return commensurate with risk, to repatriate dividends and to meet financial obligations with creditors and suppliers. In formulating the fiscal regime, the Government will aim to balance the country's interest with those of investors by ensuring that the mining taxation regime is equitable, stable and predictable, not distorted and internationally competitive.

In addition to being stable and predictable, the fiscal policy regime will be designed to be promotionally competitive for attracting and retaining investments in the mineral sector. Furthermore, it will aim to increase government spending in the supportive services and physical infrastructure required for development of the mining sector. Hence, the fiscal policy forms an important instrument for providing incentives to investors.

- Strategies for the Fiscal Regime

    Strategies for creating an attractive fiscal regime include the following:

    (i) Reviewing and revising the fiscal regime with the aim of creating a more effective taxation system with streamlined procedures and modalities to ensure an efficient, fair, stable and competitive tax regime for mining and mineral trading;

    (ii) Establishing procedures and methods that ensure that taxes owing to the Government from the mineral sector are properly assessed and collected to promote compliance;

    (iii) According favourable consideration to exploration and development expenditures with a view to minimizing front-end costs and therefore enhancing investment in mineral exploration and development;

(iv) Ensuring easy availability of pertinent information on the fiscal policy and the tax regime for mining activities;

(v) Deploying fiscal incentives to encourage domestic mineral beneficiation and establishment of local marketing infrastructure for mineral trading;

(vi) Encouraging and supporting investment in the development of social and economic infrastructure in the mining localities by providing favourable tax incentives;

(vii) Harmonizing Tanzania's fiscal regime with those of competing countries as a tool to mitigate mineral smuggling and encourage trading through official channels.; and

(viii) Providing incentives which encourage re-investment of profits earned from mining operations.

The Legal and Regulatory Framework

The policy of the Government is to establish an internationally competitive legal and regulatory framework to attract and sustain foreign and local investment in the mineral sector, and to create a stable and conducive business climate. The legal framework should also aim to deter information hoarding on new discoveries, freezing of exploration acreage for speculative purposes, transfer pricing and tax evasion. These will be reflected in the Mining Act of Tanzania.

- Strategies for the Legal and Regulatory Framework

The strategies for the legal and regulatory framework include the following:

(i) Harmonizing and consolidating all statutes under which the mineral sector operates into one Mining Act which sets out clear, simple and transparent procedures for allocation of rights, the transition from exploration to mining rights, and the transfer of these rights;

(ii) Streamlining the licensing procedures to harmonize small-scale and large-scale mining operations; ensuring transparency and fairness by conferring ownership of mineral rights on the basis of first come, first served basis;

(iii) Rationalizing the licensing system to ensure exclusivity of licensed areas;

(iv) Establishing a system of title maintenance which encourages active mineral exploration and exploitation and discourages hoarding for speculative and other purposes;

(v) Grouping minerals into categories for the purpose of facilitating targeting of incentives, penalties, specialized skills development and mineral administration;

(vi) Harmonising the mining laws with other statutes being administered by other institutions that directly or indirectly affect the development of the mineral sector;

(vii) Ensuring environmental protection and land reclamation; and

(viii) Ensuring that contractual rights and obligations are protected and providing for settlement of disputes through the Courts or by international arbitration.

10.17 USA

Bureau of Land Management – Energy and Mineral Policy[795]

This statement sets forth the Bureau of Land Management (BLM) policy tor the management of energy and mineral resources on public lands, a component of the agency's multiple use mandate. The BLM seeks 10 implement its multiple use mission to balance various uses to achieve healthy and productive landscapes, including the development of energy and minerals in an environmentally sound manner.

This Energy and Mineral Policy reflects the provisions of six important acts of Congress relating to conventional, alternative, and renewable energy; and mineral resources, as follows:

The Domestic Minerals Program Extension Act of 1953 states that each department and agency of the Federal Government charged with responsibilities concerning the discovery, development, production, and acquisition of strategic or critical minerals and metals shall undertake to decrease further, and to eliminate wherever possible, the dependency of the United States on foreign sources of supply of such material.

---

795 The following information is provided from: http://www.blm.gov/pgdata/etc/medialib/blm/wo/Information_Resources_Management/policy/ib_attachments/2008.Par.15798.File.dat/IB2008-107_att1.pdf

The Mining and Minerals Policy Act of 1970 declares that it is the continuing policy of die Federal Government to foster and encourage private enterprise hi the development of a stable domestic minerals industry and the orderly and economic development of domestic mineral resources. His act includes all minerals, including sand and gravel, geothermal, coal, and oil and gas,

The Federal Land Policy and Management reiterates that the 1970 Mining and Minerals Policy Act shall be implemented and directs that public lands be managed in a manner that recognizes the Nation's need for domestic sources of minerals and other resources. It also mandates that „scarcity of values" be considered in land use planning.

The National Materials and Minerals Policy, Research and Development Act of 1980 requires the Secretary of the Interior to improve the quality of minerals data in Federal land use decision-making.

The Energy Policy Act of 2005 encourages energy efficiency and conservation, promotes alternative and renewable energy sources, reduces dependence on foreign sources of energy, increases domestic production, modernizes the electrical grid, and encourages the expansion of nuclear energy.

The Energy independence and Security Act of 2007 to move the United States toward greater energy independence, to increase the production of dean renewable fuels, and support modernization of die nation's electricity transmission and distribution system.

The BLM recognizes that public lands are an important source of tie Nation's energy and mineral resources, including renewable energy resources such as geothermal, wind, solar, and biomass. The public lands ate also important for the siting of infrastructure facilities to support the development of energy and minerals resources. The BLM makes public lands available for orderly and efficient development of these resources under the principles of Multiple Use Management, and the concept of Sustainable Development as was defined at the World Summit on Sustainable Development in 2002, in Johannesburg, South Africa, where 192 countries, including the United States, endorsed its resolution on minerals.

The following principles will guide the BLM in managing energy and mineral resources on public lands:

1. The BLM land use planning and multiple-use management decisions will recognize that energy and mineral development can occur concurrently or sequentially with other resource uses, providing that appropriate stipulations or conditions of approval are incorporated into authorizations to prevent un-

necessary or undue degradation, reduce environmental impacts, and prevent a jeopardy opinion.

2. Land use plans will incorporate and consider energy and geological assessments as well as energy and mineral potential on public lands through existing energy, geology and mineral resource data, and to the extent feasible, through new mineral assessments to determine mineral potential. Partnerships with the National Renewable Energy Laboratory, Federal and State agencies, such as the U. S. Geological Survey and State Geologists, to obtain existing and new data will be considered.

3. Withdrawals and other closures of the public land must be justified in accordance with the Department of the Interior Land Withdrawal Manual 603 DM 1 and the BLM regulations at 43 CFR 2310. Petitions to the Secretary of the Interior for revocation of land withdrawals in favour of energy and mineral development will be evaluated through the land use planning process.

4. The BLM will work cooperatively with surface owners and mineral operators in recognizing their rights on split-estate lands. In the absence of a Surface Owner Agreement and in managing development of the Federal mineral estate on a non-federal surface, die BLM will take into consideration surface owner mitigation requests from pre-development to final reclamation.

5. The BLM endorses Sustainable Development that encourages Social, Environmental, and Economic considerations before decisions are made on energy and mineral operations. The BLM actively encourages private industry development of public land energy and mineral resources, and promotes practices and technology that least impact natural and human resources.

6. The BLM will adjudicate and process energy and mineral applications, permits, operating plans, leases, rights-of-ways, and other land use authorizations for public lands in a timely and efficient manner and in a manner to prevent unnecessary or undue degradation, The BLM will require financial assurances, including long-term trusts, to provide for reclamation of the land and for other purposes authorized by law, Prior to mine closure, reclamation considerations should include partnerships to utilize the existing mine infrastructure for future economic opportunities such as landfills wind farms, biomass facilities, and other industrial uses.

7. Energy and mineral-related permit applications will be reviewed consistent with the requirements of NEPA and other environmental laws, The BLM will work closely with Federal, Stale and Tribal governments to reduce duplication of effort while processing energy and mineral-related permit applications.

8. The BLM will monitor locatable, salable and leasable mineral operations and energy operations to ensure proper resource recovery and evaluation, production verification, diligence, and enforcement of terms and conditions. The United States will receive market value for its energy and mineral resources unless otherwise provided by statute, and royalty rates will be monitored and evaluated to protect the public interest.

9. The BLM will continue to develop e-Government solutions that will provide for electronic submission and tracking of applications and the use of GIS technology to support development of energy and mineral resources, The BLM will continue to provide public access to current mineral records, including spatial display of all types of authorizations and mineral resource and ownership data. Data systems, such as LR 2000, will be kept current and best management practices sought to reduce backlogs and to identify errors.

10. The BLM will strive to maintain a professional workforce in adjudication, energy, geology, and engineering to support energy and mineral development.

11. To the extent provided by law, regulation, secretarial order, and written agreement with the Bureau of Indian Affairs, the BLM will apply the above principles to the management of mineral resources and operations on Indian Trust lands in order to comply with its Trust Responsibilities.

Chapter 11 Appendix 2

The Brussels Commission devoted particular attention to the non-ferrous metals in its communication of 7 February 1975 Raw Materials Community (Supplement 1/75 bulletin EC). The evaluation of the supply problems of "critical non-ferrous metals" and the recommended measurements by the commission for four non-ferrous metals (chromium, copper, tungsten and zinc) can be seen in the tables below.

Table 85: Primary self-supply in developed countries, 1978–1980 (in %) (Sources: EC Raw Material balances (various years), Luxembourg)

Raw Material	EC			USA			Japan		
	1978	1979	1980	1978	1979	1980	1978	1979	1980
Al	12,4	11,4	11,3	5,9	6,5	6,7	–		
Cu	0,2	0,2	0,2	63,4	70,1	68,9	4,9	3,4	3,3
Pb	12,6	11,6	10,9	40,6	43,9	58,0	19,1	16,3	10,9
	–	–	–				2,0	0,6	0,9
Mo				196,3	195,8	255,9	2,0	0,8	0,8
Ni				5,2	7,5	7,2	–		
V	–	–	–	63,8	76,4	81,1	–	–	–
W	13,7	9,6	8,7	50,3	41,6	38,8	39,2	29,2	25,8

Table 86: Consolidated EC Raw material balances 1980

Raw Material		Sources					Usage			
		Mining	Recycling	Import	Lager-abbau	Total assets	Consumption	Export	Lager-abbau	Total assets
Al	(1000 t)	549	1021	4450		6020	4873	980	167	6020
Cu	(1000 t)	6	1010	2336	12	3364	2811	553		3364
Pb	(1000 t)	138	635	782		1555	1270	228	57	1555
Sn	(t)	3291	15.207	64614		83.112	68.323	12.610	2179	83.112
Cr	(1000 t)		51	600		651	631	17	3	651
Ni	(1000 t)		34	211		245	200	43	2	245
W	(t)	553	1513	5256	439	7761	6374	1387		7761
Ti	(1000 t)	–	2	561		563	379	180	4	563

Chapter 11 Appendix 2

Table 87: Critical non-ferrous metals (Data by Communication of the Commission dated 7. Feb. 1975 about supply of raw materials for the EC, Supplement 1/75 EC-Bulletin)

Raw Material	Dependence on imports	Possibilities		Political risk of supply	Tendency for processing in the country of origin	Security of supply	Urgency of actions
		of subsitution	of recycling				
Aluminium	56%	high	26% +	no	yes	satisfactorily	slightly
Chromium	100%	partly	22% +	yes	yes	risky	prior
Copper	95%	partly	35% +	yes	yes	risky	slightly
Iron	80%	slightly	17% +	no	yes	suffiently	slightly
Manganese	100%	slightly	slightly	yes	yes	satisfactorily	slightly
Platinum	100%	slightly	20% +	yes	yes	suffiently	slightly
Tin	83%	partly	45% −	no	yes	satisfactorily	slightly
Tungsten	95%	partly	25% −	yes	no	risky	prior
Zinc	75%	partly	20% +	no	yes	risky	prior

Appendix 2

Results of the Public Consultation on Commission Raw Material Initiative
(European Commission, DG Enterprise and Industry, 2008a)

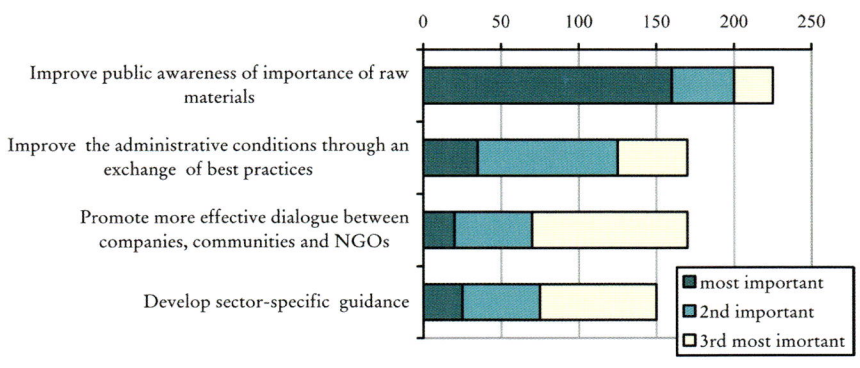

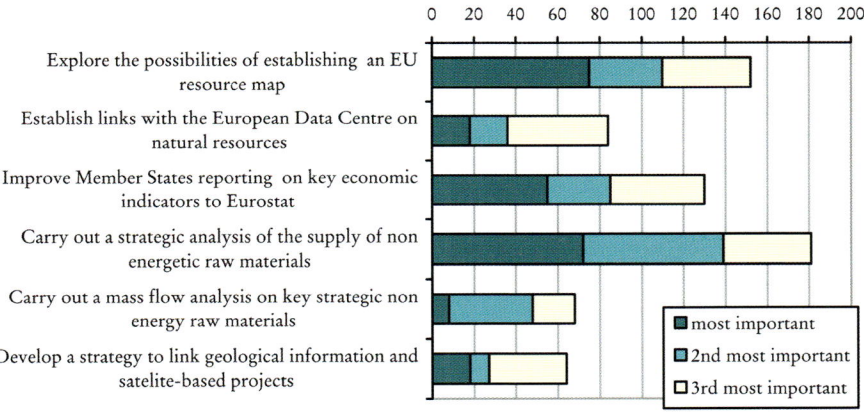

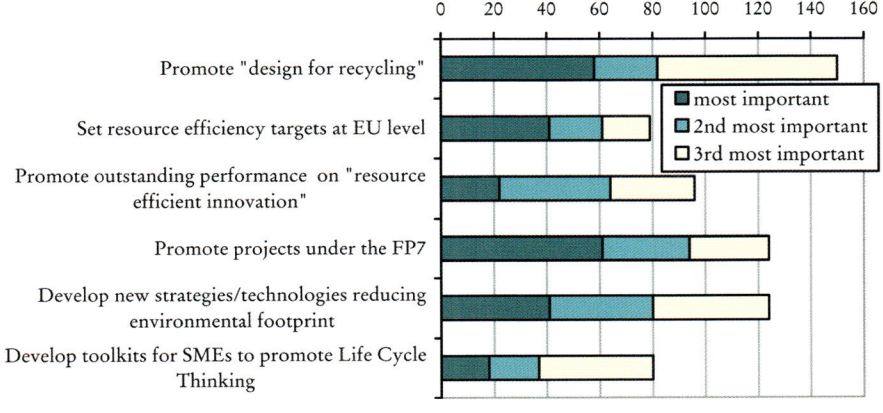

Chapter 11 Appendix 2

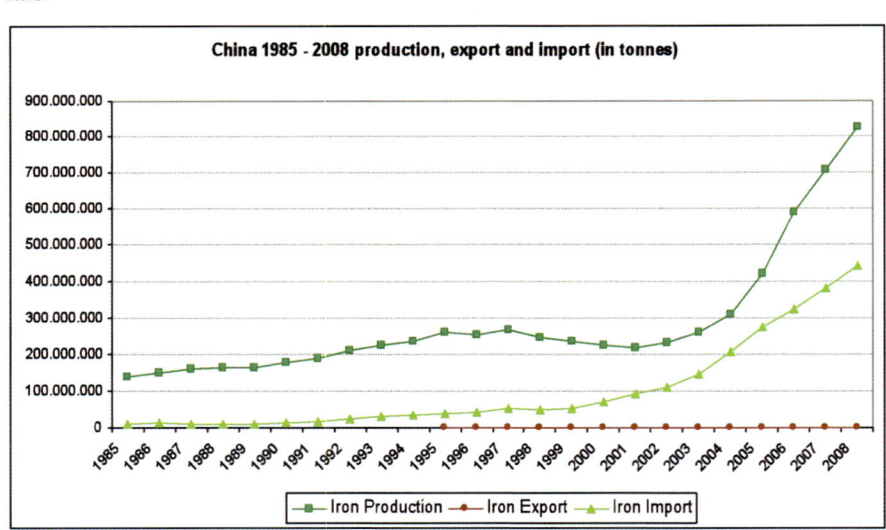

China

Some information about production, import and export of selected raw materials:

Appendix 2

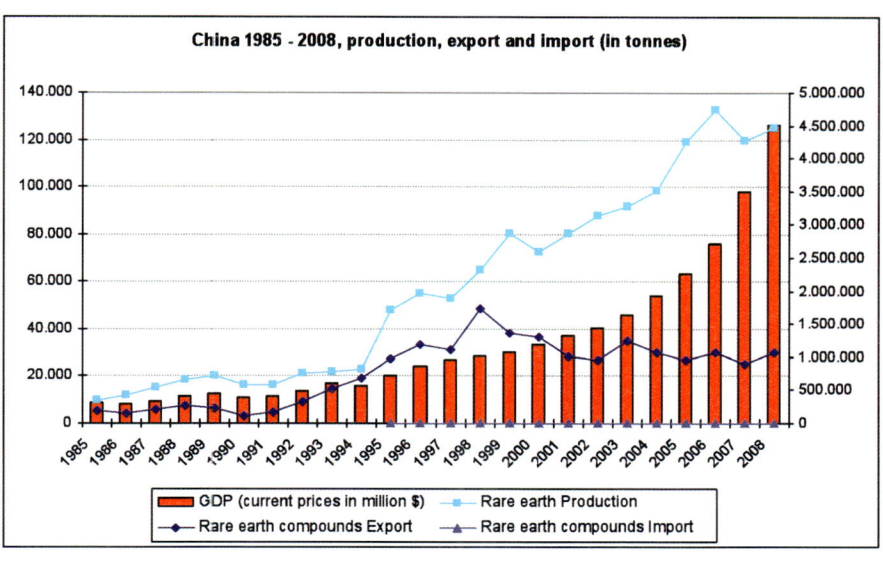

India

Some information about production, import and export of selected raw materials:

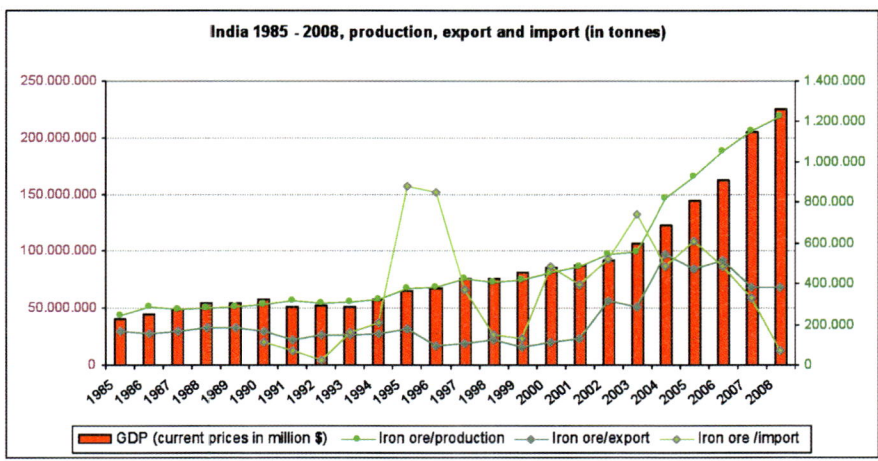

Chapter 11 Appendix 2

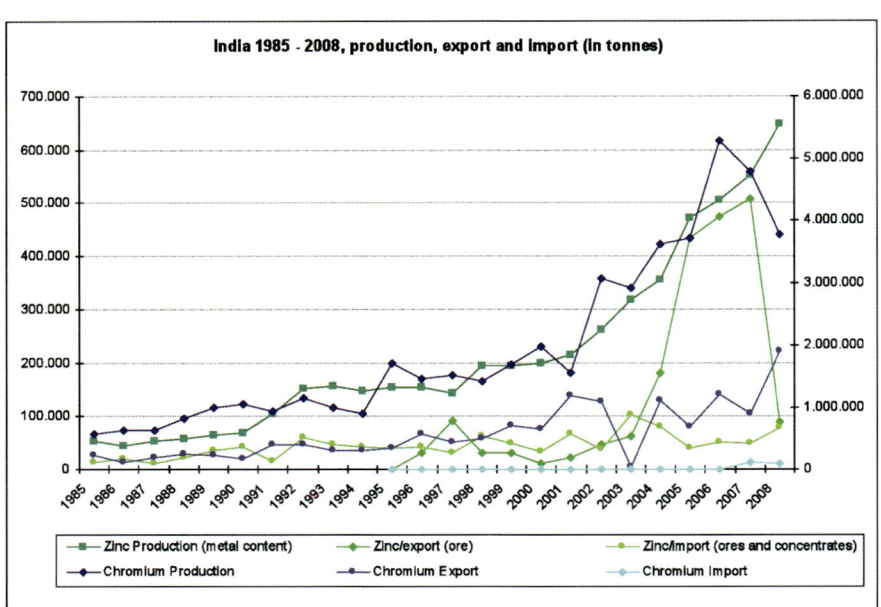

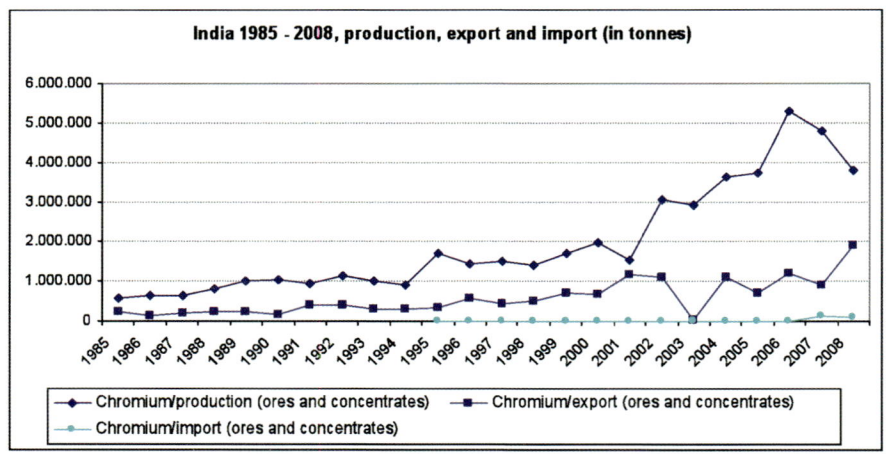

South America

Some information about production, import and export of selected raw materials:

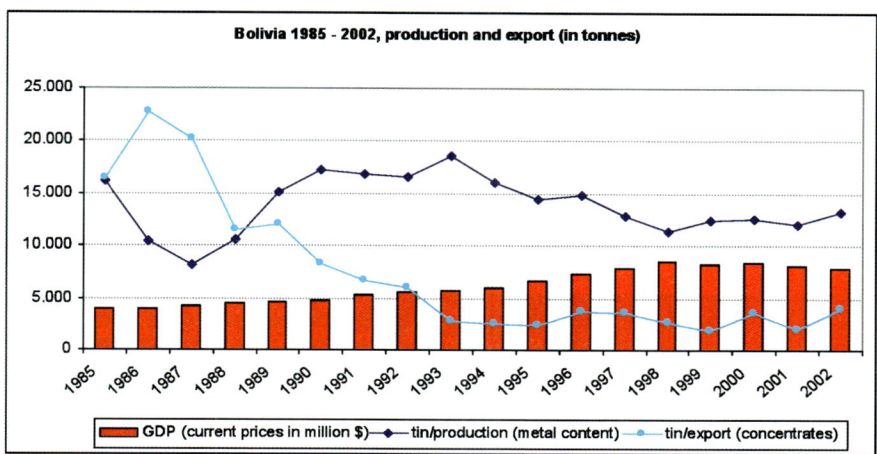

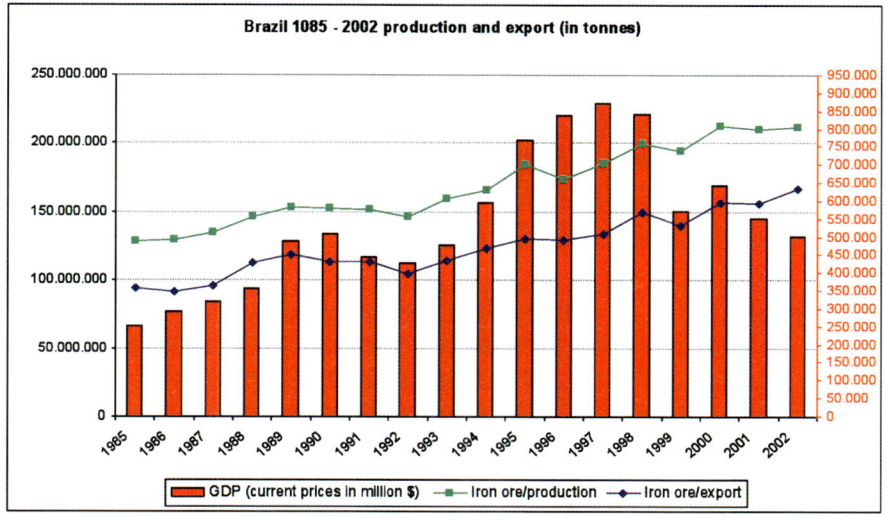

Chapter 11 Appendix 2

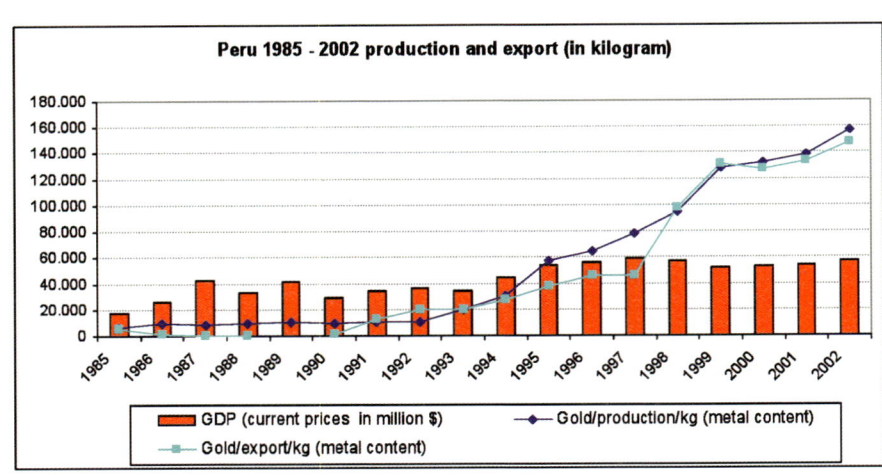

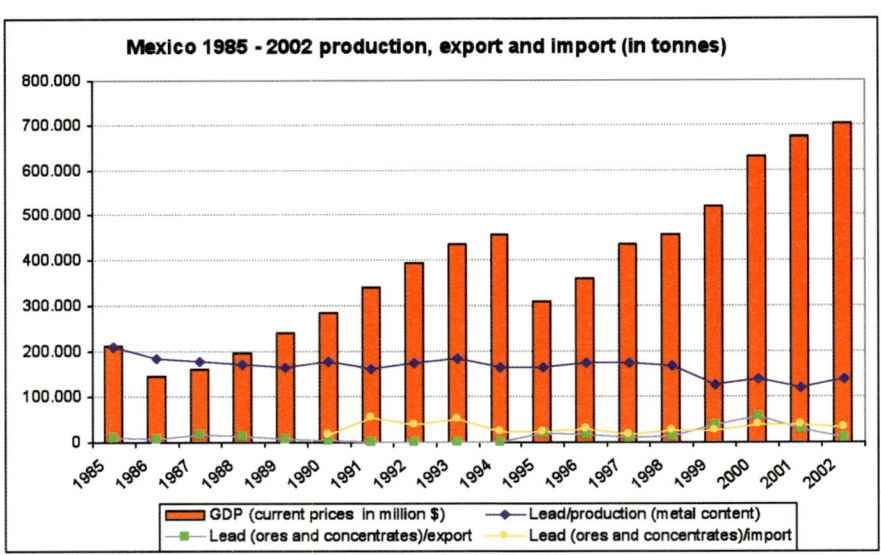